Ecotoxicology Essentials

Ecotoxicology Essentials

Environmental Contaminants and Their Biological Effects on Animals and Plants

Donald W. Sparling
Cooperative Wildlife Research Laboratory and Department of Zoology,
Southern Illinois University, Carbondale, IL, USA

AMSTERDAM • BOSTON • HEIDELBERG • LONDON
NEW YORK • OXFORD • PARIS • SAN DIEGO
SAN FRANCISCO • SINGAPORE • SYDNEY • TOKYO

Academic Press is an imprint of Elsevier

Academic Press is an imprint of Elsevier
125 London Wall, London EC2Y 5AS, UK
525 B Street, Suite 1800, San Diego, CA 92101-4495, USA
50 Hampshire Street, 5th Floor, Cambridge, MA 02139, USA
The Boulevard, Langford Lane, Kidlington, Oxford OX5 1GB, UK

Notices
Knowledge and best practice in this field are constantly changing. As new research
and experience broaden our understanding, changes in research methods, professional
practices, or medical treatment may become necessary.

Practitioners and researchers must always rely on their own experience and knowledge
in evaluating and using any information, methods, compounds, or experiments
described herein. In using such information or methods they should be mindful of
their own safety and the safety of others, including parties for whom they have a
professional responsibility.

To the fullest extent of the law, neither the Publisher nor the authors, contributors, or
editors, assume any liability for any injury and/or damage to persons or property as a
matter of products liability, negligence or otherwise, or from any use or operation of
any methods, products, instructions, or ideas contained in the material herein.

British Library Cataloguing-in-Publication Data
A catalogue record for this book is available from the British Library.

Library of Congress Cataloging-in-Publication Data
A catalog record for this book is available from the Library of Congress.

ISBN: 978-0-12-801947-4

For Information on all Academic Press publications
visit our website at https://www.elsevier.com/

  Working together
to grow libraries in
developing countries

www.elsevier.com • www.bookaid.org

Typeset by MPS Limited, Chennai, India

Contents

Section II
Chemistry and Effect of Major Families of Contaminants

Section III
Identifying and Evaluating Large Scale Contaminant Hazards

Section I

Basic Principles and Tools of Ecotoxicology

Chapter 1

An Introduction to Ecotoxicology

Terms to Know

Ecotoxicology
Anthropogenic
Sublethal Effects
Lethality
Compensatory Effects
Additive Effects
Octanol/water Coefficient (K_{ow})
Soil/water Coefficient (K_{oc})
Persistence
Photolysis
Bioavailability
Bioassimilation
Bioconcentration
Biomagnification
Hyperaccumulate
Dissolved Organic Carbon (Matter)
Environmental Matrix
Environmentally Relevant Concentration
Median Lethal Concentration (Dose)
Median Effect Concentration (Dose)
No Adverse Effects Level
Lowest Observed Adverse Effects Level

INTRODUCTION

A good starting place in this introductory chapter is to define the basic subject of this book, ecotoxicology. Although toxicology, the science of poisons to humans, has been studied for hundreds of years, ecotoxicology is relatively new. It is generally accepted that Rachel Carson's landmark book *Silent Spring* (1962) awoke the country to the potential dangers of pesticides, mostly DDT and its relatives, to the environment and that this book served as the impetus for starting the science. The term *ecotoxicology* was first coined by Rene Truhaut in 1969 to denote a natural extension of ecology and toxicology that included

Ecotoxicology Essentials.

3

the effects of chemical pollutants on any aspect of the environment (Truhaut, 1977). The discipline has received many definitions since then including from Cairns (1989): "toxicity testing on one or more components of the ecosystem" and Hoffman et al. (2003) as "the science of predicting effects of potentially toxic agents on natural ecosystems and on nontarget species." Virtually all other definitions include chemicals or contaminants, effects, and ecosystem or ecology. The overriding objective of ecotoxicology is to understand how chemicals (usually of human origin or *anthropogenic*) behave in the natural environment and how they affect organisms in that environment. Specific investigations will have objectives that refine or limit that overriding one, but essentially all ecotoxicological studies fall under that one umbrella.

As you might surmise after a few moments of thought, to accomplish that one main objective requires many different disciplines. Practitioners of the science have at least some expertise in chemistry, physiology, ecology, statistics, risk assessment, and similar areas. Very few investigators can truly be experts in all of these areas so in today's era of specialization, scientists tend to focus on one of these areas or subdisciplines.

We can identify some of these specialties knowing that any definition will meet with exceptions and that there is some degree of overlap among subdisciplines. Environmental chemists study fate and transport of contaminants in ecosystems. This includes how the contaminants get into the ecosystem in the first place, how they move from one compartment (ie, air, water, sediment, soil, organisms) to another, and how they degrade or change form in these compartments. The study of chemicals in the environment has become increasingly sophisticated with continual improvements in instrumentation and procedures to detect chemicals at the parts per billion level or even lower.

Scientists who are interested in physiological effects often use controlled studies in the laboratory or in controllable outdoor experiments. Some questions they ask include: Do the contaminants cause cancer, disrupt endocrine systems, retard growth or development, or affect behavior? At what point do these sublethal effects (ie, physiological changes short of death of the organism) become lethal and the organism dies? Statistically supported cause and effect relationships are the goal in these investigations. Historically, these controlled experiments have emphasized the effects of a given chemical in isolation from all other chemicals. However, this does not reflect most real-life exposures. In actuality, organisms are seldom exposed to only one contaminant at a time; mostly they encounter a battery of chemicals that have a variety of interactions. As a result, many studies are now looking at the interaction between two or more chemicals.

Ecotoxicologists are mostly interested in how contaminants affect populations, communities, and ecosystems. Contaminants are often thought of as additional stressors along with predation, disease, competition, and other factors. Changes at these higher levels of organization may occur through physiological effects of organisms, but they also involve numerous interactions among chemicals and the other stressors that organisms face under seminatural conditions (see Chapter 2). While statistical support of hypotheses is also critical in these tests, the complexity of chemicals interacting with other stressors may make it difficult to ascertain that a specific contaminant is having an effect on a population, leading to decreased population growth or health. This difficulty is compounded in field studies so the most informative studies are usually those that involve both field investigations and controlled experiments.

When studying chemical (or any other) stressors in general, there are three possible outcomes: (1) the stressor has no apparent effect on the population; (2) the stressor does seem to cause death or debilitation of individuals in a population but may not affect population dynamics—the concept of compensatory effects in population ecology (Burnham and Anderson, 1984); or (3) the stressor is affecting individuals and has a measurable effect on population size, growth, or health—the concept of additive effects. Under the concept of compensatory effects, organisms that die due to the contaminant would likely die from some other factor, so the overall net loss to the population is not increased. According to the additive effects concept, the mortality or harm to individuals from chemical stressors is in addition to the mortality experienced from other stressors. At times the additive effect from contaminants may exceed the effects produced by all other stressors and contaminants become the major limiting factor on the population. It is one thing to show that organisms in the field are dying from exposure to contaminants, but it is often very difficult to distinguish between compensatory and additive effects. One of the most convincing examples of additive effects were population declines of fish-eating birds due to eggshell thinning related to exposure to dichlorodiphenyldichloroethylene (DDE), a derivative of dichlorodiphenyltrichloroethylene (DDT) that was used to control insect pests from the1940s to the early 1970s (see the FOCUS section in Chapter 4 for more details).

Risk assessment assimilates information on chemical fate and transport, dose and effect relationships, and ecological factors to develop models that predict the probability that a chemical will have deleterious effects on a population or ecological entity of concern. Frequently this involves developing models that encompass chemical and other factors that may occur in a contaminant scenario and extrapolating possible outcomes by changing the values of those factors. Often risk assessment is directed toward human health issues, but the same principles of modeling pertain to animal and plant populations.

We will cover all of these topics and several others in this book. The primary objective of this book is to provide students with a solid background in the discipline of ecotoxicology. This includes an understanding of specific contaminants and how these contaminants affect individuals, populations, and communities. In addition, we hope to foster an appreciation for the nature of experimental studies in ecotoxicology. The objective of this chapter is to provide an introduction to the book by discussing several basic concepts associated with factors in the environment that influence fate and transport of contaminants, and describing key concepts of the responses of organisms to environmental contaminants.

CHARACTERISTICS OF CHEMICALS THAT AFFECT THEIR PRESENCE IN THE NATURAL ENVIRONMENT

Chemicals have intrinsic characteristics that affect how they behave under natural conditions. Questions that can be asked in this regard include:

- Are they repulsed by water or do they mix well with it?
- Do they tend to adhere to soil, sediments, or suspended particles or are they found free in the environment?
- Are they normally found as solids, liquids, or gases?
- How persistent are they in the environment?
- How likely are they to be found in the atmosphere or in any other compartment of the environment?
- Do they have very similar "cousins" or isomers?
- How does their structure influence their toxicity?

Solubility

Water has been called the "universal solvent" because many organic and inorganic molecules can dissolve in water, at least to some extent. However, water solubility from one chemical to another can vary tremendously. Solubility becomes especially important in the environment when a contaminant is present in water. Some metals have comparatively high solubility, especially under acidic conditions. Polychlorinated biphenyls (PCBs) with many chlorine atoms, high molecular weight polycyclic aromatic hydrocarbons (PAHs), and some organochlorine pesticides (OPs) are notoriously insoluble in water although they may be soluble in oils or lipids. In contrast, PCBs with two or three chlorines, PAHs with relatively low molecular weight, and some OPs are reasonably soluble in the mg/kg or parts per million (ppm) range.

Chemists use a measure called the octanol/water coefficient (K_{ow}) to rank chemicals based on their relative ability to dissolve in water or an organic solvent. The coefficient is determined by quantifying the maximum concentration of a chemical that dissolves in octanol, an organic solvent divided by the maximum concentration of that same chemical in water at standard conditions of temperature and pressure. Although octanol is polar due to a hydroxyl molecule, it does not mix (immiscible) with water. Water is strongly polar. Nonpolar and weakly polar substances preferentially dissolve in octanol. The higher the K_{ow}, therefore, the more nonpolar the compound. Log K_{ow} values are inversely related to aqueous solubility and directly proportional to molecular weight (US EPA, 2009). For example, the PAH naphthalene has a log K_{ow} of 3.3 and a water solubility of 31.8 mg/L. In comparison, another PAH, benzo[a]pyrene has a K_{ow} of 6.0 and a solubility of 0.0000000055 mg/L.

Organic molecules that have a relatively high water solubility or a log K_{ow} less than 3.5 are usually called hydrophilic ("water-loving"), whereas those with log K_{ow} greater than 3.5 are called lipophilic ("lipid or fat-loving"). Hydrophilic chemicals are also lipophobic ("lipid-fearing") and lipophilic chemicals are similarly hydrophobic. Hydrophobic chemicals, if in a body of water, are typically on the surface of the water such as PAH films which appear as an oily sheen or sheen, are attached to particulates. Polar molecules have a difficult time entering plant or animal cells because the outer nonpolar portion of the cell or plasma membrane provides a barrier to these molecules unless there are special receptors on the outer surface of the membrane (water passes through freely because of its small molecular size). Nonpolar molecules more readily pass through the membrane and into the interior of cells where they can cause many problems, as we shall see in the following chapters of this book. Passage through the plasma membrane for ions such as dissolved metals can be facilitated if they are attached to organic molecules such as methylmercury which is formed by a bond between the metal mercury with a nonpolar methyl ($-CH_3^-$) group. Organic contaminants with moderately high K_{ow} values have a potential to bioconcentrate and biomagnify because of their attraction to lipids.

Soil/Water Partition Coefficients

Contaminants in soil or water can be found attached to particulates, such as soil, or suspended matter in the water column. In general, contaminants that are attached to particulate matter are less available to organisms (ie, bioavailable) and less reactive to other environmental factors such as ultraviolet radiation than the same chemical free in water. Each molecule has a characteristic propensity to adsorb to soil, but this propensity may change with environmental conditions such as pH. The ratio between a contaminant in water and that attached to particulates under standard conditions of pH, temperature, and pressure is referred to as its *soil/water partition coefficient* (K_{oc}). This coefficient is calculated by measuring the amount of contaminant adhering to organic carbon particles and dividing that by the amount in the water phase. The K_{oc} is primarily used for soil matrices, but the values can also be used to assess the likelihood that a contaminant will be free in the water column of a lake or stream rather than attached to particles in the water. Chemicals with high K_{oc} values are more likely to adhere to particles and less likely to be bioavailable than chemicals with lower values. Most often, the log of K_{oc} is used.

Vapor Pressure

A general definition of vapor pressure is that it is the pressure exerted by a vapor in thermodynamic equilibrium with its condensed phases (solid or liquid) at a given temperature in a closed system. It is usually measured in mm Hg. Vapor pressure is relevant to ecotoxicology because it is a reflection of the tendency of a chemical to be present in the atmosphere. Contaminants with low vapor pressure are not very likely to be present in air. Vapor pressure increases as a substance is heated or if it is in a liquid rather than solid state. Vapor pressures for metals in the solid state are near zero and their melting points are too high to be environmentally relevant so, if they are in the atmosphere at all, it is mostly because they are attached to windblown particles. For, PAHs, dioxins, furans, and polybrominated diphenyl ethers (PBDEs), vapor pressures tend to decline as molecular weight or halogenation increases. Vapor pressure for PAHs, for example, ranges from 2.9×10^{-6} to $11.1\,mm\,Hg$ and for PCBs, PBDEs, and dioxins the range is 9.6×10^{-11} to $0.05\,mm\,Hg$. OPs have vapor pressures between 10^{-5} to $10^{-7}\,mm\,Hg$, as do organophosphates.

Melting Point

The melting point or the temperature in which a solid form of a contaminant becomes liquid is important for a couple of reasons. If a contaminant is solid at normal environmental temperatures (say 0°C to 40°C), it will be less mobile, less likely to vaporize, less likely to be found in the atmosphere, and less likely to be biologically available than if it is a liquid at those temperatures. Temperatures are often higher than environmentally realistic for organic molecules or metals to be gases, so that phase is not very important for most contaminants. It is axiomatic that melting points among contaminants span a huge range. Most metals have melting points exceeding several hundred degrees Celsius or more than 1000°F. However, there are the exceptions of mercury, which becomes a liquid at −38°C (−37°F), and arsenic that does not have a melting point, *per se*, but sublimates or turns into a gas at 615°C (1137°F) and never becomes a liquid. Most of the organic contaminants have low melting points at lower degrees of halogenation or molecular weight which then increases with the degree of halogenation or molecular weight. For example, the lightweight PAH naphthalene has a molecular weight of 128.2 g/mol and a melting point of 80.3°C (176.5°F) and the heavier benzo[a]pyrene at 252.3 g/mol has a melting point of 495°C (923°F).

Structure

The molecular structure of a contaminant has many ramifications for its behavior. In Chapter 4 on OPs, we will discuss stereoisomers, which are molecules with the same molecular weight and same atomic composition but different structures. These stereoisomers may have very different behavior in the environment and different toxicities. For example, endrin and dieldrin are stereoisomers of each other (Fig. 1.1), but endrin is approximately 10 times more toxic to laboratory mammals than dieldrin.

The most toxic PCBs, dioxins, and furans are called *coplanar* or *planar*, as opposed to nonplanar PCBs. PCBs are composed of two phenyl or benzene rings and if the two rings lie in the same plane, they are coplanar (Fig. 1.1) and can attach to a particular receptor at the cell level. Binding with this receptor facilitates their entrance into the cell where they can alter protein synthesis.

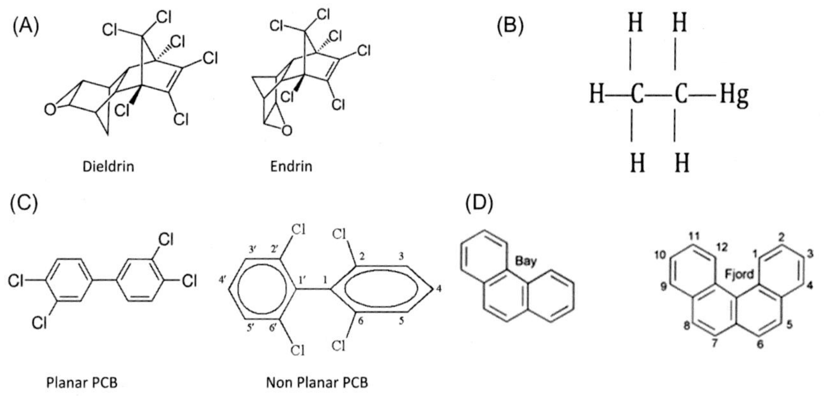

FIGURE 1.1   Differences in structure can be very important in how a contaminant behaves. (A) Dieldrin and endrin are organochlorines that are stereoisomers of each other and have different toxicities. (B) Hg is much more toxic when attached to an organic molecule such as ethane; (C) Planar PCBs are more carcinogenic than nonplanar PCBs, and (D) the Bay and Fjord regions of PAHs can increase toxicity. *From webbook.nist.gov.*

Coplanar PCBs are dioxin like and pose a serious cancer threat to humans and wildlife. The rings of nonplanar PCBs cannot lie in the same plane, cannot bind with the cellular receptor, and are much less toxic than coplanar PCBs (see Chapter 6).

Some PAHs pose serious cancer risks because one end of their molecule has an open bay or fjord region (Fig. 1.1) that allows them to attach to a receptor on the cell membrane and enter the cell readily. PAHs without this bay region are less serious of a risk (see Chapter 7).

As we have already discussed in this chapter and will spend more time on in Chapter 8, metal ions may be blocked from entering a cell by the plasma membrane. When attached to a methyl- or ethyl- group, however, they can more readily enter cells. Methylation can also increase the toxicity of PAHs. Strictly speaking, the attachment of a methyl- or ethyl- group to either a metal or any other molecule creates a different molecule so it is not just a structural change but it has a similar connotation.

Persistence

Persistence is the ability of a contaminant to stay in the environment unchanged. Metals, being elements, of course, are permanent although they may be buried deep in sediments or soil and be biologically unavailable. Persistence is typically measured in half-lives. A half-life is the constant amount of time (given constant environmental conditions) it takes half of a quantity of a contaminant to degrade into something else. It may sound a bit surprising, but it takes the same amount of time for 100 kg of a PCB to degrade to 50 kg as it does for 10 kg to degrade to 5 kg. Sometimes the degradation process results in contaminants that are more toxic than the original or parent product. For example, p,p′-DDE, which is a degradate of DDT, has a much stronger effect on producing eggshell thinning in fish-eating birds than the p,p′-DDT that was actually sprayed on a farmer's field. Both have half-lives measured in decades. Some plastics have half-lives measured in hundreds of years. Although they may be mechanically broken down into smaller and smaller pieces and become dispersed, their actual molecular composition remains unchanged. In contrast, low-molecular-weight PAHs in the atmosphere may last less than an hour. As with several of the other factors mentioned previously, persistence of many organic contaminants increases with molecular weight and/or halogenation.

ENVIRONMENTAL FACTORS THAT AFFECT CONTAMINANTS

Contaminants, or any molecule in the environment for that matter, are subject to many weathering or degrading factors that cause their eventual breakdown or increase their rate of loss into soils or sediment. For the most part, organisms live primarily in the upper few centimeters of sediment or dense clay and only a bit deeper in looser soils. If contaminants are buried deeper, they are usually out of the realm of biological organisms unless upheaval due to earthquakes, volcanoes, or human digging operations bring them back to the surface.

Ultraviolet Radiation and Oxidizers

Sunlight is a powerful force in the degradation of many different kinds of contaminants, especially in the presence of oxidizers such as hydrogen and other peroxides, oxygen, ozone, chlorate, and several others. Photochemical reactions in the environment are initiated by the absorption of a photon, typically in the wavelength range of 290–700 nm, which extends from the ultraviolet through the visible light portion of electromagnetic energy (Fig. 1.2). The energy of an absorbed photon is transferred to electrons in the chemicals and briefly changes their configurations (eg, promotes the molecule from a ground state to an excited state). Often excited state molecules are not kinetically stable in the presence of oxidizers and can spontaneously decompose (oxidize or hydrolyze) in a process called *direct photolysis*. Alternatively, molecules decompose to produce high energy, unstable fragments that can react with other molecules around them which is called *indirect photolysis*. De la Cruza et al. (2012) experimentally exposed wastewater to ultraviolet radiation at 254 nm with or without an oxidizer (hydrogen peroxide) and tested the responses of 30 pharmaceuticals and pesticides. With ultraviolet radiation alone from 0% to 100% of the contaminant was photo-oxidized within 10 min and six chemicals were completely degraded by ultraviolet (UV) light. For those chemicals that did not completely degrade under UV light, the hydrogen peroxide increased removal of the contaminant.

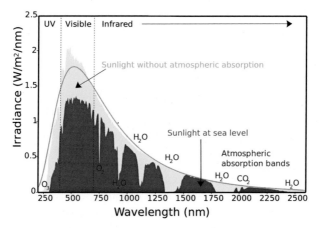

FIGURE 1.2   Solar radiation on earth. UV light is a potent factor in the degradation of organic contaminants through photolysis, but visible light at higher wavelengths also contributes to this degradation. *From Wiki Media.*

While hydrogen peroxide and other oxidizers occur naturally, they can also be artificially supplemented to enhance remediation.

Biological Organisms

Although contaminants are toxic, indeed some were created to be especially toxic to pests, biological organisms can still aid in their degradation, particularly if their concentrations are not extreme. Certain species of bacteria and fungi are noted for their ability to biodegrade specific types of contaminants, even those such as organochlorines and some PCBs that have long half-lives. Science has produced "biodegradable plastics" for shopping bags and other uses, with the idea that bacteria will decompose an organic matrix that holds plastic particles together. Some of these biodegradable plastics are not fully biodegradable in reality. Rather, while bacteria and other organisms break down the organic matrix, they do not actually break down the plastics themselves, and these remain as very small pieces in the environment.

Many organisms assimilate or *bioaccumulate* contaminants, taking them up into their tissues. Certain species of plants are particularly tolerant of contaminants and can actually help remove the contaminants from polluted soils and sediments. These plants *hyperaccumulate*—that is, they assimilate high concentrations of the contaminants and, when the growing season is over, can be harvested and burned or composted to help remove the contaminants. Plants and other organisms that take up contaminants at higher concentrations than are in the surrounding environment are said to *bioconcentrate* and the ratio of contaminant concentrations in the organism compared with the sediment, water, or soil is called a *bioconcentration factor (BCF)*. Sometimes BCFs can be on the order of several hundred times environmental levels.

Particulate Matter

We have discussed the influence of particulates on contaminants under the soil/water partitioning coefficient so we will just add a few more words here. Soil or sediment particles range in size from 0.002 mm clay to 2.0 mm coarse sand. Clay particles have slight negative charges that can form weak ionic bonds with contaminants and their collective ionic attraction can be strong enough to tightly hold contaminants in place. Larger soil particles have nooks and crannies that facilitate adsorption. In the atmosphere, these particles may carry contaminants many kilometers from their source before eventually coming back down to land or water as *dry precipitation*. When the contaminants are nitrates or sulfates produced during the combustion of fossil fuels, such as with coal-fired electrical generating plants, they may acidify lakes and streams far downwind of the source as *acid deposition* (see Chapter 9).

As we said, contaminants that are bound to particulates are less bioavailable than those that are free in the water column or in the water between soil particles and contribute less to total toxicity. One type of particulate that is well known for reducing exposure and toxicity to aquatic organisms is dissolved organic carbon (DOC), sometimes also called *dissolved organic matter (DOM)*. Similarly, soils and sediments with high organic matter tend to ameliorate contaminant toxicity compared to soils with low organic matter. Soils with higher organic matter tend to have reduced toxicity to organisms than soils with little organic matter across a wide spectrum of contaminants for similar reasons.

Measuring Environmental Contaminants

A substantial part of ecotoxicology is determining how much and what type of contaminants are present in a particular locale. To address this need, scientists and engineers have developed several very sensitive instruments to separate the contaminant from the environmental slurry, or matrix, that often accompanies field collections and to quantify the concentrations. The environmental slurry may be mud, particulates, large quantities of air, body parts or lipids if sampling organisms, or anything else that is of interest to ecotoxicologists. To ensure that results are comparable from one laboratory to another, a large battery of standardized methods or protocols have been developed and published covering just about every environmental matrix and many different types of organisms. Accurate, reliable, and repeatable protocols are especially important in regulatory situations that may involve legal issues. To this end, the US Environmental Protection Agency (US EPA) has published its own set of protocols that can be easily referred to in the methods sections of papers or reports that use these protocols—thus, interested readers can find detailed information on how the analyses were conducted. An example of a collection of EPA protocols is the Response Protocol Toolbox for water that contains methods for testing drinking water and groundwater (US EPA, 2015), but there are hundreds of other protocols for specific purposes and instruments.

Detailed descriptions of these protocols are beyond the scope of this book; interested students should consult with experienced environmental chemists for more details. We will briefly identify a few of the analytical instruments used to quantify contaminants just to provide some idea of how analyses are conducted. Note that there are several other methods including ion chromatography (IC), high performance liquid chromatography (HPLC), HPLC-MS, and others. There are some other instruments that are used to quantify contaminants; a detailed discussion of them is also beyond the scope of this book, however, so we've limited our discussion to the most widely used tools.

Gas Chromatography (GC)—This process is used in many types of analytical chemistry. Gas chromatography can be used with organic contaminants that can be vaporized at high temperatures without destroying their molecular composition. The instrument or gas chromatograph consists of two elements. A mobile element consists of a carrier gas, usually an inert gas such as helium or an unreactive gas such as nitrogen. A stationary element is a microscopic layer of liquid or polymer on an inert solid support, inside a piece of glass or metal tubing called a column.

The vaporized compounds being analyzed are carried by the gas to interact with the walls of the column, which are coated with a specific stationary element chosen for the class of contaminants of interest. Each contaminant leaves the gaseous mixture or elutes at a different time in the column, known as the retention time of the compound. The comparison of retention times is what gives GC its analytical usefulness. The final product is a chromatogram that has time as its horizontal axis and displays peaks when the various chemicals elute (Fig. 1.3). Naturally, the process is more complicated than described here. For example, the separation of mixtures with chemicals having closely spaced elution times can be improved by controlling the gas flow rate, type of carrier gas, temperature of the gas, and variation in how the sample is injected into the instrument. In a chromatogram, the area under a peak is proportional to the amount of analyte present in the chromatogram. If the investigator has an idea what chemicals may be present, their specific identities can be confirmed through the use of reference standards of chemicals and looking for similar elution times. However, the next method described can be used with greater confidence to identify chemicals even when there is limited knowledge beforehand. GC seems to be the instrument of choice for television detective shows, but the results take a lot longer to produce than the hour-long programs would allow and are not nearly as precise as suggested by Hollywood.

Mass Spectrophotometry (MS)—This process further helps identify the amount and type of chemicals present in a sample by measuring the mass-to-charge ratio and abundance of gas-phase ions. Typically a sample, which may

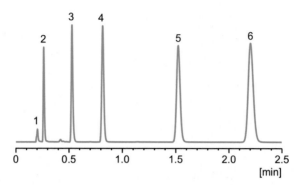

FIGURE 1.3 A chromatogram showing six spikes representing different chemicals. The height of the spike is related to the concentration of that chemical so element 1 has a low concentration and elements 3 and 4 have higher concentrations. The min refers to the number of minutes required for a chemical to be eluted. Element 6 took the longest time to elute. The tiny peak between elements 2 and 3 may be a false reading or artifact.

be solid, liquid, or gas, is ionized by bombarding it with electrons. This causes some of the sample's molecules to break into charged fragments. These ions are then separated according to their mass-to-charge ratio, typically by accelerating them and subjecting them to an electric or magnetic field: ions of the same mass-to-charge ratio will undergo the same amount of deflection while others will be deflected to a greater or lesser extent. The ions are detected by an electron multiplier or similar device. Results are displayed as spectra of the relative abundance of detected ions as a function of the mass-to-charge ratio. The atoms or molecules in the sample can be identified by correlating known masses to the identified masses or through a characteristic fragmentation pattern. There are manuals that assist in identifying unknowns. The mass spectrum (Fig. 1.4) is a plot of the ion signal as a function of the mass-to-charge ratio. Each organic chemical has a unique plot or fingerprint under a constant set of analytical methods.

When gas chromatography is used in combination with mass spectrometry, one unit (called a *GC/MS*) holds both instruments. The gas chromatograph is used to separate different compounds and shoot them into the mass spectrophotometer where it can analyze one chemical at a time.

Atomic absorption spectroscopy (AAS)—A procedure for the quantitative determination of chemical elements using the absorption of optical radiation (light) by free atoms in the gaseous state. In ecotoxicology the method is widely used with metals and metalloids. Briefly, the sample is bombarded with radiation of a given wavelength that promotes the electrons of the atoms in the atomizer to excited higher orbitals for a very short period of time. This wavelength is specific for a particular electron transition, and hence for a particular element, so the investigator has to preset the instrument for the element of interest. The

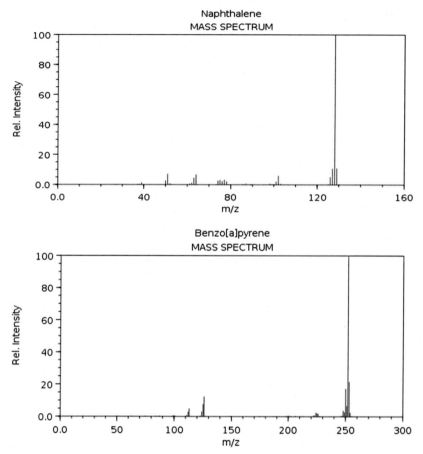

FIGURE 1.4 Mass spectrophotometry spectra for naphthalene and benzo[a]pyrene, two polycyclic aromatic hydrocarbons. The spectrum for each chemical has a characteristic number and relative height of spikes. *From NIST Chemistry WebBook (http://webbook.nist.gov/chemistry).*

radiation flux with and without a sample in the atomizer is measured using a detector, and the ratio between the two values (the absorbance) is converted to analyte concentration or mass. The technique requires standards with known analyte content to establish the relation between the measured absorbance and the sample concentration. A common way of vaporizing the sample is with a graphite furnace that combusts the sample at approximately 2700°C (4892°F).

A *cold-vapor atomization* method is specific for the determination of mercury because it is the only metal with a high-enough vapor pressure at ambient temperature. The method initiates by converting mercury into Hg^{2+} by oxidation from nitric and sulfuric acids, followed by a reduction of Hg^{2+} with tin chloride. The mercury is then swept into a long-pass absorption tube by bubbling a stream of inert gas through the reaction mixture. The concentration is determined by measuring the absorbance of this gas at 253.7 nm. Detection limits for this technique are in the parts per billion (μg/kg or μg/L) range.

This brings up the topic of units of measurement. Contaminants are frequently in very small concentrations compared with what we might be used to. A liter of water has 1000 milliliters (mL), and a milliliter has approximately 10 drops and weighs 1 g, which is not very much to the average person, but huge in the world of contaminants. The following is a brief rundown on measurements typically discussed with environmental contaminants.

Basic Units:
Liter (L)
Kilo (k)—1000; eg, 1 kg = 1000 g (1 kg)
Milli (m)—1/1000 or 10^{-3}; eg, 1 mL = 1/1000 L; 1 mg = 1/1000 g
Micro (μ)—1/1,000,000 or 10^{-6}; eg, 1 μL = 1/1,000,000 L; 1 μg = 1/1,000,000 g
Nano (n)—1/1,000,000,000 or 10^{-9}; eg, 1 nL = 1/1,000,000,000 L; 1 ng = 10^{-9} g
Pico (p)—1/1,000,000,000,000 or 10^{-12}; eg, 1 pL = 10^{-12} L; 1 pg = 10^{-12} g
For atmospheric measurements 1 cubic meter or m^3 = 1000 L.

Concentrations:
How much is one part per million or one part per billion? I, frankly, cannot imagine that without some reference. According to the National Environmental Services Center of the University of West Virginia (Satterfield, 2004), 1 ppm is comparable to 1 in. in 16 miles, 1 s in 11.5 days, 1 min in 2 years, one car in bumper-to bumper traffic from Cleveland to San Francisco, or four drops of ink in a 50-gal. barrel of water. How about one part per billion? A classic website for the toxicity of chemicals (Extoxnet, http://pmep.cce.cornell.edu/profiles/extoxnet/TIB/ppm.html) described it this way: "To give you an idea of how little this would be, a pinch of salt in 10 tons of potato chips is also one part (salt) per billion parts (chips)." Other examples include one silver dollar in a roll stretching from Detroit to Salt Lake City, one sheet in a roll of toilet paper stretching from New York to London, or 1 s in nearly 32 years (Satterfield, 2004).

1 g/kg or 1 g/L or 1 mL/L = 1 part per thousand or (sometimes) ppt (*ppt* also stands for "parts per trillion")
1 mg/kg or 1 mg/L or 1 μL/L = 1 part per million (ppm)
1 μg/kg or 1 μg/L or 1 nL/L = 1 part per billion (ppb)
1 ng/kg or 1 ng/L or 1 pL/L = 1 part per trillion (the other ppt).

Depending on the method of extraction, the care used in extracting and preparing the sample, and the instrument used in final analysis, the lower limits of quantification can be in the ppb or ppt range for water and solids.

Air concentrations are often in the μg/m^3 or even ng/m^3 range (or lower), which would be per 1000 L of air or parts per trillion or quadrillion. Special sampling devices are used to scavenge large quantities of air over a period of time in order to have enough material to quantify. Chemists sometimes make a distinction for all matrices between limits of detection and limits of quantification with the former signifying that they are pretty sure the substance is present, but at a level too low to be certain of concentration.

BASICS OF ASSESSING TOXICITY

In any given species and any given contaminant, there are going to be individuals that are highly sensitive to the chemical and others that are quite tolerant of it. Take humans, for example. From 3% to 5% of the population has a sensitivity to aspirin (salicylic acid) where taking a single pill can induce allergy-like symptoms and even anaphylaxis. Other individuals can take several times the recommended dose without any harmful effects or even any pain-relieving effects, for that matter. Most people experience pain relief with the recommended dose of 650 mg (two tablets or pills), but because aspirin is relatively safe compared with many medications, it has a huge standard daily dose range from 50 to 6000 mg. Another example is gluten, a protein commonly found in wheat and some other grains. Most people can eat breads and cereals containing gluten without concern. For others with a gluten sensitivity or celiac disease, ingestion of just a small amount can cause bloating, vomiting, abdominal pain, flatulence, stomach noises, and problems associated with decreased eating including weight loss, tiredness, and anemia. Similar ranges of tolerance are common in other organisms exposed to contaminants.

To accommodate this range of tolerance within a population, scientists have devised standard protocols of conducting laboratory experiments to assess the toxicity of a given compound. Again, the US Environmental Protection Agency has developed and published many of these protocols over the years. Specific details of toxicity testing are covered in Chapter 2, but we will cover the very basics here.

Toxicity tests are defined by their end points, duration, and mode of exposure. For many years, the so-called gold standard was to determine the *median lethal dose* (or concentration) or the amount that statistically kills 50% of the test population. The term *dose* (LD50) is used when the amount of contaminant an animal is exposed to is precisely known as when they are given an injection or when the contaminant is applied to the skin as a cream or oil, or when it is force fed to an animal. *Concentration* (LC50) is used when the contaminant is presented in water, food, or some other means, and the concentration actually taken in is not precisely known and may vary somewhat among individuals.

Statistical methods including *probit* analysis or *logistic regression* were developed to provide an estimate with confidence intervals of the median lethal dose and other reference points. For example, a hypothetical probit analysis of an experiment on ducks and cadmium using several doses and replicates at each dose may result in an estimate of a median lethal concentration of 15 mg cadmium/kg of duck body weight with a 95% confidence interval of 5–25 mg/kg. This means that the best estimate we have considering the variability within the experiment is that 50% of the ducks will die at 15 mg/L, but we are 95% confident that the true median lethal concentration is somewhere between 5 and 25 mg/L.

Early experiments quickly revealed that the duration of the exposure was very important to determine lethal concentrations. A concentration that showed no effect after a day might result in 100% mortality by the end of a week. Thus, an essential part of the experimental description is the duration of the test. There are several global institutions concerned with standardizing such tests including the American Public Health Association, US EPA, American Society for Testing and Materials, International Organization for Standardization, Environment Canada, and Organization for Economic Cooperation and Development. Standardized tests offer the ability to compare results between laboratories and are useful in the regulatory component of contaminants. Some specific definitions include

Acute—Tests requiring 1–4 days (24–96 h)
Subacute—A standard 28-day test
Subchronic—Standardized to 90 days
Chronic—More than 90 days

The 96-h acute test, undoubtedly the most commonly used aquatic toxicity test, was established by the US EPA not because 96 h has any special relevance to the toxicity of a compound, but because such studies could be set up and completed within a standard work week. An investigative team can begin exposing their animals on Monday morning and complete the study on Friday. Similar acute toxicity tests have been developed for mammals and birds in which animals are exposed once and observed for a predefined period or exposed repeatedly for a set number of days followed by a period of observation without dosing.

Outside of the regulatory arena and more into the experimental side of things, acute usually means 1–96 h, subacute from 4 to 7 days, and *chronic* can mean anything more than 7 days and is often related to the life cycle of the species of interest. For example, tests on amphibians may focus on the embryonic or larval stages of the test organisms and span that entire period, even if it requires many weeks as with some species of frogs. For birds and mammals, chronic tests may also cover a particular life stage, such as from hatching or birth until the first molt or weaning, or may occur in known aged animals.

Experiments conducted in the 1970s and later showed that the median lethal concentration or dose for an animal was often higher than what that animal would normally encounter. In other words, the concentrations necessary to kill an animal, especially in acute tests exceeded the *environmentally relevant* or expected concentrations. Such tests revealed what could potentially happen under very severe conditions, but not necessarily under natural conditions. Much lower concentrations over longer periods of time were found to produce sublethal effects such as reduced growth or development, endocrine disruption, histopathology, or other physiological changes. Thus, the terms *median effective concentration* (EC50) or *dose* (ED50) were developed and estimated statistically using the same methods as with LC50, but with a sublethal marker used as an endpoint. Along with the EC50 concept, other benchmark concepts were developed, including the No Adverse Effects Level (NOAEL), defined as the highest tested dose or concentration of a chemical at which no harmful effect is found in exposed test organisms where higher doses or concentrations did result in adverse effects. A similar concept, the Lowest Observed Adverse Effects Level (LOAEL), is the lowest concentration of a contaminant found by experiment or observation that causes an adverse alteration of morphology, function, capacity, growth, development, or lifespan of a target organism. A few studies have used other measures of effect including the mean time for effects to be observed. Chapter 2 discusses each of these concepts in greater detail and addresses how to calculate these benchmarks.

STUDY QUESTIONS

1. In your own words, distinguish the terms *ecotoxicology* and *toxicology*.
2. At this early stage in the course, do you have any preconceived ideas of what aspect of ecotoxicology—environmental chemistry, physiological effects, influences at ecological scales of reference, or risk assessment—sound most interesting?
3. Why might the investigation of sublethal effects due to contaminants be more realistic than investigating lethal effects?
4. What is the difference between compensatory and additive mortality in population studies? Discuss how you might go about distinguishing whether a contaminant has either one of those effects.
5. Discuss with your class recent events involving contaminant exposure in real life situations.
6. In general, what influence does increased halogenation or molecular weight have on the solubility, vapor pressure, melting point, and persistence of PCBs?
7. What is the importance of vapor pressure in predicting whether a contaminant will be found in the atmosphere?
8. What is the difference among bioavailability, bioaccumulation, and bioconcentration?
9. What is the role of dissolved organic matter on bioavailability?
10. What instrument would you use if you:
    a. Wanted to determine the concentration of a given metal in an environmental sample?
    b. Had to positively identify an unknown chemical from an environmental sample?
    c. Wanted to estimate the number of chemicals present in a sample?
11. What is the difference between:
    a. Median lethal dose and median lethal concentration?
    b. Median lethal dose and median effective dose?
    c. Lowest Observed Adverse Effects Level and No Observed Effects Level?

REFERENCES

Allison, J.D., Allison, T.L., 2005. Partition Coefficients for Metals in Surface Water, Soil, and Waste. U.S. Environmental Protection Agency Office of Research and Development, Washington, DC.

Burnham, K.P., Anderson, D.R., 1984. Tests of compensatory vs additive hypotheses of mortality in mallards. Ecology 65, 105–112.

Cairns Jr., J., 1989. Will the real ecotoxicologist plese stand up? Environ. Toxicol. Chem. 8, 843–844.

De la Cruza, N., Giménez, J., Esplugas, S., Grandjean, D., de Alencastro, L.F., Pulgarín, C., 2012. Degradation of 32 emergent contaminants by UV and neutral photo-fenton in domestic waste-water effluent previously treated by activated sludge. Water Res. 46, 1947–1957.

Hoffman, D.J., Rattner, G., Burton Jr., G.A., Cairns Jr., J., 2003. Introduction. In: Hoffman, D.J., Rattner, G., Burton Jr., G.A., Cairns Jr., J. (Eds.), Handbook of Ecotoxicology Lewis Publ, Boca Raton, FL, pp. 1–15.

Satterfield, Z., 2004. What Does ppm or ppb Mean? National Environmental Services Center., <http://www.nesc.wvu.edu/ndwc/articles/ot/fa04/q&a.pdf>.

Truhaut, R., 1977. Eco-toxicology: objectives, principles and perspectives. Ecotoxicol. Environ. Saf. 1, 151–173.

US EPA U.S. Environmental Protection Agency, 2015. Response Protocol Toolbox (RPTB). <http://yosemite.epa.gov/water/owrccatalog.nsf/SingleKeyword?Openview&Keyword=Response+Protocol+Toolbox+(RPTB)&count=2000&CartID=null> (accessed 01.12.15.).

Chapter 2

Basics of Toxicity Testing

Terms to Know

Toxicant
Static Test
Static Renewal Test
Flow-through Test
Gavage
Dietary Test
Categorical Data
Continuous Data
Discrete Data
Interval Data
Experimental Unit
Sampling Unit
Good Laboratory Practices (GLPs)
Quality Assurance (QA)

INTRODUCTION AND TERMS OF THE TRADECRAFT

Simply stated, many of the questions that ecotoxicologists face focus on exposure and effects. More often than not, when considered within the context of field conditions—regardless of the type of environment—exposure and effects become entangled and must be considered as an integrated outcome. A major function of ecotoxicology is thus to bring scientific findings to regulatory agencies that can then make informed decisions on how to interpret that data and decide resource management. Another area of ecotoxicology involves scientific investigations with no direct interest in management decisions that seek to obtain clearer understandings of the physiological or ecological effects of chemicals on biological organisms. Implicit in nearly all studies regarding the toxicity of chemicals, whether the studies occur under controlled laboratory settings or in the field, is an interest in how the chemicals affect organisms under so-called natural conditions.

In general, toxicity refers to the potential for a substance to produce adverse effects on biota. A *toxicant*, then, is simply an agent, most often a chemical acting alone or as a constituent in a complex environmental mixture, that produces an adverse effect in a biological system, causing death or damage to an organism's structure or function. When considering higher levels of biological organization, such as communities or ecosystems, the harm is often expressed as affecting the system's sustainability. For individuals within a species or as members of categories within a species (eg, adult versus neonate, male versus female), adverse effects are most readily characterized by a biological response defined in terms of a measurement that is outside the normal range for healthy organisms. Most often, adverse effects are characterized as departures from normal ranges for endpoints related to survival, growth, and reproduction. Even these endpoints may be subject to much discussion, especially when viewed from various perspectives of resource management. In the following sections, we will briefly survey toxicity testing, starting with a very brief historical setting and then continue to an overview of the process of toxicity testing that relates to its implementation in order to address questions critical to ecotoxicologists.

TOXICITY TESTING: HISTORICAL PERSPECTIVES AND ITS ROLE AS ONE TOOL IN THE ECOTOXICOLOGY TOOLBOX

Historical Perspectives

Concern about the effects of toxicants on organisms (usually humans) is ancient, stemming from as far back as 2700 years BC with Shen Nung, the father of Chinese medicine. Supposedly he sampled many different herbs until 1 day he overdosed and died (De Carvalho, 2012). As with much of history, many inroads have contributed to the toxicity testing practiced today, particularly related to its role in characterizing biological responses potentially linked to the adverse effects of chemicals when released into the environment. Although early accounts of toxicity were based on observation and anecdotal evidence linking chemical exposures to adverse biological outcomes, early efforts to develop methods that assess chemical toxicity may be illustrated by examples of toxicity-test methods development. In Chapter 1, we discussed some of this with regard to the concern over pesticides in the 1960s and today. Concern over polychlorinated biphenyls, polycyclic aromatic hydrocarbons, and the many other chemicals that are developed and released annually has also expanded our understanding of ecotoxicology. For example, throughout the 1970s and early 1980s, the United States faced an energy crisis characterized by increased energy use, dwindling domestic supplies of oil, and increasing reliance on foreign oil. To develop alternative fuels intended to relieve the country's sole reliance on imported oil, development of "synfuels" was undertaken by major petroleum companies whose aim was to fulfill the country's energy needs by developing the country's supply of oil shale and tar sands, or from liquefaction or gasification of coal by exposure to high temperatures and pressures (Probstein and Hicks, 1990). As these technologies were developed, however, concern for their environmental impacts heightened because conversion of oil shale and tar sands, or the liquefaction or gasification of coal, yielded effluents and production byproducts that were complex mixtures of organic and inorganic compounds and potentially toxic, particularly when released to surface waters during or following the manufacturing process. Development of these energy technologies, and the disposal and management of their derivative complex chemical mixtures, gained the attention of regulatory and science organizations such as the U.S. Environmental Protection Agency (US EPA), academic researchers, national laboratories such as Oak Ridge National Laboratory, and industry. This represented the beginning of the test methods development community, whose efforts, combined with those of the energy-development community, yielded technologies and tools for evaluating hazards and risks associated with these technologies, which were beginning to be understood.

For example, aquatic toxicity tests were developed because resource management concerns focused on releases of synfuels-processed water to surface waters, and the many other contaminants that enter water bodies. Some of these tests relied on macroinvertebrates such as *Daphnia magna* or *Daphnia pulicaria*, (eg, Biesinger and Christensen, 1972; Geiger et al., 1980), fishes such as fathead minnow (*Pimephales promelas*), and the coldwater rainbow trout (*Oncorhynchus mykiss*) (Sprague, 1969, 1970). The US EPA and the American Society for Testing and Materials (ASTM) were instrumental in developing these standards (eg, E-729 on the Standard Practice for Conducting Acute Toxicity Tests with Fishes, Macroinvertebrates, and Amphibians, ASTM, 2015a). While toxicity tests were developed for fishes or aquatic invertebrates, Dumont and colleagues (Dumont et al., 1979; Dumont and Schultz, 1980) were developing amphibian test methods in Oak Ridge National Laboratory (Oak Ridge, Tennessee) using amphibian embryos to examine the toxicity of coal and oil-shale conversion technologies. Other investigators applied similar toxicity tests in various laboratory, field, or integrated field-laboratory investigations for amphibians (eg, Linder, 2003) throughout the 1990s. These efforts continue today to address data needs (US EPA, 2002a; Bleiler et al., 2004).

Similarly, test methods to evaluate sediments and soils with respect to a chemical's toxicity to biota exposed in the laboratory tests or in field settings were developed. Sediment test methods are numerous, using aquatic and wetland plants (algae, nonvascular, and vascular plants), benthic and epibenthic invertebrates, fish, and amphibians. For soils, a range of test methods are available, including those focused on soil biota, terrestrial invertebrates, and vertebrates. These methods were developed throughout the 1990s. A list of available representative test methods to evaluate the range of exposure matrices potentially encountered in field settings is summarized in Tables 2.1–2.3.

TABLE 2.1 Consensus-Based Tests Developed by the ASTM Are Representative of Those Available to Evaluate Toxicity of Exposure Matrices of Interest to Ecotoxicologists

Aquatic Toxicity Tests: Freshwater

ASTM E1218-04(2012)	Standard Guide for Conducting Static Toxicity Tests With Microalgae
ASTM E1415-91(2012)	Standard Guide for Conducting Static Toxicity Tests With *Lemna gibba* G3
ASTM E1711-12	Standard Guide for Measurement of Behavior During Fish Toxicity Tests
ASTM E1604-12	Standard Guide for Behavioral Testing in Aquatic Toxicology
ASTM E1241-05(2013)	Standard Guide for Conducting Early Life-stage Toxicity Tests With Fishes
ASTM E1192-97(2014)	Standard Guide for Conducting Acute Toxicity Tests on Aqueous Ambient Samples and Effluents With Fishes, Macroinvertebrates, and Amphibians
ASTM E2455-06(2013)	Standard Guide for Conducting Laboratory Toxicity Tests With Freshwater Mussels
ASTM E2122-02(2013)	Standard Guide for Conducting In-situ Field Bioassays With Caged Bivalves
ASTM E1366-11	Standard Practice for Standardized Aquatic Microcosms: Fresh Water
ASTM E1193-97(2012)	Standard Guide for Conducting *Daphnia magna* Life-cycle Toxicity Tests
ASTM E1768-95(2013)	Standard Guide for Ventilatory Behavioral Toxicology Testing of Freshwater Fish
ASTM E1439-12	Standard Guide for Conducting the Frog Embryo Teratogenesis Assay-*Xenopus* (FETAX)
ASTM E1242-97(2014)	Standard Practice for Using Octanol-water Partition Coefficient to Estimate Median Lethal Concentrations for Fish Due to Narcosis
ASTM D3978-04(2012)	Standard Practice for Algal Growth Potential Testing With *Pseudokirchneriella*
ASTM E1841-04(2012)	Standard Guide for Conducting Renewal Phytotoxicity Tests With Freshwater Emergent Macrophytes
ASTM E729-96(2014)	Standard Guide for Conducting Acute Toxicity Tests on Test Materials With Fishes, Macroinvertebrates, and Amphibians

(Continued)

TABLE 2.1 Consensus-Based Tests Developed by the ASTM Are Representative of Those Available to Evaluate Toxicity of Exposure Matrices of Interest to Ecotoxicologists *Continued*

ASTM E1440-91(2012)	Standard Guide for Acute Toxicity Test With the Rotifer Brachionus
ASTM E1295-01(2013)	Standard Guide for Conducting Three-brood Renewal Toxicity Tests With *Ceriodaphnia dubia*
Aquatic Toxicity Tests: Saltwater	
ASTM E1498-92(2012)	Standard Guide for Conducting Sexual Reproduction Tests With Seaweeds
ASTM E1563-98(2012)	Standard Guide for Conducting Static Acute Toxicity Tests With Echinoid Embryos
ASTM E1463-92(2012)	Standard Guide for Conducting Static and Flow-through Acute Toxicity Tests With Mysids From the West Coast of the United States
ASTM E1022-94(2013)	Standard Guide for Conducting Bioconcentration Tests With Fishes and Saltwater Bivalve Mollusks
ASTM E724-98(2012)	Standard Guide for Conducting Static Acute Toxicity Tests Starting With Embryos of Four Species of Saltwater Bivalve Molluscs
ASTM E1191-03a(2014)	Standard Guide for Conducting Life-cycle Toxicity Tests With Saltwater Mysids
ASTM E2317-04(2012)	Standard Guide for Conducting Renewal Microplate-based Life-cycle Toxicity Tests With a Marine Meiobenthic Copepod
ASTM E1367-03(2014)	Standard Test Method for Measuring the Toxicity of Sediment-associated Contaminants With Estuarine and Marine Invertebrates
ASTM E1562-00(2013)	Standard Guide for Conducting Acute, Chronic and Life-cycle Aquatic Toxicity Tests With Polychaetous Annelids
Sediment Toxicity Tests: Saltwater and Freshwater	
ASTM E1706-05(2010)	Standard Test Method for Measuring the Toxicity of Sediment-associated Contaminants With Freshwater Invertebrates
ASTM E1688-10	Standard Guide for Determination of the Bioaccumulation of Sediment-associated Contaminants by Benthic Invertebrates
ASTM E1525-02(2014)	Standard Guide for Designing Biological Tests With Sediments
ASTM E2591-07(2013)	Standard Guide for Conducting Whole Sediment Toxicity Tests With Amphibians

(Continued)

TABLE 2.1 Consensus-Based Tests Developed by the ASTM Are Representative of Those Available to Evaluate Toxicity of Exposure Matrices of Interest to Ecotoxicologists *Continued*

ASTM E1391-03(2014)	Standard Guide for Collection, Storage, Characterization and Manipulation of Sediments for Toxicological Testing and for Selection of Samplers Used to Collect Benthic Invertebrates
ASTM E1611-00(2013)	Standard Guide for Conducting Sediment Toxicity Tests With Polychaetous Annelids
Soil and Terrestrial Toxicity Tests	
ASTM E2172-01(2014)	Standard Guide for Conducting Laboratory Soil Toxicity Tests With the Nematode *Caenorhabditis elegans*
ASTM E1676-12	Standard Guide for Conducting Laboratory Soil Toxicity or Bioaccumulation Tests With the Lumbricid Earthworm (*Eisenia fetida*) and the Enchytraeid Potworm (*Enchytraeus albidus*)
ASTM E1197-12	Standard Guide for Conducting a Terrestrial Soil-core Microcosm Test
ASTM E1163-10	Standard Test Method for Estimating Acute Oral Toxicity in Rats
ASTM E1619-11	Standard Test Method for Chronic Oral Toxicity Study in Rats
ASTM E2045-99(2014)	Standard Practice for Detailed Clinical Observations of Test Animals
ASTM E857-05(2012)	Standard Practice for Conducting Subacute Dietary Toxicity Tests with Avian Species
ASTM E981-04(2012)	Standard Test Method for Estimating Sensory Irritancy of Airborne Chemicals
ASTM E1963-09(2014)	Standard Guide for Conducting Terrestrial Plant Toxicity Tests
General Guidance for Toxicity Testing	
ASTM E943-08(2014)	Standard Terminology Relating to Biological Effects and Environmental Fate
ASTM E1847-96(2013)	Standard Practice for Statistical Analysis of Toxicity Tests Conducted Under ASTM Guidelines
ASTM E2186-02a(2010)	Standard Guide for Determining DNA Single-strand Damage in Eukaryotic Cells Using the Comet Assay
ASTM E2385-11	Standard Guide for Estimating Wildlife Exposure Using Measures of Habitat Quality

TABLE 2.2 Representative US EPA Tests Available to Evaluate the Toxicity of Exposure Matrices of Interest to Ecotoxicologists[a] (see also EPA 2002b–d)

Guideline Designation	Title
Aquatic Fauna Test Guidelines	
850.101	Aquatic Invertebrate Acute Toxicity Test, Freshwater Daphnids
850.102	Gammarid Acute Toxicity Test
850.1025	Oyster Acute Toxicity Test (Shell Deposition)
850.1035	Mysid Acute Toxicity Test (Shrimp, Saltwater)
850.1045	Gammarid Acute Toxicity Test (Shrimp, Freshwater)
850.1055	Bivalve Acute Toxicity Test (Embryo Larval)
850.1075	Fish Acute Toxicity Test, Freshwater and Marine
850.1085	Fish Acute Toxicity Mitigated by Humic Aid
850.13	Daphnid Chronic Toxicity Test
850.14	Fish Early-life Stage Toxicity Test
850.171	Oyster BCF
850.173	Fish BCF
850.1735	Whole Sediment Acute Toxicity Invertebrates, Freshwater
850.174	Whole Sediment Acute Toxicity Invertebrates, Marine
850.179	Chironomid Sediment Toxicity Test
850.18	Tadpole/Sediment Subchronic Toxicity Test
850.185	Aquatic Food Chain Transfer
850.19	Generic Freshwater Microcosm Test, Laboratory
850.1925	Site-specific Aquatic Microcosm Test, Laboratory
850.195	Field Testing for Aquatic Organisms
Beneficial Insects and Invertebrates Test Guidelines	
850.302	Honey Bee Acute Contact Toxicity
850.303	Honey Bee Toxicity of Residues on Foliage
850.304	Field Testing for Pollinators
Nontarget Plants Test Guidelines	
850.4	Background-nontarget Plant Testing
850.4025	Target Area Phytotoxicity

(*Continued*)

TABLE 2.2 Representative US EPA Tests Available to Evaluate the Toxicity of Exposure Matrices of Interest to Ecotoxicologists[a] (see also EPA 2002b–d) *Continued*

Guideline Designation	Title
850.41	Terrestrial Plant Toxicity, Tier I (Seedling Emergence)
850.415	Terrestrial Plant Toxicity, Tier I (Vegetative Vigor)
850.42	Seed Germination/Root Elongation Toxicity Test
850.4225	Seedling Emergence, Tier II
850.423	Early Seedling Growth Toxicity Test
850.425	Vegetative Vigor, Tier II
850.43	Terrestrial Plants Field Study, Tier III
850.44	Aquatic Plant Toxicity Test Using Lemna spp. Tiers I and II
850.445	Aquatic Plants Field Study, Tier III
850.46	Rhizobium-legume Toxicity
850.48	Plant Uptake and Translocation Test
Toxicity to Microorganisms Test Guidelines	
850.51	Soil Microbial Community Toxicity Test
850.54	Algal Toxicity, Tiers I and II
Other Test Guidelines	
850.62	Earthworm Subchronic Toxicity Test

[a]*Series 850—Ecological Effects Test Guidelines for US EPA, Office of Pesticide Programs and Toxic Substances; see http://www.epa.gov/test-guidelines-pesticides-and-toxic-substances/series-850-ecological-effects-test-guidelines, last accessed December 15, 2015.*

TABLE 2.3 In Europe, Organisation for Economic Co-operation and Development (OECD) Tests Are Available to Evaluate Toxicity of Exposure Matrices of Interest to Ecotoxicologists as Exemplified by These Methods

Guideline Designation	Title
201	Alga, Growth Inhibition Test
202	*Daphnia* sp. Acute Immobilization Test
203	Fish, Acute Toxicity Test

(Continued)

TABLE 2.3 In Europe, Organisation for Economic Co-operation and Development (OECD) Tests Are Available to Evaluate Toxicity of Exposure Matrices of Interest to Ecotoxicologists as Exemplified by These Methods *Continued*

Guideline Designation	Title
204	Fish, Prolonged Toxicity Test
205	Avian Dietary Toxicity Test
206	Avian Reproduction Test
207	Earthworm, Acute Toxicity Tests
208	Terrestrial Plant Test: Seedling Emergence and Seedling Growth Test
209	Activated Sludge, Respiration Inhibition Test
210	Fish, Early-life Stage Toxicity Test
211	*Daphnia magna* Reproduction Test
212	Fish, Short-term Toxicity Test on Embryo and Sac-fry Stages
213	Honeybees, Acute Oral Toxicity Test
214	Honeybees, Acute Contact Toxicity Test
215	Fish, Juvenile Growth Test
216	Soil Microorganisms: Nitrogen Transformation Test
217	Soil Microorganisms: Carbon Transformation Test
218	Sediment-water Chironomid Toxicity Using Spiked Sediment
219	Sediment-water Chironomid Toxicity Using Spiked Water
220	Enchytraeid Reproduction Test
221	*Lemna* sp. Growth Inhibition Test
222	Earthworm Reproduction Test (*Eisenia fetida/Eisenia andrei*)
223	Determination of the Inhibition of the Activity of Anaerobic Bacteria Reduction of Gas
224	Production From Anaerobically Digesting (sewage) Sludge
227	Terrestrial Plant Test: Vegetative Vigour Test

OPENING THE ECOTOXICOLOGY TOOLBOX: COMMON ELEMENTS OF TOXICITY TESTS

Toxicity tests determine the level of biological response indicated by adverse effects displayed by test organisms following exposure. Acute toxicity tests determine whether chemical concentrations of a sample will produce adverse effects in a group of test organisms during a short-term exposure under controlled conditions. In one experiment, a 50% lethal response (concentration at which 50% of the test organisms die or median-lethal concentration or LC50) is the most frequently measured endpoint used to characterize acute toxicity. An LC50 for a specific chemical or a percent effluent lethal to 50% of the test organisms during the test period is calculated from the survival data (number of animals alive upon test termination) using one of the several statistical methods available (see Finney, 1978; Newman, 2012; ASTM, 2015b). Results from acute tests are frequently used as guides for the design of chronic toxicity tests.

Chronic toxicity tests allow for the evaluation of adverse effects of a sampled exposure matrix—water, sediment, or soil—under relatively long-term exposure compared with acute toxicity tests. Lengthening test duration to include one or more complete life cycles of the test organism or performing the test during a sensitive life stage allows for the detection of subtler adverse effects, such as reduction in growth or reproduction. Evaluation of these effects from long-term exposure to the sampled matrix provides a direct estimate of the toxicant's effect threshold. For example, during life-cycle tests with fishes and invertebrates, certain developmental stages, usually the younger ones, have consistently been shown to be more sensitive than other stages. Use of short-term tests with the early developmental stages may also serve as a method to predict chronic toxicity. Chronic toxicity tests are generally more sensitive than acute tests, primarily because the measures of toxicity focus on endpoints other than survival, which are better used to define no adverse effects levels. In addition, chronic tests provide a better estimate of responses for a population in the field. Acute tests are less sensitive measures of toxic conditions (relative to chronic tests or biomarkers); thus, the absence of an acute toxic response cannot be interpreted as the absence of toxicity. Chronic tests, however, may not detect all sublethal effects.

Regardless of the test matrix—water, sediment, or soil—biota exposed in toxicity tests are generally placed in test chambers for a prescribed exposure period under specified exposure conditions to various concentrations of the sample. For example, beakers holding environmental samples are often used for aquatic or sediment toxicity tests (Fig. 2.1), whereas pots or other containers are used for testing soils or exposed sediments. Toxicity tests are most often conducted as experiments completed under controlled laboratory settings where exposure concentration is a primary variable of concern, particularly with respect to adverse biological effects linked to chemicals. Tests are standardized to maximize comparability and reproducibility. Test methods frequently

FIGURE 2.1 Toxicity tests with aquatic vegetation or invertebrates are often conducted using small beakers.

specify exposure as a series of test concentrations of a chemical or mixture of chemicals for a defined period of time. In a multiple-concentrations test, design data provide a toxicity estimate in terms of an LC50 or a no-observed-adverse effects-concentration that is defined in terms of mortality, growth, or reproduction. Such numeric endpoints are most often determined as outcomes from statistical analysis. Tests commonly consist of a control and a minimum of five exposure concentrations, although the specific test conditions will depend on the objectives of the study and expected range of toxicity.

Selection of test organism is one element that defines a specific toxicity testing procedure. Test species are selected based upon the ease of laboratory culture, the availability of adequate background information such as physiology, genetics, and behavior and sensitivity to a wide range of toxicants. For instance, for toxicity tests of freshwater samples, a species of water flea (*Ceriodaphnia dubia* or *D. magna* or relative) and a fish, such as the fathead minnow (*P. promelas*) are often used to evaluate chemical effects in aquatic habitats (Fig. 2.2). Other species used include the amphipod *Hyalella* sp. and the rainbow trout may also be used to address specific concerns. Regardless of the test species, toxicity tests are often *static*, meaning that the organisms are maintained in the original test solutions for the duration of the test, but test conditions may vary depending on the length of exposure or life history attributes of the species used in the test. For example, a *static renewal test* may be used, such as a three-brood static-renewal test using the cladoceran *C. dubia*. In this test, the solutions are renewed periodically by transferring the test organisms to chambers with freshly prepared test solutions. For the static-renewal using

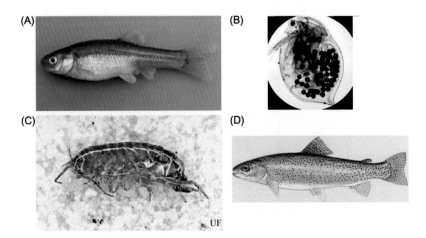

FIGURE 2.2 Some common species used in aquatic toxicity tests: (A) the fathead minnow (*P. imephales promelas*), (B) the copepod *D. magna,* (C) the amphipod *Hyallela* sp., and (D) the rainbow trout (*O. mykiss*).

C. dubia, the test is initiated with organisms that are less than 24 h old and born within 8 h of each other. There is also a *flow-through test* in which treated water flows through the test chamber constantly and is mainly used for organisms that live in streams and other flowing bodies of water.

Test endpoints measure survival and reproduction during the test period. The original neonates are introduced into each test container at the beginning of the test and monitored for survival and for the number of offspring they produce. Similarly, the *P. promelas* chronic-toxicity test measures both survival and growth during the test period. In this particular fish toxicity test, individual fathead minnows introduced into each test container at the beginning of the test are monitored for survival, and at test termination, the weight of the control groups of fish is compared to that of the test exposure groups.

Some commonly used terrestrial species are the mallard (*Anas platyrhynchus*), northern bobwhite (*Colinus virginianus*), Japanese quail (*Coturnix* sp.), and field mice (*Peromyscus* sp.). The traditional lab rats, mice, and rabbits are also used, often in a human toxicology framework. A standard method of toxicity is a dietary exposure wherein a relatively stable, nonvolatile, contaminant is added to food or dissolved in drinking water. Other methods include gavage, which places a known amount of chemical or bolus into the animals' stomach or proventriculus, and injection.

Whether toxicity tests are used to characterize water, sediment, or soil samples, all test designs will include a control (or untreated sample) to ensure that the effects observed are associated with or are attributable to exposure to the test material. As such, experimental design provides a basis for interpreting the test results. For example, water samples collected to evaluate toxicity of an effluent discharge (such as that from a water treatment facility) would be prepared for testing by thorough mixing, and specified limits for test conditions (minimum and maximum) would need to be fulfilled for any test to be valid. The sample is then diluted with control water. For example, five concentrations may be used (with the appropriate number of replicates) to span the concentration range from a 0% to 100% sample or five concentrations of a single chemical may be used in a concentration series. Test organisms would then be randomly assigned to test chambers with a sufficient volume of test material. For aquatic toxicity tests, initial conditions are recorded, such as temperature, dissolved oxygen (DO), and pH, in separate vessels. When completed following the standard guidelines, quality control components of toxicity tests are incorporated into experimental design, which underscores their critical role in the interpretation of test outcomes.

Regardless of the test matrix, laboratory conditions are specified for exposure with a particular focus on environmental conditions potentially influencing outcomes other than chemical(s) in the exposure matrix. For example, in soil toxicity tests using earthworms such as *Eisenia fetida*, all studies rely on the specification of incubation temperatures for exposure chambers as that which might naturally occur and the photoperiod (eg, 16:8 h of light:dark) would be specified along with other exposure conditions, with the primary variable of concern to the ecotoxicologist being chemical concentrations in the test matrix.

Simply stated then, toxicity tests measure effects of chemicals in contaminated soils, sediments, waters collected from areas where chemicals have been or may have been released to the environment. Alternatively, toxicity tests may be conducted proactively, wherein chemicals may be characterized with respect to their adverse effects on biota exposed in water, sediment, soil, and very occasionally to air to better understand hazards associated with that chemical if it were released to the environment as part of its specified use. For instance, during the development of agricultural chemicals or a new product for commercial use, a company may evaluate toxicity of the chemical as part of a regulatory process or as part of a research study. Often, samples of soil, sediment, or water are collected from the field and returned to the laboratory for testing with several standard laboratory test species. Toxicity tests can also be conducted in mobile laboratories or in situ, and investigating resident species from the area rather than standard test species (Anderson et al., 2004; McGee et al., 2004).

Acute effects are those that occur rapidly as a result of short-term exposure. Exposure is considered relative to the organism's life span. The most commonly measured acute effect in aquatic organisms is death. Chronic effects occur when a toxicant produces adverse effects as a result of a repeated or long-term exposure. Chronic effects include lethal and sublethal responses (such as abnormal growth and/or reproduction). Criteria for effects, such as mortality and reproduction, are evaluated by comparing those organisms exposed to different dilutions of the sample chemical with those organisms (controls) exposed only to nontoxic dilution water.

Tests in combination with chemical characterization of exposure matrices—surface water, groundwater, sediments, soils, and air—provide data critical to understanding the possible effects and toxicity of these chemicals to target and nontarget organisms. As a side note, the toxicity of airborne contaminants is seldom tested because atmospheric concentrations of contaminants are generally very low and short-lived compared with the other matrices. The possible exceptions might include gases known to be present in smog, second-hand cigarette smoke in the human context, or highly volatile organic molecules. While chemical analyses of a sample of soil or water from an environmental source provide a measure of the total concentration of specific chemicals potentially available for exposure to organisms, toxicity tests provide estimates of actual bioavailable chemical contaminants and what happens if an organism is exposed. There can be a substantial difference between the concentration of a chemical in an environmental source and what an organism is actually exposed to because of the complexity of the matrix or analytical detection limits; thus, toxicity tests play an important role and potentially link the occurrence of chemical contamination to biological effects.

Toxicity tests also indicate potential for population- or community-level effects, although demonstration and quantification of ecological effects at these levels of biological organization also require field surveys. As such, environmental sampling of exposure matrices and evaluation of toxicity are required critical components if causal analysis is a chief objective for any integrated field and laboratory investigation. Results from such integrated chemical and biological investigations incorporate measures of chemicals' bioavailability in the field, whereas reliance only on measuring chemical concentrations reveals little about whether the chemicals are affecting organisms or higher levels of organization. Furthermore, outcomes of such integrated field and laboratory studies may be specific to the location from which the sample was collected, thus they can be used to develop maps of the extent and distribution of bioavailable contamination and toxic conditions. However, depending on the methods applied to evaluating toxicity, results of toxicity tests cannot always be directly translated into an expected magnitude of effects on populations in the field. Interpretation of results from standardized toxicity tests reflect uncertainties dependent on factors such as test species, water conditions, sediment or soil quality, test duration, and others related to how and where field samples were collected and specific methods inherent to the implementation of toxicity tests. Field survey data, however, may provide insights that benefit interpretation of analytical chemistry and toxicity data. The two are only marginally comparable because field conditions have far more complexities than what can be designed in laboratory studies. It is difficult, for example, to design laboratory studies that contain all of the chemicals that organisms in the field may be exposed to, not including the multitude of predators, competitors, and possible disease organisms. As we will see in Chapter 10, however, some experimental designs can help bridge the gap between laboratory and field exposures.

Toxicity Data Analysis

Statistical analyses and mathematical modeling can provide a quantitative basis to summarize the data collected during a toxicity test. Data garnered from toxicity tests are varied, yet may fit into categories that strongly influence which analytical tools are conveniently available for data analysis and subsequent interpretation within the context of exposure and effects. In general, toxicity data occur as one of two data types—categorical data or measurement data. The term *categorical data* refers to objects grouped together based on some qualitative trait. In general, categorical data consider attributes such as color (eg, flower color, pelage), sex, age classifications (eg, adult, juvenile, neonates), or survival (eg, dead, alive). Such sampling of these everyday examples indicates that categorical data can also be classified based upon the number of categories that are potentially characteristic of all members of the population. Categorical data are classified as nominal, ordinal, or binary (dichotomous) in character (Fig. 2.3). Nominal data are a type of categorical data in which objects fall into unordered categories (eg, flower colors), whereas ordinal data are categorical data in which order is important, such as how the developmental stages of some invertebrates are an ordered set referred to as eggs, larvae, juveniles, and adults, or pathological states such as morbidity may be scored as none, mild, moderate, and severe. *Binary* or dichotomous data are categorical data that occur as one of two possible states; that is, there are only two independent categories: survival data assigns outcomes as "dead" or "alive." Binary data can either be nominal or be ordinal. Whereas ordinal data must be analyzed using statistical methods classified as nonparametric, the data can be transformed into frequencies which may meet the assumptions of parametric statistics.

Not surprisingly, *measurement data* are those that are measured based on some quantitative trait and the resulting data are set of numbers (eg, height, weight, age, or stream velocity) (Fig. 2.4). Measurement data may be classified as discrete or continuous, where discrete measurement data occur only as

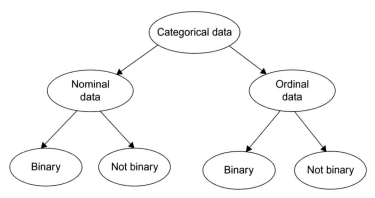

FIGURE 2.3 A diagram showing the different kinds of categorical data, all of which might be collected during an ecotoxicological investigation.

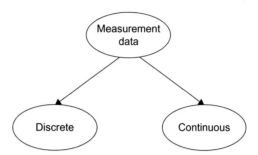

FIGURE 2.4 Measurement data can be either discrete, such as three mice, or continuous, as in a weight of 0.245 mg.

integers (eg, as count data), such as population counts for fish in a pond. In contrast to discrete measurement data, continuous measurement data may occur as any whole number and take on any value in the interval between whole numbers, such as distance, height, and age. For instance, there may be 100 fish in a pond, but not 100.56 though a single fish may be 5.321 cm long. Categorical data are also commonly summarized using percentages (or proportions or frequencies), and measurement data are typically summarized using averages (or means) or some descriptive statistic that characterizes a particular attribute of a sample of numbers taken from a population of interest.

Some data obtained from toxicity tests may be derived as *interval data* or as *frequency data*, which may be handled in various ways depending on the number of organisms included in the experimental design. Interval data are simple ratio estimators (eg, number of eggs or offspring per parent), which allows the ranking of items measured as part of data collection. Interval data are often analyzed as if the data were continuous, although certain statistical evaluations should be completed prior to following through with data analysis. Frequency data are similar in mathematical origins, arising as a ratio estimator most often encountered when exposed organisms are counted or measured, then compared with the comparable outcomes observed in controls. Binary data, when considered relative to control data may also follow a similar data analytical path. Typically, binary data are presented as the number of individuals observed in a specified condition (eg, number dead or number alive) relative to the total number of individuals observed in each experimental unit.

A further distinction in data analysis is an *experimental unit* compared to a *sampling unit*. Experimental units, also referred to as replicates, are the smallest unit of an experiment to which a treatment can be allocated independently of all other units. For most ecotoxicologists, these terms simply translate to plant pots, beakers, or aquaria. This term is usually not the same as a sampling unit, which is generally the biota unit for which the measurement is taken (eg, the plant or animal being tested, such as fish, daphnid, or plant). For instance, in laboratory studies scientists often set up a set of beakers, aquaria, or cages and place a few to several organisms into that container. Organisms in the same container share many things including food, competition, social interactions (for animals), etc. Thus, the individual organisms cannot be considered independently of each other. However, the separate containers that house these organisms can be considered independently. You are probably aware of the concept of sample size, which is very important in statistics. If you had 25 containers, each holding five organisms, the sample size would be 25, not 125.

Data Quality and Good Laboratory Practices (GLPs). Regardless of the type of data derived from toxicity tests, the data must be collected and analyzed in accordance with a scientifically sound research design that will allow analysis of data through appropriate statistical methods—as the adage states, "garbage in, garbage out (GIGO)." Although the adage had its origins in mathematical modeling and computer science, GIGO reinforces the need for principals of data quality evaluation, at least from a resource management and scientific basis, to be incorporated into the evaluation of biological effects potentially linked to exposures of environmental chemicals. For example, based on our brief overview of toxicity testing and the role that statistical data analysis plays in characterizing toxicity endpoints such as LC50, scientists must be certain that the data that is collected is of the highest quality that can be attained given the context of the study. In laboratory studies, the investigator can control almost all variables, including environmental factors and dosages. Field situations are more complicated, but still require that data meet the highest standards possible, which requires a throughly evaluated study plan that includes the type of statistical analyses that will be conducted even before the first data point is measured. Consultation with a statistician is highly recommended. In addition, the investigator must assure throughout the study that the data are of the highest quality possible and that any deviations from the original study design are carefully documented. Although the process of developing data quality standards continues to evolve to address the needs of science and the regulatory community, consensus-based standards have been developed by agencies such as the US EPA, but future work is continually needed, particularly given the increasing application of toxicity assessment to the environmental hazards assessment (HA) or ecological risk assessment (ERA) process (more on this in Chapter 13).

It is better to collect more data than needed than to collect not enough data to answer key questions of the study. After all data are collected, the investigator can evaluate the quality of the data and determine to what extent the original data and its accompanying statistical analyses are of high quality, relevant, and representative of the environmental questions being evaluated within the context of toxicity. For instance, let's say that you were tasked to do a study on the effects of contaminant x on the endangered red-breasted cockinfluff. If you used too few animals to arrive at a statistically valid sample then all the information you collected would be wasted. If instead you collected a few more cockinfluffs to meet statistical rigor, you could have a meaningful study. Indeed, for regulatory applications, each agency generally establishes data-quality guidance and criteria, which are then reflected in standard guidance and procedures for conducting toxicity tests (Tables 2.1–2.3, ASTM, 2015b; US EPA, 2002b,c,d). Such efforts should be taken to determine if any errors in data collection or analysis occurred and to what extent these errors might affect test outcomes. From data quality evaluation and toxicity test development perspectives, such efforts improve the quality of overall data collection, analysis, and interpretation processes, which inevitably leads to outcomes that contribute to improved decision-making efforts on the part of resource management or valid scientific interpretations by scientists.

In studies conducted for the regulatory agencies, their protocols for conducting the tests and data handling must be followed exactly. Studies conducted outside of that context may develop their own methodology and analysis as long as they meet sound scientific processes and can get through a strict peer-review process prior to publication of the study in scientific literature. However, we cannot understate the importance of data quality evaluation either way. Various tools are deployed in these evaluations, particularly within regulatory contexts, and include data evaluation by categorical descriptors that define the suitability of ecotoxicological data for subsequent analysis and interpretations, such as those suitable for inclusion in risk analysis and assessment activities that ultimately support more informed risk-management decisions.

GOOD LABORATORY PRACTICES

Often, the data quality evaluation is completed in a gradual manner that covers the entire process of a study. The general rules dictating the conditions of the tests, accuracy checks and double checks, identification of the protocols that will be employed, and rules for data collection down to the smallest details such as what to do when an error is made in writing down a data value (cross it out once and rewrite the correct value next to the error) are all part of GLPs. Companies that want to register a new compound or consulting firms investigating a potential breach in environmental laws must follow GLPs when existing data are gathered (or may be submitted from a toxicity testing laboratory as part of a regulatory requirement in a chemical's evaluation), then reviewed relative to data acceptance criteria (eg, control treatments were reported within acceptable ranges of test performance, duration and environmental conditions were within accepted ranges, replication was included as part of the toxicity-test design, toxicity endpoints such as survival or growth, and species used in the toxicity test were identified and exposed following specific guidance for toxicity testing). All of this is done because millions of dollars might be at stake either from a research and development context or from a litigious matter.

GLPs have been promulgated through regulations that became part of the regulatory landscape in the latter part of the 1970s in response to the poor business practices supporting applied research and development activities by laboratories working in toxicological and pharmaceutical companies and contract facilities conducting work for these organizations.

In the United States, GLP regulations were first instituted by the US Food and Drug Administration (US FDA, 1981), and then by the US EPA (1997). GLP regulations established rules for good practice and required that work be performed in compliance with an organization's preestablished plans and standardized procedures. GLP guidance stresses: (1) resources, (2) characterization, (3) rules, (4) results, and (5) quality assurance (QA).

Under GLP specifications in a regulatory context, guidance is focused on *resources* and centers on *organization and personnel*. GLP regulations specify structure and responsibilities of research and development businesses at all levels of their organization, including qualifications and the training of staff. Documentation of these organizational practices is required. At the same time, *facilities and equipment* are emphasized and, in particular, the organization must assure sufficient facilities and equipment are available to perform the studies and that the equipment and instrumentation satisfy requirements related to its qualification, calibration, and maintenance. When scientific studies are conducted outside of a regulatory context, greater freedom is allowed to meet the objectives of the study; nevertheless, attempts to follow GLPs as closely as possible can help assure the most reliable results in these contexts as well.

Within the context of GLP, *characterization* refers to an organization's ability to perform a study correctly, to essentially know as much as possible about the materials and processes required to successfully implement and complete any specific study. To complete toxicity tests, the organization must be knowledgeable of biota involved in the tests (often an animal or plant) and the specific requirements of test implementation. In part, this expert knowledge of a test system, its biological components, and instrumentation or equipment necessary to conduct the test are represented in the rules formulated and written by the organization in test protocols and written procedures. Specific documentation in the form of study plans or protocols must be developed by the organization that details the principal steps implemented as part of conducting research studies. A central tenet of the scientific method is the ability to repeat studies and obtain similar results; hence, detailed routine procedures must also be available to scientists involved in the study to assure comparability and acceptance of data generated by the same protocol regardless of the staff involved or time of implementation. Protocols provide experimental design and general timelines for a study and written standard operating procedures (SOPs) subsequently provide the *technical details* necessary to conduct any specific study. As such, a given protocol and accompanying SOPs should assure that any study could be repeated exactly as originally completed, if necessary.

Given the data needs driving GLP, results remain an outcome of primary interest. For example, results of any given toxicity test will reflect *raw data* derived from the study and provide a basis for establishing scientific interpretations and arriving at conclusions. As such, raw data should be presented as part of a *final report* that include a written account of a study's performance with the scientific interpretation of the data. Final reports are subject to review by regulatory authorities, and may become part of the submission for registration and marketing approval, if regulatory mandates are behind the test's implementation. Once a given study is completed and final report revised and accepted following review, GLP requires that study reports be placed in an organization's archives for storage to ensure safekeeping for many years and allow for prompt retrieval.

A measure of success for GLP is determined, at least in part, from outcomes to the fifth component of the process—namely, QA. Under the auspices of GLP, QA is a process implemented by members of the testing or research organization (eg, a toxicity-testing laboratory) that are responsible for assuring compliance with the guidance, specifications, and rules of GLP-governing operations within a laboratory or research facility. As such, QA must be independent from the scientists involved in the operational aspects of the study as part of the overarching research or testing process.

FOCUS—GLPs IN OPERATION

To illustrate the roles that GLPs and data quality play within ecotoxicology, especially with respect to toxicity testing and evaluation of chemicals released to the environment, we will offer a brief consideration of an integrated laboratory and field approach to assessing effects of metals in a large watershed in Montana that had been impacted by historic mining operations. Ambient metal levels—copper, cadmium, and zinc, among others—and arsenic were the principal chemicals of concern to resource management and people living in the area. Over a period of many years, ranging from the mid- to late-1980s through early 2000s, the watershed, and in particular the sediments and soils, surface waters, and groundwater within the system, were the subject of a series of designed studies focused on (1) characterizing adverse effects associated with arsenic and metals released to riparian and wetland habitats, and (2) restoring the system to assure its long-term sustainability.

As typically implemented, integrated laboratory and field studies relied on many tools to inform resource management throughout their implementation of the evaluation and restoration process. These tools consisted of a variety of steps: (1) laboratory toxicity tests to evaluate effects of arsenic and metals on test species under controlled exposure conditions, (2) field studies principally focused on recording "laboratory-to-field" observations that might influence interpretation of toxicity tests, and (3) physical-mathematical models intended to characterize current states of the watershed and forecast future conditions potentially linked to resource management practices.

Toxicity Testing Under Field and Laboratory Conditions

Our case study focuses on the Clark Fork River in western Montana (Fig. 2.5), which was initially recognized as a Superfund site in 1989 after reconnaissance investigations in the early 1980s. Those reconnaissance investigations of the

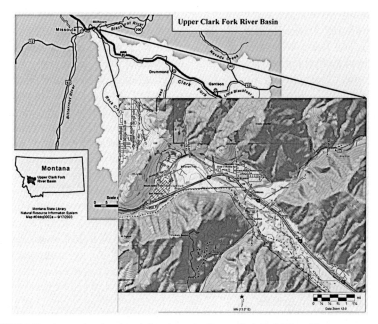

FIGURE 2.5   A map showing the mining site at the Upper Clark Fork River, Montana.

river and Milltown Reservoir, located as the confluence of the Blackfoot and Clark Fork rivers near Missoula, Montana, identified chronic exposure to fish and wildlife from metal-contaminated sediment that was deposited in aquatic and terrestrial habitats along riparian and wetland habitats in the watershed.

During the early 1990s, a series of studies evaluated aquatic, wetland, and terrestrial habitats across multiple trophic levels, and the results of these comprehensive studies informed remedial actions for the river and wetlands in the upper Clark Fork River watershed. Soil contamination evaluations and ecological assessments were completed for wetlands linked to the river, and biological evaluations, principally toxicity tests, were included as part of the field and laboratory investigations. Depending upon habitat type, field and laboratory methods directly lent themselves to the assessment process and considered bioavailability and subtle expressions of adverse biological effects associated with chronic exposures to arsenic and metals. A variety of biological test methods (eg, terrestrial and aquatic tests) were critical to the evaluation of aquatic and wetland habitats. For evaluating heavy-metal effects throughout the Clark Fork River and wetland system, field and laboratory methods included toxicity tests for plants, particularly those species found in emergent and upland habitats, soil macroinvertebrate (earthworm) tests, preliminary studies using amphibian and bacterial test systems, fish and aquatic invertebrate tests, as well as sediment toxicity tests using benthic and epibenthic invertebrates (eg, rainbow trout, *D. magna*, *Hyalella azteca*, and *Chironomus riparius*).

Detailed summaries of technical findings were reported in the scientific literature (eg, Pascoe and DalSoglio, 1994; Linder et al., 1999; Morey et al., 2002), and when considered in conjunction with soil characterizations and chemical analysis of soils, sediments, and biological materials, these biological and ecological evaluations yielded an integrated evaluation of ecological effects and exposure occurring in riparian wetland habitats within the impacted areas of the western reaches of the Clark Fork River watershed. Overall results of the aquatic, wetland, and terrestrial studies suggested that acute, adverse biological effects were largely absent from the wetland; however, adverse effects to reproduction, growth, and physiological endpoints of various terrestrial and aquatic species were related to metals exposures in more highly contaminated depositional areas. For example, feeding studies with contaminated diets collected from the upper reaches of the Clark Fork River indicated that trout were at high risk from elevated metals concentrations in surface water, sediment, and aquatic invertebrates. Integration of chemical analyses with ecotoxicological and ecological evaluations of metal effects on the wetlands, fish, and aquatic- and sediment-dwelling invertebrates provided an important foundation for environmental decisions and long-term plans for restoring the river and adjacent habitats.

Today, the wetlands at Milltown are recognized as a state park in Montana. During the early 1990s, a series of studies evaluated aquatic, wetland, and terrestrial habitats across multiple trophic levels and results of these comprehensive studies informed remedial actions for the river and wetlands in the upper Clark Fork River watershed.

STUDY QUESTIONS

1. According to this chapter, how did the study of synfuels lead to aquatic toxicology testing?
2. What distinguishes an acute test from a chronic one? Which of the two is more conservative—that is, likely to find some effect from a chemical?
3. Discuss with your class what conditions would lend themselves to a static test, static renewal test, or flow-through test.
4. How many experimental units do you have under the following conditions:
   a. Twenty-five *Daphnia,* each housed in separate beakers
   b. One hundred mallards placed as pairs in separate cages
   c. Five tadpoles collected with one swipe of a collecting net from the same pond
5. Review the material from Chapter 1. What environmental or chemical factors might reduce the bioavailability of a chemical?
6. What information can be gained from an integrated field and laboratory study versus a study that only uses one of the two approaches?

REFERENCES

Anderson, B.S., Hunt, J.W., Phillips, B.M., Nicely, P.A., Tjeerdema, R.S., Martin, M., 2004. A comparison of in situ and laboratory toxicity tests with the estuarine amphipod *Eohaustorius estuarius*. Arch. Environ. Contam. Toxicol. 46, 52–60.

ASTM, 2015a. ASTM Book of Standards, Volume: 11.06. Developed by Subcommittee: E50.47, American Society for Testing and Materials. West Conshohocken, Pennsylvania.

ASTM, 2015b. ASTM E1847—96(2013) Standard Practice for Statistical Analysis of Toxicity Tests Conducted Under ASTM Guidelines. ASTM Book of Standards, Volume: 11.06. Developed by Subcommittee: E50.47, American Society for Testing and Materials. West Conshohocken, Pennsylvania, p. 10.

Biesinger, K.E., Christensen, G.M., 1972. Effects of various metals on survival, growth, reproduction, and metabolism of *Daphnia magna*. J. Fish. Res. Board Canada 29, 1691–1700.

Bleiler, J., Pillard, D., Barclift, D., Hawkins, A., Speicher, J., 2004. Development of a Standardized Approach for Assessing Potential Risks to Amphibians Exposed to Sediment and Hydric Soils. Tech. Rept. TR-2245-ENV. Naval Facilities Engineering Service Center (NFESC). Port Hueneme, CA. Sectional pagination, p. 340.

De Carvalho, J.P., 2012. Toxicology Timeline. <http://www.toxipedia.org/display/toxipedia/Toxicology+Timeline>.

Dumont, J.N., Schultz, T.W., 1980. Effects of coal gasification sour water on *Xenopus laevis* embryos. J. Environ. Sci. Health A 15, 127–138.

Dumont, J.N., Schultz, T.W., Jones, R.D., 1979. Toxicity and teratogenicity of aromatic amines to *Xenopus laevis*. Bull. Environ. Contam. Toxicol. 22, 159–166.

Finney, D.J., 1978. Statistical Method in Biological Assay Mathematics in Medicine Series, third ed. Macmillan Publishing Company, Inc., New York, p. 505.

Geiger, J.G., Buikema Jr., A.L., Cairns Jr., J., 1980. A tentative seven-day test for predicting effects of stress on populations of *Daphnia pulex*. In: Eaton, J.G., Parrish, P.R., Hendricks, A.C. (Eds.), Aquatic Toxicology. ASTM STP 707 American Society for Testing and Materials, Philadelphia, PA, pp. 13–26.

Linder, G., 2003. Integrated field and laboratory tests to evaluate effects of metals-impacted wetlands on amphibians: a case study from Montana. In: Linder, G., Krest, S., Sparling, D., Little, E. (Eds.), Multiple Stressor Effects in Relation to Declining Amphibian Populations. ASTM STP 1443 ASTM International, West Conshohocken, PA, pp. 184–204.

Linder, G., Pascoe, G.A., DalSoglio, J.A., 1999. Case study: an ecological risk assessment for the wetlands at Milltown Reservoir, Missoula, Montana. In: Lewis, M.A., Mayer, F.L., Powell, R.L., Nelson, M.K., Klaine, S.J., Henry, M.G., Dickson, G.W. (Eds.), Ecotoxicology and Risk Assessment for Wetlands. SETAC Press, Pensacola, FL, pp. 153–190.

McGee, B.L., Fisher, D.J., Wright, D.A., Yonkos, L.T., Ziegler, G.P., Turley, S.D., et al., 2004. A field test and comparison of acute and chronic sediment toxicity tests with the estuarine amphipod *Leptocheirus plumulosus* in Chesapeake Bay, USA. Environ. Toxicol. Chem. 23, 1751–1761.

Morey, E.R., Breffle, W.S., Rowe, R.D., Waldman, D.M., 2002. Estimating recreational trout fishing damages in Montana's Clark Fork River basin: summary of a natural resource damage assessment. J. Environ. Manage. 66, 159–170.

Newman, M.C., 2012. Quantitative Ecotoxicology, second ed. CRC Press, Boca Raton, FL, p. 592.

Pascoe, G.A., DalSoglio, J.A., 1994. Planning and implementation of a comprehensive ecological risk assessment at the Milltown Reservoir Clark-Fork River Superfund site, Montana. Environ. Toxicol. Chem. 13, 1943–1956.

Probstein, R.F., Hicks, R.E., 1990. Synthetic Fuels. McGraw-Hill Book Company (New York)/pH Press, Cambridge, MA, p. 490.

Sprague, J.B., 1969. Measurement of pollutant toxicity to fish, I—bioassay methods for acute toxicity. Water Res. 3, 793–821.

Sprague, J.B., 1970. Measurement of pollutant toxicity to fish, II—utilizing and applying bioassay results. Water Res. 4, 3–32.

US EPA, 1997. Good Laboratory Practice Standards. Title 40—Protection Of Environment Part 160—Good Laboratory Practice Standards. CFR Title 40, Volume 14, Parts 150–189.

US EPA, 2002a. Methods for Evaluating Wetland Condition: Using Amphibians in Bioassessments of Wetlands. EPA-822-R-02-022. Office of Water, U.S. Environmental Protection Agency, Washington, DC, p. 41.

US EPA, 2002b. Methods for Measuring the Acute Toxicity of Effluents and Receiving Waters to Freshwater and Marine Organisms, fifth ed. EPA-821-R-02-012. U.S. Environmental Protection Agency, Washington, DC, p. 275.

US EPA, 2002c. Short-term Methods for Estimating the Chronic Toxicity of Effluents and Receiving Waters to Freshwater Organisms, fourth ed. EPA-821-R-02-013. U.S. Environmental Protection Agency, Washington, DC, p. 350.

US EPA, 2002d. Short-term Methods for Estimating the Chronic Toxicity of Effluents and Receiving Waters to Marine and Estuarine Organisms, third ed. EPA-821-R-02-014. U.S. Environmental Protection Agency, Washington, DC, p. 486.

US FDA, 1981. Guidance for Industry: Good Laboratory Practices Questions and Answers. U.S. Department of Health and Human Services, Washington, DC.

Chapter 3

Bioindicators of Contaminant Exposure

Terms to Know

Biomarker
Bioindicator
Reactive Oxygen Species (ROS)
Oxidative Stress
Monooxygenase
EROD (Ethoxyresorufin-O-dealkylase)
Glutathione
Lactate Dehydrogenase
Alanine Aminotransferase
Aspartate Aminotransferase
Creatinine kinase
Aminolevulinate dehydratase
Acetylcholinesterase
Vitellogenin
Flow Cytometry
Micronucleus
Heinz Body
Necropsy
Genomics
Proteomics
Teratogenesis

INTRODUCTION

If an investigator wants to know if an animal under natural conditions was exposed to certain contaminants, there are several things that he or she can do. Presumably, the investigator already has some idea that the contaminant(s) of interest are present in the environment; whether in air, water, sediment, or soil; and that they are at concentrations that can cause problems based on information from scientific literature. The investigator should certainly know if the animal is in potential contact with the contaminants. For example, if the contaminants were only found in sediments or water and the animal in question was terrestrial, the risk that the animal would be exposed is likely very small.

Ecotoxicology Essentials.

From there, the investigator has a few decisions to make. He or she can choose to collect some of the animals and measure contaminant concentrations in selected tissues. This usually requires euthanizing the animals and sampling liver, kidney, lipids, or whole bodies through sophisticated processes and instrumentation. Sacrificing animals may be a problem if the species is threatened, endangered, or otherwise of concern. In addition, for many contaminants, especially organic ones, processing and analyzing samples can be very time consuming and expensive. For-profit contract labs often charge hundreds of dollars per sample and several samples are almost always necessary to provide an adequate idea of contaminant exposure. Even university labs have considerable costs in terms of equipment maintenance, reagents, and student wages for running analyses. There may also be competition for use of instruments and analyses can be delayed.

In addition, the presence of contaminants in selected tissues or whole bodies does not necessarily mean that the animals are adversely affected by these contaminants. Certain inferences can be made from the literature, especially if the contaminant concentrations are high, but some species can be very tolerant to contaminants before showing any adverse reactions that may not be known beforehand. Others are very sensitive to contaminants and may experience problems at concentrations not predicted by previous research. Physiological signs or behavior can be used to assess if an animal is affected by the contaminants but these signs are most effective if prior studies have confirmed that they reliably indicate effects due to exposure. Through repeated laboratory testing, certain signs or health conditions reoccur. If these signs are repeated in a variety of species they begin to form a pattern of dose and effects. Once an effect is identified as being consistent it can be used to help the diagnosis of future situations.

Identification and use of these signs are often less invasive than direct measurement of contaminant concentrations, reduce or eliminate the need to sacrifice animals, are less expensive, require less or simpler preparation procedures, and lead to direct understanding of potential biological effects. These methods fall under the category of *bioindicators.*

The quantification of contaminant residues often requires the use of several grams of sample tissue. If an investigator focuses on relatively small animals such as insects, cladocerans, or even amphibian larvae, several grams of tissue could require many organisms. In contrast, bioindicators often require only a drop or two of plasma and a reliable microplate reader (Fig. 3.1). Microplate readers allow tens of samples to be analyzed with duplicates and standards within the span of a few minutes.

After extensive associations between contaminant concentrations and physiological or behavioral effects, bioindicators might be used in absence of contaminant measurements. Confirmation of contaminant residues in the environment of a small subsample of organisms is necessary to substantiate cause and effect. In other words, bioindicator and contaminant measurements work best together. Bioindicators might even be used in the absence of residue determinations when there is good correspondence between the strength of the bioindicator and

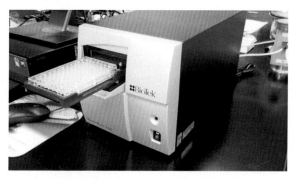

FIGURE 3.1   A microplate reader with a 96-well microplate in its drawer. Microplate readers are used in many bioindicator assays.

the degree of environmental contamination or when the bioindicator is clearly affected by specific contaminants. For instance, exposure to lead in soil, food, or water is clearly related to reduced levels of the enzyme ALAD (aminolevulinate dehydratase, see next and Chapter 8). Depression of another enzyme, acetylcholinesterase, is a good indicator of exposure to carbamates, organophosphorus, and, to a lesser degree, pyrethroid pesticides.

Bioindicators versus Biomarkers

There are two terms, biomarkers and bioindicators, that have been extensively used and interchanged in the toxicological literature. According to Adams et al. (2001), the term *biomarker* refers to any functional measure of exposure that is characterized at an organizational level below the total organism, such as physiological responses. In contrast, the authors limited *bioindicator* to those structures or processes that reflect exposure or effects at higher levels of organization (eg, entire organism, population, community, or ecosystem). This would include behavior of organisms, growth rates, population growth or decline, and interspecific interactions. However, even Adams (2003) used the term "bioindicator" to refer to biological responses expressed at biomolecular levels. In practice "biomarker" and "bioindicator" have been used interchangeably. I mention this to reduce student confusion if they see these terms used in different contexts. However, I imagine that some practitioners would disagree with lumping these terms. We will use "bioindicator" regardless of the organizational level the observation occurs.

Bioindicators and Diagnoses

Much of the initial information on bioindicators was developed during controlled experiments in the laboratory and then taken into the field. For example, the use of the enzyme acetylcholinesterase (AChE) activity for detecting exposure to organophosphate and carbamate pesticides (see Chapter 5 for more information on these pesticides and AChE) was first promoted in the 1960s (Beam and Hankinso, 1964) and was employed through a long history of laboratory studies on the response of AChE to these pesticides (eg, Bai and Reddy, 1977; Hayes et al., 1980; Hill and Fleming, 1982). The results of those studies have led to the finding that a 50% depression in brain acetylcholinesterase activity in moribund birds is usually sufficient to diagnose pesticide poisoning (Hill, 2003), even in court cases.

Some bioindicators such as AChE and others that we will discuss are diagnostic for a specific group of contaminants. Others respond to a wider variety of chemicals, but their diagnostic capability can be enhanced by using multiple bioindicators. In any case, the use of bioindicators can be very effective in conducting screenings to see if any effects are expressed before conducting extensive chemical analyses.

TYPES OF BIOINDICATORS

Bioindicators can be classified into several categories. We will examine the following:

1. Indicators of oxidative stress—P450 system and glutathione
2. Plasma enzymes—lactate dehydrogenase, creatinine kinase, aspartate aminotransferase, alanine aminotransferase, AChE, and aminolevulinate dehydratase
3. Use of cellular contents of plasma blood cells
4. Evidence of endocrine disruption—specific hormones, vitellogenin
5. Evidence of chromosomal damage—Heinz bodies, flow cytometry, karyotyping
6. Histopathology—liver, kidney, other organs
7. Genomic inducible DNA, microsatellite arrays, proteomics
8. Reproductive and developmental—teratogenesis, malformations, eggshell thinning

In Chapter 10 and 11 we will discuss some effects on populations and communities.

OXIDATIVE STRESS

Oxidative stress is a result of an imbalance between the production of molecules that are called reactive oxygen species (ROS) and a biological system's ability to readily detoxify the reactive elements or repair the resulting damage. In toxicology, certain environmental contaminants including polycyclic aromatic hydrocarbons (PAH, see Chapter 7), polychlorinated biphenyls (PCBs, see Chapter 6), and chlorinated hydrocarbons (see Chapter 4) are metabolized with the aid of the P450 system within cells. During metabolism, the contaminants produce radicals and peroxides, collectively known as ROS. Examples of ROS include OH^- (hydroxyl ion), O_2^- (superoxide), H_2O_2 (hydrogen peroxide), and $OONO^-$ (peroxynitrile). Oxidative stress can cause toxic damage to all components of the cell. In humans, oxidative stress is thought to be involved in the development of Asperger's syndrome, attention deficit hyperactive disorder (ADHD), cancer, Parkinson's disease, Alzheimer's disease, atherosclerosis, heart failure and myocardial infarction, sickle cell anemia, autism, and chronic fatigue syndrome. The most studied effects of ROS in animals involve various cancers (Sies, 2013).

Ecotoxicologists often use two methods of assessing oxidative stress in a cell: measurement of components of the P450 system and differences in glutathione.

Cytochrome P450

Cytochrome P450s (which are discussed in more detail in Chapter 7) are a class of proteins present in all tissues, but most notably in liver. They are induced, which means that their respective genes are activated to produce more enzyme, by the presence of certain chemicals such as the contaminants listed previously. While the normal function of cytochromes P450 are to break down naturally occurring toxins and organic molecules, they are "duped" by these contaminants into producing ROS in excess. When cellular repair mechanisms are unable to keep up with this production of ROS, damage occurs in the tissue.

The cytochromes of the P450 system, while inducible, often do not reach levels that mark exposure to contaminants at environmentally realistic concentrations. However there are several other associated enzymes called *monooxygenases* that are sensitive to contaminant exposure and whose activities can be measured to assess P450 induction. The most widely evaluated monooxygenase is EROD (ethoxyresorufin-O-dealkylase), other related enzymes include PROD and BROD.

Antioxidants Including Glutathione

If oxidative damage does occur in cells the body uses several ways to repair this damage. Antioxidants can serve as bioindicators of this damage and subsequent repair. One commonly studied antioxidant is *glutathione*. Glutathione is a tripeptide comprised of three amino acids (cysteine, glutamic acid, and glycine) present in most animal tissues, but is particularly important in livers. Glutathione acts as a free radical scavenger that combines with these free radicals (ROS) and detoxifies them. When reduced glutathione (GSH) reacts with these radicals it serves as an electron donor and, in the process, glutathione is converted to its oxidized form: glutathione disulfide (GSSG). Once oxidized, glutathione can be reduced back to GSH by glutathione reductase. The ratio of

GSH to GSSG within cells can be used as a measure of cellular toxicity. The investigator compares the ratios in animals collected from a reference site and one or more potentially contaminated sites. The absolute concentrations of GSH and GSSG can be important, but the ratio of the two is often more informative. Other antioxidants that can be used to indicate oxidative damage include catalase, superoxide dismutase, peroxidases, ascorbate, and alpha-tocopherol. These can all be used as biomarkers of oxidative stress, but they are not specific for a given source of the stress.

To provide an example of how glutathione might be used, Świergosz-Kowalewska et al. (2006) measured total GSH and GSSG in tissues of bank voles (*Clethrionomys glareolus*) that were collected at various distances from a zinc smelter. They also determined the activity of various enzymes associated with glutathione activity. The most sensitive parameter of metal toxicity for animals living in a chronically contaminated environment was the GSH/GSSG ratio, which decreased in the liver of animals with high Cd levels.

PLASMA ENZYMES

If your doctor has ever had your blood panel done, you may be familiar with several of the plasma constituents that they use; however, you may not know what information these constituents provide. Just as your doctor relies on LDH, AST, and ALT to provide information on the health of your liver, so too does the ecotoxicologist rely on this information to diagnose if animals in contaminated sites are being affected. Because we do not have data on most wild species to tell us what is a "normal range," scientists compare data among reference and contaminated sites to determine if there is a difference. In addition to the normal panel, we added two other enzymes that can serve as bioindicators.

As a side note, there is a difference between plasma and serum and some tests may recommend one over the other. Plasma is the liquid fraction of uncoagulated blood after it has been centrifuged. Serum is the liquid fraction of blood after it is coagulated. If plasma is required then blood must be drawn into a vial that has been coated with an anti-coagulant such as heparin to prevent coagulation, but blood for serum is drawn into untreated vials (see Fig. 3.2). Although they are both products of whole blood, concentrations of enzymes, contaminants, and other factors can often differ markedly between serum and plasma. We will provide a brief description of commonly sampled blood constituents next.

Lactate Dehydrogenase (LDH)

This enzyme is found in animals, plants, and microbes. A dehydrogenase is an enzyme that transfers a hydride (H^-) from one molecule to another. Lactate dehydrogenase catalyzes the conversion of pyruvate to lactate. Elevated levels of LDH compared to samples from reference sites may indicate several things

FIGURE 3.2   Taking a blood sample from the jugular of the Harlequin duck (*Histrionicus histrionicus*). *Source: Courtesy Sea Duck Joint Venture and Tim Bowman photographer.*

including cancer, heat stress, anemia (better determined through other measures), and lung or liver disease. In wildlife, elevated LDH is usually used to diagnose liver toxicity because the liver is the primary organ for metabolizing contaminants and one of the first to be affected.

Alanine Aminotransferase (ALT)

ALT is another enzyme that may indicate liver problems if animals from contaminated sites have higher activities than reference animals. ALT is a transaminase enzyme, meaning that it is involved in amino acid metabolism. It catalyzes two parts of the alanine cycle. ALT is found in serum and in various body tissues, but is most common in the liver. Serum ALT level, serum AST (aspartate transaminase) level, and their ratio (AST: ALT ratio) are commonly measured as biomarkers for liver health. Elevated concentrations of ALT and elevated ALT: AST ratios are indicative of liver pathology.

Aspartate Aminotransferase (AST)

AST catalyzes a reaction between the amino acids aspartate and glutamate and is an important enzyme in amino acid metabolism. AST is found in the liver, heart, skeletal muscle, kidneys, brain, and red blood cells. Since it is more widespread than ALT, elevated AST may be indicative of heart, pancreas, anemia, kidney, and musculoskeletal pathologies and is usually associated with other tests to support a diagnosis. As previously mentioned, the ALT: AST ratio is rather specific for liver toxicity but elevated AST combined with elevated activity of the next constituent may be specific for kidney disorders.

Creatinine Kinase (CK)

Creatinine kinase (CK, or creatinine phospho-kinase) is an enzyme found in several tissues and cell types that catalyzes the degradation of the waste product creatinine. Elevated CK is an important indicator of a few problems. Muscle cells are rich in the enzyme and acute smooth, cardiac, or striated muscle damage will release CK into the blood stream. Thus, an abundance of the enzyme in serum can indicate muscle damage. In wildlife testing, it is frequently used to indicate renal failure, especially when it is accompanied by elevated AST. Kidneys can be damaged by heavy metals and other contaminants because they are the first-line method of excreting water-soluble contaminants.

Aminolevulinate Dehydratase (ALAD)

Aminolevulinate dehydratase is not part of a normal blood panel, but it is highly sensitive to lead exposure. ALAD is an important enzyme in the formation of blood. Even a small amount of lead will severely depress ALAD activity. Unfortunately, while low levels of ALAD clearly show lead exposure, the enzyme is so sensitive that it cannot be used to estimate the degree of lead toxicity (See Chapter 8 for more information on ALAD).

Acetylcholinesterase (AChE)

This enzyme, which is discussed in greater detail in Chapter 5, is diagnostic of exposure to carbamate and organophosphate pesticides when depressed. Pyrethroid pesticides may also depress AChE, but not as strongly as the other two groups of pesticides. AChE is an enzyme that breaks down acetylcholine, an important neurotransmitter in the brain and in neuromuscular junctions. Without AChE, acetylcholine remains in the synapses and causes continual, unregulated neural discharges that result in asphyxiation.

Other Blood Panel Measurements

Bioindicators in this category can be analyzed through several methods, but are not routinely applied to wildlife toxicity. We won't focus on these measurements, but suffice to say that elevated alkaline phosphatase and bilirubin can indicate liver disorders; elevated levels of blood urea nitrogen (BUN), and potassium suggest kidney pathology, as do low levels of protein and albumin.

Examples of Blood Enzymes in Practice

There are a great many examples that illustrate the use of plasma enzymes in ecotoxicology—we will present two of these examples. The Couer d'Alene River Basin of Idaho has been a major lead and zinc mining area for many decades. The U.S. Fish and Wildlife Service became interested in the area when it was reported that large numbers of waterfowl—ducks, geese, and swans—were dying. Both laboratory and field investigations were undertaken to determine the cause(s) of this mortality and to assess sublethal harm to the waterfowl. In one study, Hoffman et al. (2006) fed mallards (*Anas platyrhynchos*) diets mixed with 12% clean sediment and treatments with 12% sediment from three different areas of Couer d' Alene (4520–6990 mg/kg lead). They found that blood lead concentrations were higher in all three treatments compared to the birds on clean sediments. In addition, plasma ALAD activity was depressed by 90% or more with lead-contaminated sediment from all sites. Plasma enzyme activities for ALT, CK, and LDH-L were elevated by as much as 2.2-fold. Other blood changes also occurred and are reported below.

In another study, Hoffman et al. (1996) combined several of the tests described above to test the effects of a polychlorinated biphenyl (PCB 160) on the development of American kestrels (*Falco sparverius*) nestlings. Birds were gavaged (force fed) for 10 days with 5 μL/g body weight of corn oil (controls) or PCB 126 at concentrations of 50, 250, or 1000 ng/g body weight. Dosing with 50 ng/g of PCB 126 resulted in a hepatic concentration of 156 ng PCB/g wet weight, liver enlargement, and mild clotting of blood causing local necrosis. They also saw more than 10-fold increases in hepatic microsomal EROD and BROD (part of the P450 system). At 1000 ng/g, the liver concentration of PCB 126 was 1098 ng/g, elevated PCB levels were accompanied by decreased bursa weight, and increased ALT, AST, and LDH-L activities in addition to the previous effects. Highly significant positive correlations were noted between liver concentrations of PCB 126 and the ratio of oxidized to reduced glutathione.

OTHER BLOOD COMPONENTS

In addition to plasma or serum enzymes, other physical constituents and conditions in blood indicate adverse effects to wildlife.

Hematocrit

This is the volume of red blood cells or erythrocytes in uncoagulated blood compared with the volume of plasma (Fig. 3.3). Heavy metals such as lead and mercury are major contaminants that effect hematocrit.

Hemoglobin

This pigment is tightly correlated to hematocrit because erythrocytes contain it and transport oxygen through the body. Again, heavy metals are a primary group of contaminants that may reduce hemoglobin and blood oxygen.

Protoporphyrin IX

This is a biochemical widely used as a carrier molecule for divalent cations. Together with iron (Fe^{2+}), the body of the heme- group of hemoglobin, myoglobin, and many other heme-containing enzymes like cytochrome c and catalase are formed. It can also be affected by heavy metal contamination.

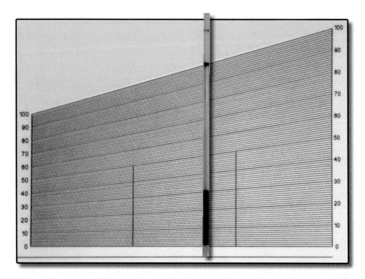

FIGURE 3.3 A hematocrit is the proportion of the total blood that is composed of red blood cells or erythrocytes. It can be easily measured using a special card developed for this purpose. *Source: qeluna82 http://blogs.rediff.com/qeluna82/2015/02/23/hematocrit/.*

In the Couer d'Alene study cited above (Hoffman et al., 2006) free erythrocyte protoporphyrin (FEP) concentrations were elevated by contaminated sediment from all sites. Hematocrit values and hemoglobin concentrations were up to 30% lower for all birds exposed to lead site sediments. Dietary amendments with phosphorous-restored hematocrit, hemoglobin, and plasma enzyme activities were lower compared to control levels. Although amendments of phosphorus substantially reduced the bioavailability of lead and alleviated many of the adverse hematological effects, lead concentrations in the blood of mallards given the amended sediments were still above those believed to be harmful to waterfowl under the conditions at that time.

Differential White Blood Cell Counts

The cellular portion of blood consists of platelets, red blood cells (erythrocytes) that carry oxygen, and white blood cells (leukocytes) that are part of the immune system. The proportion of specific white blood cells in whole blood can provide information on whether a particular contaminant has an *immunotoxicological* effect—that is, does it alter an organism's ability to defend itself against bacterial or viral infections? There are automated ways of making counts, but the typical university lab will have students manually count the different kinds of white blood cells. This manual version directs the researcher to place a drop of blood onto a microscope slide and count the number of leukocytes (Fig. 3.4)

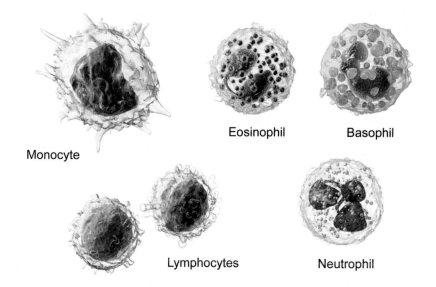

Eosinophil   Basophil

Monocyte

Lymphocytes   Neutrophil

White Blood Cells

FIGURE 3.4   Types of white blood cells or leukocytes. Frequencies of these cells relative to erythrocytes can provide information on immune status of an animal. *Source: Blausen.com staff. "Blausen gallery 2014." Wikiversity Journal of Medicine. http://commons.wikimedia.org/wiki/ File:Blausen_0909_WhiteBloodCells.png#/media/File:Blausen_0909_WhiteBloodCells.png.*

and erythrocytes and calculate the leukocyte: erythrocyte ratio. A standard technique is to count all the leukocytes for 1000 erythrocytes. If immune functions are impaired, the total ratio of leukocytes:erythrocytes will be lower in affected sites than in reference areas. Similarly, abnormal percentages of specific leukocytes indicate certain conditions are present. For example, an elevated percentage of neutrophils suggest acute infection or stress whereas a low percentage indicates anemia, exposure to radiation, or severe bacterial or viral infection. Elevated lymphocyte percentage indicates a chronic bacterial or viral infection and a lower percentage suggests exposure to radiation. High monocyte percentages are associated with parasitic infection. High eosinophil counts could mean allergies, cancer, or parasites. High basophil counts suggests allergies, low counts are related to acute infections or cancers (Medline Plus, 2015). In most vertebrates, both leukocytes and erythrocytes have nuclei; only mature mammalian erythrocytes lack a nucleus.

Kundu and Roychoudhury (2009) examined the effects of malathion, a commonly used organophosphorus pesticide on cricket frogs (*Fejervarya limnocharis*) by adding 6 μg/L malathion into the water of their aquaria for up to 10 days. They found a significant decrease in the total erythrocyte count in malathion-exposed animals after 24–96 h of exposure as compared to controls. They also observed significant decreases in hemoglobin and a significant increase in leukocyte counts. Lymphocytes and eosinophils were elevated and the authors suggested that this was a result of direct stimulation of the immunological defense due to the presence of a toxic substance.

EVIDENCE OF ENDOCRINE DISRUPTION

Many organic contaminants and organometals (metals attached to an organic molecule, such as methylated mercury, selenomethionine, or tributyltin) can disrupt hormone function or concentration. Polychlorinated biphenyls, polybrominated diphenyl ethers, DDT, bisphenol A, and phthalates are among known endocrine disruptors. The three groups of hormones that have received the greatest attention are those involved with sex (testosterone and estrogen), thyroid (thyroxine and triiodothyronine), and stress (cortisol, epinephrine and corticosterone). Overt signs such as feminization of males or masculinization of females, sterility, or other such evidence are rare among healthy wildlife species, but can increase due to contaminant exposure. Histological examination of gonads may expose problems but requires sacrificing animals, which, as previously stated, can cause problems for rare or endangered species. Titers of specific hormones can be determined through blood samples but these hormones often cycle throughout the day, making it difficult to get representative samples consistently. Hormone titers can also be affected by many other factors and great care must be taken to assure that animals are exposed to the same level of stress and handling for stress hormones, and come from the same age brackets and genders for sex and thyroid hormones.

Vitellogenin

This is a glycolipoprotein that has the properties of a sugar, fat, and protein. It belongs to a family of several lipid transport proteins. In egg laying animals, vitellogenin is essential in the deposition of yolk in the egg and induction of its gene is gender specific; it is essentially limited to females. Any trace of the protein in males suggests that the gene has been induced by contaminants and further investigations may be warranted.

As an example of using vitellogenin, tris (1,3-dichloro-2-propyl) phosphate (TDCPP) is an organophosphate flame retardant that is detectable in the environment and biota that prompts concern over its risk to wildlife and human health. Wang et al. (2015) investigated whether long-term exposure to low concentrations of TDCPP could affect reproduction in zebrafish (*Danio rerio*). Embryos were exposed to low concentrations of TDCPP from 2 h after fertilization of the eggs until sexual maturation. Exposure to TDCPP significantly increased plasma estradiol (an estrogen) and testosterone levels in females, but had no effect in males. TDCPP also reduced fecundity as evidenced by decreased egg production. Furthermore, hepatic vitellogenin was increased in both females and males, suggesting TDCPP has estrogenic activity. Histology showed that oocyte maturation in females was accelerated and that sperm formation was hindered in males. The authors also observed smaller eggs and increased malformation rates in the offspring. They concluded that long-term exposure to low concentrations of TDCPP can impair fish reproduction.

Hypothalamic-Pituitary-Thyroid Axis and Endocrine Disruption

As an example of thyroid disruption, Pandey and Mohanty (2015) examined the effects of the neonicotinoid pesticide imidacloprid and the carbamate mancozeb on the red munia (*Amandava amandava*), a bird that breeds in tropical Asia. Endocrine disruption can occur at several points along the hypothalamic-pituitary-thyroid axis (HPT, Fig. 3.5) and the authors investigated elements along this axis.

Briefly, the HPT is a chain-of-command system wherein the hypothalamus regulates the pituitary through secretion of releasing factors, in this case thyroid

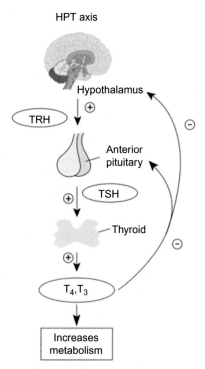

FIGURE 3.5 Diagram of the hypothalamic-pituitary-thyroid axis. Hormones and releasing factors from these organs control the production of thyroid hormones and general metabolism. *Source: http://embryology.med.unsw.edu.au/embryology/index.php/BGD_Lecture_-_Endocrine_Development#Hormones.*

releasing factor or TRF (TRH in the figure). The pituitary in turn secretes thyroid stimulating hormone which stimulates the thyroid to produce the thyroid hormones thyroxin and triiodothyronine. These, in turn, regulate metabolism and provide a negative feedback loop on the hypothalamus. When the thyroid hormones reach critical concentrations in blood, they inhibit the hypothalamus' production of TRF.

Pandey and Mohanty (2015) exposed adult male birds to very small concentrations of both imidacloprid and mancozebin food for 30 days during the pre-breeding and breeding seasons. Weight, volume, and histopathology of the thyroid gland were distinctly altered. Histologically, several changes were observed in thyroid follicular cells. Impairment along the thyroid axis varied between the pesticides and stage as determined by plasma levels of thyroid (thyroxin) and pituitary (thyroid stimulating hormone, TSH) hormones. In the pre-breeding phase, mancozeb decreased thyroxin and increased plasma TSH, showing responsiveness of the hypothalamic-pituitary-thyroid (HPT) axis to feedback regulation. On imidacloprid exposure, however, plasma levels of both thyroxin and TSH decreased, which indicated that the normal negative feedback mechanism was not functioning. During breeding, impairment of the HPT axis was more pronounced as plasma thyroxin and TSH were significantly decreased in response to both pesticides. According to the authors, it seems that low doses of these pesticides could affect thyroid homeostasis.

GENETIC AND CHROMOSOMAL DAMAGE

PAHs, PCBs, dioxins, furans, and some organometals or their metabolites can enter a cell's nucleus and form adducts or attachments with its DNA. These adducts can alter gene expression, break chromosomes, damage DNA, and cause mutations. The resulting genetic damage can lead to tumors, cancers, and many other problems associated with genetic functions. Cells can repair some of this damage, but in doing so they usually produce additional DNA that can be detected by various means. Fortunately, the vast majority of the damage is done in somatic cells so that only the exposed animals are affected. Rarely does damage extend to germ cells, ova, and spermatozoa, which could result in reduced fertility or genetic problems in subsequent generations. We will discuss three of several methods to analyze genetic damage next.

^{32}P-Postlabeling Technique

This is a very sensitive procedure for detecting and quantifying carcinogenic DNA adducts. The method utilizes microgram quantities of DNA to detect adducts at frequencies of 1 in 10^{10}, making it applicable to realistic environmental exposures (Phillips and Arlt, 2007). However, the method is complex compared with some of the others cited in this chapter and requires sophisticated instrumentation. ^{32}P-postlabeling technique has been used to identify adducts in several species of marine fish living in polluted waters (Stein et al., 1989; Varanasi et al., 1989) and in a host of other species since.

Flow Cytometry

Flow cytometry has been around since the 1950s and its instrumentation has been frequently updated. Flow cytometry uses lasers to count and sort cells and detect several biomarkers associated with cells traveling up to thousands of particles per second. One parameter is the measurement of the total amount of DNA in at least 10,000 cells at once. When chromatids break due to various stress factors, the parts can reattach to each other or to sister chromatids with the addition of small amounts of new DNA used in the repair or reattachment process (like a glue or patching compound). When the total DNA is compared among sites thought to be clean and contaminated, a difference is associated with increased chromosomal damage. For example, Sparling et al. (2015) reported that tadpoles of the frog *Pseudacris regilla* inhabiting mid-elevation wet meadows in Sequoia National Park had significantly more DNA than those in Lassen or Yosemite National Parks. Elevated DNA corresponded to greater concentration of several organic contaminants including organochlorine pesticides and flame retardant chemicals in Sequoia National Park.

Visual Inspection

Three methods rely on observation through the microscope. A *micronucleus* (Fig. 3.6A) is a fragment of DNA from the nucleus of somatic cells that is formed during the anaphase stage of mitosis. They are cytoplasmic bodies having a portion of chromosome or even an entire chromosome that was not carried to the opposite poles during this stage. Their formation results in the daughter cell lacking a part or all of a chromosome. They are a good indicator of genotoxicity.

Eggshell Thinning

An early biomarker of contaminants on wildlife was eggshell thinning produced by DDT and its relatives (see Chapter 4 for details). Thin eggshells were used extensively as bioindicators of DDT exposure in fish-eating birds such as brown pelicans (*Pelecanus occidentalis*), peregrine falcons (*Falco peregrinus*), and bald eagles (*Haliaeetus leucocephalus*). Since DDT and virtually all other organochlorine pesticides that could cause thin eggshells have been banned, eggshell thinning due to organochlorines is no longer a concern.

Heinz Bodies

These bodies appear as small round inclusions within red blood cells (Fig. 3.6B). They are formed by damage to hemoglobin, usually through oxidative stress. Heinz bodies indicate that a reactive oxygen species was created in hemoglobin and that can cause cell damage leading to premature cell death. They may be induced by organic contaminants such as polychlorinated biphenyls, organochlorine pesticides, dioxins, and furans.

Chromosomal Breakage

This and other forms of damage can be caused by radiation and several contaminants (Fig. 3.6C). A visual inspection of a karyotype can reveal many of these damages.

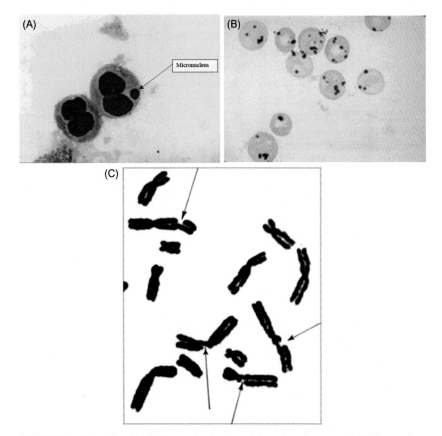

FIGURE 3.6 Examples of bioindicators that be viewed through a microscope. (A) Micronucleus; (B) Heinz body; (C) Chromosomal breakage.

HISTOLOGY

Direct observations of tissues in selected organs can indicate problems occurring at the cellular level. For larger animals, biopsies of liver, kidney, and other tissues can be taken without sacrificing the animals. As part of laboratory studies, however, a complete necropsy (ie, a complete examination of a deceased animal; referred to as an autopsy for humans) can provide extensive information on the effects of contaminants and allow for the development of diagnostic bioindicators.

There are so many things that can be examined through histology that we cannot hope to discuss them all here. We already have seen how viewing blood samples can be useful. Primary organs worthy of inspection in many cases are the liver, kidneys, spleen, and pancreas. Precancerous or cancerous tumors can also be detected through histopathological examination. These are the primary organs of metabolism or excretion for many chemicals and, as a result, often become damaged. For instance, white phosphorus (WP) has been associated with military target practice because it produces a white vapor that helps pinpoint where the howitzer and mortar shells might be going. The substance will vaporize in air but in cold, wet sediments WP can exist for years. When massive numbers of waterfowl were dying in a tidal marsh within an U.S. Army fort in Alaska, investigators determined that WP was killing the birds. Birds that survive are often emaciated and display characteristic fatty livers (Fig. 3.7,

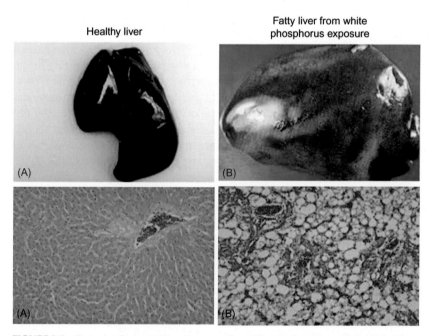

FIGURE 3.7   Example of a normal liver from a mallard (A) and a liver from a duck that had been gavaged with white phosphorus (B). The top photos are the livers and the bottom are histograms of liver sections.

Sparling, 1997). Liver biopsies can be used as bioindicators of effect in areas where WP is suspected.

In Chapter 5, we will discuss studies on the feminization of male frogs that are associated with the widespread use of the herbicide atrazine. Feminization is detected by the presence of ovotestes. These are gonads that are predominately testes with cells designed for the production of spermatozoa but also have regions that are distinctly ovarian with defined follicles. Ovotestes have been found in the African clawed frog (*Xenopus laevis*) and northern leopard frog (*Rana [Lithobates] pipiens*) (Hayes et al., 2003, 2010). In agricultural areas, particularly in the Midwest where corn propagation is extensive and atrazine use is pervasive, these ovotestes could serve as another type of bioindicator of exposure and effect. However, the examination for ovotestes generally results in the death of the animal.

GENOMICS AND PROTEOMICS

Genomics is a highly specialized and rapidly developing field, a part of which includes genotoxic effects of contaminants and development of bioindicators. It is an aspect of molecular genetics that applies recombinant DNA, DNA sequencing methods, and bioinformatics to sequence, assemble, and analyze the function and structure of genomes (ie, the complete set of DNA within an organism). This science holds considerable promise in developing our understanding of the relationships between contaminant exposure and genetics, and is already developing useful markers.

One method that is already in use is *gene microarrays* (Fig. 3.8). These arrays consist of coated chips or plates onto which DNA sequences representing different genes are affixed. The plates are exposed to contaminants, or alternatively to disease agents. If a chemical forms an adduct with a strand of DNA or otherwise induces the gene, the spot turns green under fluorescent light and reveals that the chemical can result in genotoxicity. Microarrays serve as excellent screening tools, but further tests will be necessary to demonstrate the actual toxicity of a chemical and dose/response relationships for the species. Thomas et al. (2002, p. 920) stated: that "one would have to make one overriding assumption—that most, if not all, toxic chemical exposures will alter gene expression at some level. In support of this assumption, toxicity by its very nature results from some form of cellular dysfunction or cell death that will result in changes in gene expression at some level." The specific expressions can be archived and serve as a microfingerprint of toxicity based on more extensive studies.

Several years ago, gene arrays were restricted to humans, mice, and rats. Today the method is much more available. For example, Hou et al. (2015) used microarrays on tomato plants (*Lycopersicon esculentum*) and cadmium-, chromium-, mercury-, and lead-spiked soil. The soil concentration of each metal was limited to 1/10 of the median lethal dose (LC50) for 7 days. They found expression changes in 29 Cd-specific, 58 Cr-specific, 192 Hg-specific,

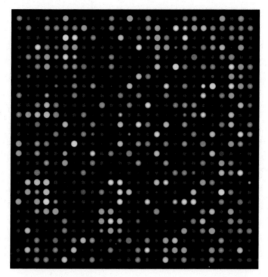

FIGURE 3.8   A gene microarray. DNA representing specific genes are applied to a back plate and dilute concentrations of suspected contaminants are applied to the DNA. If the gene is activated by the contaminant the spot turns green under fluorescent light.

and 864 Pb-specific genes. Statistical analysis of the data revealed that the characteristic gene expression profiles induced by each metal was unique. They were also able to pinpoint the function of the expressed genes and found that they were related to ion transport, structural organization, reproduction, lipid metabolism, and diverse other cellular processes. The authors concluded that microarray-based analysis of tomatoes was a sensitive tool for the early detection of potential toxicity of heavy metals in soil, an effective tool for identifying specific genes, and could be useful for assessing risk due to heavy metals in agricultural soil.

We made a partial scan of the scientific literature for a single year, 2015, and found that gene microarrays have been used on African clawed frog with silver, rainbow darters (*Etheostoma caeruleum*) and municipal wastewater, the plant rock cress (Arabidopsis) and a PAH, water fleas (*Daphnia magna*) and heavy metals, marine mussels and pyrene, dolphins and PCBs, and silkworms (*Bombyx mori*) and an insecticide. There seems to be no limit to the species or type of contaminant that can be used to indicate potential toxicity. One major advantage of microarrays is that only a small sample of DNA is necessary and for all but very small animals (such as insects) a simple biopsy of tissue, perhaps a toe clip, is sufficient for a sample. Another benefit of microarrays is that they are often sensitive to pg/L or ng/L concentrations of chemicals.

Gene microarrays are just one element of genomics. As previously mentioned, this very rapidly growing field of investigation promises much in our understanding of mechanisms of contaminant action, cross species variation, and the prediction of risk to humans and other species (Thomas et al., 2002).

A parallel field of investigation is *proteomics*, the high throughput separation, display, and identification of proteins that involves the total protein picture of an organism (Anderson and Anderson, 1998; Blackstock and Weir, 1999). Some of the same approaches are used in proteomics as in genomics, which makes sense because there is a direct relationship between genes and proteins produced by an organism. The basic concept is to identify unique relationships between proteins and toxicity and to use this knowledge to recognize bioindicators. Recent advances in high throughput techniques to identify specific proteins have greatly aided this field. Advantages of proteomics is that they can use tissue biopsies or even body fluids such as urine for their analysis. In 2015, proteomics studies included cadmium tolerance in aquatic plants, toxicity of ammonia to chickens, zebrafish and the insecticide chlorpyrifos, small crustaceans *Gammarus fossarum* to endocrine disruptors, *Daphnia magna* and copper, and oysters and carbon dioxide. See Kennedy (2002) for a good review of proteomics and toxicology.

REPRODUCTIVE AND DEVELOPMENTAL BIOINDICATORS

Teratogenesis

Abnormal development of body structures and organs can occur in embryos of all animals, although the greatest concern is for vertebrates. In mammals, birds, and reptiles, this abnormal development is usually observed at birth or hatching. In amphibians and fish, which also have a larval stage malformations, it can occur either during embryogenesis or during the larval period. The term "teratogenesis" is specific to abnormal development that occurs during embryonic development; abnormal development that occurs during the larval period is called malformations. Some forms of teratogenesis are diagnostic of specific kinds of contaminants.

Perhaps the most compelling example of teratogenesis occurred among American coots (*Fulica americana*), mallards, eared grebes (*Podiceps nigricollis*), and black-necked stilts (*Himantopus mexicanus*) in the Kesterson National Wildlife Refuge in the 1980s. The refuge collected water from Californian agricultural fields that had accumulated selenium and salt. This selenium caused a host of malformations including excess fluid in the cranium causing swelling, reduced sizes of eyes, and malformed feet and crossed bills. Laboratory studies confirmed that selenium was the cause of these defects (Hoffman, 1990) and that seleno-L-methionine, which was the form most common in plants (Heinz et al., 1987), was far more potent than elemental selenium in producing teratogenesis. Among nests, 13–53.8% contained chicks that were malformed and between 8.6% to 16.4% of the eggs investigated had malformed embryos (Ohlendorf et al., 1988). In Chapter 8, we will talk about malformed frogs due to lead, however this is not a case of teratogenesis since the animals had already hatched before the abnormalities occurred.

STUDY QUESTIONS

1. List at least four benefits of using bioindicators or biomarkers.
2. Describe in a general way how a biomarker for a particular contaminant or family of contaminants might be developed.
3. Discuss how oxidative stress occurs. What diseases or conditions have been linked to oxidative stress?
4. What is ROS?
5. You suspect that a population of animals living in a contaminated pond may be suffering oxidative stress. What would you look for to confirm that suspicion?
6. Suppose you make the hypothesis that mallard ducks feeding in a site contaminated by lead may have liver problems. How could you do to test that hypothesis without sacrificing animals?
7. What types of health conditions can you detect with a differential white blood cell count?
8. What are some ways that an investigator might determine that genetic damage has occurred to animals living in polluted environments?

REFERENCES

Adams, S.M., 2003. Establishing causality between environmental stressors and effects on aquatic ecosystems. Hum. Ecol. Risk Assess. 9, 17–35.

Adams, S.M., Giesy, J.P., Tremblay, L.A., Eason, C.T., 2001. The use of biomarkers in ecological risk assessment: recommendations from the Christchurch conference on biomarkers in ecotoxicology. Biomarkers 6, 1–6.

Anderson, N.L., Anderson, N.G., 1998. Proteome and proteomics: new technologies, new concepts, and new words. Electrophoresis 19, 1853–1861.

Bai, A., Reddy, C.C., 1977. Inhibition of acetylcholinesterase as a criterion to determine degree of insecticide poisoning in *Apis-cerana-indica*. J. Apicult. Res. 16, 112–114.

Beam, J.E., Hankinso, D.J., 1964. Application of acetylcholinesterase inhibition method for detecting organophosphate residues and related compounds in milk. J. Dairy Sci. 47, 1297–1305.

Blackstock, W.P., Weir, M.P., 1999. Proteomics: quantitative and physical mapping of cellular proteins. Trends Biotechnol. 17, 121–127.

Hayes, Al, Wise, R.A., Weir, F.W., 1980. Assessment of occupational exposure to organophosphates in pest-control operators. Am. Indust. Hygiene Assoc. J. 41, 568–575.

Hayes, T., Haston, K., Tsui, M., Hoang, A., Haeffele, C., Vonk, A., 2002. Atrazine-induced hermaphroditism at 0.1 ppb in American leopard frogs (*Rana pipiens*): laboratory and field evidence. Environ. Health Perspect. 111, 568–575.

Hayes, T.B., Khoury, V., Narayan, A., Nazir, M., Park, A., Brown, T., et al., 2010. Atrazine induces complete feminization and chemical castration in male African clawed frogs (*Xenopus laevis*). Proc. Natl. Acad. Sci. USA. 107, 4612–4617.

Heinz, G.H., Hoffman, D.J., Krynitsky, A.J., Weller, D.M.G., 1987. Reproduction in mallards fed selenium. Environ. Contam. Toxicol. 6, 423–433.

Hill, E.F., 2003. Wildlife toxicology of organophosphorus and carbamate pesticides. In: Hoffman, D.J., Rattner, B.A., Burton Jr., G.A., Cairns Jr., J. (Eds.), Handbook of Ecotoxicology, 2nd ed. Lewis Pub, Boca Raton, FL, pp. 281–312.

Hill, E.F., Fleming, W.J., 1982. Anticholinesterase poisoning of birds field monitoring and diagnosis of acute poisoning. Environ. Toxicol. Chem. 1, 27–38.

Hoffman, D.E., 1990. Embryotoxicity and teratogenicity of environmental contaminants to bird eggs. Rev. Environ. Contam. Toxicol. 115, 39–45.

Hoffman, D.J., Melancon, M.J., Klein, P.N., Rice, C.P., Eisemann, J.D., Hines, R.K., et al., 1996. Developmental toxicity of PCB 126 (3,3′,4,4′,5-pentachlorobiphenyl) in nestling American kestrels (*Falco sparverius*). Fund. Appl. Toxicol. 34, 188–200.

Hoffman, D.J., Heinz, G.H., Audet, D.J., 2006. Phosphorus amendment reduces hematological effects of lead in mallards ingesting contaminated sediments. Arch. Environ. Contam. Toxicol. 50, 421–428.

Hou, J., Liu, X.H., Wang, J., Zhao, S.N., Cui, B.S., 2015. Microarray-based analysis of gene expression in *Lycopersicon esculentum* seedling roots in response to cadmium, chromium, mercury, and lead. Environ. Sci. Technol. 49, 1834–1841.

Kennedy, S., 2002. The role of proteomics in toxicology: identification of biomarkers of toxicity by protein expression analysis. Biomarkers 7, 269–290.

Kundu, C.R., Roychoudhury, S., 2009. Malathion-induced sublethal toxicity on the hematology of cricket frog (*Fejervarya limnocharis*). J. Environ. Sci. Health B 44, 673–680.

Medline Plus, 2015. Blood differential. <http://www.nlm.nih.gov/medlineplus/ency/article/003657.htm> (accessed 30.06.15.).

Ohlendorf, H.M., Kilness, A.W., Simmons, L.J., Stroud, R.K., Hoffman, D.E., Moore, J.F., 1988. Selenium toxicosis in wild aquatic birds. J. Toxicol. Environ. Health 24, 67–92.

Pandey, S.P., Mohanty, B., 2015. The neonicotinoid pesticide imidacloprid and the dithiocarbamate fungicide mancozeb disrupt the pituitary-thyroid axis of a wildlife bird. Chemosphere 122, 227–234.

Phillips, D.H., Arlt, V.M., 2007. The 32P-postlabeling assay for DNA adducts. Nat. Protoc. 2, 2772–2781.

Sies, H. (Ed.), 2013. Oxidative Stress Elsevier.

Sparling, D.W., 1997. Ecotoxicology of white phosphorus in an Alaskan tidal marsh. <www.pwrc.usgs.gov/resshow/sparl1rs/sparl1rs.htm>. (accessed 14.06.15.).

Sparling, D.W., Bickham, J., Cowman, D., Fellers, G.M., Lacher, T., Matson, C.W., et al., 2015. In situ effects of pesticides on amphibians in the Sierra Nevada. Ecotoxicology 24, 262–278.

Stein, J.E., Reichert, W.L., Nishimore, M., Varanasi, U., 1989. 32P-postlabelling of DNA: a sensitive method for assessing environmentally induced genotoxicity. Proc. Oceans 7, 385–390.

Swiergosz-Kowalewska, R., Bednarska, A., Kafel, A., 2006. Glutathione levels and enzyme activity in the tissues of bank vole *Clethrionomys glareolus* chronically exposed to a mixture of metal contaminants. Chemosphere 65, 963–974.

Thomas, R.S., Rank, D.R., Penn, S.G., Zastrow, G.M., Hayes, K.R., Hu, T., 2002. Application of genomics to toxicology research. Environ. Health Perspect. 110 (Suppl. 6), 919–923.

Varanasi, U., Reichert, W.L., Stein, J., 1989. 32P-postlabelling of DNA adducts in liver of wild English sole (*Parophyrs vetulus*) and winter flounder (*Pseuopleuronectes americanus*). Cancer Res. 49, 1171–1177.

Wang, Q.W., Lam, J.C.W., Han, J., Wang, X.F., Guo, Y.Y., Lam, P.K.S., et al., 2015. Developmental exposure to the organophosphorus flame retardant tris (1,3-dichloro-2-propyl) phosphate: estrogenic activity, endocrine disruption and reproductive effects on zebrafish. Aquat. Toxicol. 160, 163–171.

Section II

Chemistry and Effect of Major Families of Contaminants

Chapter 4

Organochlorine Pesticides

Terms to Know

Aliphatic
Aromatic
Hydrocarbon
Gamma-Aminobutryic Acid (GABA)
Cyclodienes
Persistent Organic Pollutant (POPs)
Legacy Compounds
Stockholm Convention
Stereoisomer
Enantiomer
Diastereomer
Maternal Transfer
Xenobiotic
Apoptosis
Minimal Risk Level (MRL)
Maximum Concentration Level (MCL)
Permissible Exposure Level (PEL)

INTRODUCTION

Organic molecules are those that have carbon atoms serving as the backbone or major structural component of the molecule. The simplest group of organic molecules are *hydrocarbons* that contain only carbon, hydrogen, and oxygen. The carbons in these hydrocarbons can be arranged in straight lines and are called *aliphatics* (eg, ethane or pentane) (Fig. 4.1), or they can be in *aromatic* ring structures (eg, benzene). The hydrogen and oxygen atoms can also be arranged to form alcohols, organic acids, and other structures. Other elements such as nitrogen, chlorine, bromine, or sulfur can be attached to hydrocarbons to form more complex organic molecules. Organic molecules form the basis for the chemistry of living organisms although are also found in many nonliving sources, such as petroleum.

    Organochlorine pesticides (OCPs) are molecules that have at least one ring structure and multiple chlorine atoms, thus the name organochlorine. They include many different compounds including dichlorodiphenyltrichloroethane (DDT)

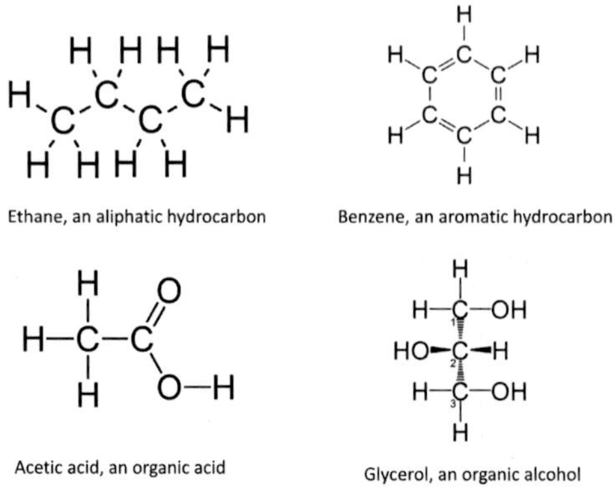

Ethane, an aliphatic hydrocarbon

Benzene, an aromatic hydrocarbon

Acetic acid, an organic acid

Glycerol, an organic alcohol

FIGURE 4.1 Types of organic molecules: the two lines between carbons in benzene and between carbon and oxygen in acetic acid represent double bonds. Aliphatics have carbons arranged in straight lines, aromatics have a ring structure in their solid form. Organic acids have a double-bonded carbon and oxygen with an OH or hydroxyl group on the same carbon, and alcohols have at least one hydroxyl group without the double-bonded oxygen.

and its relatives, gamma-hexachlorocyclohexane (γ-HCH or lindane), methoxy-chlor, endrin, endosulfan, heptachlor, chlordane, toxaphene, and mirex, to list just a few. Almost all organochlorines have been banned from use in the United States, Canada, and Europe, as well as in some other areas or countries. The intended function of organochlorines was to control insect pests by interfering with their nervous systems. Specifically, DDT and related compounds operate on sodium channels along axons of neurons so that the passage of an "action potential" along the nerve is disrupted. It causes uncontrolled, repetitive, and spontaneous discharges along the axon of a neuron, which most often leads to loss of control in the muscles used for breathing and consequently asphyxiation. A subclass of OCPs called *cyclodienes* includes endrin, aldrin, endosulfan, etc., and acts on the *gamma-aminobutryic acid (GABA) receptors* of the neuron. GABA is a very common neurotransmitter in the central nervous system, especially in the cerebral cortex. Cyclodienes bind to the GABA receptors and interfere with the ability of GABA to convey the action potential to the next neuron. Typical symptoms in animals exposed to these pesticides include convulsions, agitation, excitability, and other nervous disorders. In addition to the intended function of OCPs, extensive studies have shown that they have many negative effects including endocrine disruption, cancer, disruption of enzyme functions, and numerous other physiological disorders.

Most OCPs are *lipophilic*, which means that they readily dissolve in fats and oils, and *hydrophobic*, which means they do not mix well in water. Due to their lipophilic nature, OCPs bioaccumulate through bioconcentration and biomagnification. Most are very persistent compounds in the environment, remaining for years or decades after application and are part of a larger group of contaminants called *persistent organic pollutants (POPs)* that also includes dioxins, furans, polychlorinated biphenyls (PCBs), and brominated diphenyl ethers (BDEs) used in fire retardants. Although most OCPs have either been banned or their usage greatly restricted in the United States by the EPA, they are still found in the environment due to their persistent nature and thus, are also called *legacy compounds* and are used for limited purposes.

SOURCES AND USE

OCPs became popular as insecticides in the 1930s. The most famous, or perhaps infamous, of these pesticides is *DDT*. The chemical was initially synthesized in 1874 but wasn't used as a pesticide until 1939. At that time, it was one of the first chemicals to be widely used on farm crops. From then through the 1960s, other OCPs were manufactured in the United States. Based primarily on its risk to human health and persistence, the EPA delisted or banned DDT in 1972, Canada did so a few years later and Mexico banned it in 2000. The US EPA banned other OCPs through 2010; endosulfan is the most recent OCP to be have its registration revoked.

In 1995, the United Nations called for an international examination on the use of POPs. This led to the *Stockholm Convention* in 2001 which was an international agreement calling for a cessation of 12 of the most noxious pesticides included in five OCP groups: toxaphene, chlordane, and heptachlor; DDT; aldrin, dieldrin, endrin; hexachlorohexane (HCH) and lindane; and pentachlorophenol. The convention also curtailed the production of dioxins and furans. Initially, 128 nations signed the convention and it was instated in 2004. Since then, the total number of signatories is 179. Interestingly, the United States did not internally ratify the convention.

DDT continues to be used for one important human health factor—malaria. There does not seem to be any pesticide currently available that is as cheap and effective in controlling the mosquito vectors of malaria as DDT. Malaria-prone countries in South America, India, and Africa are therefore allowed to use this compound under the UN's World Health Organization guidelines to treat mosquito netting and spray the walls of their huts or houses, but not to use it in a broad application to wetlands. Unfortunately, the mosquitoes in these countries are developing the same resistance to DDT as was seen in crop pests in the United States prior to its being banned. Additionally, India remains highly contaminated by DDT and HCH, in part due to the increasing use of OCPs in mosquito control (Sharma et al., 2014).

GENERAL CHEMICAL CHARACTERISTICS

Some general chemical characteristics of OCPs are presented in Table 4.1. A notable feature of many OCPs is their low solubility in water and a high lipophilicity, which is shown by the relatively high log K_{ow}s. Recall from Chapter 1

TABLE 4.1 Some Characteristics of Organochlorine Pesticides

Common Name	Log K_{ow}	Log K_{oc}	MW^1	Solubility (mg/L)	Field Half-Life	US Status
Aldrin	6.5	4.24	365	0.03	365 days	Banned 1974
Chlorbenzilate	4.6	3.45	325	13	NA	Banned in 1999
Chlordane	6	4.78	410	0.06	283–1387 days	Banned 1988
DDD	5.0–6.2	5.36	320	0.05	15 years	Banned 1972
DDE	5.6–6.9	5.94	318	0.06	15 years	Banned 1972
DDT	6.0	5.63	355	0.001	15 years	Banned 1972
Dicofol	4.3	3.79	370	0.8	16 days	In use as a mitacide
Dieldrin	3.6–6.2	4.08	381	0.186	1000 days	Banned 1987
Endosulfan	3.1	4.09	407	0.5	4–200 days	Phased out starting 2010
Endrin	3.2–5.3	4.00	381	0.24	4300 days	Banned 1987
Heptachlor	4.4–5.5	4.38	373	0.06	250 days	Used only for fire ant control in transformers
HCB	3.9–6.4	4.75	285	0.04	1000 days	Banned 1966
Lindane	3.7	1355	291	7	100–1424 days	Used only to control lice and scabies
Methoxychlor	4.8–5.1	3.13	345	0.1	128 days	Banned in 2003
Mirex	6.9	6.0	546	0.00007	3000 days	Banned 1976
Pentachlorolphenol	3.3–5.9	4.30	266	14	48 days	Used only to preserve wood
Toxaphene	5.9	5.0	414	3	9 days	Banned 1990

that K_{ow} or octanol-water coefficient is the ratio of a compound's concentration in *n*-octanol relative to its concentration in an equal volume of water after the octanol and water have reached equilibrium and that the higher the K_{ow} value, the more soluble chemicals are in fats and oils and the less soluble they are in water. Molecules with a $K_{ow} > 3.5$ are generally considered to be lipophilic and hydrophobic. Lindane, pentachlorophenol, and chlorbenzilate have relatively high water solubilities and may be found dissolved in water at toxic concentrations. Mirex, DDT, and aldrin with lower water solubilities would be less likely to be in the water column of lakes, ponds, or rivers. Another guideline, K_{oc} or soil-water partitioning coefficient, reflects the tendency of the compound to adhere to soil particles with high K_{oc} values. Mirex and DDE, for example, have high K_{oc} values which suggest that they will readily bind with soils and sediments. Binding may reduce their availability to organisms in water. The melting points of all of the listed OCPs show that they are solids at room temperature and most are white or colorless crystalline structures.

Structure

Structurally, OCPs fall into five classes (Blus, 2003, Fig. 4.2): (1) DDT and its analogs including DDT and dichlorodiphenyldichloroethylene (DDE); (2) hexachlorocyclohexane (HCH), such as lindane; (3) cyclodienes including aldrin, dieldrin, endrin (sometimes referred to as "drins" in the literature), heptachlor, chlordane, and endosulfan; (4) toxaphene; and (5) mirex and chlordecone.

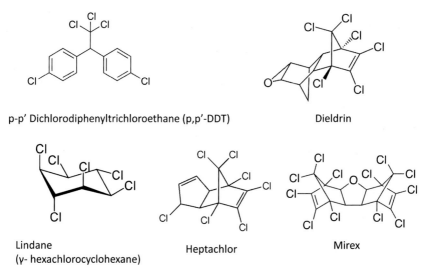

p-p' Dichlorodiphenyltrichloroethane (p,p'-DDT)   Dieldrin

Lindane   Heptachlor   Mirex
(γ- hexachlorocyclohexane)

FIGURE 4.2   Structural formulas for representatives of the five major classes of OCPs.

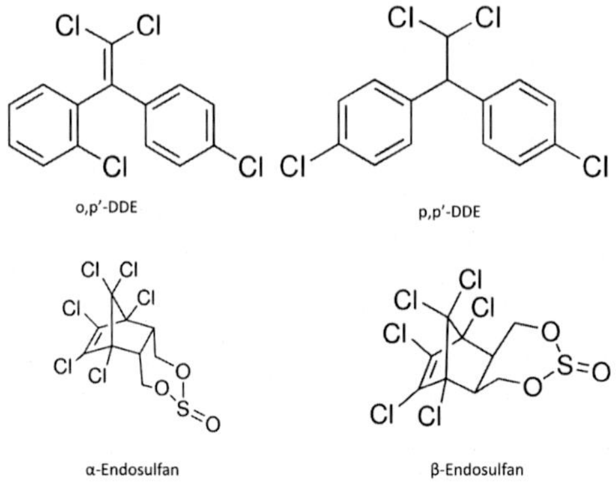

o,p'-DDE

p,p'-DDE

α-Endosulfan

β-Endosulfan

FIGURE 4.3 Examples of stereoisomers. The two forms of DDE are enantiomers, note the position of the bottom chlorine atoms. α and β endosulfan are diastereomers and mirror images of each other.

Toxaphene is made up of approximately 200 chemicals, many of which are proprietary and not identified by the manufacturer. Technical-grade chlordane is also a mixture of more than 120 structurally related chemical compounds but is primarily composed of *cis*- and *trans*-chlordane, heptachlor, and *cis*- and *trans*-nonachlor (Bondy et al., 2000). Mirex and chlordecone (also called kepone) have only one form.

Many OCPs are represented by different stereoisomers including enantiomers and diastereomers. Briefly, *stereoisomers* (Fig. 4.3) are chemicals with the same molecular formulas—they have the same molecular weight and the same number of carbon, hydrogen, oxygen, and chlorine atoms, but the atoms are rearranged. Stereoisomers can be further divided into *enantiomers* (also called *chiral* structures) which are nonsuperimposable mirror images of each other and *diastereomers* that can be superimposed. These stereoisomers can radically differ in their chemical and biological properties. For instance, the DDT family includes DDT, DDD, and DDE; all have enantiomeric ortho- and para-isomers with DDE and DDD being metabolites of the parent compound DDT. Thus, we have o,p'-DDT, p,p'-DDT, and similar configurations for DDD and DDE. In contrast, when you see *cis*- and *trans*- as in *cis*-chlordane and *trans*-chlordane or the chemical name is preceded by a Greek letter as in α-endosulfan or β-endosulfan we are discussing diastereomers. Diastereomers can be transformed into each other during metabolism whereas enantiomers cannot.

Persistence

While Table 4.1 lists approximate half-lives of the compounds, actual persistence in the environment is affected by many different factors. Typically, the covalent C-Cl (or carbon with fluorine or bromine for that matter) bond is very

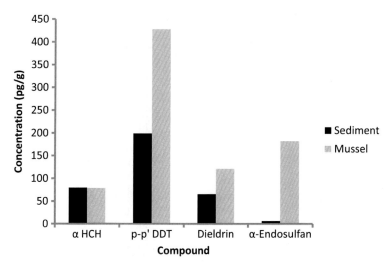

FIGURE 4.4   An example of bioconcentrations of a few organochlorine pesticides. Note that the concentrations in the mussels are generally higher than in the environment. *Okay et al. (2011).*

stable and resistant to microbial degradation, although a few strains of bacteria and fungi have the ability to break down the parent compounds. In addition, certain plants can metabolically degrade OCPs and earthworms can enhance the degradation of DDT in soil (Lin et al., 2012). Physical factors that help degrade OCPs include light, pH, and moisture. Low pH and the presence of radicals such as CO_3^- and HCO_3 accelerate degradation. Ultraviolet radiation is very significant in degrading OCPs.

Persistence also varies by chemical. DDT and its derivatives are very stable in the environment. However, DDT is metabolized chiefly into DDE. For that reason, residue studies often find higher DDE than DDT concentrations in the environment, but the DDE/DDT ratio can be even higher in organisms. For example, Okay et al. (2011) measured the concentrations of OCPs in mussels and sediments in Turkey and found that the ratio of DDE/DDT was 3.5 and 7.1 in sediment and mussels, respectively (Fig. 4.4). They also found that the concentrations of OCPs were generally higher in mussels than in sediments because of bioaccumulation. Heptachlor and aldrin are rapidly metabolized by the body and have a short half-life in organisms. However, their metabolites may be just as, or sometimes even more, toxic than the parent compound. Heptachlor epoxide, the major breakdown product of heptachlor is far more stable in the body and the environment; both forms have approximately the same toxicity (ATSDR, 2007). Aldrin breaks down into dieldrin and its stereoisomer, endrin. Aldrin is not toxic to insects (Blus, 2003), but endrin has one of the highest toxicities among OCPs. Organochlorines volatize from soil and travel through

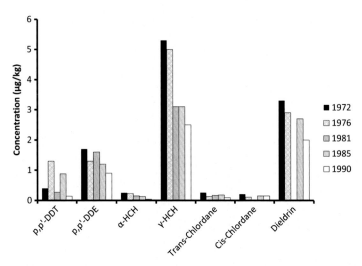

FIGURE 4.5 Concentrations of various OCPs in United Kingdom soils, showing a gradual decline overtime. *Meijer et al. (2001).*

aerial transport. Thus, countries that still permit the use of OCPs may provide a source of "new" OCPs to the rest of the world (Bidleman et al., 2013).

Eventually, these factors of degradation cause even the most persistent POPs to breakdown and their concentrations in environmental sources and organisms diminish. Meijer et al. (2001) measured OCPs in soils from the United Kingdom that had been archived from 1972 until 1990. They found that p,p′-DDT, dieldrin and γ-HCH were consistently higher than other OCPs. However, there was a general downward trend for all of the OCPs they examined (Fig. 4.5).

As an example in animals, Braune et al. (2007) studied ivory billed gull (*Pagophila eburnea*) eggs from the Canadian Arctic from 1976 to 2004 and reported that dieldrin concentrations in 2004 were 56% of what they were in 1976. They also found that p,p′-DDD concentrations were only 8% of what they were at the start (Fig. 4.6). Note that that not all the OCPs declined. In particular, the total HCH, heptachlor epoxide, and especially *trans*-nonachlor were all higher in 2004 than in 1987. The reasons for higher concentrations in 2004 compared to those in 1976 were not explained by the study.

Another example is that of peregrine falcon (*Falco peregrinus*) plasma on San Padre Island in Texas from 1978 to 2004 (Henny et al., 2009). The plasma's p,p′-DDT dropped from 0.879 µg/g wet weight in 1978 to 0.013 µg/g, in 2004, a decrease of 98%. Moreover, the frequency of occurrence for several OCPs including p,p′-DDT, p,p′-DDE, heptachlor epoxide, dieldrin, oxychlordane, and mirex went from 10 to 43% of the birds sampled in 1978 to 0% in 2004. Thus, given enough time, we may eventually see the end of these POPs.

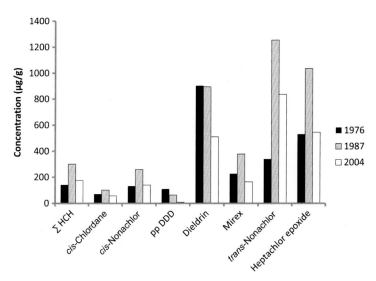

FIGURE 4.6   Concentration of selected OCPs in ivory billed gull eggs in 1976, 1987, and 2004. Here the general pattern is an increase from 1976 to 1987 as environmental OCPs were still abundant but then a decline from 1987 to 2004 as the OCPs were slowly degraded in the environment. *Braune et al. (2007).*

Bioaccumulation, Bioconcentration, and Biomagnification

OCPs are known to bioaccumulate in organisms and this process can involve bioconcentration and biomagnification. Bioaccumulation means that OCPs enter the tissues of organisms; another term for this is that they are *assimilated*. They may be found in higher concentrations in some tissues, such as liver or kidney than in others like muscle because of their lipophilic nature; this is bioconcentration. For example, refer back to Fig. 4.3. Note that most of the OCPs were higher in the mussels than they were in the sediment that contained the mussels. The exception to this was α-HCH, which is more water soluble than the other OCPs and has a lower tendency to be stored in organisms. Organochlorine pesticides can also biomagnify in that the concentrations of the compounds increase at higher trophic levels. For instance, concentrations of DDT, DDE, or other OCPs may be in the parts per billion in water. Animals such as zooplankton assimilate these OCPs. Zooplankton are eaten by insects which are consumed by small fish that are ingested by bigger fish, etc. and, at each step, the concentrations of OCPs increase through biomagnifications (Fig. 4.7). Because the concentrations of OCPs are very low initially, no negative effects may be seen until the highest trophic levels are reached. At that point concentrations may be 10 million times those in water. This process of biomagnification was very important in the population declines seen in many fish-eating raptors in the 1980s and 1990s (see FOCUS section at the end of this chapter).

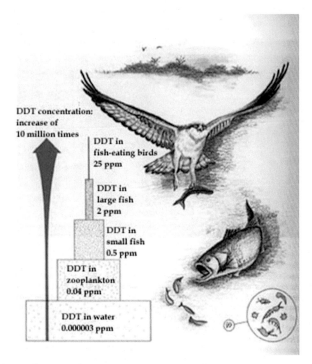

FIGURE 4.7   A diagrammatic representation of biomagnification of DDT in an aquatic food chain. *Courtesy of the U.S. Fish and Wildlife Service.*

EXAMPLES OF OCP CONCENTRATIONS IN ENVIRONMENTAL SOURCES

Concentrations of OCPs in various parts of the environment depend on many factors. OCPs that are more water soluble, such as HCHs, can be found in higher concentrations in water than those that are more lipophilic. Those that have high K_{oc} values are likely to adhere to soil or sediment particles where they can remain relatively stable for many years. These and other POPs are controlled by three main factors (Miglioranza et al., 2013): (1) physiochemical properties of the compound; (2) environmental factors such as pH, temperature, Ultraviolet radiation-B, wind, soil texture, and organic carbon; and (3) characteristics of the organisms inhabiting the area of concern.

Initially, OCPs were introduced into the environment through pesticide application, usually in the form of sprays. In areas where OCPs are still used, this continues to be the major point of entry into the environment. Spills, leaching of stock piles or dump sites, and direct discharge into effluents or emissions from industrial processes are other means. From there, the contaminants can be recycled through various parts of the environment. If discharged into water through runoff or direct spray, they are most likely to adhere to sediments than attach to particulates or move freely in the water column. OCPs can volatilize from dry soil, which is a major way that they are released into the atmosphere in areas where they are no longer used but still exist. Eventually OCPs can return to water or soil as wet or dry precipitation. Wet precipitation occurs with rain or snow and dry precipitation occurs when the OCPs attach to dust or other atmospheric debris. Weather factors such as snowmelt or rain can cause runoff into streams or wetlands, whereas wind erosion can resuspend the contaminants in the atmosphere.

Of the four major environmental compartments—air, water, sediments, or soil—sediments generally contain the highest concentrations of OCPs followed by soil; water and air have low but often measureable concentrations. Units of measurement for tissues, soil, or sediment are usually expressed as ng/g or µg/kg. In water, the concentrations are usually much lower and measured as ng/L (Table 4.2). In air, the concentrations are even lower and are measured as pg/m^3. Because there are $1000 L$ in $1 m^3$, a 1-pg concentration in air would be the equivalent of 1 over 1 quintillion, which is the same as $1/1 \times 10^{18}$. Air concentrations are often too small for even the best laboratory instruments to measure directly, so air is passed through a collector in an air sampler to "scavenge" the OCPs. Here, the amount of air that flows through the device is measured or the sampler is positioned for a set period of time, and the pesticide is collected and condensed with the concentration determined by final concentration/amount of air passed or unit of time.

CONCENTRATIONS OF OCPs IN ANIMALS

Many factors affect the concentrations of OCPs in living organisms. Location is a very important factor. Although OCPs are widespread globally, their concentrations vary by orders of magnitude from one site to another. An animal must have contact with OCPs for some period of time to assimilate them from the environment. For somewhat sedentary species, this means that the OCPs have to be in their home range. Migratory species can take in OCPs from their winter or their summer ranges, or both. Aquatic organisms such as freshwater and marine fish and turtles and marine mammals such as seals, walruses, whales, and dolphins may be constantly bathed in contaminated waters. Many of the fish and mammals also ingest contaminated prey, leading to biomagnification. For the most part, terrestrial species ingest OCPs with prey organisms but aquatic organisms may also be exposed through dermal absorption or by assimilation through gills.

Diet is another important factor that can affect the concentration of OCPs, but this is often closely associated with focal species. For example, birds of prey collected in the same area can vary by at least 100 times in their OCP concentrations

TABLE 4.2 Concentrations of Selected OCPs in Environmental Sources

Matrix	Location	Contaminant	Date	Concentration	Units	Comments	Reference
Atmosphere	Lake Ontario	cis-Chlordane	2000–2001	6.9	pg/m^3	2000–2001	Shen et al. (2005)
	Lake Huron			1.2			
	Lake Superior			2.1			
	Cornbelt			14			
	Lake Ontario	α-Endosulfan		93			
	Lake Huron			18			
	Lake Superior			12			
	Cornbelt			48			
	Lake Ontario	p,p′-DDT		ND			
	Lake Huron			0.59			
	Lake Superior			0.6			
	Cornbelt			12			
	Lake Ontario	p,p′- DDE		24			
	Lake Huron			2.7			
	Lake Superior			2			
	Cornbelt			16			
	Lake Ontario	Hexachlorobenzene		99			

Media	Location	Compound	Period	Concentration	Site	Units	Statistic	Reference
	Lake Huron			66				
	Lake Superior			84				
	Cornbelt			131				
Precipitation	Brule River, Wisconsin	γ-HCH	1997–2003	860 ± 150	On Lake Superior	pg/L	Mean ± SE	Sun et al. (2006)
		Chlordane		180 ± 41				
		trans-Nonachlor		16 ± 2.2		Mean ± SE		
		p,p'-DDT		77 ± 21				
		p,pDDE		36 ± 5.7				
		methoxychlor		150 ± 35				
	Chicago, Illinois	γ-HCH		1100 ± 270	On Lake Michigan	pg/L		
		Chlordane		1100 ± 230		Mean ± SE		
		trans-Nonachlor		100 ± 12				
		p,p'-DDT		570 ± 93				
		p,p'-DDE		210 ± 37				
		methoxychlor		370 ± 56				
Snow	High elevations in western U.S. National Parks	α-Endosulfan	2001–2003	0–0.21	2001–2003	ng/L	Range	Mast et al. (2007)
		β-Endosulfan		0–1.2				
		Endosulfan sulfate		0–0.21				

(Continued)

TABLE 4.2 Concentrations of Selected OCPs in Environmental Sources *Continued*

Matrix	Location	Contaminant	Date	Concentration	Units	Comments	Reference
Water	High elevations in Sierra Nevada, California	α-Chlordane	2001	1.20	ng/L	2001	Fellers et al. (2013)
		p,p′-DDE		0.41		Median	
		p,p′-DDT		1.28			
		α-Endosulfan		0.05			
		β-Endosulfan		0.07			
		Endosulfan sulfate		0.06			
		trans-Nonachlor		0.02			
		α-HCH		0.06			
Water	South China	α-HCH	2002–2003	38–596	pg/L	Range	Yu et al. (2008)
		β-HCH		48–877			
		γ-HCH		13–1656			
		p,p′-DDE		43–886			
		p,p′-DDD		66–783			
		p,p′-DDT					
Freshwater Sediments	High elevations in Sierra Nevada, California	α-Chlordane	2001	0.55	ng/g	Median	Fellers et al. (2013)
		p,p′-DDE		1.25			
		α-Endosulfan		0.35			

		Compound	Year	Value	Units	Notes	Reference
		β-Endosulfan		0.17			
		Endosulfan sulfate		0.28			
		α-HCH		17.98			
		Heptachlor epoxide		0.19			
		cis-Nonachlor		0.87			
		trans-Nonachlor		0.53			
Freshwater Sediments	New Zealand	∑ Endosulfans	2007	0.04 ± 0.02	ng/g	Near an organic sheep/cattle ranch	Shahpoury et al. (2013)
		α-HCH		0.02 ± 0.01			
		γ-HCH		0.17 ± 0.08			
		∑ DDTs		0.34 ± 0.10			
		Endrin		0.06 ± 0.3			
		∑ Chlordanes		0.90 ± 0.52			
Marine Sediments	Southern CA	p,p´-DDE	2009	127–18437	ug/Kg dry wt	2009, 18 sites	Chen et al. (2012)
		p,p´-DDD		9–18743		Highest values outside of an active effluent system	
		p,p´-DDT		17–25114			

(Continued)

TABLE 4.2 Concentrations of Selected OCPs in Environmental Sources *Continued*

Matrix	Location	Contaminant	Date	Concentration	Units	Comments	Reference
Marine sediments	Turkey	α-HCH	2007	ND-358	pg/g	17 sites	Okay et al. (2011)
		β-HCH		4.7–316			
		p,p′-DDT		1.31–6650			
		p,p′-DDD		5.39–5725			
		p,p′-DDE		12.2–1567			
		Dieldrin		0.52–181			
		α-Endosulfan		nd-37			
Ag soils	British Columbia	γ-HCH	1999–2000	8–57	ng/g dw	Muck	Bidleman et al. (2006)
				<LOD		Tilled Loams	
				<LOD		Orchard loams	
		Dieldrin		410–800		Muck	
				<LOD		Tilled Loams	
				<LOD-8		Orchard loams	
		trans-Chlordane		55–715		Muck	
				<LOD		Tilled Loams	
				<LOD		Orchard loams	

Region	Compound	Soil/medium	Units	Period	Value	Reference
	p,p'-DDT	Muck			590–990	
		Tilled Loams			< LOD–145	
		Orchard loams			530–6900	
	p,p'-DDE	Muck			350–395	
		Tilled Loams			5–165	
		Orchard loams			2200–7100	
	p,p'-DDT	Muck			520–685	
		Tilled Loams			< LOD–5	
		Orchard loams			200–1400	
Southern U.S.	α-HCH	Farm soils	ng/g dw	1996–2000	0.05	Bidleman and Leone (2004)
	γ-HCH				0.05	Geometric means
	Σ Chlordanes				0.16	
	p,p'-DDE				17	
	p,p'-DDD				0.57	
	p,p'-DDT				8.5	
	Σ DDTs				31	
Ag Soils	Toxaphene				92	

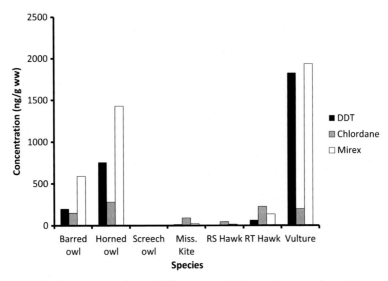

FIGURE 4.8 Mean concentrations of OCPs measured in livers of raptors collected from coastal South Carolina, 2009–2010. The concentrations reflect the diets of the raptors (see text). *Miss,* Mississippi; *RS,* Red-Shouldered; *RT,* Red-Tailed; *Vulture,* turkey vulture. *Yordy et al. (2013).*

(Yordy et al., 2013, Fig. 4.8). All of the predators examined in Yordy et al.'s study were opportunistic, so diets differ more by degree than absolutes; from screech owls (*Asio otus*) and Mississippi kites (*Ictinia mississippiensis*) that consume a fair amount of insects and small mammals, to great horned owls (*Bubo virginianus*) and hawks that feed more on medium-sized mammals and birds. The turkey vulture (*Cathartes aura*) is, of course, a scavenger, but red-tailed hawks (*Buteo jamaicensis*) also scavenge. Ospreys (*Pandion haliaetus*) are piscivores or fish eaters. One great horned owl (not reported in the figure due to its very high contaminant levels) had 168 times the amount of DDT in its liver than the mean for barred owls (*Strix varia*), 510 times more than red-tailed hawks, and 10 times higher than the highest turkey vulture concentration reported.

In another example, the green sea turtle (*Chelonia mydas*) and Kemp's ridley sea turtles (*Lepidochelys kempii*) inhabit many of the world's oceans including the Gulf of Mexico. Green sea turtles, however, are primarily herbivores, feeding on sea grasses and algae (Swarthout et al., 2010) and have significantly lower concentrations of several OCPs than Kemp's ridley sea turtles that are more carnivorous and feed on crabs and mollusks (Fig. 4.9).

Beyond location and diet, other factors play variable and case-specific roles in affecting OCP concentrations. Age is sometimes important due to the bioaccumulation of pesticides or because older animals have more fat and less lean muscle than juveniles or young adults. Gender can often be a factor, but whether the male or female has the higher concentration depends on species.

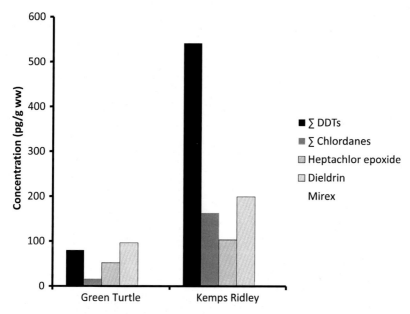

FIGURE 4.9   Concentrations of OCPs in blood of green turtles and Kemps Ridley turtles collected in the Gulf of Mexico, 2001–2002. Green turtles are primarily herbivores whereas Kemps Ridley turtles are mostly carnivores. *Swarthout et al. (2010).*

For example, the blubber of adult male common bottlenose dolphins (*Tursiops truncates*) living off the coast of Sarasota Florida had significantly higher concentrations of several OCPs than juvenile males or females and 10-40 times the amount found in adult females (Yordy et al., 2010, Fig. 4.10). The dolphins were in the process of, or recently completed weaning, young and had recommenced eating fish, which contributed to their contaminant body burdens. The authors attributed the lower concentrations of OCPs in the adult females to their ability to offload some of them into their milk.

Maternal Transfer

This brings up the concept of *maternal transfer*. Contaminants including OCPs can be passed from mother to young in many species of vertebrates. This has been of great concern for nursing human mothers in many parts of the world (eg, Azeredo et al., 2008; Hernik et al., 2014). Milk mostly consists of water, but also contains lipids. In mammals, lipophilic compounds can be transferred with the fat component of milk. Often maternal transfer can relieve mothers of substantial percentages of their contaminant body burdens, but it also means that their young are subsequently exposed to them.

Bird eggs are around 9–10% lipid by weight, which is concentrated in the yolk where the early embryo will develop. Other egg laying vertebrates have

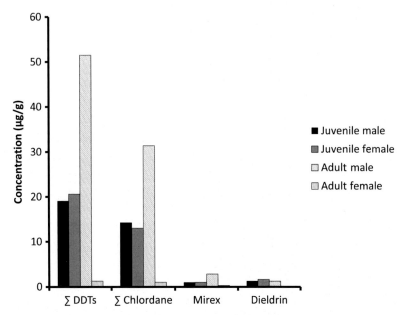

FIGURE 4.10 Concentration expressed as lipid weight of organochlorines in bottlenose dolphin blubber and milk by age and class. Note the high concentrations overall higher concentrations in males and low concentrations in adult females. The low concentrations in adult females are a factor of maternal transfer. *Yordy et al. (2010).*

roughly the same lipid concentration or more in their eggs. Thus, lipophilic compounds can be transferred from the female when she is producing eggs to the developing embryos. Amphibians and fish have far less lipid in their rapidly developing eggs and are unlikely to contain nearly the same amount of contaminants that occur in shelled eggs.

Some studies have shown that the order of laying can make a big difference in the concentration of contaminants in egg. In species such as the blue tit (*Parus caeruleus*), contaminant concentrations dropped drastically through the egg laying sequence (Van den Steen et al., 2009); for example, the last eggs contained only 68% of the total OCPs found in the first egg. Similarly, the first five eggs laid by female snapping turtles (*Chelydra serpentina*) had two to three times the concentration of OCPs as the last five eggs (Bishop et al., 1995); clutch size ranges from 12 to approximately 40 in this species. This difference due to laying sequence can be important for a couple of reasons. First, since the early eggs receive a larger dose of contaminants, they may be at greater risk for harm. Second, it is important to try to obtain the same egg in sequence among different nests for biomonitoring purposes. However, this pattern does not hold for all species of birds. In some species there is no significant difference due to sequence of laying and for other species, like the common tern (*Sterna hirundo*), the first egg may have lower concentrations than subsequent eggs (Nisbet, 1982).

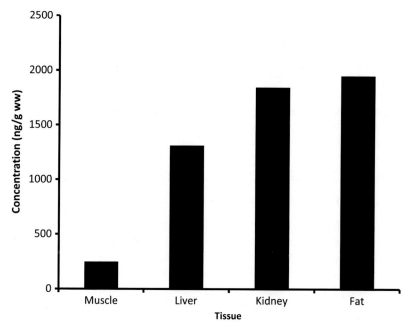

FIGURE 4.11   Concentrations of p,p′-DDE in various organs of bald eagles showing the relatively low concentrations in muscle compared to organs that contain higher levels of lipids. *Kumar et al. (2002).*

OCP Concentrations Vary Among Tissues

Organochlorine pesticides are not equally distributed through an organism. In most cases, organs and tissues with greater lipids will have higher OCP concentrations than lean compartments. For this reason, body concentrations of OCPs are often measured in terms of lipid percentage. Other methods of measurement are unadjusted from fresh, wet, or dry tissue. Animal tissue is 70–90% water, which means that after water is removed only 10–30% of the body tissue remains but, at least for chemicals that are not very volatile, most of the contaminant remains thus concentrations based on dry weights will typically be considerably higher than the other two. Concentrations based on fresh and wet weights are indicative of what the animal experiences. Lipid-adjusted standardized concentrations from one tissue to another and among animals and dry weights represent what a predator would consume if it ate the entire animal. Muscle and plasma tend to be lower in lipid content than liver or kidney, where we often see OCPs sequestered. For example, the sequence for p,p′-DDE in bald eagles (*Haliaeetus leucocephalus*) from Michigan using wet weights was fat > kidney > liver > muscle > plasma (Fig. 4.11, Kumar et al., 2002). Plasma is not shown on the figure because it had only 9.92 ng/g, below the level that the graph could resolve.

In the Patagonian silverside (*Odontesthes hatcheri*), a South American fish, p,p'-DDE had the highest concentrations among the DDTs and, based on concentration in lipids, had the following sequence in tissues: liver > gill > muscle > gonads (Ondarza et al., 2014). The authors wanted to determine if eating the fish could be harmful because the silverside is commonly eaten by humans. Muscle is the most commonly consumed tissue and the authors found that the concentrations of OCPs were not of concern in this area; however, PCBs had higher concentrations than organochlorines and were consider potentially harmful to human health.

BIOLOGICAL EFFECTS OF ORGANOCHLORINE PESTICIDES

Considerations with Plants

Few studies have been conducted on the effects of OCPs on plants. Liu et al. (2009) exposed white clover (*Trifolium repens*) to relatively high concentrations of endosulfan. At the highest concentration, 10 mg/L, significant chromosomal damage was observed in somatic (nonreproductive) tissues. The authors did not speculate on what that might mean to the individual plants. DDT derivatives at concentrations of 100–600 mg/kg inhibited photosynthesis activity and reduced cell density and shoot growth in three species of algae. These derivatives also affected the growth of roots, but had no effect on germination in ryegrass (*Lolium perenne*, Chung et al., 2007). To produce toxic effects, however, the DDT derivatives were several times higher than those necessary to induce toxicity in most animals or could be expected in the environment.

However, the interaction of OCPs and plants have other more significant concerns. Plants can bioaccumulate OCPs from the environment. This can either be a good thing or a bad thing, depending on your perspective. The ability of plants to absorb OCPs means that they can be used in remediation of contaminated sites. Plants that have high affinities for OCPs can be planted in contaminated sediments, harvested at the end of the growing season, and burned along with their pesticide burdens. On the negative side, plants used for human or livestock consumption also assimilate chlorinated pesticides.

An Interesting Relationship

One other rather interesting anecdote about plants and OCPs is that plants seemed to help grizzly bears (*Ursus arctos horriblis*) reduce their body burdens of OCPs (Christensen et al., 2013). This study demonstrated some of the complexity in trying to understand the exposure of populations of wild animals to chlorinated hydrocarbons and the type of information that should be considered for risk assessment. Coastal or intertidal grizzlies in British Columbia, Canada, feed substantially on pink salmon (*Oncorhynchus gorbuscha*) during their fall spawning, whereas inland bears consume more sockeye salmon (*O. nerka*) when they are available. The mean concentration of total OCPs in pink salmon was 444 ng/g based on lipid weight while mean sockeye concentration was 1890 ng/g. Sockeyes also had significantly higher concentrations of PCBs compared to pink salmon (Fig. 4.12A). Throughout the year intertidal grizzlies can also consume mussels (*Mytilus* sp.) and Dungeness crabs (*Cancer magister*) that have 4–26 times the OCP concentrations as plants, respectively. During the spring salmon are not available so both populations turn to a greater percentage of plant material in their diets. In their 2013 study, Christensen et al. found that the favorite vegetation of grizzlies in the Koeye River region of British Columbia had 0.3–1% of the total OCPs as fall-running salmon. In a model based on estimated food consumption, the authors determined that, during the fall, adult grizzlies living in interior areas were exposed to the highest concentrations of organochlorines due to the large amount of sockeye salmon in their diets. For example, daily exposures for total DDTs ranged from 83–170 µg for intertidal bears, but were 290–760 µg for interior bears. However, in the spring when salmon were not available, inland bears had the lowest daily exposure to POPs including OCPs, PCBs, and polybrominated diphenyl ethers (PBDEs). Total DDTs dropped to 27 µg/day for intertidal bears, but only 0.31 µg/day for interior grizzlies. Subadult bears, which are not as successful in obtaining salmon as adults, tended to have lower OCP concentrations in both areas and seasons. This can be seen in seasonal and regional differences in the feces of the bears (Fig. 4.12B). Following meals of either pink or sockeye salmon, bears excreted 1.5% or less of the ingested POPs. However, POP excretion increased substantially after a meal of vegetation or berries. From 51% to 510% of the ingested PCBs and 360% of the PBDEs were eliminated. Most vegetation is indigestible to grizzlies, which would explain the relatively high excretion concentration up to the amount consumed, but not in excess. In humans, a high water-insoluble fiber fraction in the diet will enhance POP excretion, further assisting elimination of POPs and a similar phenomenon may at work for bears.

The overall mechanics strongly suggest that increased consumption of vegetation reduces the assimilation of POPs in grizzlies, and aids in removing stored body tissue concentrations of the contaminants. This is finding is further supported by the fact that many PCB and PBDE congeners that were excreted after the vegetation meals were not detected in the vegetation itself. Maybe there are a couple of good reasons why your mother told you to eat your vegetables!

Effects in Animals

The effects of OCPs on animals including invertebrates and vertebrates are very broad indeed. The most infamous of these effects include cancer and endocrine disruption, which are found in all vertebrates, including humans. However, many other sublethal effects due to OCPs have been reported in an extensive literature. The acute toxicity of most OCPs generally occurs at concentrations that are higher than those considered environmentally realistic so death under natural conditions may be slow and often seen as a general wasting away or

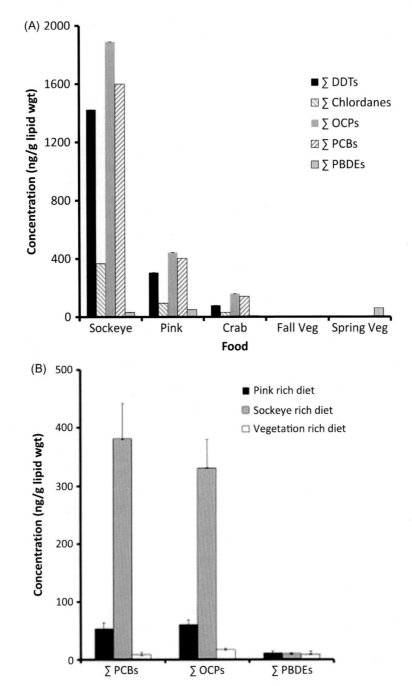

FIGURE 4.12   Relationships in POP concentrations between grizzly bears in Alaska and their foods. (A), Concentrations of POPs in food; (B), Concentrations in grizzly bear feces.

chronic illness. The lipophilic and persistent nature of most OCPs can lead to long-term storage in adipose tissue, followed by a rerelease into the circulatory system during harsh environmental conditions. This can cause delays from the time of first exposure to the onset of effects. DDT can remain in the human body for 50 years or more (Mrema et al., 2013). As previously mentioned, the ability of OCPs to biomagnify leads to greater toxicity at higher trophic levels, such as in fish-eating birds and mammals. They can also be passed from mother to offspring through lipids in eggs or lactation in mammals.

Field Effects

Most OCPs are highly toxic to aquatic organisms, but are less toxic to birds or mammals. At times, OCPs have been directly involved in the deaths of wild animals including fish and birds. During a 16-month period in 1996 and 1997, Stansley and Roscoe (1999) documented chlordane poisoning in six species of songbirds and four species of raptors in New Jersey. At one roost, they recovered 425 dead or sick birds; including common grackles (*Quiscalus quiscula*), European starlings (*Sturnus vulgaris*), and American robins (*Turdus migratorius*). Sick birds displayed signs that were consistent with chlordane poisoning including convulsions, muscle spasms, and excessive vocalizing. Chlordane poisoning was also diagnosed in Cooper's hawks (*Accipiter cooperi*). The timing of the Cooper's hawk mortalities coincided loosely with the July peak in songbird mortalities, and may have been due to hawks feeding on other birds debilitated by chlordane.

Aldrin, which has very low toxicity on its own, is readily converted to dieldrin or endrin in the field and in organisms, and both of these are very toxic to birds and mammals. An early example of wildlife mortality due to aldrin/dieldrin occurred in Illinois when aldrin was applied on farm fields to control Japanese beetles (*Popillia japonica*, Labisky and Lutz, 1967). The application was effective on the beetles, but it also wiped out 25–50% of the pheasants living in sprayed areas and severely depressed reproduction; more than half of the pheasant hens lost broods that year. Fortunately, diminished reproduction only lasted the year of application.

Endrin was probably the chief OCP responsible for the extirpation of brown pelicans in Louisiana during the early 1960s. A large discharge of endrin into the Mississippi River from a plant in Tennessee resulted in the chemical moving downstream towards the Gulf of Mexico (Mount and Putnicki, 1966). Along the way, millions of fish perished with mortality increasing as the pesticide approached the Gulf. Gulf menhaden (*Brevoortia patronus*), a baitfish, took one of the biggest hits. Various estimates of the preexposed Louisiana population of brown pelicans at this time ranged from 12,000 to 85,000 birds but the entire population was extirpated. Based on endrin residues and reproductive declines in the pelicans, Blus et al. (1979) made a convincing argument that endrin was responsible for the extirpation of this species through the decrease in fish.

Chronic Effects

OCPs disrupt certain hormones, enzymes, growth factors, and neurotransmitters; they also induce key genes involved in metabolism of steroids and *xenobiotics* (substances that are foreign to the body). These disruptions cause complex and incompletely understood changes in many different parts of the body. A direct, nongenetic result of these influences is an alteration in the homeostatic condition of cells, leading to oxidative stress and accelerated *apoptosis* or cell death. In healthy individuals, apoptosis is a way of removing damaged or unnecessary cells without causing damage to surrounding cells. The abnormal apoptosis caused by OCPs is related to various pathologies including immunodeficiency, autoimmune diseases, cancer, and reproductive problems (Alison and Sarraf, 1995; Mrema et al., 2013).

OCPs also have genetic effects involving DNA functioning and gene transcription. The details of this are complex and more detailed than what we can fully explain in this chapter. Briefly, OCPs may suppress some genes or reactivate genes that have been naturally suppressed; these can then lead to uncontrolled cell proliferation and cancer. Depending on the stage of animal development, these gene alterations can also result in permanent changes in developmental patterns, leading to endocrine disruption and reduced immunosuppression. This is an area of active research and many questions remain to be answered, but it is likely another major way that OCPs cause serious chronic conditions. Mrema et al. (2013) cautioned that much of the ongoing research with OCPs is done on animal models at environmentally unrealistically high concentrations which confounds making inferences to real-world exposures.

Under laboratory conditions, specific effects of OCPs can be determined and the effects of one might be distinguished from the effects of other OCPs. For example, Bondy et al. (2000) tested the responses of laboratory rats to *cis*-nonachlor, *trans*-nonachlor, and technical-grade chlordane. Both nonachlors are common components of technical-grade chlordane. The researchers treated rats with single chemicals via gavage for 28 days and found differences among the three OCPs. Only *trans*-nonachlor caused overt mortality with 43% of the rats dying on the highest dose. Residues of *trans*-nonachlor accumulated in fat more than the cis-nonachlor. *Trans*-nonachlor and chlordane but not *cis*-nonachlor increased kidney weights and depressed ion transport. *Trans*-nonachlor had the greatest effect on liver histology and some differences were observed between males and females. The major metabolite for all chemicals was oxychlordane. In terms of overall toxicity, the authors decided that: *trans*-nonachlor> chlordane > *cis*-nonachlor. Under field conditions, it is likely that all of these effects would have to be ascribed as a combined effect.

A huge problem with finding the "smoking gun" or specific cause for wildlife declines under field conditions due to contaminants is that populations of wildlife are seldom exposed to only one contaminant. Many studies on residue analysis show that there are high correlations among OCPs, other organochlorines such as PCBs and heavy metals, for example. Thus, when wildlife mortality or debilitation is seen, it is often very difficult to identify a single contaminant as the cause. For example, the black-footed albatross (*Phoebastria nigripes*) is one of three northern hemisphere albatrosses. It nests on the Hawaiian chain, primarily on the Leeward Islands halfway between Japan and the west coast of the United States, but feeds on the open oceans in the central Pacific. Finkelstein et al. (2007) sampled blood of adult birds on Midway Island for PCBs; OCPs; trace concentrations of silver, cadmium, tin, lead, chromium, nickel, copper, zinc, arsenic, selenium, and total mercury. They found all of these contaminants in the birds and a very high correlation among the concentrations of organochlorine compounds. When they compared these contaminant concentrations against sensitive markers of the immune systems, they found significant relationships between organochlorines and mercury with increased immune responses. So, the immune systems of these fish-eating birds were indeed stressed, but identifying the relative importance of the organochlorines or mercury causing this stress was not possible.

Whereas laboratory experiments might generate more specific results than field studies, not all animals are suitable candidates for laboratory studies. Their size, status (threatened or endangered), or propensity to cope with laboratory conditions may prevent them from being used under such controlled environments. Moreover, stress is an important factor in how animals respond to treatment, and stress conditions differ considerably between laboratory and field.

Endocrine Disruption

The endocrine disruption effect of OCPs and other chlorinated hydrocarbons, such as PCBs, occurs through the contaminants interfering with the endocrine system at several points. One way this occurs is at the site of target cells—those that typically respond to a given hormone. Hormones attach to specialized binding receptors on specific types of cells that are sensitive to that hormone. In the case of reproductive hormones, cells in the ovary, testes, or some other organ possesses these receptors, while those in other organs such as the heart do not have them or have them in much lower frequencies. DDT and its derivatives, endosulfan, and lindane mimic the effects of hormones by binding to these receptors and triggering the effects normally produced by the hormones. Other OCPs interfere with various enzyme pathways responsible for the synthesis of hormones. Clinical tests with in vitro exposure of cell cultures show that DDT, lindane, and endosulfan block a specific pathway for biotransformation of estrogen and enhance the levels of a potent estrogen (Bradlow et al., 1995). Thus, DDT and particularly o,p'-DDT mimics 17-β-estradiol, the most common form of estrogen in mammals. p,p'-DDE acts as an antiandrogen, counteracting the effects of testosterone (Kelce et al., 1995). In rats, treatment with p,p'-DDT over a course of days decreased the number and motility of spermatozoa and induced testosterone metabolism, leading to lower testosterone concentrations in rats (Ben Rhouma et al., 2001). The same exposure led to increased thyroid hormone T3, decreased T4 in the blood, and produced hypothyroidism (Tebourbi et al., 2010).

Sometimes when wild animals are not available, suitable surrogates of other species can be tested for their responses to OCPs and other contaminants. Theoretically, closely related species should have similar responses to the same chemicals. While that is not always the case, using surrogate species may be the best we can do. Domestic sheep (*Ovis aries*) are in the same genus as bighorn sheep (*O. canadensis*) and Dall sheep (*O. dalli*) and are not too distant from other ungulates, such as deer. Therefore, experiments with domestic sheep can help us predict what could happen to wildlife exposed to OCPs. Beard and Rawlings (1999) administered lindane and pentachlorophenol (PCP) to ewes through their diets to determine what effects the pesticides might have on reproduction and thyroid functioning. Ewes were given 1 mg/kg bw/day for 5 weeks prior to mating and then through pregnancy and lactation. The only reproductive effect observed was that pregnancy rate for ewes on PCP were lower than those on control or lindane diets. PCP also significantly reduced serum T4 concentrations compared with controls. T4 (thyroxin) is a prohormone directly produced by the thyroid and converted into triiodothyronine or T3, through the mediation of thyroid stimulating hormone (TSH). T3 is the more biologically active hormone and regulates metabolism and other activities throughout the body. Although the researchers found that serum T4 concentrations were significantly reduced compared with controls, the ability of TSH to convert T4 into T3 was not altered. It is not uncommon for the cells that synthesize hormones to enlarge if insufficient hormone is being released and the thyroid follicular cells were enlarged under both lindane and PCP. The authors suggested that the effects on the thyroid could influence reproduction and general performance in the long run.

Effects of Organochlorines on Humans

Due to ethical considerations it is usually not possible to conduct controlled laboratory studies on humans with chemicals known to be toxic. Occasionally with widely used substances such as tobacco, defined populations can be followed and compared to control groups to determine what differences may occur between the two groups. However, with most organochlorines human exposure has not been sufficient to develop test groups that would satisfy statistical constraints. Therefore, most of what we state about OCP effects in humans is based on extrapolation from animal tests, often rats and mice. These tests must be carefully evaluated to ascertain that the concentrations are used in environmentally realistic dosages. Tests that use, for example, concentrations that are 1000 times greater than what we would expect humans to be exposed to are not very predictive. They might suggest what could potentially happen, but would not have much value from a risk assessment perspective.

Given those caveats, several OCPs are probable or potential carcinogens in humans. According to the Centers for Disease Control (CDC, 2013), human health effects from DDT at low environmental doses are unknown; at higher doses the chemical can cause vomiting, tremors or shakiness, and seizures.

The CDC considers DDT a possible human carcinogen. The US EPA (2015a) lists the following human health hazards from DDT and its derivatives: probable human carcinogen; damage to the liver; temporarily damage to nervous systems; reduced reproductive success; and damaged reproductive system (in both males and females). In a review paper, Rogan and Chen (2005) stated that exposure to DDT at amounts that would be needed in malaria control might cause preterm birth and early weaning; that toxicological evidence shows endocrine-disrupting properties; and that human data also indicate possible disruption in semen quality, menstruation, gestational length, and duration of lactation.

In a 1940s study (summarized in ATSDR, 2002), Velbinger exposed volunteers to oral doses ranging from 250 to 1500 mg DDT, which was equivalent of up to 22 mg/kg bw. At the lowest doses, volunteers described variable sensitivity or tingling of the mouth. Six hours after exposure to 750 or 1000 mg DDT, they noted numbness of the lower part of the face, uncertain gait, malaise, cold moist skin, and hypersensitivity to contact. Prickling of the tongue and around the mouth and nose, loss of balance, dizziness, confusion, tremors, malaise (nausea), headache, fatigue, and severe vomiting occurred in all volunteers exposed to 1500 mg DDT. All volunteers exposed to any dose achieved almost complete recovery from these acute effects within 24 hours after exposure.

The *Minimal Risk Level* (MRL) or the estimated daily human exposure to a hazardous substance that is likely to be without appreciable risk of adverse, noncancer health effects over a specified duration of exposure. For DDT the MRL is 0.0005 mg/kg/day (ATSDR, 2002). The "safe" concentration of DDT and its derivatives in drinking water or *maximum concentration limit* (MCL), as determined by the US EPA is 0.0000072 µg/L (US EPA, 2014). Environmental air in the work place should not exceed OSHA's *Permissible Exposure Limit* (PEL) of 1 mg/m^3.

Toxaphene also affects the human central nervous system, resulting in convulsive seizures (US EPA, 2015b). Chronic inhalation exposure to toxaphene in humans results in reversible respiratory toxicity. Chronic oral exposure in animals has negatively affected the liver, kidneys, spleen, adrenal and thyroid glands, central nervous system, and the immune system. Due to animal studies that have reported an increased incidence of thyroid gland tumors and liver tumors via ingestion, the US EPA has classified toxaphene as a probable human carcinogen (US EPA, 2015b). The MCL for toxaphene is 0.003 mg/L and its PEL is 0.5 mg/m^3.

FOCUS—EGGSHELL THINNING

This is a classic story concerning the effects of organochlorines, especially DDT and its derivatives, on the populations of several species of fish-eating birds. As mentioned in the beginning of this chapter, DDT became a very popularly used pesticide soon after World War II. DDT was actually synthesized in 1874, but its insecticidal properties were not identified until 1939 and its serious use began

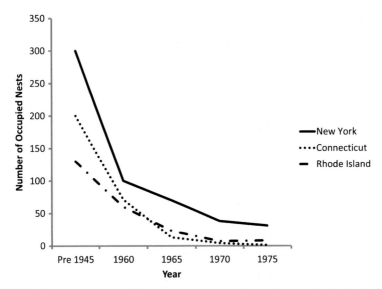

FIGURE 4.13 Number of occupied osprey nests in selected areas along the North Atlantic Coast in the United States. Note the dramatic decline in nests in 1975 compared to the period before the extensive use of DDT (1945). *Henny et al. (2010).*

in 1945 (Henny et al., 2010). Other OCPs including aldrin, dieldrin, toxaphene, endosulfan, and methoxychlor were marketed later. Some of these were capable of affecting eggshell thickness, but we will focus on DDT and DDE, the most widely known culprits of this phenomenon. DDT was known to cause incidental mortality to nontarget organisms in the late 1940s (Cottam and Higgins, 1946; Bartnett, 1950), but major public attention was aroused when members of more than 90 species of birds, mostly American robins were killed following treatment of forests for Dutch elm disease in the 1950s (Carson, 1962) and bald eagles in the 1950s (Broley, 1952).

The role of DDT in eggshell thinning and the effects on bird populations gradually became apparent. Many populations of fish-eating birds including brown pelicans American bald eagles, peregrine falcons, and ospreys (Fig. 4.13) were suffering severe depletions.

Due to these population declines, brown pelicans and peregrine falcons were declared endangered by the Endangered Species Conservation Act and the bald eagle was listed as endangered in 1978 under the Endangered Species Act, which replaced the Conservation Act. Ospreys have a cosmopolitan distribution that spared the species from being depleted and off the endangered species list.

The problem was not confined to North America. Raptors in Norway, such as the merlin (*Falco columbarius*) were experiencing thinning eggshells (Fig. 4.16, Nygård, 1999). In England, Ratcliffe (1967, 1970) was the first to describe thinning eggshells as a possible problem and to link the thinning to organochlorines. He reported on eggshell thinning in golden eagles (*Aquila chrysaetos*), peregrines, and sparrowhawks (*Accipter nissus*). Among these, peregrines went from an average of 3.5 eggs and 2.5 young per nest in England prior to 1950 to an average of 1.5–1.7 young per nest after 1950. The number of nests that had lost young peregrines increased from 3.7 to 38.9% over the same time periods. Chemical residue analyses revealed that raptor eggs contained measurable concentrations of several OCPs with DDE having the highest concentrations in peregrines and sparrow hawks. Total OCP concentrations negatively correlated to an eggshell index that reflects overall egg size and thickness. Concentrations of these OCPs and associated PCBs are highly correlated in the environment and in tissues, so identifying the specific chemical was difficult. Hickey and Anderson (1968) found similar relationships in North American raptors: (1) population declines became noticeable following widespread use of DDT; (2) eggshell thinning was prevalent in these disappearing populations; and (3) eggshell thickness was negatively correlated with OCP concentrations, particularly p,p′-DDT. Virtually all field studies on eggshell thinning, however, were confounded by the presence of other POPs.

Several experimental laboratory studies confirmed that p,p′-DDE is unique in its ability to affect eggshell thickness and is the most likely cause for effects observed under field conditions. Much of this research was conducted by C.E. Lundholm and summarized in a thorough review (Lundholm, 1997). DDT was the most common form of the OCP family applied to agricultural crops, but the pesticide used was actually an 80%/20% mixture of p,p′-DDT and o,p′-DDT. Of these, the mechanism of p,p′-DDT is primarily neurotoxic and that of o,p′-DDT is primarily estrogenic. Both isomers are readily metabolized to the corresponding DDE isomers. A few long-term studies (eg, Cooke, 1973) concluded that DDT could cause thinning, but these studies were so long-term that DDT was most likely metabolized to DDE. Direct application of p,p′-DDT did not cause eggshell thinning in ducks, but comparable treatment with p,p′-DDE was effective in causing thinning after a couple of days (Lundholm, 1985). o,p′-DDE wasn't involved because it produced a slight but nonsignificant eggshell thinning in ducks (Lundholm, 1976). Other contaminants can temporarily (a few days to weeks) thin eggshells (Haegle and Tucker, 1974) by reducing feeding behavior (called *inappetance*), causing a generalized listlessness, or affecting calcium uptake, but the effects of p,p′-DDE are extremely longlasting. Substantial thinning has been observed for up to 2 years following exposure (Longcore and Stendell, 1977). Other OCPs can reduce the size or weight of eggs or interfere with egg production, but are either far less potent than p,p′-DDE in causing thinning or produce other overt signs not seen with this contaminant. For example, very high lethal doses of p,p′-DDE can restrict egg laying, but the thinning that occurred at those DDE concentrations did not cause other overt signs of toxicity.

As with other toxins, not all avian species react to p,p′-DDE in the same way or at the same concentrations. An early classification (Cooke, 1973) that is still generally valid today placed birds into three groups: (1) insensitive species—Galliforms (ie, quail, grouse, and chickens) that do not show eggshell thinning at environmentally relevant concentrations; (2) moderately sensitive—species such as American kestrels (*Falco sparverius*) and several waterfowl species, including mallards—that show eggshell thinning in 5–15% of normal eggs; and (3) sensitive species such as brown pelicans and peregrine falcons. Table 4.3 lists some tested species and the lowest known p,p'-DDE concentrations that cause significant eggshell thinning.

There is a disconnect between most of the laboratory studies on the effects of p,p'-DDE and other OCPs on eggshell thinning and what can be done under natural conditions. This disconnect pops up now and then when comparing laboratory and field studies and other endpoints of interest. Most of the laboratory studies report on the duration of the exposure and the concentrations being tested, but fewer report contaminant levels in eggs. The amount of contaminant exposure can be controlled because they are conducted in the laboratory. However, since OCPs bioaccumulate, residue levels in eggs and other tissues are a factor of concentration and duration. Longer studies with higher concentrations can, in theory, attain the same residue concentrations as shorter studies with lower concentrations. In the field, determining the actual exposure is virtually impossible. Studies have measured the concentration of OCPs in food items or in eggs and related that to concentrations in predators, and some have used models to estimate intake but it is impossible to determine how much contaminant a particular predator has actually consumed over a period of time, let alone the variation within a population of predators. Thus, assessing MRLs in a tissue such as egg membranes and relating that to eggshell thinning is realistic for field studies. Ideally, laboratory studies should have quantified contaminant concentrations in eggs and relate those to exposure and effects but, as mentioned, this

TABLE 4.3 Lowest DDT Concentrations That Have Been Known to Cause Eggshell Thinning in Selected Species of Avian Raptors

Species	Scientific Name	p,p'-DDE Concentration	Source
Brown pelican	*Pelecanus occidentalis*	3 µg/g	Blus (1982)
Peregrine falcon	*Falco peregrinus*	30 µg/g	Ambrose et al. (1988)
Bald eagle	*Haliaeetus leucocephalus*	3–16 µg/g	Wiemeyer et al. (1984)
Osprey	*Pandion halietus*	A	Wiemeyer et al. (1988)
Spanish imperial eagle	*Aquila adalberti*	3.5 µg/g	Hernández et al. (2009)
White-faced ibis	*Plegadis chihi*	4 µg/g	King et al. (2003)

was rarely done. To be most effective, laboratory studies should consider what is relevant under field conditions and provide information that might be used in risk assessment. That is not to say that laboratory experiments should be completely dictated by field studies, but it is very helpful when there is some correspondence.

A mechanism for how p,p'- DDE thins eggshells is summarized in Lundholm (1997). The process is complicated, so we will only briefly summarize it here. The creation of eggs is finalized in the eggshell gland within the fallopian tubes of female birds. Here albumin, yolk, shell membranes and shell are developed. Calcium, bicarbonate, magnesium, sodium, and potassium are secreted onto the egg to form the shell which is mostly calcium carbonate, $CaCO_2$. The presence of sodium and bicarbonate in the shell gland stimulates the transfer of Ca from the blood to the gland, and this transfer is regulated by a calcium-binding protein called calbindin. Calbindin, in turn, may be regulated by hormones like progesterone. It appears that DDE may uniquely (among OCPs) interfere with progesterone activity by affecting steroid hormone receptors, which in turn interferes with calbindin. Another hypothesis is that DDE interferes with the synthesis of prostaglandins, which are lipids that behave like hormones. One of their functions is to help regulate egg laying or oviposition timing. If timing is delayed, eggs will have thin or missing shells.

Since the EPA banned DDT and other OCPs the populations of the most affected species grew remarkably. For example, the European merlin has seen a gradual restoration of eggshell thickness (Fig. 4.14). The shells have not reached the thickness they had prior to the DDT era because residues of the both

FIGURE 4.14   The long-term decline and partial return of mean eggshell thickness in merlins in Norway before and after the widespread use of DDT in Europe. *From Nygård (1999).*

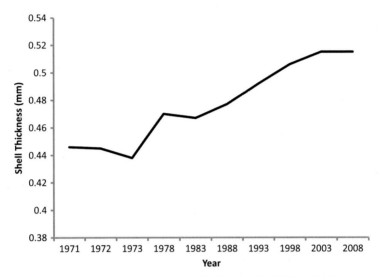

FIGURE 4.15   Eggshell thickness in osprey from Sweden since the DDT era. In this study, a 10% decrease in shell thickness (eg, 0.5–0.45) means that maximum number of young decreases 6.8% and a 20% thinning results in 28% fewer chicks. *Odsjö and Sondell (2014).*

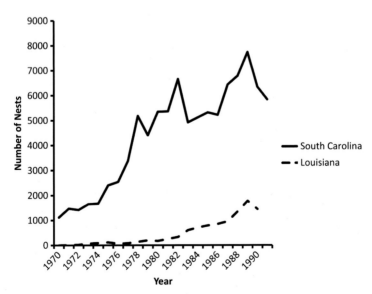

FIGURE 4.16   Number of brown pelican nests shortly before and after the ban of DDT in South Carolina and Louisiana. The species had been virtually extirpated from Louisiana prior to the ban. *Wilkinson et al. (1994).*

DDT and DDE still persist in their foods, but the eggs no longer break (Nygård, 1999). Eggshells of ospreys have become significantly thicker in many parts of the world, such as Sweden (Fig. 4.15). The mean concentrations of p,p′-DDE in peregrine falcons have dropped; for example migrating peregrine falcons in Texas during 2004 had only 2% of the p,p′-DDE in their blood plasma as they had in 1978 (Henny et al., 2009). Their blood plasma still contained DDE due to the persistence of the compound, but the concentrations were far below those considered to cause thinning. Similar decreases were seen in other OCPs from the same study. Brown pelican populations along the eastern seaboard have grown exponentially since the 1970s and even the extirpated population along Louisiana's coasts has responded dramatically with the assistance of transplanted birds from other populations (Fig. 4.16, Wilkinson et al., 1994). Bald eagles were taken off of the endangered species list in 2007 although they are still protected by the stringent Bald Eagle Protection Act; both the brown pelicans and peregrine falcons were delisted in 1999.

STUDY QUESTIONS

1. Are organochlorines aromatic or aliphatic organic compounds? Why?
2. What do we mean when we say that a particular chemical is a POP?
3. What was the purpose of the Stockholm Convention?
4. True or False: The Stockholm Convention put a global ban on DDT and prohibits its use for any purpose whatsoever.
5. Why, from a biochemical perspective, are organochlorines so persistent?
6. Describe the general trend in global organochlorine concentrations over the past decade or so.
7. In what way is the lipophilic and persistent nature of organochlorines related to bioconcentration and biomagnification?
8. What factors are associated with the occurrence of OCPs in organisms?
9. What are the benefits and risks of maternal transfer? Who gets the benefits and who gets the risks?
10. What are the relative advantages or disadvantages of measuring concentrations of organochlorines via wet weight, live weight, dry weight, or percent lipid?
11. Can you think of any other situation that might be similar to grizzly bears, salmon, and POPs? It would require a diet for part of the year that contains little POPs and substantially more contaminants in another part of the year.
12. How might laboratory studies that use dosages of OCPs that are considerably higher than expected environmental conditions be misleading?
13. Why might it be difficult to identify a given organochlorine as the chief cause for animal mortality under field conditions?
14. What signs (clinical conditions) would you look for if you suspected that a living animal was suffering toxicity from endrin or endosulfan?
15. What are the relationships among PEL, MCL, and MRL?

REFERENCES

Alison, M.R., Sarraf, C.E., 1995. Apoptosis: regulation and relevance to toxicology. Hum. Exp. Toxicol. 14, 355–362.

Ambrose, R.E., Henny, C.J., Hunter, R.E., Crawford, J.A., 1988. Organochlorines in Alaskan peregrine falcon eggs and their current impact on productivity. In: Cade, T.J., Enderson, J.H., Thelander, C.G., White, C.M. (Eds.), Peregrine Falcon Populations: Their Management and Recovery Peregrine Fund, Boise, ID, pp. 385–393.

ATSDR Agency for Toxic Substances and Disease Registry, 2002. Toxicological Profile for DDT, DDE and DDD. U.S. Dept Health Human Services. Agency for Toxic Substances and Disease Registry, Atlanta GA.

ATSDR Agency for Toxic Substances and Disease Registry, 2007. Toxicological Profile for Heptachlor and Heptachlor Epoxide. U.S. Dept Health Human Services. Agency for Toxic Substances and Disease Registry, Atlanta GA.

Azeredo, A., Torres, J.P.M., de Freitas Fonseca, M., Britto, J.L., Basto, W.R., Azevedo e Silva, C.E., et al., 2008. DDT and its metabolites in breast milk from the Madeira River basin in the Amazon, Brazil. Chemosphere 73, 5246–5251.

Bartnett, D.C., 1950. The effect of some insecticide sprays on wildlife. Proc. Ann. Conf. West Assoc. State Game Fish Comm. 30, 259.

Beard, A.P., Rawlings, N.C., 1999. Thyroid function and effects on reproduction in ewes exposed to the organochlorine pesticides lindane or pentachlorophenol (PCP) from conception. J. Toxicol. Environ. Health A 58, 509–530.

Ben Rhouma, B.K., Tebourbi, O., Krichah, R., Sakly, M., 2001. Reproductive toxicity of DDT in adult male rats. Hum. Exp. Toxicol. 20, 393–397.

Bidleman, T.F., Leone, A.D., 2004. Soil-air exchange of organochlorine pesticides in the Southern United States. Environ. Poll. 128, 46–57.

Bidleman, T.F., Leone, A.D., Wong, F., van Vliet, L., Szeto, S., Rpiley, B.D., 2006. Emission of legacy chlorinated pesticides from agricultural and orchard soils in British Columbia, Canada. Environ. Toxicol. Chem. 25, 1448–1457.

Bidleman, T.F., Kurt-Karakus, P.B., Wong, F., Alegria, H.A., Jantuen, L.M., Hung, H., 2013. Is there still "new" DDT in North America? An investigation using proportions of DDT compounds. In: McConnell, L.L., Dachs, J., Hapeman, C.J. (Eds.), Occurrence, Fate and Impact of Atmospheric Pollutants on Environmental and Human Health. ACS Symposium Series 1149 American Chemical Society, Washington DC, pp. 153–181.

Bishop, C.A., Lean, D.R.S., Brooks, R.J., Carey, J.H., Ng, P., 1995. Chlorinated hydrocarbons in early life stages of the common snapping turtle (*Chelydra serpentina*) from a coastal wetland on Lake Ontario, Canada. Environ. Toxicol. Chem. 14, 421–426.

Blus, L., Cromartie, E., McNease, L., Joanen, T., 1979. Brown pelican—population status, reproductive success and organochlorine residues in Louisiana, 1971–1976. Bull. Environ. Contam. Toxicol. 22, 128–135.

Blus, L.J., 1982. Further interpretation of the relationship of organochlorine residues in brown pelican eggs to reproductive success. Environ. Poll. 28, 15–33.

Blus, L.J., 2003. Organochlorine pesticides. In: Hoffman, D.J., Rattner, B.A., Burton Jr, G.A., Cairns Jr, J. (Eds.), Handbook of Ecotoxicology, seconded. Lewis Publishers, Boca Raton, pp. 313–340.

Bondy, G.S., Newsome, W.H., Armstrong, C.L., Suzuki, C.A.M., Doucet, J., Fernie, S., et al., 2000. Trans-nonachlor and cis-nonachlor toxicity in Sprague-Dawley rats: Comparison with technical grade chlordane. Toxicol. Sci. 58, 386–398.

Bradlow, H.L., Davis, D.L., Lin, G., Sepkovic, D., Tiwari, R., 1995. Effects of pesticidess on the ratio of 16-Alpha/2-hydroxyestrone - A biologic marker of breast cancer risk. Health Perspect. 103, 147–150.

Braune, B.M., Mallory, M.L., Gilchrist, H.G., Letcher, R.J., Drouillard, K.G., 2007. Levels and trends of organochlorines and brominated flame retardants in ivory gull eggs from the Canadian Arctic, 1976 to 2004. Sci. Total Environ. 378, 403–417.

Broley, M.J., 1952. Eagle Man. Pellegrini and Cudahy, New York.

Carson, R., 1962. Silent Spring. Houghlin Mifflin, New York.

Center for Disease Control and Prevention, 2013. Factsheet: Dichlorodiphenyltrichloroethane (DDT). Avaiabe from: <http://www.cdc.gov/biomonitoring/DDT_FactSheet.html> (accessed 10.07.15.).

Chen, Z.S., Chen, L., Liu, Y., Cui, L., Tang, C.L., Vega, H., et al., 2012. Occurrence of DDA in DDT-contaminated sediments of the Southern California Bight. Mar. Poll. Bull. 64, 1300–1308.

Christensen, J.R., Yunker, M.B., MacDuffee, M., Ross, P.S., 2013. Plant consumption by grizzly bears reduces biomagnification of salmon-derived polychlorinated biphenyls, polybrominated diphenyl ethers and organochlorine pesticides. Environ. Toxicol. Chem. 32, 995–1005.

Chung, M.K., Hu, R., Wong, M.H., Cheung, K.C., 2007. Comparative toxicity of hydrophobic contaminants in microalgae and higher plants. Ecotoxicology 16, 393–402.

Cooke, A.S., 1973. Shell thinning in avian eggs by environmental pollutants. Environ. Poll. 4, 85–152.

Cottam,C., Higgins,E.,1946. DDT: Its effects on fish and wildlife. U.S. Fish and Wildlife Service Circ 11, Washington DC.

Fellers, G.M., Sparling, D.W., McConnell, L.L., Kleeman, P.M., Drakeford, L., 2013. Pesticides in amphibian habitats of Central and Northern California, USA In: McConnell, L.L. Dachs, J. Hapeman, C.J. (Eds.), Occurrence, Fate and Impact of Atmospheric Pollutants on Environmental and Human Health. ACS Symposium Series, 1149, pp. 123–150.

Finkelstein, M.E., Grasman, K.A., Croll, D.A., Terhsy, B.R., Keitt, B., Jarman, W.M., et al., 2007. Contaminant-associated alteration of immune function in black-footed albatross (*Phoebastria nigripes*), a north Pacific predator. Environ. Toxicol. Chem. 26, 1896–1903.

Haegle, M.A., Tucker, R.K., 1974. Effects of 15 common environmental pollutants on eggshell thickness in mallards and coturnix. Arch. Environ. Contam. Toxicol. 11, 98–102.

Henny, C.J., Kaiser, J.L., Grove, R.A., Johnson, B.L., Letcher, R.J., 2009. Polybrominated diphenyl ether flame retardants in eggs may reduce reproductive success of ospreys in Oregon and Washington, USA. Ecotoxicology 198, 802–813.

Henny, C.J., Grove, R.A., Kaiser, J.L., Johnson, B.L., 2010. North American osprey populations and contaminants: historic and contemporary perspectives. J. Toxicol. Environ. Health 13B, 579–603.

Hernàndez, M.K., Gonzàlez, L.M., Oria, J., Sànchez, R., Arroyo, B., 2009. Influence of contamination by organochlorine pesticides and polychlorinated biphenyls on the breeding of the Spanish imperial eagle (*Aquila adalberti*). Environ. Toxicol. Chem. 27, 433–441.

Hernik, A., Góralczyk, K., Struciński, P., Czaja, K., Korcz, W., Minorczyk, M., et al., 2014. Characterising the individual health risk in infants exposed to organochlorine pesticides via breast milk by applying appropriate margins of safety derived from estimated daily intakes. Chemosphere 94, 158–163.

Hickey, J.J., Anderson, D.W., 1968. Chlorinated hydrocarbons and eggshell changes in raptorial and fish-eating birds. Science 162, 271–273.

Kelce, W.R., Stone, C.R., Laws, S.C., Gray, L.E., Kemppainen, J.A., Wilson, E.M., 1995. Persistent DDT metabolite p,p′-DDE is a potent androgen receptor antagonist. Nature 375, 581–585.

King, K.A., Zaun, B.J., Schotborgh, H.M., Hurt, C., 2003. DDE-induced eggshell thinning in white-faced ibis: a continuing problem in the western United States. Southwest Nat. 48, 356–364.

Kumar, K.S., Kannan, K., Giesy, J.P., Masunaga, S., 2002. Distribution and elimination of polychlorinated dibenzo-p-dioxins, dibenzofurans, biphenyls and p,p′-DDE in tissue of bald eagles from the Upper Peninsula of Michigan. Environ. Sci. Technol. 36, 2789–2796.

Labisky, R.F., Lutz, R.W., 1967. Responses of wild pheasants to solid-block application of aldrin. J. Wildl. Mange 31, 13–24.

Lin, Z., Li, X.M., Li, Y.T., Huang, D.Y., Dong, J., Li, F.B., 2012. Enhancement effect of two ecological earthworm species (Eisenia foetida and Amynthas robusta E. Perrier) on removal and degradation processes of soil DDT. J. Environ. Monitor 14, 1551–1558.

Liu, W., Zhu, L.S., Wang, J., Wang, J.H., Xie, H., Song, Y., 2009. Assessment of the genotoxicity of endosulfan in earthworm and white clover plants using the Comet assay. Arch. Environ. Contam. Toxicol. 56, 742–746.

Longcore, J.R., Stendell, R.C., 1977. Shell thinning and reproductive impairment in black ducks after cessation of DDE dosage. Arch. Environ. Contam. Toxicol. 6, 293–304.

Lundholm, C.E., 1976. Comparison of p,p′DDE and o,p′-DDE on eggshell thickness and calcium binding activity of shell gland in ducks. Acta. Pharmacol. Toxicol. 47, 377–384.

Lundholm, C.E., 1985. Studies of the effects of DDE on the calcium metabolism of the eggshell gland during formation of the eggshell in ducks and domestic fowls. Linköping University Medical Dissertations No. 205. Dept of Pharmacol, Fac Health Science, S-581 85 Linköping, Sweden.

Lundholm, C.E., 1997. DDE-induced eggshell thinning in birds: effects of p,p′-DDE on calcium and prostaglandin metabolism of the eggshell gland. Comp. Biol. Physiol. 118C, 113–128.

Mast, M.A., Foreman, W.T., Skaates, S.V., 2007. Current-use pesticides and organochlorine compounds in precipitation and lake sediment from two high-elevation national parks in the Western United States. Arch. Environ. Contam. Toxicol. 52, 294–305.

Meijer, S.N., Halsall, C.J., Harner, T., Peters, A.J., Ockenden, W.A., Johnston, A.E., et al., 2001. Organochlorine pesticide residues in archived UK soil. Environ. Sci. Technol. 35, 1989–1995.

Miglioranza, K.S.B., Gonazlez, M., Ondarza, P.M., Shimabukuro, V.M., Isla, F.I., Fillmann, G., et al., 2013. Assessment of Argentinean Patagonia pollution: PBDEs, OCPs and PCBs in different matrices from the Rio Negro basin. Sci. Total Environ. 452, 275–285.

Mount, D.I., Putnicki, G.J., 1966. Summary report of the 1963 Mississippi fish kill. Trans. N. Am. Wildl. Nat. Resour. Conf. 31, 177–184.

Mrema, E.J., Rubino, F.M., Brambilla, G., Moretto, A., Tsatsakis, A.M., Colosio, C., 2013. Persistent organochlorinated pesticides and mechanisms of their toxicity. Toxicology 307, 74–88.

Nisbet, I.C.T., 1982. Eggshell characteristics and organochlorine residues in common terns: variation with egg sequence. Colonial Waterbirds 5, 139–143.

Nygård, T., 1999. Long term trends in pollutant levels and shell thickness in eggs of merlin in Norway, in relation to its migration pattern and numbers. Ecotoxicology 8, 23–31.

Odsjö, T., Sondell, J., 2014. Eggshell thinning of osprey (Pandion haliaetus) breeding in Sweden and its significance for egg breakage and breeding outcome. Sci. Total Environ. 470, 1023–1029.

Okay, O.S., Karacik, B., Henkelmann, B., Schramm, K.W., 2011. Distribution of organochlorine pesticides in sediments and mussels from the Istanbul Strait. Environ. Monit. Assess. 176, 51–65.

Ondarza, P.M., Gonzalez, M., Fillmann, G., Miglioranza, K.S.B., 2014. PBDEs, PCBs, and organochlorine pesticides distribution in edible fish from Negro River, Argentinian Patagonia. Chemosphere 94, 135–142.

Ratcliff, D.A., 1967. Decrease in eggshell weight in certain birds of prey. Nature 215, 208–210.

Ratcliff, D.A., 1970. Changes attributable to pesticides in egg breakage frequency and eggshell thickness in some British birds. J. Appl. Ecol. 7, 67–72.

Rogan, W.J., Chen, A., 2005. Health risks and benefits of bis(4-chlorophenyl)-1,1,1-trichloroethane (DDT). Lancet 366, 763–773.

Shahpoury, P., Hageman, K.J., Kimberly, J., Matthaei, C.D., Magbanua, F.S., 2013. Chlorinated pesticides in stream sediments from organic, integrated and conventional farms. Environ. Poll. 181, 219–225.

Sharma, M.B., Baharat, G.K., Tayal, S., Nizzetto, L., Cupr, P., Larssen, T., 2014. Environment and human exposure to persistent organic pollutants (POPs) in India: a systematic review of recent and historical data. Environ. Intet. 66, 48–64.

Shen, L., Wania, F., Lei, Y.D., Teixeria, C., Muir, D.C.G., Bidleman, T.F., 2005. Atmospheric distribution and long-range transport behavior of organochlorine pesticides in North America. Environ. Sci. Technol. 39, 409–420.

Stansley, W., Roscoe, D.E., 1999. Chlordane poisoning of birds in New Jersey, USA. Environ. Toxicol. Chem. 18, 2095–2099.

Sun, P., Blanchard, P., Brice, K., Hites, R.A., 2006. Atmospheric organochlorine pesticide concentrations near the Great Lakes: temporal and spatial trends. Environ. Sci. Technol. 40, 6587–6593.

Swarthout, R.F., Keller, J.M., Peden-Adams, M., Landry, A.M., Fair, P.A., Kucklik, J.R., 2010. Organohalogen contaminants in blood of Kemp's ridley (*Lepidochelys kempii*) and green sea turtles (*Chelonia mydas*) from the Gulf of Mexico. Chemosphere 78, 731–741.

Tebourbi, O., Hallegue, D., Yacoubi, M.T., Sakly, M., Ben Rhouma, K., 2010. Subacute toxicity of p,p'-DDT on rat thyroid: Hormonal and histopathological changes. Environ. Toxicol. Pharmacol. 29, 271–279.

US EPA U.S. Environmental Protection Agency, 2014. Draft Update of Human Health Ambient Water Quality Criteria: 4,4′-DDT. Available from: <http://water.epa.gov/scitech/swguidance/standards/criteria/current/upload/Draft-Update-of-Human-Health-Ambient-Water-Quality-Criteria-4-4-DDT-50-29-3.pdf>.

US EPA U.S. Environmental Protection Agency, 2015a. DDT. Availabe from: <http://www.epa.gov/pbt/pubs/ddt.htm> (accessed 10.07.15.).

US EPA U.S. Environmental Protection Agency, 2015b. Toxaphene. Available from: <http://www.epa.gov/airtoxics/hlthef/toxaphen.html>.

Van den Steen, E., Jaspers, V.L.B., Covaci, A., Neels, H., Eens, M., Pinxten, R., 2009. Maternal transfer of organochlorines and brominated flame retardants in blue tits (*Cyanistes caeruleus*). Environ. Intl. 35, 69–75.

Wiemeyer, S.N., Lamont, T.G., Bunck, C.M., Sindelar, C.R., Grandich, F.J., Fraser, J.D., et al., 1984. Organochlorine pesticide, polychorobiphenyl, and mercury residues in bald eagle eggs, 1969-79, and their relationships to shell thinning and reproduction. Arch. Environ. Contam. Toxicol. 13, 529–549.

Wiemeyer, S.N., Bunck, C.M., Krynitsky, A.J., 1988. Organochlorine pesticides, polychlorinatedbiphenyls, and mercury in osprey eggs 1970-79 and their relationships to shell thinning and productivity. Arch. Environ. Contam. Toxicol. 17, 767–787.

Wilkinson, P.M., Nesbitt, S.A., Parnell, J.F., 1994. Recent history and status of the eastern brown pelican. Wildl. Soc. Bull. 22, 420–430.

Yordy, J.E., Wells, R.S., Balmer, B.C., Schwacke, L.H., Rowles, T.K., Kucklick, J.R., 2010. Life history as a source of variation for persistent organic pollutant (POP) patterns in a community of common bottlenose dolphins (*Tursiops truncatus*) resident to Sarasota Bay, FL. Sci. Total Environ. 408, 2163–2172.

Yordy, J.E., Rossman, S., Ostrom, P.H., Reiner, J.L., Bargnesi, K., Hughes, S., et al., 2013. Levels of chlorinated, brominated and perfluorinated contaminants in birds of prey spanning multiple trophic levels. J. Wildl. Dis. 49, 347–354.

Yu, Z.Q., Chen, L.G., Maw, B.X., Wu, M.H., Sheng, G.Y., Fu, J., et al., 2008. Diastereoisomer- and enantiomer-specific profiles of hexabromocyclododecane in the atmosphere of an urban city in South China. Environ. Sci. Technol. 42, 3996–4001.

Chapter 5

Current Use Pesticides

Terms to Know

FIFRA
MSDS
Integrated pest management
Target organism
Anilide
Azole
Carbamate
Acetylcholinesterase
Dinitroaniline
Neonicotinoid
Organophosphorus pesticide
Pyrethroid
Triazine
Ovotestes
Phosphonoglycine

INTRODUCTION

Chapter 4 discussed organochlorine pesticides (OCPs), a group of chemicals that was widely used in the past, but had many problems including broad spectrum toxicity that affected both target pests and nontarget species along with a very long persistence in the environment. In this chapter, we discuss pesticides that are currently in production and use in the United States. These pesticides are chemically very diverse and often designed for a particular group of pests and have relatively short half-lives. Before we get into any specific group of pesticides, however, let us develop some basic understanding.

WHAT IS A CURRENT USE PESTICIDE?

By the roots of the word, a "pesti-cide" is something that kills pests. In this broad definition, a pesticide could include mechanical killing, burning, biological control, chemical destruction, or any other method of permanently removing a pest. However, in this chapter we will focus on manufactured chemical control agents that are currently being produced and are registered for use by the U.S. Environmental Protection Agency (US EPA).

Pesticide Use Is Controversial

The subject of pesticides and their use can be controversial. Some argue that all pesticides are dangerous and that we are poisoning our streams and rivers with pesticide runoff from agricultural fields. They also argue that pesticides are not species specific and that many beneficial insects and plants are victims to spraying. These are valid concerns.

Proponents of chemical pesticides, however, are also correct when they state that agriculture in its current form within North America, consisting of hundreds to thousands of acres of a single crop, would not be sustainable without pesticides. They point out that only with the aid of pesticides can we produce sufficient food for the world's human population. Without pesticides, production of farm commodities would suffer and the global problems of famine would be far worse.

Regulation of Pesticides

The bottom line is that pesticides are neither bad nor good in and of themselves. It is how humans use pesticides that can be the problem. Indiscriminate spraying can cause substantial damage to the environment, so use of pesticides should follow carefully prescribed instructions to minimize the damage they can cause. In the United States, the Federal Insecticide, Fungicide, and Rodenticide Act (FIFRA, 1947 substantially amended in 1972) is the chief federal law affecting the production and use of pesticides. Under FIFRA, all commercially manufactured pesticides must be registered by the US EPA. To have a pesticide registered, the manufacturer must provide documentation on the research conducted to assess the toxicity of the compound to a variety of plant and animal models, along with all other information it has gathered that might be relevant. Also included in the package that the company submits to the US EPA are detailed directions on how the pesticide will be applied. These directions include what the active ingredient pesticide will be mixed with (called the *formulation*), where it will be used, when it will be used, methods of application, and application concentrations. The EPA has separate processes for conventional chemicals, biological pesticides, and antimicrobials. Each process is complex and the reports are detailed. Often, specific toxicity tests involving a cadre of representative species and evidence that the chemical is not carcinogenic to humans must be performed. While the process of registration is usually adequate, there are concerns because only a select handful of species are approved for testing and these are not necessarily the ones that will be exposed to the chemical; endangered species are a special example of this (Racke and McGaughey, 2012). In addition, the EPA allows the manufacturers to do their own testing for the most part and report the results rather than having objective, independent labs do the toxicity evaluations (Boone et al., 2014). This suggests a possible opportunity for a conflict of interest to occur.

Further, the EPA mandates that each container of pesticide must have a *Material Safety Data Sheet* attached to it that describes the chemical, its manufacturer, toxicity, permitted crop use (for agriculture or home use), application rates, clean up procedures, and other relevant information for that material. For commercial uses, certified applicators are required to be licensed. Fines can be assessed for any misuse of a pesticide. In Canada, the *Pest Management Regulatory Agency of Health Canada* serves parallel functions to the U.S. EPA. In addition, states and provinces have their own regulations which can be more, but not less strict, than federal regulations. All of these agencies aim to make sure that the use of pesticides is as safe as possible, but the system is not fool proof and harm to the environment and nontarget organisms sometimes occurs.

Alternatives to Conventional Uses of Pesticides

Methods other than broad usage of chemical pesticides exist to increase the sustainability of agriculture. One possible solution is to grow crops organically. This means that only naturally occurring substances are used in crop production. Manure or nutrient-enriched soils are used as fertilizers, weeds can be mechanically turned over or removed, and insects may be controlled through natural products such as sulfur or biological agents, including bacteria, viruses, and even predators such as ladybug beetles or praying mantises. Organic agriculture is labor intensive and organically grown foods are more expensive than nonorganic foods. Not too long ago, organic foods were essentially unavailable from mainstream grocery stores. When their availability first increased, organic foods were substantially more expensive than conventional foods. As demand for organic foods increased, however, their prices dropped to just above that of conventional foods of the same type.

Another potential solution is for science to develop chemicals with a high degree of specificity and limited environmental persistence. These would target a certain species or small group of species, exert their controlling effect, and dissipate into the environment in very short time. Such chemicals are the objectives of much of the research and development in the pesticide industry, but further work is necessary. A similar approach is to develop insecticides that are species- or group-specific, but become incorporated in plant tissues to provide a longer period of protection and only target those organisms that infect or feed on the plants. Some companies are taking a different approach—through genetic modification, they are designing crops that are resistant to herbicides so that the herbicide can be used with increased safety on noxious weeds.

Integrated pest management (IPM) can be an effective and environmentally sensitive approach to pest management. It relies on a combination of practices including comprehensive knowledge about the life cycles of pests, their interactions with the environment, and effective means of interrupting their life cycles. IPM integrates organic means of control with careful application of synthetic pesticides when necessary. IPM can be used in agricultural and home settings but it has generally been applied to smaller scale operations, not the large mono-cultures that provide most of our crops.

ECONOMICS OF CURRENT USE PESTICIDES

The pesticide industry is economically huge. Economics include research and development, production of formulations, and marketing; enforcement of regulations; and actual purchases of the chemicals. The economic savings due to crop protection and disease reduction are formidable. On the other hand, pesticides can inflict damage to various components of the environment and this damage should be considered in an overall assessment of pesticide economics.

Factors involved in the production and distribution of pesticides include:

- According to an industry-sponsored website: pesticide development and testing by the crop protection industry, along with EPA registration take an average of 9 years and costs pesticide manufacturers \$152–\$256 million for each crop protection product introduced to market (US EPA, 2014; Crop Life America undated).
- The costs of manufacturing the pesticides are tightly held secrets by the companies so this remains unknown.
- Enforcement of regulation costs vary, but the 2016 U.S. EPA budget request includes \$47 million "to ensure the safety of chemicals in our environment, reduce the risk, and prevent pollution at the source" (US EPA, 2015c). This does not include the costs of enforcement by states.
- Actual cost of pesticide purchases within the United States alone was \$11.78 billion in 2006 and \$12.45 billion in 2007, the most recent year that data are available online. Globally, sales amounted to more than \$35 billion in both years (US EPA, 2015d). In general, the sale of conventional pesticides has shown a slow decline since a peak in 1979 (Fig. 5.1).

The cost of ecological effects of pesticides is an entirely different set of issues. Pimental (2005) estimated that the annual major economic and environmental losses due to the application of pesticides in the United States were:

- Public health, \$1.1 billion;
- Pesticide resistance in pests, \$1.5 billion;
- Crop losses caused by pesticides, \$1.4 billion;
- Bird losses due to pesticides, \$2.2 billion;
- Groundwater contamination, \$2.0 billion.

Benefits from the use of pesticides include (Crop Life America undated):

- Increased food production
- Decreased cost of food
- Enhanced quality of foods
- Household pest control

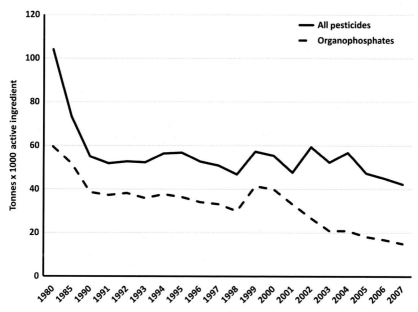

FIGURE 5.1   Total sales of conventional pesticides in the United States, 1970–2006. *Source: US EPA 2015. Pesticide Industry Sales and Usage 2006 and 2007 market estimates. http://www.epa. gov/opp00001/pestsales/07pestsales/market_estimates2007.pdf.*

- Protection of industry and infrastructure due to clogged waterways, obstructed highways and access to highway, utility, and railroad rights of way.
- Enhanced recreation areas including lawns, gardens, public parks, athletic fields, lakes, ponds, and other areas for public enjoyment.
- Human health, livestock, and pet concerns; such as diseases, allergies, parasites, and other factors.

The costs of these benefits are diffuse—occurring all over the country—and therefore difficult to pin down.

A BRIEF HISTORY OF PESTICIDE USE

Since we covered OCPs in Chapter 4 and this chapter will look at many more pesticides, it might be valuable to reflect again on the history of pesticides and their use. Pest control through chemicals extends into very early human history. There are records of Sumerians (5000–2000 BC) using sulfur compounds to control weeds as early as 4500 years ago. The Chinese also used sulfur and later mercury and arsenic compounds to reduce body lice and household insect pests. Sometimes the cure was worse than the problem, however, because both mercury and arsenic are highly toxic and undoubtedly had their own adverse effects. Ancient Romans also used sulfur to control insects and salt to kill weeds. Much of pest control was conducted through mechanical means—having slaves pick off insect larvae or manually pull weeds for example.

The use of chemicals to control insects exploded after World War II with the introduction of the organochlorines (OCs). dichlorodiphenyltrichloroethane (DDT), lindane (gamma-hexachlorocyclohexane, or γ-HCH), aldrin, dieldrin, endrin, toxaphene, and several others were easily applied on crop fields, relatively in expensive, and best of all, effective in controlling insect pests on a variety of plant crops. The herbicide 2,4-D (2,4-dichlorophenoxyacetic acid) was also developed at this time. Later, the insecticides were used to control insects in homes and were very effective in controlling mosquitoes that served as vectors for malaria. We refer you to the FOCUS section at the end of Chapter 4 for more information on the problems that developed from persistence, buildup of tolerance, and biomagnification demonstrated by these pesticides.

Starting in 1972 with DDT, the US EPA began banning OCPs. This process took some time, however, with the most recent OC banned, endosulfan, being phased out in the United States starting in 2010. As we will see, some less persistent OCs are still in use.

As OCs were being phased out, new pesticides came on the market. These "new generation" pesticides included organophosphates (OPs) and carbamates which were first discovered in the 1930s and 1940s but were not extensively used until OC usage was cut back. In the 1960s, pyrethroids were synthesized from pyrethrum, a natural insecticide derived from the chrysanthemum flower (*Chrysanthemum cinerariifolium* or *C. coccineum)*. All three of these groups are still in use today and account for a large share of pesticides applied to crops and for household use. Subsequently, the science of pesticide development has grown tremendously as new chemicals and even entire families of chemicals are synthesized. Active ingredient formulations were developed for specific purposes such as herbicides, acaricides, fungicides, or nematicides. Half-lives of these new pesticides (the time that it takes for ½ of the original amount of pesticide to degrade) are typically measured in days, weeks, or a few months rather than years or decades.

Pesticide research continually seeks increasingly efficacious pesticides—that is, those that are very effective at low concentrations, very specific to a group of pests, and rapidly decompose. Some control agents with very high species-specificity include synthetic pheromones or sex attractants that lure male insects of a given species to traps where they die before mating. The use of pheromones is an important factor in the control of boll weevils (*Anthonomus grandis)*, a serious pest on cotton in the South. Biological organisms such as *Bacillus thurigensis* or Bt, a bacterium that is specific for certain insect larvae such as caterpillars, have been used with success in controlling pests. This field continues to provide many new opportunities.

TYPES OF PESTICIDES

There are many different classes of chemical pesticides based on chemical composition and on the type of pest they control. Chemicals may fall into broad classes of organic molecules (not to be confused with organic farming) including OCs, OPs, carbamates, pyrethroids, neonicotinoids, anilides, triazines, and several others. Explaining the nature of these chemicals is a principal objective of this chapter.

Another way of classifying pesticides is by the *target organism*, the group of plants or animals that the chemical and its application were designed to control—that is, the plants designated as weeds, or the animals as crop depredators or disease vectors. Both can be serious economic and health risks and have had a long history with humans. Some pesticides may have broad action and can be toxic to both plants and animals; although, because of the differences in physiology, chemical characteristics, and application instructions, most pesticides are designed to be more toxic to one group more than others. Thus, we can conveniently group pesticides into the following categories based on their target organisms (Table 5.1).

Certainly there can be inadvertent effects from pesticides but these effects can be reduced through proper application. Warfarin is an anticoagulant often used to control rats and mice—it is a common rodenticide. It can also be toxic to any other vertebrate that consumes it, so methods of application often dictate what the most common target organism will be. Placing warfarin in attics and

TABLE 5.1 Types of Pesticides and their Target Organisms

Type of Pesticide	Target Organisms
Acaricide	Mites, ticks, spiders
Microbiological	Bacteria, viruses, other microbes
Avicide	Birds
Fungicide	Fungi
Herbicide	Plants
Insecticide	Insects
Molluscicide	Snails, slugs
Nematicide	Nematodes
Piscicide	Fish
Rodenticide	Rodents
Sanitizer	Microbes

crawl spaces will generally keep it away from pets and increase the probability of exposure to vermin. Interestingly, warfarin is often used in medicine to reduce the possibility of blood clot formation. As another example, many of the nematicides are also registered as pesticides and vice versa, but nematodes inhabit soil whereas many insects live on plants or soil surfaces; thus, integrating the pesticide into the soil will target nematodes and protect many insect species that can be beneficial.

The number of different kinds of pesticides is mind boggling. The Pesticide Action Network (2014), a clearing house of pesticide information, lists the following types of pesticides and the number of chemical families for each. Each family may have one to many individual pesticides: insecticides—20 families; herbicides—36; natural—8; fungicides—12; microbiological—11; inorganic/metals—16; soaps, solvents and adjutants (surfactants, carriers, etc.)—11; other—7. Altogether, the EPA registers more than 1375 pesticides (Scorecard, 2011). Included in the registered products, in addition to synthetic pesticides, are biologicals, organically derived products, and oils often included as solvents or adjutants but that may have biological properties of smothering insect eggs or preventing fungal spores from contacting living plants. Next are descriptions of some of the more common pesticide families; however the list is not exhaustive.

Anilides

Anilides are aromatic molecules with the chemical formula of $C_6H_5NH_2$ (Fig. 5.2A). They are under the broader class of anilines which are often classified as amines due to the—NH_2 group. Different molecular groups may be attached to the phenol group (which makes the molecule an aromatic) and determines the product and defines their biological activity. Depending on their structure, anilides may be fungicides, microbiologicals, or herbicides. The US EPA has registered five anilides and several more are being considered. Those registered include boscalid (fungicide), fenhexamid (fungicide), flufenacet (herbicide), flutolanil (fungicide), milfluidide (herbicide), and propanil (fungicide). In addition, the US EPA lists propanil as one of the 25 most extensively used pesticides in the United States with 1800–2700 tonnes (metric tons) of active ingredient used per year in 2007 (US EPA, 2015a).

We will use boscalid, a fungicide developed by the company BASF in 2003 as an example anilide. Boscalid is marketed under the name Endura®. Boscalid seems to be a very stable fungicide with an assessed low acute toxicity to birds and mammals and moderate toxicity to aquatic organisms, including plants. It is not considered to pose significant cancer risk to humans. It is not readily degraded by anything including sunlight or biological organisms. Boscalid has a dissipation time in excess of 100 days, and perhaps more than a year in soil. In water, boscalid adsorbs readily to sediments which reduces its bioavailability and risk to animals. The mode of action of boscalid and other anilides is inhibition of succinate dehydrogenase, an enzyme in both the electron transport

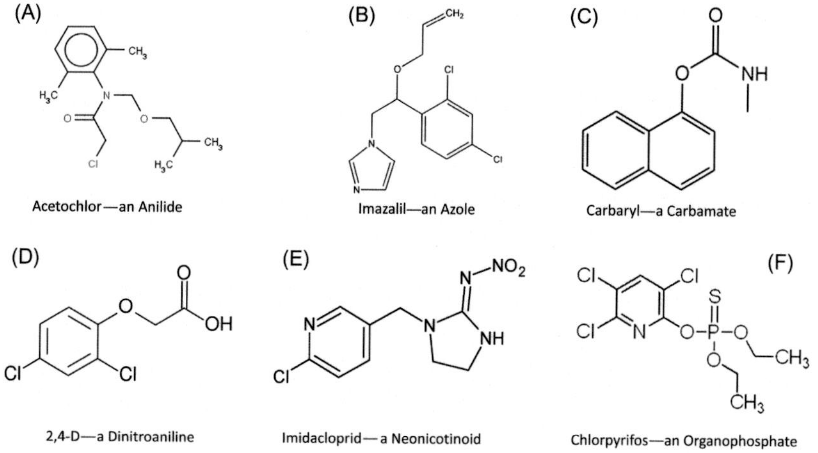

(A) Acetochlor—an Anilide

(B) Imazalil—an Azole

(C) Carbaryl—a Carbamate

(D) 2,4-D—a Dinitroaniline

(E) Imidacloprid—a Neonicotinoid

(F) Chlorpyrifos—an Organophosphate

FIGURE 5.2   Structural formulas for common pesticides.

system and citric acid cycle that converts glucose into Adenosine triphosphate (ATP) (SDHI Fungicides http://www.frac.info/work/work_sdhi.htm).

Azoles

An azole is a group of chemicals with a single five-membered heterocyclic ring containing nitrogen and at least one other noncarbon atom of either nitrogen, sulfur, or oxygen (Fig. 5.2B). They are all fungicides with some used in agriculture and others used in medicine to treat fungal infections. The EPA has registered 21 of these chemicals as agricultural fungicides, none of which made the top usage list in 2007 (US EPA, 2015a). Of these, 13 have slight to moderate acute toxicity to birds and mammals. However, some such as tetraconazole, propiconazole, cypraconazole, and imazalil are possible human carcinogens as determined from animal testing. Other azoles are possible endocrine disruptors, reproductive toxins, or developmental toxins in mammals. All azoles target the same active site in a particular fungal enzyme (Hof, 2001). They interfere with the activity of fungal lanosterol 14α-demethylase, a member of the cytochrome P450 family. Blockage of the enzyme results in disorganization of the cell walls and ultimately cessation of fungal growth. Azoles tend to be longlasting in soils and sediments, often remaining months to a year or more. For humans, that means that azole residues may linger on the fruits and vegetables that we eat (Hof, 2001).

Carbamates

A carbamate is an organic compound derived from carbamic acid (NH_2COOH). Carbamates, along with a related group of pesticides called OPs, largely replaced the OCs in the 1970s. Carbamates are a diverse group that includes herbicides (eg, thiobencarb, phenmedipham), fungicides (mancozeb, ziram), and insecticides (carbaryl, carbofuran, methomyl, oxamyl). Several carbamates are extensively used in the United States and have familiar trade names. For instance, carbaryl (Fig. 5.2C), the first carbamate developed, is currently made by Bayer CropScience and is sold under the trade name Sevin®, a very common insecticide around homes. Carbaryl, was introduced in 1956 and more of it has been used globally than all other carbamates combined (Fishel, 2015a). Carbaryl has been widely used in lawn and garden settings because of its relatively low mammalian oral and dermal toxicity and the broad spectrum of insects in its register. Aldicarb, also manufactured by Bayer, is sold as Temik®. Among the top 25 pesticides used in the United States that are carbamates are metam sodium (22,700–25,000 tonnes) and metam potassium (3900–4100 tonnes), both fumigants. Other widely used carbamates include the fungicide mancozeb (1800-2700 tonnes) and the insecticide aldicarb (1400–1800 tonnes).

Early signs of carbamate poisoning include muscle weakness, dizziness, sweating, and slight body discomfort. These can be accompanied or followed by headache, excess salivation, nausea, vomiting, abdominal pain, and diarrhea at higher levels of exposure. Other signs include contraction of the pupils with blurred vision, incoordination, muscle twitching, and slurred speech (Fishel, 2015a).

Aldicarb is currently a chemical of concern due to its very high toxicity. It has been banned in the European Union and listed as a limited-use pesticide by the US EPA (Gilbert, 2014). Another carbamate of concern is carbofuran, which was banned by the European Union and Canada in 2008 and the US followed suit a year later because of its very high toxicity to birds and mammals and its known cause of many avian die-offs.

Mode of Action for Carbamates and Organophosphorus Pesticides

In all animals, carbamates and OPs are neurotoxins that operate on the synapse of neurons, especially at the neuromuscular junctions where neurons innervate muscles. These chemicals deactivate an enzyme called *acetylcholinesterase* (AChE) that hydrolyzes the neurotransmitter *acetylcholine* (ACh) (Fig. 5.3). Let us step back a bit and explain the process. An action potential of an innervating neuron reaches the terminal end of the neuron where it would otherwise stop because it cannot cross the gap or *synapse* between neuron and muscle or between neurons in the brain. However, in the terminus of the neuron are vesicles that contain ACh. If the action potential is sufficiently strong, it causes several of these vesicles to rupture, allowing the ACh to diffuse across the synapse. ACh opens calcium channels in the muscle junction which ultimately results in contraction of the muscle. As long as ACh remains in the synapse it can continue to elicit contractions in the muscle. Therefore AChE is released into the synapse to destroy the ACh and prohibit further contractions. All of this happens in a normal neuromuscular junction and in the brain.

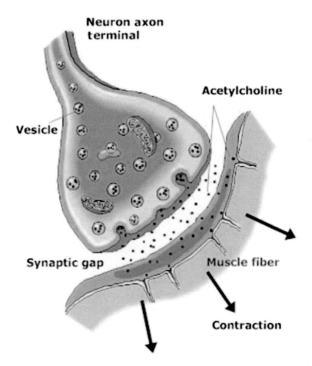

FIGURE 5.3   A neuromuscular junction. The innervating neuron is on top and the muscle is displayed on the bottom. Exocytosis occurs when the vesicle membrane dissolves, releasing ACh into the synapse. Endocytosis is when the vesicle reforms and contains new ACh. Normally, AChE will hydrolyze the ACh, preventing spontaneous contractions of the muscle fibers. When a carbamate or OP pesticide is in the system, the pesticide blocks the action of the ACh esterase and allows spontaneous firing of the muscle. *Source:* https://kin450-neurophysiology.wikispaces.com/ Myofascial+Referred+Pain.

Now suppose the animal has been exposed to either a carbamate or an organophosphate pesticide. These chemicals bind with the AChE and prevent its action on ACh. Thus, the muscles continue to contract uncontrollably and, if the dose is sufficiently high, the animal dies usually from asphyxiation since the respiratory system becomes disordered. These pesticides can also cause sublethal problems leading to convulsions, spasms, reduced growth, teratogenicity, reproductive effects, endocrine disruption, and a variety of other functions during prolonged exposures. Ethyl carbamate or urethane has well know carcinogenic potential in laboratory animals (Schlatter and Lutz, 1990).

A major difference exists between carbamates and OPs in that carbamates can spontaneously reverse their binding with AChE which allows the enzyme to function once again and, if it survives the initial shock, the animal may spontaneously recuperate comparatively quickly. OPs, on the other hand, permanently bind with AChE and the animal must produce new enzyme once the pesticides have dissipated. A prolonged recovery may be more than the animal can survive or it may make it more vulnerable to predators.

Individual chemicals vary in their ability to suppress AChE and some carbamates are extremely toxic while others have mild to low toxicity to birds and mammals, if not in other animals as well. For example, 1 mg/kg aldicarb is sufficient for the LC50 in rats as is 8 mg/kg of carbofuran and 5.4 mg/kg of oxamyl. The same chemicals are lethal to birds in the 51–500 mg/kg; 10–50 mg/kg; and <10 mg/kg ranges, respectively, depending on the species of bird. Other carbamates are not lethal until several 100 mg/kg.

A Tragedy Due to Carbamate Production

What is arguably the worst pesticide-related industrial accident ever occurred in Bhopal, India in 1984. A massive leak consisting of 30 tonnes of methyl isocyanate and other gases occurred at a plant owned by Union Carbide India Limited. Methyl isocyanate is an intermediate chemical in the manufacturing of carbamates and is extremely toxic and irritating. The gas poured out during the night over a shanty town in Bhopal, gassing thousands of people. The official death toll of the number of initial deaths was approximately 3000 people, but even today the numbers vary and range from 2000 to 4000; some estimates even approach 8000. Over the years, it is estimated that over 16,000 people died prematurely due to the leak. Hundreds of other people were blinded and many children suffered other complications (Malik, 2014).

Carbamates used as fungicides or herbicides affect the enzymes of these organisms in various ways. Some reduce the growth and formation of mycorrhizae that are the main portion of the fungus, others disrupt mitosis by interfering with the filaments which separate sister chromatids and others affect cellular functions in different ways. The mode of action is unclear for some of these fungicides and herbicides.

Dinitronanilines

Dinitroanilines are a class of chemical compounds with the chemical formula $C_6H_5N_3O_4$ (Fig. 5.2D). They are derived from both aniline and dinitrobenzenes. They are a small group of primarily pre-emergent herbicides that include trifluralin (3200–4100 tonnes in production), pendimenthalin (3200–4100 tonnes), and metolchlor-S (13,600–15,500 tonnes) on the EPA's top selling list (US EPA, 2015a). Other examples include ethalfluralin and oryzalin. Dinitroaniline herbicides act by inhibiting cell mitosis. Specifically, they inhibit the microtubulin synthesis necessary in the formation of cell walls and in chromosome movement to daughter cells during mitosis. Affected cells do not complete division and remain as single cells with multiple nuclei. Death of the plant occurs when roots no longer grow, causing the plant to dehydrate. Dinitroanilines have half-lives measured in months in soil and can sometimes carry over through the winter to affect spring crops. They are lipophilic, volatile, and not very water soluble, so they do not pose a substantial threat in aquatic systems. In human and animal welfare, the synthesis of trifluralin can result in the production of nitrosamines, harmful substances known to cause cancer in laboratory animals (Saghir et al., 2008).

Inorganics and Metals

This is a very broad group of largely unrelated compounds that display toxic properties to pests of all kinds. There are hundreds of chemicals in this group. Major pesticides are compounds that contain sulfur; copper; bromine; elemental chlorine, chlorates, or chlorites; boric acid; nonanoic acid; hydrogen peroxide; zinc compounds, but no organic components such as ethylene, ethanol, benzene rings, etc. From our previously presented short history of pesticides we saw that sulfur has been used for centuries and is still used today as a fungicide and insecticide. Copper compounds also have many target organisms among fungi and insects. They are sometimes mixed with other pesticides to give them extra effect. Chlorine and its related compounds are used as fumigants or in other applications as microbiologicals. Bromide is used in compounds primarily as a fungicide or microbiological. Many of these chemicals are found in the environment but are still registered by the US EPA. Chloropicrin (fumigant with 4100–5000 tonnes annual usage) and copper hydroxide (fungicide with multiple other purposes and 2700–3600 tonnes usage) are on the US EPA's list of most commonly used pesticides (US EPA, 2015a). The risk of these compounds to nontarget organisms varies considerably from one to another but, in general, this group has lower acute toxicity to vertebrates than several other pesticide groups.

Neonicotinoids

The insect resistance in tobacco has been recognized for decades. Many insects and larger herbivores avoid tobacco due to the presence of nicotine, which is a natural repellent. Researchers have synthesized this natural compound and modified it into neonicotinoids that are more effective in repelling and killing insects than nicotine itself (Fig. 5.2E). Neonicotinoids were designed in the 1980s (Laurino et al., 2011) and first registered for use in the mid-1990s. They are systemic pesticides, meaning that they are incorporated into plant tissues and they persist in the environment. Seven neonicotinoids are commonly found in pesticide formulations including imidacloprid, thiamethoxam, clothianidin, acetamiprid, thiacloprid, dinotefuran, and nitenpyram. Some of these are applied to seeds so that as the plant grows the pesticide becomes incorporated into the developing tissues. Neonicotinoids have essentially cornered the market of pesticides used as seed coatings. Note that while these neonicotinoids are widely used in the United States, imidacloprid, clothianidin, and thiamethoxam are no longer allowed in the European Union.

Neonicotinoids, like nicotine, bind to nicotinic ACh receptors of a cell and trigger a response by that cell. In mammals, nicotinic ACh receptors are located in cells of both the central and peripheral nervous systems. In insects, these receptors are limited to the central nervous system (Fishel, 2015b). As you might guess, nicotinic ACh receptors are activated by the neurotransmitter ACh. While low to moderate activation of these receptors causes nervous stimulation, high levels over stimulate and block the receptors, causing paralysis and death. AChE cannot break free from neonicotinoids and their binding is irreversible (Gervais et al., 2010). Based on a lack of carcinogenic activity in animals, EPA classifies them as Group E with respect to carcinogenicity (ie, no evidence of carcinogenicity).

Neonicotinoids are widely used pesticides in the world. In the United States, neonicotinoids are currently used on about 95% of corn and canola crops; the majority of cotton, sorghum, and sugar beets; and approximately half of all soybeans (Fishel, 2015b). They are also used on the vast majority of fruit and vegetable crops, including apples, cherries, peaches, oranges, berries, leafy greens, tomatoes, and potatoes; they are applied to cereal grains, rice, nuts, and wine grapes. Neonicotinoids are applied at much lower rates than other insecticides (Fishel, 2015b) which may contribute to their absence from the US top 25 list. Only a few neonicotinoids have been widely used; these include imidocloprid and thiomethoxam, both with broad ranges of application. Acetamiprid is used against sucking insects such as aphids and whiteflies on citrus, cotton, and some vegetables. Clothiamidin is widely used in corn and canola seed treatments and it can be directly applied to rice, tobacco, and grapes (Fishel, 2015b).

Costs and Benefits of Neonicotinoid Use

Environmentally, neonicotinoids have strengths and weaknesses. One factor that is both positive and negative is that neonicotinoids remain in plant tissues for months to even more than a year (Maus et al., 2005). This means that the chemical is sequestered from nontarget animals and application can be widely spaced, which is valuable. However, their half-lives in the environment range from days to years (eg, 1–3 years in soil, 1 year in water under darkness for imidacloprid, 200–300 days for thiamethoxam, 148–1155 days for clothianidin, 4.9–25.1 days for acetamiprid, 68 days for thiacloprid) (Hopwood et al., 2012). While it might be good for the grower and ultimately require fewer applications, this persistence may increase risk to nontarget organisms.

These pesticides have received a lot of attention concerning their possible involvement with honeybee (*Apis mellifera*) die-offs or what is called *Colony Collapse Disorder* (CCD) in the United States and Europe (Suchali et al., 2000). Beginning in 2006, beekeepers in the United States began to notice unusual declines and extinctions of their honey bee colonies. A third of honey bee colonies in the United States were lost during each winter between 2006 and 2009 (Mullin et al., 2010). Approximately 29% of 577 beekeepers across the United States reported CCD and losses of up to 75% of their colonies according to

Stokstad (2005). The economic loss in terms of decreased pollination of crops amounts to millions of dollars. The causes of CCD remain unclear, but a variety of suggestions have been postulated including pesticides, parasites, fungus, other pathogens, beekeeping practices (such as the use of antibiotics, or long-distance transportation of beehives), malnutrition, and immunodeficiency.

There has been great concern over imidacloprid which is very toxic to bees under laboratory conditions (Suchali et al., 2000). Foraging bees can bring the substance back to the colony where the queen can be poisoned and the colony dies out or collapses (Scholar and Krischik, 2014). As little as 0.25 µg/kg of imidacloprid in honey might cause premature death in older workers and queens (Anderson et al., 2015). Intoxication can also reduce the foraging efficiency of bees. Studies on honeybees in field situations suggest that other factors such as simultaneous exposure to other pesticides and diseases may be as or more important in these die-offs than neonicotinoids alone (Sandrock et al., 2014; Thompson et al., 2014; Fairbrother et al., 2014).

Some positives are that these pesticides are very effective at low dose concentrations in controlling noxious insects. They are also considerably more toxic to insects than to mammals, birds, and most fish. For example, imidacloprid, the most widely studied neonicotinoid, is considered moderately toxic with the acute oral LD50 in rats between 424 and 475 mg/kg bw; for mice it is between 130 and 170 mg/kg bw (WHO, 2004). Both ranges are well above environmentally realistic concentrations. If imidacloprid is applied on the skin of mice and rats, it is essentially nontoxic—concentrations up to 5000 mg/kg bw do not cause mortality, indicating that the pesticide is not absorbed through the skin. In fish, imidacloprid has LC50s around 211 mg/L in rainbow trout (*Oncorhynchus mykiss*) and 280 mg/L in carp (*Cyprinus carpio*), also greater than environmentally realistic concentrations (Tisler et al., 2009). In contrast, LC50s for sensitive aquatic in vertebrates are typically less than 20 µg/L and for sensitive terrestrial invertebrates it can be less than 10 µg/kg. Neonicotinoids are often used in combination with other insecticides to produce synergistic effects.

Sublethal effects of imidacloprid in rats under laboratory conditions include reduced weight, increases in liver enzymes as measured in plasma, decreased AChE activity in brain and plasma, and liver and thyroid gland histopathology. Some studies have shown genotoxicity in invertebrates with this chemical (Kavi et al., 2014). Imidacloprid is placed in class E of carcinogen assessment by the US EPA, indicating little to no risk of cancer. Perhaps one of the more serious sublethal effects seen in male rats is atrophy of the testes and eventual loss of spermatogenesis in chronic exposures of rats (Koshlukova, 2006) and reptiles (Cardone, 2015).

Organochlorines

A few OCs are still in use. As mentioned, endosulfan was the most recent OC to be phased out by the EPA with the last permissible use to occur in 2016.

In developing nations afflicted with malaria, DDT is still used as a spray on screens and walls to prevent the mosquito vector from entering residences. Lindane is still used in scalp treatments for lice and scabies even though the pesticide has carcinogenic properties. Interestingly, the terminology seems to have changed sometime over the last several years. Instead of "organochlorine" many sources are listing compounds as "halogenated organics" or "chlorinated hydrocarbons."

Dichloropropenes are listed as soil fumigants for nematodes and 1,3-dichlorpropene, the most commonly used isomer, is on the EPA's top 25 pesticides in terms of use. From 12,200 to 14,500 tonnes of this chemical were used in the United States during 2007 (US EPA, 2015a). The pesticide has been shown to be carcinogenic to animals and is suspected of being a carcinogen to humans. In the United States, the Department of Health and Human Services has determined that 1,3-dichloropropene may reasonably be anticipated to be a carcinogen. The International Agency for Research on Cancer has determined that 1,3-dichloropropene is possibly carcinogenic to humans. The EPA has also classified 1,3-dichloropropene as a probable human carcinogen. We always have to keep in mind whether a chemical can cause cancer or any other ailment at environmentally realistic concentrations. Since we cannot ethically conduct controlled experiments with humans as we might with other animals, variation in classifying currently used OCs will persist.

1,3-dichloropropene has been approved for 16 crops in the United States and is a primary nematicide for crops such as potatoes, sugar beets, carrots, onions, and others. Unlike the traditional OCs discussed in Chapter 4, 1-3-dichloropropene is relatively short-lived. Half-lives in air are approximately 7–50 h before photo-oxidation degrades it (ATSDR, 2008). In water, half-lives are approximately 10 days and stabilizers are used to keep the pesticide bound to the soil to increase effectiveness. Most of the chemical sprayed onto soil volatizes and enters the air.

Laboratory experiments with commercial treatments of 1,3-dichloropropene, which also contains the pesticide chloropicrin, caused swelling in the nasal epithelium, urinary bladder tissues, and stomach lining of rats. Neurological effects included ataxia (loss of coordination) and inability to stand. Reduced litter size was seen in rats given chronic doses of 150 mg/kg bw. Little is known about the effects of this pesticide on humans (ATSDR, 2008). The use of 1,3-dichloropropene is restricted and being phased out by the European Union.

The release of pentachlorophenol to the environment is sizable (ATSDR, 2001). An estimated 620 tonnes is released to the atmosphere each year by wood preservative plants and cooling towers where it is used as a microbiological. An additional 890 tonnes are released to the land and 17 tonnes to water through direct runoff and water treatment plants.

Pentachlorophenol was developed in the 1930s and, at one time, was one of the most widely used pesticides in the United States. Since 1984, its use is restricted to commercial applications in wood preservatives for fence posts, telephone poles, railroad ties, and the like (ATSDR, 2001). Pentachlorophenol causes many adverse effects in humans and, presumably, other animals. These effects include liver, thyroid, and immune system deficiencies as well as cardiac, reproductive, and neurological disorders. The chemical can cause cancer in rats although even technical-grade pentachlorophenol is contaminated with other OCs and polychlorinated biphenyls (PCBs) which many augment the carcinogenicity (ATSDR, 2001). The US EPA has classified pentachlorophenol as a Group B2, probable human carcinogen.

In mink, 1 mg/kg bw/day over 3 weeks caused reproductive effects. A dose of 2 mg/kg/day administered twice per week over 43 days increased the frequency of ovarian cysts in sheep. The single dose, 96-h LD50 in rats is 50 mg/kg bw and 129–134 mg/kg in mice (ATSDR, 2001).

Chronic exposure by inhalation to pentachlorophenol in humans has resulted in inflammation of the upper respiratory tract and bronchitis; blood effects such as anemia; effects on the kidney and liver; immunological effects; and irritation of the eyes, nose, and skin. Chronic oral exposure to pentachlorophenol in animals has resulted in effects on the liver, kidney, blood, endocrine, immune system, and central nervous system (US EPA, 2015b). The half-life of this pesticide in soil or sediment varies from weeks to months, considerably shorter than the banned OCs. In water and air it is subject to photo-oxidation and has half-lives usually less than a week. The Stockholm Convention has proposed placing a ban on pentachlorophenol as it has with many persistent organic pollutants (see Chapters 4 and 6).

Organophosphates

OPs became popular in the mid-1970s and 1980s along with carbamates. Their most common use is as insecticides but some are registered as herbicides or fungicides. Like carbamates, OP insecticides are neurotoxins that interfere with the breakdown of ACh in the synapse between neurons and the neuromuscular junction. They are similar in structure and function to nerve gases, such as sarin. OPs were the most commonly used insecticides in the world but are being replaced by neonicotinoids in the United States. OP insecticides of primary concern include: azinphos-methyl, chlorpyrifos (Fig. 5.2F), diazinon, dichlorvos, dimethoate, malathion, methamidophos, naled, and oxydemeton-methyl. Ethephon (3200–4090 tonnes used per year), a plant growth regulator; chlorpyrifos (3200–4090 tonnes), the most extensively used insecticide; and acephate (900–1800 tonnes), another insecticide, are on the EPA's top 25 list of pesticides for 2007.

OP pesticides are generally regarded as safe for use on crops and animals due, in part, to their relatively fast degradation rates. Their degradation varies as a function of microbial composition, pH, temperature, and availability of sunlight. Under laboratory conditions at 25°C and pH 7, biodegradation is about 10 times faster than chemical hydrolysis, which in turn is roughly 10 times faster than photolysis (Ragnarsdottir, 2000). OPs have relatively low log

K_{ow} values (2–3) and thus have moderate to high solubility (10–10,000 mg/L). Hydrolysis is the primary way OPs degrade in water. Estimates of actual half-lives due to hydrolysis at 25°C and pH 7 range from 1.3 days for disulfoton to 230 days for chlorpyrifos. In contrast, hydrolysis nearly stops in cold, acidic waters and half-lives can be measured in thousands of years (Ragnarsdottir, 2000). Half-lives in neutral pH soils ranges from 1 day for methamidophos to 60–120 days for chlorpyrifos. Other means of degrading OPs include photolysis in air and surfaces, volatilization from soils, and microbiological activity in soil and water. Many different plants and animals metabolize OPs so bioconcentration can occur, but biomagnification does not.

Metabolism of OPs Can Increase Their Toxicity

Some OPs such as diazinon, profenfos, and dimethoate are direct toxins that need no transformation to be effective. Most, however, must be converted to their oxon or sulfon derivatives for maximum toxicity (Fig. 5.4). Some OP pesticides that are converted to an oxon form include chlorpyrifos, azinphos-methyl, fonofos, and malathion. Among those that are activated to sulfones are disulfoton and terbufos. This conversion can take place in the environment but also in exposed organisms through enzymes associated with the P450 system. The conversion process makes the molecule much more efficient in binding with AChE and the converted forms may be 10–100 times more toxic to amphibians than the parent compound (Sparling and Fellers, 2007). These same pesticides have been linked to amphibian declines in pristine mountainous areas in California (Sparling and Fellers, 2009; Sparling et al., 2015). As we mentioned in Chapter 3, the P450 system functions as one of the major ways animals detoxify adverse chemicals in their bodies so it is somewhat ironic that species with less well-developed P450 systems are less prone to being intoxicated by anthropogenic toxins than those with well-developed systems. We see similar

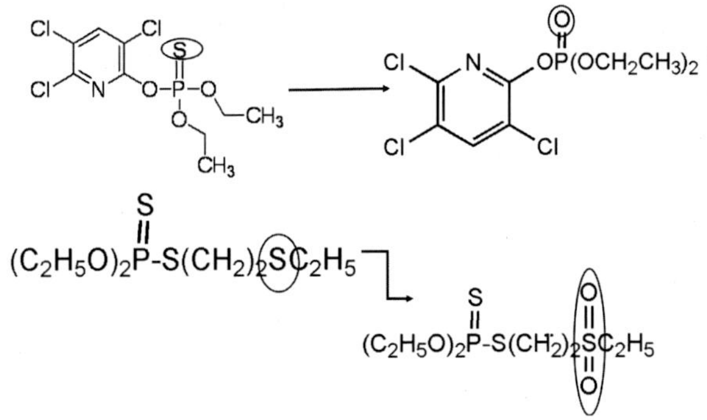

FIGURE 5.4 Structural formulas showing the oxon and sulfon forms of representative OP pesticides. The oxon and sulfon forms are degraded pesticides, but are often more toxic than the parent compounds.

relationships between the P450 system and polychlorinated biphenyls, dioxins, and furans (see Chapter 6).

While very toxic to insects, most OPs have moderate to low toxicity to birds and mammals. Table 5.2 presents some lethal dose concentrations of OPs to birds and mammals. Note the major exceptions to mammalian toxicity for azinphos-methyl, methyl parathion, and phorate. These same OPs have high toxicity to birds. Fish and bees tend to be more sensitive to these pesticides than mammals. Despite the comparatively low toxicity to mammals for most OPs, these pesticides are either the cause or have been linked to many human diseases including the exotically named Ginger Jake syndrome which was caused by adding an OP to the "medicine" Jamaica Ginger to hide its high alcohol content during the period of alcohol prohibition in the United States. Approximately 600 people died from the Spanish toxic oil syndrome in 1981 when they drank OP-contaminated colza oil that should have been used for industrial purposes. OPs have been linked to attention deficit disorder, Parkinson's disease, motor

TABLE 5.2 Some Representative Acute Toxicity Values of Organophosphorus Pesticides to Rats (oral), Rabbits (dermal), Birds, Fish and Bees

Pesticide	Rat Oral LD50 (mg/kg bw)	Rabbit Dermal LD50 (mg/kg bw)	Bird Acute oral LD50[a]	Fish Acute in Water LC50[b]	Bee[c]
Azinphos-methyl	4	150–200	HT	VHT	HT
Chlorpyrifos	96–270	2000	HT	HT	HT
Diazinon	1250	2020	HT	HT	HT
Disulfoton	2–12	3.6–15.9	HT	MT	HT
Malathion	5500	>2000	MT	HT	HT
Methyl Parathion	6	45	NA[d]	ST	NA
Naled	191	360	MT	HT	HT
Phorate	2–4	20–30 (guinea pig)	VHT	ST	MT
Phosmet	147–316	>4640	ST	HT	HT

Source: Fischel (2014).
[a]*Bird toxicity LC50 (mg/kg bw): ST = 501-2000; MT = 51-500; HT = 10-50; VHT = <10.*
[b]*Fish toxicity LC50 (mg/L):ST = 10-100; MT = 1-10; HT = 0.1-1; VHT = <0.1.*
[c]*Bee toxicity MT, moderately toxic (kills if applied on bees); HT, highly toxic (kills upon contact as well as residues).*
[d]*Data not available.*

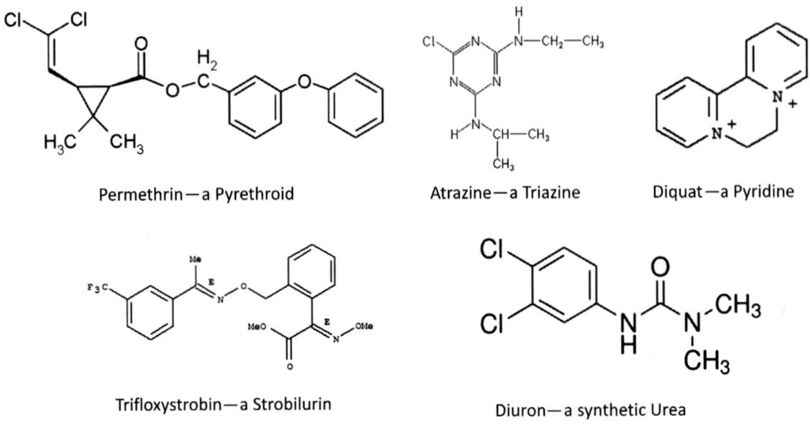

Permethrin—a Pyrethroid   Atrazine—a Triazine   Diquat—a Pyridine

Trifloxystrobin—a Strobilurin   Diuron—a synthetic Urea

FIGURE 5.5   Structural formulas for three other common pesticides.

neuron disease, various neuropathies, and multiple sclerosis. Farmers, pesticide applicators, and those working in pesticide plants are the groups at highest risk for these chronic disorders (Ragnarsdottir, 2000).

Phosphonoglycine

There are only two chemicals of any concern in this family of pesticides but one at least commands a huge market. Both are herbicides. In 1970, Monsanto developed the herbicide glyphosate (Fig. 5.5A) which has become the most widely used herbicide in North America, if not the world. In 2007, the US EPA (2015a) estimated that between 180 and 185 million pounds (between 81.8 and 84.1 thousand tonnes) were sold in the United States alone. Glyphosate tops the US EPAs list of the 25 most widely used pesticides.

Glyphosate is formulated as an isopropylamine salt of N-(phosphonomethyl)-glycine, a diammonium salt of glyphosate, or a trimethylsulfonium salt of glyphosate, but its most widely used formulation is the isopropylamine salt. Monsanto held the registration for glyphosate until 2000, at which time the formulation became available to other manufacturers. However, by that time, the trade name Roundup® was widely known as was the name for the aquatic formulation Rodeo®. In recent years, Monsanto delved into the genetic modification business, producing corn, canola, sugar beets, cotton, and soybeans that are glyphosate resistant, thus allowing the herbicide to be used as a postemergent product to control weeds without damaging resistant crops. Monsanto benefits twice, first by selling the genetically modified seeds and then by marketing the herbicide. Glyphosate interferes with enzymes essential in the production of chlorophyll, cytochromes, and peroxidases.

Glufosinate is marketed as its ammonium salt, glufosinate-ammonium. It is a broad spectrum contact herbicide used to control a wide range of weeds postemergence. It is a natural compound isolated from two species of *Streptomyces* fungi. Glufosinate was first produced in Japan in 1984 and is sold as Liberty® in North America. Bayer CropScience has begun genetically modifying some crops such as corn, soybeans, and cotton to be resistant to this herbicide; similar to how Monsanto is working with glyphosate. It inhibits the activity of an enzyme which is necessary for the production of glutamine and for ammonia detoxification, allowing for toxic buildup of ammonia in cells. Glufosinate also inhibits the same enzyme in animals.

Both products have a high propensity to particulates in soil. Both are also highly water soluble with a solubility of 786 g/L for isopropylamine glyphosate and 1370 g/L for glufosinate. Glyphosate readily and completely biodegrades in soil even under low temperature conditions. Its average half-life in soil is about 60 days, but may be longer in cold environments. Biodegradation in foliage and litter is somewhat faster (NPIC, 2010). In field studies, residues are often found the following year and may have residual effects. Glyphosate may enter aquatic systems through accidental spraying, spray drift, or surface runoff. It dissipates rapidly from the water column as a result of adsorption and possibly biodegradation. The half-life in water is usually a few days, but may take weeks depending on pH and temperature (NPIC, 2010). Based on its water solubility, glyphosate is not expected to either bioaccumulate or bioconcentrate in aquatic organisms. Glyphosate has a very low volatility so atmospheric concentrations are usually negligible.

Glufosinate-ammonium has similar characteristics to glyphosate. Its half-life in soils varies from 3 to 43 days but may be 70 under certain conditions (Jewell, 1998). The herbicide is more long-lived than glyphosate in water and half-lives can be approximately 300 days. Like glyphosate, glufosinate-ammonium does not bioaccumulate, is very soluble in water, and has low vapor pressure.

Commercial glyphosate-based formulations generally consist of a mixture of the isopropylamine (IPA) salt of glyphosate, a surfactant, and various minor components including antifoaming and color agents, biocides, and inorganic ions to produce pH adjustment (Bradberry et al., 2004). This makes field estimates of the toxicity of glyphosate formulations complicated. Many laboratory studies have used technical-grade glyphosate without the rest of the formulation ingredients. Further, glyphosate can be used as any of five different salts. As a result, human, wildlife, and inadvertent vegetative poisoning with this herbicide can be due to the complex and variable mixtures. Therefore, it is difficult to separate the toxicity of glyphosate in the field from that of the formulation as a whole or to determine the contribution of surfactants to overall toxicity.

Glyphosate and Amphibians

Ronald Relyea and his students have done extensive research on the effects of glyphosate on amphibians, mostly in mesocosms studies. Their work can serve as a learning experience to demonstrate the complexity of understanding toxicity of chemicals to living organisms. An early study (Relyea, 2005) showed that

Roundup® was lethal to both larval (ie, tadpoles) and juveniles of three species of frogs. Subsequently, Relyea and Jones (2009) determined that the 96-h LC50 for nine species of larval frogs using Roundup Original® ranged from 2.2 to 5.6 mg a.i./L while that for three species of larval salamanders ranged from 7.6 to 8.9 mg a.i./L. A typical in-field concentration of glyphosate is around 4.2 mg/L, about midway for the LC50s in frogs but lower than that for salamanders. Recall, however, that longer exposures require lower concentrations to have similar effects so chronic exposures at these concentrations are probably lethal. Jones et al. (2010) found that the timing and quantity (ie, a single dose or multiple dose at equivalent total concentrations) can affect toxicity and that glyphosate tended to be at higher concentrations near the top of the water column than deeper. Increased competitive stress, produced by higher stocking densities, elevated toxicity of glyphosate formulations to bullfrog (*Rana catesebeina*) tadpoles (Jones et al., 2011). In contrast, the presence of dragonfly nymphs, a natural predator on frog larvae, reduced the apparent toxicity of glyphosate compared to tanks without predators, likely because the higher surface concentrations of glyphosate forced the larvae to go deeper where cover was thicker and the chances of being eaten were lower.

Complicating the understanding of glyphosate toxicity is that the specific formulation may be very important. Working with Australian frogs, Mann and Bidwell (1999) determined the acute (48-h) toxicity of technical-grade glyphosate acid, glyphosate isopropylamine, and three glyphosate formulations tadpoles of four species. The LC50 values for Roundup® herbicide tested ranged between 8.1 and 32.2 mg a.i./L. Touchdown® herbicide was slightly less toxic than Roundup® with LC50 values ranging between 27.3 and 48.7 mg/L. Roundup® Biactive was practically nontoxic to the tadpoles with LC50 values ranging from of 911 mg/L to more than 1000 mg/L. Technical-grade glyphosate isopropylamine was also practically nontoxic. The formulations differed in the form of glyphosate and the surfactant used.

Glyphosate has very low acute toxicity to most animals. The acute oral LD50 in rats exceeds 4500 mg/kg by weight in several studies (NPIC, 2010) and the toxicities of the technical acid isopropylamine and Roundup® are nearly the same. The oral LD50 for the trimethylsulfonium salt was somewhat more toxic at 750 mg/kg. Oral LD50 values for glyphosate are greater than 10,000 mg/kg in mice, rabbits, and goats. It is practically nontoxic by skin exposure and is not even irritating to the skin of rabbits or guinea pigs. Glyphosate is slightly toxic to wild birds. The dietary LC50 in both mallards and bobwhite quail is greater than 4500 ppm (Extoxnet, 1996). It is also practically nontoxic to fish and may be slightly toxic to aquatic invertebrates.

Glufosinate has low to moderate toxicity to mammals. Median lethal doses in rats were approximately 1600 mg/kg by weight in mice weighing 450 mg/kg. Dogs (beagles) were somewhat more sensitive with LD50s around 300 mg/kg (Jewell, 1998). Chronic effects, however, have included problems with brain development in rats. Exposure to 3.75, 7.5, and 15 mg/L glufosinate in water

FIGURE 5.6   *Chrysanthemum cinerariifolium*, the flower from which natural pyrethrum is derived.

increased the frequency of micronuclei in red blood cells of larvae of the toad *Rhinella arenarum* (Lajmanovich et al., 2014) which indicates chromosomal damage. The LC50 for the medaka (*Oryzias dancena*), a species of marine fish, was 8.76 mg/L and lower concentrations produced sublethal effects of blood congestion, gill histopathology, and lipid degeneration and necrosis in livers (Kang et al., 2014). Other than those studies and a couple more, not much is known about the effects of glufosinate on wildlife or fish.

Pyrethroids

The chrysanthemum plant is native to Eastern Europe and Asia. It produces a natural insecticide, *pyrethrum*, that has been used for thousands of years as a mosquito repellent and as a remedy for lice in ancient Persia and elsewhere. Pyrethrum is actually a natural mixture of six chemicals called *pyrethrins* (ATSDR, 2003a). You can make that same insecticide today by drying chrysanthemum flowers, extracting, and crushing the seed cases and rinsing them thoroughly in a small quantity of water. The water then contains the pyrethrins and can be rubbed on the skin. *Chrysanthemum cinerariifolium* and *C. cineum* are the commercial varieties for producing the natural insecticide (ATSDR, 2003a). *C. cinerariifolium* looks much like the common daisy (Fig. 5.6). Its flowers are typically white with yellow centers and are really quite attractive. The commercial pyrethrum is in the form of an oleoresin or cream.

In addition to pyrethrum, there have been two generations of *synthetic pyrethroids*. Pesticides in both generations are still used, although the second generation ones have larger sales volumes. The first generation pyrethroids, developed in the 1960s, include bioallethrin, tetramethrin, resmethrin, and bioresmethrin. They are more active than the natural pyrethrum but unstable in sunlight. The second generation pyrethroids were introduced in the mid-1970s and are substantially more resistant to degradation by light and air, thus improving their suitability for agriculture. They include permethrin (Fig. 5.5A), cypermethrin, and deltamethrin. However, these second generation chemicals have significantly higher toxicities to mammals than the earlier formulations. Over the subsequent decades, these pyrethroids were followed by other compounds such as fenvalerate, lambda-cyhalothrin, and beta-cyfluthrin. Over 1000 synthetic pyrethroids have been manufactured, but only a small number have been registered for use in the United States (ATSDR, 2003a).

The EPA listed 33 pyrethroids in its 2011 cumulative risk assessment of the chemical family (US EPA, 2011). None of these made the top 25 user list in 2007, in part perhaps because application rates are usually substantially lower than for other pesticides and in part because pyrethroids are not as popular as they once were for agricultural purposes. Rather, pyrethroids find a niche in the home pesticide arena. More than 3500 registered products contain pyrethroids (US EPA, 2013). Allethrin, is the active ingredient in Raid® insect spray, bifenthrin is an active ingredient in Ortho Home Defense Max®, and cyfluthrin and other pyrethroids are used in Baygon®, a home insecticide. Flea and tick powders for dogs and cats typically contain first generation pyrethroids that are effective against the ticks and safe for the pets. Caution should be used with cats, however, for they do not have one of the enzymes necessary for rapid breakdown of pyrethroids.

Pyrethroids are synthetic esters derived from the naturally occurring pyrethrins. With the exception of deltamethrin, pyrethroids are complex mixtures of isomers rather than one single compound. It is the mixture and the ratio of isomers that distinguish one from another. All of these chemicals have very low vapor pressure (10^{-5}–10^{-7} mm) which equates with low volatilization and, with the exception of two or three, they are nearly insoluble in water although they are soluble in a variety of solvents and oils. Accordingly, their log K_{ow} values are approximately 5–8 (ATSDR, 2003a).

Technical-grade pyrethroids are usually formulated for use in commercial products. The formulated products may contain up to 99% inert ingredients that improve storage, application handling, or effectiveness. The actual identity of these additives are largely unknown and fall into those categories of "trade secret" or proprietary chemicals. Piperonyl butoxide, which has its own insecticidal characteristics, is often one of these "inert" ingredients even though it is not truly inert.

Pyrethroids tend to be very short-lived. Because they have low vapor pressures, air pollution is not much of a concern. Those that do enter the atmosphere typically last for a day or two (ATSDR, 2003a). Also, because they are generally insoluble, pyrethroids do not mix within the water column, rather, they tend to adhere to dissolved organic carbons and other particulates and end up in sediments where bioavailability to free-swimming organisms is low. Bioavailability tends to decrease as sediment organic matter concentration increases. Early laboratory experiments failed to recognize these attributes and, under controlled situations with no sediment present, they concluded that pyrethroids had higher toxicities than generally observed under more natural conditions (Palmquist et al., 2012). That is not to say that pyrethroids are not toxic to aquatic organisms, they are highly toxic to aquatic insects, and moderately high to highly toxic to fish and amphibians. They are just not as toxic as originally thought.

Half-lives of these chemicals in water are measured in hours to days and are reduced even more by photo-oxidation and microbial degradation. In soils and sediments, typical half-lives may be chemical-specific but most degrade in a few weeks or less. The ATSDR (2003) cites half-lives in soil for fenvalerate (88–280 days), cyfluthrin (56–63 days), permethrin (4–40 days), and cypermethrin (6–60 days).

Accurate estimates of quantities of pyrethroids produced and used in the United States are not available except on a state by state basis and not all states keep good records. For example, in California, cypermethrin (one of the pyrethroids that is used extensively for agriculture) sales varied from 36 to 58 tonnes between 2003 and 2012. Although annual sales varied, there was a general decline from the peak years of 2005 and 2006. In total, more than 460 tonnes of pyrethroids were sold in California in 2012.

Pyrethroids are neurotoxins whose outward signs may be similar to organophosphate or carbamate exposure, but the mechanisms are different. Rather than affecting synapses, pyrethroids react with voltage-gated sodium channels along nerve axons, prolonging the time during which the channels are open. Opened channels extend depolarization of the nerve which leads to either a series of short bursts or a prolonged burst. Outward signs of toxic exposures include incoordination, convulsions, and paralysis (Soderlund and Bloomquist, 1989).

Based on specific signs and mode-of-action on sodium channels, pyrethroid insecticides can categorized into type I and type II classes. Type I pyrethroids include permethrin, resmethrin, tetramethrin, and others. These produce repetitive nerve discharges that cause restlessness, hyperexcitation, and body tremors. Type II pyrethroids, such as cypermethrin, deltamethrin, and esfenvalerate produce nerve depolarization and blockage of impulses (Soderlund and Bloomquist, 1989) that induce hyperactivity, incoordination, convulsions, and writhing.

Pyrethroids are particularly toxic to insects and other invertebrates. Fortunately, metabolism of pyrethroids is high for most animals and sublethal toxicity tends to be unimportant at environmentally realistic concentrations (ATSDR, 2003a).

Fish and aquatic organisms are highly susceptible to pyrethrins and pyrethroids. Among five species of fish from different orders, 96-h LC50s ranged from 24.6 to 114 μg/L (Mauck and Olson, 1976). Velisek et al. (2011) determined that the LC50 in rainbow trout was 1.0 μg/L for deltamethrin, 1.98 μg/L for cypermethrin, and 1.47 μg/L for bibenthrin. In carp, the median lethal concentrations were slightly higher: 3.25 μg/L, 2.91 μg/L, and 5.75 μg/L, respectively. For two species of aquatic macroinvertebrates, the 24-h LC50s of deltamethrin and cypermethrin were 0.89–1.51 μg/L, respectively (Lutnicka et al., 2014). For *Ceriodaphnia dubia*, the LC50 value for permethrin of 0.19 μg/L was 223 times lower than that of the fungicide chlorothalonil and 3700 times lower than that for the herbicide atrazine (Phyu et al., 2013). Microgram per liter (ppb) concentrations are all environmentally relevant. Note, however, that these laboratory tests do not include sediments or dissolved organic carbons which could reduce bioavailability and hence toxicity of pyrethroids through adsorption onto particles.

For humans, industrial exposures rarely cause significant problems (ATSDR, 2003a). The most common symptoms of pyrethroid exposure in humans is a temporary numbing of the skin call *paresthesia*. Pyrethrins have been classified as possible human carcinogens because long-term studies on rats resulted in elevated frequencies of liver cancer (ATSDR, 2003a). Neither pyrethroids nor pyrethrins have been shown to produce genotoxic, endocrine, or reproductive effects in mammals. Mild immunological suppression occurred in rats and mice with permethrin, deltamethrin, and cypermethrin, usually following weeks of exposure in food (ATSDR, 2003a).

Triazines

Triazines consist of a benzene or phenol ring with three nitrogen atoms substituting for three carbons (Fig. 5.5B); the prospective hydrogen atom sites can also be substituted with other molecules to increase the diversity of these compounds. They were first developed in the early 1950s. All triazines are synthetic. Due to limited substitution sites on the benzene ring, there are three isomers of each of these compounds. Triazines that have the nitrogen atoms evenly spaced on the benzene ring such as the 1,3,5 locations are called symmetric (s)-triazines. Those with nitrogen atoms at other locations such as 1,2,4 are asymmetric (as)—triazines or triazones. One of the s-triazines is the basis for melamine, a resin that has been used for making Melmac® dishware and other products which were widely popular in the 1940s through 1960s. (For TV buffs, you might also remember that this was the home planet for ALF.) For our purposes, however, we will focus on those triazines that are used as herbicides.

Metribuzon is the primary asymmetrical triazine herbicide. It is a pre- and postemergent herbicide used on asparagus, carrots, corn, sugarcane, potatoes, sorghum, soybeans, barley, tomato, garbanzo, peas, recreational turf areas, wheat, alfalfa, hay, and pastures (Patterson, 2004). It inhibits photosynthesis primarily by interfering with the electron transport system.

There are several symmetric triazine herbicides and they can be combined with amines, chlorines, or thio (sulfur) groups. Among these are two very major herbicides, atrazine and simazine. Atrazine is the second most widely used in the United States, only surpassed by glyphosate. Both atrazine and simazine were listed among the top 25 pesticides used in the United States by the US EPA (2015a). Due to environmental and human health concerns, atrazine was banned by the European Union in 2003. Between 33,000 and 35,000 tonnes of atrazine are applied in the United States each year. Simazine came in as number 16 of the top 25 with an annual usage of 2300–3200 tonnes (US EPA, 2015a). Atrazine is a selective herbicide used to prevent pre and postemergence broadleaf weeds in crops. It is very widely used on corn and on turf, such as golf courses and residential lawns; more than 80% of the herbicide is used on corn and, as a result, its use is concentrated in the Midwest. Simazine is also a selective herbicide at lower application rates and used to control broad-leaved weeds and annual grasses in field, berry fruit, nuts, vegetable and ornamental crops, turfgrass, orchards, and vineyards. At higher rates, it is used for nonselective weed control in industrial areas (Extoxnet, 1996). Other s-triazines include cyanzine, terbuytlazine, atraton, and several others.

Molecular weights for herbicide triazines range from 174 to 240 mg/mole (Prosen, 2012). They tend to have low log K_{ow} values ranging from 1.1 to 3.3, which indicates that they have relatively high solubility. Simazine has a solubility of 5 mg/L, atrazine's solubility is 33 mg/L, but that of atraton is 1800 mg/L, and desethylatrazine's is 3200 mg/L (Prosen, 2012). Triazines are solid at room temperature and have low volatilization pressures, making it less likely that they will occur in the atmosphere to any great extent. Triazine herbicide's log K_{oc} values vary from 2.1 to 2.9 (Chafetz et al., 2004; Prosen, 2012), suggesting that at least some triazines have a high propensity to adsorb to particulates. If triazines are found in air, they are very likely to be attached to aerosols and if they are in water, they will either be dissolved or attached to dissolved organic carbons.

Both atrazine and simazine are relatively persistent. In soil, half-lives are measured as a few weeks to several months. However, carry-over of triazine residues through the winter has been known to damage plants in the year following application. In surface water they may persist for more than 200 days and in air they last 14 days. Photo-oxidation is not important in breaking down these chemicals in the atmosphere or in surface waters but in air ionic oxidations, it reduces half-lives to several hours (ATSDR, 2003b).

Acute toxicity of triazines is not particularly high to birds and mammals. In rats, the oral LD50s (96 h) are as follows: atrazine—1900–3000 mg/kg bw; simazine—>5000 mg/kg; cyanazine—180–380 mg/kg; atraton—1465–2400 mg/kg; and terbutyrin—2000–2980 mg/kg. The LD50 for rabbits and atrazine was 750 mg/kg. Sheep are more sensitive to simazine than are rats—their LD50 was 500–1400 mg/kg for that compound (Prosen, 2012). The dietary 8-day LD50 for atrazine in mallards (*Anas platyrhynchos*) is 19,650 mg/kg and for the northern bobwhite, it was 5600 mg/kg (Beste, 1983).

Acute toxicity of triazines can be higher in aquatic invertebrates than in mammals and birds and run in the low ppm region. For example, 96-h LD50s of six triazine herbicides to the crayfish *Pacifastacus leniusculus* ranged from 12.2 mg/L for atrazine to 77.9 mg/L for simazine (Velisek et al., 2013). For the water flea, *Daphnia magna*, the 48-h LC50 for atrazine was 6.9 mg/L, 5.7 mg/L for the scud *Gammarus fasciatus*, and 0.7 mg/L for the midge *Chironomus tentans* (Macek et al., 1976). Fish have a similar sensitivity as aquatic invertebrates, at least for atrazine. LC50s (96 h) for rainbow trout (*Onchorynchus mykiss*), bluegill (*Lepomis marcrochirus*), and zebrafish (*Danio rerio*) are 4.5–24 mg/L, 8–42 mg/L, and 37 mg/L, respectively (reviewed by Eisler, 2000).

Atrazine and Human Health Concerns

Whereas triazines are not as acutely toxic as some of the other pesticides we have discussed, continuing research is revealing that some, such as atrazine, may produce a wide range of sublethal effects. Atrazine targets the reproductive system of mammals and developing animals. It disrupts estrus (the reproductive cycle of all higher mammals except humans), alters plasma concentrations of hormones, increases the rate of miscarriages or spontaneous abortions in rats, decreases fetal body weights, and causes incomplete ossification of the fetal skeleton in rats (ATSDR, 2003b). Many of the studies that have produced these results tested subjects with atrazine concentrations more than what might be expected under normal environmental conditions, but longer term exposures at lower concentrations may produce similar results. In humans that have been exposed to atrazine in the field or industry, there is a statistically significant decrease in the term of pregnancy, an increase in the rate of prostitis (inflammation of the prostate gland), and some embryonic malformations (ATSDR, 2003b).

Reproductive effects are likely to occur through what is called the *gonadal/hypothalamic/pituitary axis* (GHPA, Fig. 5.7). The GHPA is an avenue for causing sex-related endocrine disruption in vertebrates. The interaction within this axis is complex so we'll try to keep our explanation to simple terms. The hypothalamus, which has a direct connection to the pituitary, produces *releasing factors* that stimulate the pituitary to produce and release hormones (in this case, luteinizing hormone and prolactin) into the circulatory system. These hormones stimulate cells in the testes to produce spermatozoa and to begin the process of follicular development in the ovary. The gonads produce testosterone in males and estrogen in females due to stimulation by the pituitary hormones. As the circulatory concentration of the gonadal hormones increase, they provide negative feedback to the hypothalamus and pituitary to inhibit the secretions of their releasing factors and hormones. This, in turn, inhibits sperm production in males and triggers ovulation in females. Chemicals such as atrazine that disrupt the GHPA can affect the sexual development and performance of affected animals at several locations within the axis.

Atrazine and Feminization of Male Frogs

In Chapter 3, we briefly mentioned that atrazine has been involved in a heated controversy for the past several years regarding amphibians. Bishop et al. (2010) provided an excellent summary of the effects of atrazine on amphibians and reptiles and of the controversy. They stated that between 1980 and 2009, 44 studies appeared in the peer-reviewed literature focusing on amphibians and atrazine. Between the publication of that article and 2015, more than

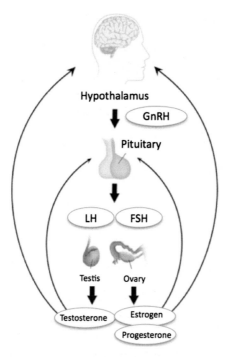

FIGURE 5.7   The gonadal/hypothalamic/pituitary axis (GHPA) demonstrating the connections among the gonads, pituitary, and hypothalamus. This axis is stimulated by environmental cues such as day length, controls gonadal hormones, and influences sexual drives. *Courtesy of Wikimedia Commons.*

45 additional papers were published on that topic—a very impressive showing for a single-issue topic. Although atrazine is not persistent, it is used often and widely enough that amphibians at various life stages can be exposed. Giddings et al. (2005) found that atrazine is found primarily in the water column and not adsorbed to sediments, thereby increasing the risk of exposure to aquatic life stages of amphibians. Very high environmental concentrations of the herbicide in water are approximately 200–300 µg/L, but effective concentrations are orders of magnitude less. Frogs and toads have received more attention as to the effects of atrazine than have salamanders. In their review of atrazine toxicity to frogs, Bishop et al. (2010) found that LC50s typically run in the 7–25 mg/L range, considerably higher than would be expected under typical environmental conditions, so sublethal effects may be more important to free-ranging anurans than acute lethal effects.

The main focus of the controversy exists between studies that have found endocrine disruption effects at very low (ppb) concentrations and those that do not. Several studies have shown that atrazine can have either estrogenic or anti-estrogenic effects at comparatively high concentrations. The crux of the matter is whether the herbicide can show these effects at environmentally realistic levels. We cannot review all of these studies, but we will present some of the positive and negative findings.

A leader on showing endocrine disruption has been Dr Tyrone Hayes from the University of California, Berkeley. In a study published in 2002, Hayes et al. (2002) exposed tadpoles of the African clawed frog (*Xenopus laevis*) to atrazine concentrations ranging from 0.01 μg/L (0.01 ppb) to 200 μg/L. At concentrations greater or equal to 0.1 μg/L, the researchers observed oocytes present in what were identified as testes. These abnormal organs or *ovotestes* are characteristic of males that have been exposed to estrogen. At 1 μg/L or higher, the cross-sectional diameters of the male frog larynges were reduced compared with control males—in males, the larynx size is affected by testosterone. In addition, plasma testosterone was lower in exposed males than controls and similar to females that had been repeatedly dosed with atrazine. Combining data from repeated studies and doses resulted in an overall rate of 10% for ovotestes compared with around 2% for controls (Hayes et al., 2006). Testicular degeneration in *Xenopus laevis* at 21 μg/L was also found by Tavera-Mendoza et al. (2002). Ovotestes or degenerated testes were also identified in leopard frog (*Rana [Lithobates] pipiens*) males subjected to 0.1 μg/L atrazine (Hayes et al., 2006) and frogs collected from farm ponds in areas with intense corn cultivation (Hayes et al., 2003; Hecker et al., 2004) had similar signs of ovotestes and other endocrine disruption characteristics. Whether these testicular abnormalities have an effect on the reproductive capability of affected males is not known.

Several other studies have found no effects at environmentally relevant concentrations opposing these observations of atrazine affecting amphibian testicular development. Carr et al. (2003) found a slight elevation in the frequency of ovotestes in *Xenopus laevis* at 25 μg atrazine/L which is on the high end of environmental concentrations. Many other studies have shown no ovotestes or overt effects of endocrine disruption with *Xenopus laevis*, even though concentrations of atrazine ranged from 0.1 to 250 μg/L (reviewed by Bishop et al., 2010).

The debate goes on. Can atrazine be expected to exert negative effects on amphibian testes? Why do we not see these changes consistently? How widespread are these effects across species and conditions? Are there any mitigating factors—positive or negative that influence this expression? Does the presence of ovotestes impair reproduction in affected individuals? These question beg that further studies be conducted, especially since so many of the world's species of amphibians are threatened or endangered.

OTHER INORGANIC AND BIOLOGIC PESTICIDES

The pesticides we have previously discussed are mostly complex, synthetic chemicals that are created in large manufacturing plants. We would be remiss if we did not spend at least some time presenting other, simpler pesticides or repellents that are more "natural" because they can occur in nature as rather simple inorganic elements or molecules or as derived from biological sources. Some of these chemicals and products may be used in certified organic farming but there are tight restrictions on a list of such chemicals (eg, USDA, 2015) and we do not attempt to make the distinction except to state that most "conventional" pesticides would preclude an farm from being organic.

For our purposes, we will distinguish these simpler chemicals as inorganic, biologically derived, and biological control organisms. The inorganic category includes elements such as sulfur and zinc and simple inorganic molecules that can occur naturally, even though they may be manufactured for high production. The biologically derived category includes various oils and other products that are byproducts from living organisms. Biological control agents include bacteria and viruses that target specific pests. We obtained our list of these chemicals from the state of California's remarkable database that lists more than 900 chemicals used in pesticide control and their amount sold or used in a particular year down to a half pound (0.2 kg) of product (California, 2014). For the most part, these chemicals do not pose risks to vertebrates under normal use.

Inorganic Chemicals Used in Pest Control

There are more than 50 chemicals that fit into this category listed in the California database. Chief among these in terms of quantity sold is sulfur and sulfur dioxide (40 million pounds or 18,200 tonnes sold in 2012). Sulfur has widespread use as a fungicide on many different crops and as a miticide. Alcohols have wide use in controlling bacteria, fungi, and insects. They are often employed in organic farming. The two most extensively used, ethyl and isopropyl alcohols, sold 2000 and 1800 tonnes in California, respectively. Petroleum distillates include mineral spirits, kerosene, white spirits, naphtha, and paraffins. They are used as insecticides and herbicides by themselves and serve as solvents for other pesticides. Petroleum oil, such as kerosene, has been widely used for controlling aphids, smothering insect eggs, and controlling weeds for decades and is still popular; 2028 tonnes of petroleum oil and more than 6400 tonnes of mineral oil were sold for pest control in California during 2012. A variety of insecticidal soaps are also sold each year as ingredients with other pesticides to reduce surface tension and facilitate the transport of active ingredients into plant and invertebrate cells.

Biologically Derived Pesticides

High among these products in use are an assortment of plant-derived oils. Margosa oil or Neem oil is a popular garden oil from an extract of the seeds of *Azadirachta indica*, native to India and Sri Lanka. In low doses, margosa oil has been a traditional remedy for centuries in India and Southeast Asia used in treating asthma, intestinal parasites, arthritis, and leprosy. As an insecticide, Neem oil has many complex, active ingredients which interfere with natural hormones and causes insects to stop laying eggs; larvae cease molting; and adults stop eating, mating, or flying. Other oils that are used to repel insects include coconut oil, soybean oil esters, garlic, eucalyptus, rosemary, thyme, general vegetable oil, tall oil—a by-product of the Kraft process of coniferous tree pulp manufacturing—, balsam fir, castor bean, cottonseed, black pepper, cedarwood, citronella, geranium, lemongrass, and peppermint. Some of these oils have defined insect repelling or killing properties, whereas others coat fungal spores and insect eggs, thus interrupting gaseous exchange and resulting in death. Urea, a by-product of animal wastes, is used extensively as a high nitrogen containing fertilizer. It is also used in the composition of pre-emergence herbicides, fungicides, insecticides, and bactericides. In 2012, California sold 288 tonnes of urea. Some perhaps unexpected compounds used in pest control include sugar, coyote urine, fox urine, hydrolyzed corn, plant hormones, red cabbage color, yeast buffalo gourd root powder, and putrescent whole egg solids.

Biological Control Agents

One of the earliest and most widely used biological insecticides is the bacterium *Bacillus thuringiensis* or Bt. During the process of development, the bacterium produces spores that contain crystals of proteinaceous insecticidal δ-*endotoxins*. When insects ingest the spores, the spores break apart within their alkaline digestive tracts and the crystals are denatured, making them soluble and thus amenable to breakage with proteases found in the insect gut. The toxin is thereby released from the crystal and enters the cells of insect guts. This paralyzes the digestive tract and the insect stops eating and starves to death. Different strains of Bt are selectively effective on specific insect larvae, including lepidopterans (caterpillars), dipterans (flies and mosquitoes), coleopterans (beetles), and nematodes. Bt has been used for decades in outbreaks of gypsy moths (*Lymantria dispar dispar*), which are serious exotic pests on oaks and aspen (Roh et al., 2007).

The δ-endotoxins or their crystals have been chemically identified and incorporated into sprays. The genes responsible for coding the crystals have also been isolated and are now injected into plant seeds as part of genetic modification (Klocko et al., 2014). Seventy-six tonnes of Bt spores were applied in California during 2012.

Streptomyces griseoviridis or Strain K61, a close relative of the *Streptomyces* that is used to produce the antibiotic streptomycin, is a naturally occurring soil bacterium initially isolated from peat in Finland. It seems to act against disease-causing fungi in at least two ways. By colonizing plant roots before the disease organisms get there, it deprives them of space and nourishment. It also produces several kinds of chemicals that may attack the harmful fungi (US EPA, 2000).

Among viruses the California listing includes is "polyhedral occlusion bodies" of the nuclear polyhedrosis virus of *Helicoverpa zea* (corn earworm). This virus preferentially infects caterpillars and the particular product is highly specific for the target species.

Metarhizium anisopliae strain F52 is a fungus that infects insects, primarily beetle larvae. It has been approved as a microbial pesticide for nonfood use in greenhouses and nurseries, and at limited outdoor sites not close to bodies of water. Many strains of *Metarhizium anisopliae* have been isolated worldwide from insects, nematodes, soil, river sediments, and decomposing organic material (US EPA, 2003).

Myrothecium verrucaria is another species of fungus which, in this case, is a plant and nematode pathogen. It is common throughout the world, often occurring on materials such as paper, textiles, canvas, and cotton. It is a highly potent cellulose decomposer. *M. verrucaria* has been formulated into a pesticide for control of nematodes. The pesticide's active ingredient is a mixture of the killed fungus and the liquid in which the fungus was grown. The active ingredient is specific only to nematodes that parasitize plants.

FOCUS—EXAMPLES OF PESTICIDE USE

Pesticides are expensive and it would not be economically prudent, or ecologically intelligent, for a grower to use pesticides that are unnecessary. Therefore, in actual use, a team consisting of growers, applicators, and extension experts in a region, and perhaps even the pesticide companies, will convene to determine what the most effective pesticides will be in the area. Thus, one can see some pesticides that have very broad application because they are efficacious on major crops and others that are used only in some regions and years. To illustrate this, let us take a look at pesticide use in two states with adequate to excellent records of pesticide use. These states track both sales and usage but note that sales and use may be unequal; some users may purchase pesticides to be used in other states or decide to store them for subsequent years, but companies that sell pesticides are held to tighter control than in-field users and overall, sales arguably provide a more accurate picture of state pesticide activity. The following analysis is based on sales.

The two states we will examine are California and Minnesota. For this comparison, we used several sources of information but relied heavily on each state's pesticide web sites (California Dept. of Pesticide Regulation, 2013; Minnesota Dept. of Agriculture, 2014). California, which started its program in 1990, has had the most complete and long standing record of pesticide use and sales of all states. All agricultural pesticide sales must be reported monthly to county agricultural commissioners, who in turn, report the data to Department of Pesticide Regulation.

Minnesota has a shorter history of keeping records. It started a pilot project in 2001 and the full project has continued since 2003. In 2006, the databases for sales were expanded to cover all crops. We use 2011 data for both states.

The two states substantially differ in their type and amount of principal agricultural crops (Table 5.3) and this is the major determinant for the types and amounts of pesticides used. For example, Minnesota, like many Midwestern states, focuses on corn and soybeans and the herbicides atrazine for corn and glyphosate for soybeans and corn are extensively used in that state. Due to its

TABLE 5.3 Number of Acres (1000's) Planted in Various Crops in California and Minnesota, 2011

Crop[a]	California	Minnesota
Corn	150	8750
Soybeans	0	7100
Wheat	535	1580
Other grains	15	282
Sugar beets	25.1	479
Peas and beans	54	207
Sweet corn	33	24.4
Potatoes	36.6	49
Apples	17.5	2.2
Grapes	796	0
Mixed Truck[b]	1037.4	0
Tree Fruit	468	0
Nuts	1446.6	0
Olives	41.5	0
Rice	580	0
Cotton	454	0
Tomatoes	285	

Source: USDA National Agricultural Statistics Service.
[a]*Crops considered as "dirty"—using extensive pesticides—are indicated in bold print.*
[b]*Several types of vegetables typically harvested by manual labor.*

northern climate, Minnesota has a smaller variety of major commercial crops than California. California is a huge agricultural state with a wide range of crops. In addition to those listed in the table, California grows rice, almonds, cotton, citrus, avocados, and pistachios and its tree and bush varieties of berries and fruits are much broader than apples, which are mainstays in Minnesota. The category "mixed truck" are those vegetables, melons, and fruit that were traditionally harvested by hand and loaded onto trucks rather than mechanically harvested. Included in this group are cantaloupe, watermelons, carrots, eggplant, strawberries, greens and lettuce, and similar commodities. A wider variety of crops implies a wider variety and more extensive use of pesticides.

In addition to the variety of crops we also have to be cognizant of the pesticide requirements for some so-called "dirty" crops. Among the common "dirty"

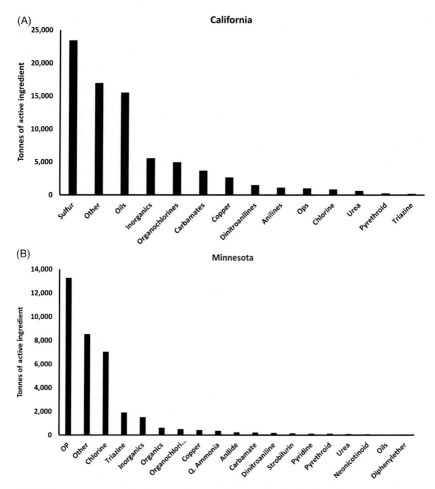

FIGURE 5.8   (A) Tonnes of active ingredient pesticides sold in California during 2011. (B) Tonnes of active ingredient pesticides sold in Minnesota during 2011.

crops are celery, peaches, strawberries, apples, blueberries, nectarines, bell peppers, greens of various types, cherries, potatoes, and grapes. "Clean" crops—those that use less pesticides than the norm—include onions, avocados, sweet corn, sweet potatoes, asparagus, kiwi fruit, cabbage, eggplant, cantaloupe, watermelon, grapefruit, and sweet peas according to the Environmental Working Group, a nongovernment organization promoting healthy foods (EWG, 2015).

By all measures, California sells and uses more pesticides than Minnesota (Fig. 5.8A,B). In 2011, California businesses sold a total of 281,522 tonnes (619,348,642.23 lbs, or 309,674 tons) of active ingredient pesticides. In contrast, Minnesota agribusinesses sold 50,340 tonnes (110,748,435 lbs, or 55,374 tons) of active ingredient during the same period; this is a difference of 5.6 times more total pesticides in California. Similarly, the state of California recorded

921 different chemicals, while Minnesota recorded only 653. Many of these chemicals in both cases were variations of the same active component. California keeps records of a greater variety of chemicals than Minnesota, which does not provide data on sulfur or adjuvants (ie, carrier chemicals, solvents, and nontoxic products in the formulations such as surfactants and soaps) but even with these dismissed, there is little comparison between the states. One other way that demonstrates the more extensive use of pesticides in California was that we limited our analyses to chemicals that sold more than 50,000 lbs (22.7 tonnes or 25 tons) or .045% of the total pesticide amount in Minnesota and 0.008% in California. It took the top 100 chemicals in Minnesota to reach this minimal limit but 200 chemicals in California before that mark was reached.

We used the Pesticide Action Network database (PAN, 2014) to determine the chemical family and the usage of a given pesticide. The largest category in California was "Other" which included those chemical families that were represented only a few times such as strobilurins, unidentified, amides, and a host of others. The exception to this group was glyphosate, which PAN lists as a phosphonoglycine but so much of it was used that we included it separately; glyphosate was eighth on the list in California. In Minnesota it was the number one pesticide followed by "Other." Despite the scale differences between the two states, Minnesota sold more glyphosate than California due to the large amounts of soybeans and corn produced in Minnesota. In both states, chlorine products, mostly elemental chlorine, follows in sales. Chlorine is used extensively as a fumigant, sterilizer, and in control of fungi.

From there, priority of use begins to differ between the two states. In California, inorganic chemicals which include several ammonia, potassium, and sodium inorganic products, as well as bentonite, cryolite, come next, followed by sulfur. Sulfur is apparently used more as a fertilizer than as a pesticide in Minnesota and we could not find records of sales or use for this element. Triazines, largely simazine, and atrazine herbicides follow. For California, oils are mostly petroleum products including distillates and mineral oil that are used as carriers and have some toxic properties in smothering eggs and spores. OCs are primarily used as herbicides or nematicides and don't include the old-time persistent chemicals covered in Chapter 4. Copper is used in many molecular formulations as herbicides, insecticides, fungicides, and microbials. Biologics include bacteria used against specific species of insect pests, insect pheromones, or products such as vegetable oil that are directly derived from living organisms. 2,6-Dinitroanilines are a small family of mostly herbicides; the OP pesticides included both insecticides and herbicides; anilides are fungicides, herbicides, and microbials; and azoles are fungicides. Pyrethroids are a group of insecticides and round out our list for this state.

In Minnesota after chlorine products, triazines are next on the priority list followed by inorganics, and biologics. OC herbicides, OP pesticides, and copper compounds are followed by anilides, carbamates, 2,6-dinitroanilines,

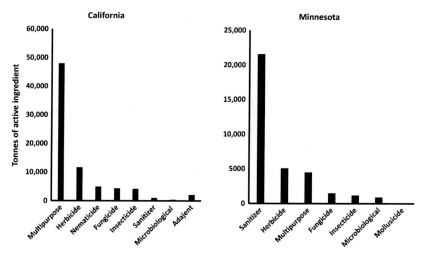

FIGURE 5.9   Tonnes of active ingredient pesticides classified to target organism sold in California and Minnesota during 2011.

azoles, and pyrethroids. Oils come in last but, as mentioned, Minnesota is not as comprehensive in its list of that group as California.

In terms of target organism, microbiologics led the list of uses in both states (Fig. 5.9). In California, this category (minus additives so we can make a fairer comparison) accounted for 53% of the total compared to 65% in Minnesota. Major elements in this category included quaternary ammonium compounds which have ammonia as the primary active ingredient, chlorine, and alcohols. Multipurpose pesticides include chemicals that are registered for three or more uses. They accounted for 20% of the chemicals by weight in California and 13% in Minnesota. Many inorganics such as phosphoric acid, as well as oils, copper, and sulfur compounds fit into this use. Herbicides came in third in Minnesota (14%) and fourth in California (8%). Despite the lower ranking in California, dealers in that state sold nearly four times as much herbicide as those in Minnesota. This category included many chemicals but glyphosate, simazine, atrazine, and 2,4-D clearly led the sales. Other leaders included acetochlor, and metolachlor and its isomer S-metolachlor. As might be expected based on the warmer, more humid climate of California and its extensive fruit and grape acres, fungicides sold well in California, coming in third place at 15% of total sales. Products in this category, besides several of the multipurpose chemicals include mancozeb, chlorothalonil (an OC), glycols (as in ethylene glycol or radiator antifreeze), strobilurins, and acrolein (an aldehyde).

Perhaps surprisingly given the amount of attention paid to them, insecticides accounted for the lowest of all the usages in both states. In California, only 4% of the total weight of active ingredient pesticides were dedicated expressly for insects and in Minnesota it was even lower at 3%. Certainly, several of the multipurpose chemicals are approved for insects. Perhaps the relatively low ranking of insecticides is because many of them have high toxicities which translates to a high "knockdown" effect and therefore comparatively less is needed to control these pests compared with weeds or fungi. The most commonly sold insecticide in Minnesota was chlorpyrifos at 790,578 lbs or 359 tonnes. In California, chlorpyrifos was also widely sold at 1,630,920 lbs or 741 tonnes. Ethyl and isopropyl alcohols are approved by the US EPA as insecticides and command large sales for agricultural use. Other inorganics, OP, and carbamate pesticides are in this category.

So, what is the take-home message to this review? I believe there are several lessons. Pesticides come in many different forms than chemicals that are especially synthesized for particular purposes to various inorganics and biologics that have been adopted to pest control. The number of products for pest control is surprising. California lists more than 900 chemicals and, even if one omits additives that have no direct lethality, there are hundreds of choices available. When DDT was popular, the available options were few but today's market has exploded in options.

The individual grower/applicator is in a competitive market and must maximize his/her profit margin; therefore, being highly selective when it comes to pesticides can make a difference. There are very significant regional differences in the types of pesticides that are used. In large part, this is due to the climate and what can be grown in a region. The whole business of pesticides has become a science both in the knowledge of chemicals and in economics.

STUDY QUESTIONS

1. What are some of the benefits that humans derive from the wise use of pesticides?
2. What are some ecological problems associated with pesticides?
3. Describe the various economic factors associated with pesticides.
4. What federal agency has the chief responsibility for registering pesticides and enforcing regulations concerning these chemicals? Do you have an opinion on whether the rules and enforcement are adequate?
5. What are some alternatives to using broad scale, nonspecific pesticides in agriculture or home use?
6. True or False. Ancient civilizations did not have any chemicals to control pests such as weeds or insects.
7. Of all the modern groups of pesticides in use, what three groups or families are the oldest (do not include OCs in your answer)?
8. What pesticide families are: (1) mostly herbicides; (2) mostly insecticides, nematicides; or (3) fungicides? What families have all three types of pesticides?

9. Describe the primary mechanism of action shared by carbamates and OP pesticides. Why might an animal recover more quickly from carbamate exposure than from OP exposure?
10. What happened in Bhopal, India?
11. What is CCD and why is it important to agriculture?
12. True or False. There are no OC pesticides in use within the United States or Canada.
13. Why does the toxicity of many OP pesticides increase once they are in an animal's body?
14. True or False. Glyphosate, the most extensively used herbicide in the United States, has very low toxicity to humans.
15. Based on half-lives in soil at typical conditions, place the following pesticides into one of three categories: Short-lived, Moderately Persistent, Persistent. The pesticides are Boscalid, Imidacloprid, Pentachlorophenol, Metamidaphos, Chlorpyrifos, Glyphosate, Permethrin, Fenvalerate, Atrazine. What cut-off points did you use to distinguish these categories?

REFERENCES

Agency for Toxic Substances and Disease Registry [ATSDR], 2001. Toxicological Profiles for Pentachlorophenol. U.S. Dept Health Human Serv, Atlanta GA.

Agency for Toxic Substances and Disease Registry [ATSDR], 2003a. Toxicological Profile for Pyrethrins and Pyrethroids. U.S. Dept Health Human Serv, Atlanta GA, Available from: <http://www.atsdr.cdc.gov/toxprofiles/tp155.pdf>.

Agency for Toxic Substances and Disease Registry [ATSDR], 2003b. Toxicological Profile for Atrazine. U.S. Dept. Health Human Serv, Atlanta GA, Available from: <http://www.atsdr.cdc.gov/toxprofiles/tp153.pdf>.

Agency for Toxic Substances and Disease Registry [ATSDR], 2008. Toxicological Profile for Dichloropropenes. US Dept Health Human Services, Atlanta GA.

Anderson, J.C., Dubetz, C., Palace, V.P., 2015. Neonicotinoids in the Canadian aquatic environment: a literature review on current use products with a focus on fate, exposure, and biological effects. Sci. Total Environ. 505, 409–422.

Beste, C.F. (Ed.), 1983. Herbicide Handbook of the Weed Science Society of America Weed Sci Soc Am, Champaign IL.

Bishop, C.A., McDaniel, T.V., de Solla, S., 2010. Atrazine in the environment and its implications for amphibians and reptiles. In: Sparling, D.W., Linder, G., Bishop, C.A., Krest, S.K. (Eds.), Ecotoxicology of Amphibians and Reptiles, second ed. SETAC, CRC Press, Pensacola FL, pp. 225–260.

Boone, M.D., Bishop, C.A., Boswell, L.A., Brodman, R.D., Burger, J., Davidson, C., et al., 2014. Pesticide regulation amid the influence of industry. Bioscience 64, 917–922.

Bradberry, S.M., Proudfoot, A.T., Vale, J.A., 2004. Glyphosate poisoning. Toxicol. Rev. 23, 159–167.

California Department of Pesticide Regulation, 2013. Summary of pesticide use report data 2011 indexed by chemical. Availabe from: <http://www.cdpr.ca.gov/docs/pur/pur11rep/chmrpt11.pdf> (accessed 15.04.15.).

Cardone, A., 2015. Imidacloprid induces morphological and molecular damages on testis of lizard (*Podarcis sicula*). Ecotoxicology 24, 94–105.

Carr, J.A., Gentles, A., Smith, E.E., Goleman, W.L., Urquidi, L.J., Thuett, K., et al., 2003. Response of larval *Xenopus laevis* to atrazine: assessment of growth, metamorphosis, and gonadal and laryngeal morphology. Environ. Toxicol. Chem. 22, 396–405.

Chafetz, B., Bilksb, Y.I., Polubesovaa, T., 2004. Sorption–desorption behavior of triazine and phenylurea herbicides in Kishon river sediments. Water Res. 38, 4383–4394.

Crop Life America. Undated. Pesticide regulation. Available from: <http://www.croplifeamerica.org/crop-protection/pesticide-regulation> (accessed 20.07.15.).

Eisler, R., Atrazine. Handbook of Chemical Risk Assessment: Health Hazards to Humans, Plants and Animals. Volume 2. Organics. Lewis Publ., Boca Raton FL, pp. 767–798.

Environmental Working Group [EWG], 2015. EWG's 2015 Shopper's Guide to Pesticides in Produce. Available from: <http://www.ewg.org/foodnews/dirty_dozen_list.php>.

Extoxnet, 1996. Pesticide Information Profiles: Simazine. Available from: <http://extoxnet.orst.edu/pips/simazine.htm> (accessed 20.07.15.).

Fairbrother, A., Purdy, J., Anderson, T., Fell, R., 2014. Risks of neonicotinoid insecticides to honeybees. Environ. Toxicol. Chem. 33, 719–731.

Fischel, F.M., 2015. 2014. Pesticide Toxicity Profile: Organophosphate Pesticides. University of Florida/IFAS Extension. Available from: <http://edis.ifas.ufl.edu/pi087>.

Fishel, F.M., 2015a. Pesticide Toxicity Profile: Carbamate pesticides. Inst Food Ag Sci, Univ FL. Available from: <http://edis.ifas.ufl.edu/pi088>.

Fishel, F.M., 2015b. Pesticide Toxicity Profile: Neonicotinoids. Inst Food Ag Sci, Univ FL. Available from: <http://edis.ifas.ufl.edu/pi117>.

Gervais, J.A., Luukinen, B., Buhl, K., Stone, D., 2010. Imidacloprid Technical Fact Sheet. National Pesticide Information Center., Available from: <http://npic.orst.edu/factsheets/imidacloprid.pdf>.

Giddings, J.M., Anderson, T.A., Hall Jr, L.W., Hosmer, A.J., Kendall, R.J., Richards, R.P., et al., 2005. Atrazine in North American Surface Waters: A Probabilistic Aquatic Ecological Risk Assessment. Society for Environmental Toxicology and Chemistry, Pensacola, FL.

Gilbert, S., 2014. Aldicarb. Toxipedia. Available from: <http://www.toxipedia.org/display/toxipedia/Aldicarb>.

Hayes, T.B., Collins, A., Lee, M., Mendoza, M., Noriega, N., Stuart, A., et al., 2002. Hermaphroditic, demasculinized frogs after exposure to the herbicide atrazine at low, ecologically relevant doses. Proc. Natl. Acad. Sci. USA 99, 5476–5480.

Hayes, T.B., Haston, K., Tsui, M., Hoang, A., Haeffele, C., Vonk, A., 2003. Atrazine-induced hermaphroditism at 0.1 ppb in American leopard frogs (*Rana pipiens*): laboratory and field evidence. Environ. Health Perspect. 111, 568–575.

Hayes, T.B., Stuart, A.A., Mendoza, M., Collins, A., Noriega, N., Vonk, A., et al., 2006. Characterization of atrazine-induced gonadal malformations in African clawed frogs (*Xenopus laevis*) and comparisons with effects of an androgen antagonist (Cyproterone acetate) and exogenous estrogen (17β estradiol): support for the demasculinization/feminization hypothesis. Environ. Health Perspect. 114 (Suppl.), 134–141.

Hecker, M., Geisy, J.P., Jones, P.D., Jooste, A.M., Carr, J.A., Solomon, K.R., et al., 2004. Plasma sex steroid concentrations and gonadal aromatase activities in African clawed frogs (*Xenopus laevis*) from South Africa. Environ. Toxicol. Chem. 23, 1996–2007.

Hof, H., 2001. Critical annotations to the use of azole antifungals for plant protection. Antimicrob. Agents Chemother. 45, 2987–2990.

Hopwood, J., Vaughan, M., Shepherd, M., Biddinger, D., Mader, E., Black, S.H., et al. 2012. Are neonicotinoids killing bees? A review of research into the effects of neonicotinoid insecticides on bees, with recommendations for action. The Xerces Society for Invertebrate Conservation. Available from: <www.xerces.org> (accessed 10.07.15.).

Jewell, T., 1998. Glufosinate fact sheet. Pest News 42, 20–22.

Jones, D.K., Hammond, J.I., Relyea, R.A., 2010. Roundup (R) and amphibians: the importance of concentration, application time, and stratification. Environ. Toxicol. Chem. 29, 2016–2025.

Jones, D.K., Hammond, J.I., Relyea, R.A., 2011. Competitive stress can make the herbicide Roundup (r) more deadly to larval amphibians. Environ. Toxicol. Chem. 30, 446–454.

Kang, G.R., Song, J.Y., Kim, D.S., 2014. Toxicity and effects of the herbicide Glufosinate-Ammonium (Basta) on the Marine Medaka *Oryzias dancena*. Fish. Aquat. Sci. 17, 105–113.

Kavi, L.A.K., Kaufman, P.E., Scott, J.G., 2014. Genetics and mechanisms of imidacloprid resistance in house flies. Pest. Biochem. Physiol. 109, 64–69.

Klocko, A.L., Meilan, R., James, R.R., Viswanath, V., Ma, C., Payne, P., et al., 2014. Bt-Cry3Aa transgene expression reduces insect damage and improves growth in field-grown hybrid poplar. Can. J. Forest Res. 44, 28–35.

Koshlukova, S.E., 2006. Imidacloprid. Risk Characterization Document: Dietary and Drinking Water Exposure. Calif Dept. Pest Reg; Calif Environ Protect Agency., Available from: <http://www.cdpr.ca.gov/docs/risk/rcd/imidacloprid.pdf> (accessed 21.04.15.).

Lajmanovich, R.C., Cabagna-Zenklusen, M.C., Attdemoa, A.M., Jungesa, C.M., Peltzera, P.M., Basso, A., et al., 2014. Induction of micronuclei and nuclear abnormalities in tadpoles of the common toad (*Rhinella arenarum*) treated with the herbicides Liberty® and glufosinate-ammonium. Mutat. Res. Genet. Toxicol. Environ. Mutagen. 769, 7–12.

Laurino, D., Porporato, M., Patetta, A., Manino, A., 2011. Toxicity of neonicotinoid insecticides to honey bees: laboratory tests. Bull. Insectol. 64, 107–113.

Lutnicka, H., Fochtman, P., Bojarski, B., Ludwikowska, A., Formicki, G., 2014. The influence of low concentration of cypermethrin and deltamethrin on phyto- and zooplankton of surface waters. Folia Biologica-Krakow 62, 251–257.

Macek, K.J., Buxton, K.S., Sauter, S., Gnilka, S., Dean, J.W., 1976. Chronic Toxicity of Atrazine to Selected Aquatic Invertebrates and Fishes. US EPA 600/3-76-047. Washington, DC.

Malik, A., 2014. 30 Years after the Bhopal Disaster, India has not learned the lessons of the world's worst industrial tragedy. Intl. Bus. News Available from <http://www.ibtimes.com/30-years-after-bhopal-disaster-india-has-not-learned-lessons-worlds-worst-industrial-1731816> (accessed 10.04.15.).

Mann, R.M., Bidwell, J.R., 1999. The toxicity of glyphosate and several glyphosate formulations on four species of southwestern Australian frogs. Arch. Environ. Contam. Toxicol. 36, 193–199.

Mauck, W.L., Olson, L.E., 1976. Toxicity of natural pyrethrins and five pyrethroids to fish. Arch. Environ. Contam. Toxicol. 4, 18–29.

Maus, C., Schoening, R., Doering, J., 2005. Residues of imidacloprid WG 5 in blossom samples of shrubs of different sizes of the species *Rhododendron* sp. after drenching application in the field. Application 2004, Sampling 2005. Bayer CropScience AG. Report No. G201813.

Minnesota Dept. of Ag., 2014. Pesticide Use in Minnesota. Available from: <http://www.mda.state.mn.us/chemicals/pesticides/pesticideuse.aspx> (accessed 08.04.15.).

Mullin, C.A., Frazier, M., Frazier, J.L., Ashcraft, S., Simonds, R., van Engelsdorp, D., et al., 2010. High levels of miticides and agrochemicals in North American apiaries: implications for honey bee health. PLoS One 5, e9754.

National Pesticide Information Center [NPIC], 2010. Glyphosate technical fact sheet. Available from: <http://npic.orst.edu/factsheets/glyphotech.html> (accessed 21.07.15.).

Palmquist, K., Salatas, J., Fairbrother, A., 2012. Pyrethroid insecticides: use, environmental fate, and ecotoxicology, insecticides. In: Farzana, P. (Ed.), Advances in Integrated Pest Management, pp. 251–278. Available from: <http://www.intechopen.com/books/insecticides-advances-in-integrated-pest-management/pyrethroidinsecticides-use-environmental-fate-and-ecotoxicology>.

Patterson, M., 2004. Metribuzin Analysis of Risks to Endangered and Threatened Salmon and Steelhead. U.S. EPA, Off. Pest. Prog.., Available from: <http://www.epa.gov/oppfead1/endanger/litstatus/effects/metribuzin/metribuzin_analysis.pdf>.

Pesticide Action Network [PAN], 2014. Pesticides Database—Chemicals. Available from: <www.pesticideinfo.org> (accessed 18.04.15.).

Phyu, Y.L., Palmer, C.G., Warne, M.S., Dowse, R., Mueller, S., Chapman, J., et al., 2013. Assessing the chronic toxicity of atrazine, permethrin, and chlorothalonil to the cladoceran *Ceriodaphnia cf. dubia* in laboratory and natural river water. Arch. Environ. Contam. Toxicol. 64, 419–426.

Pimental, D., 2005. Environmental and economic costs of the application of pesticides particularly in the United States. Environ. Develop. Sustain 7, 229–252.

Prosen, H., 2012. Fate and determination of triazine herbicides in soil. In: Hasaneen, N.H. (Ed.), Herbicides—Properties, Synthesis and Control of Weeds InTech, pp. 43–58. Available from: <http://www.intechopen.com/books/herbicides-properties-synthesis-and-control-of-weeds/fateand-determination-of-triaPesticides>in the Nation's Streams and Ground Water, 1992–2001.

Racke, K.D., McGaughey, B.D., 2012. Pesticide regulation and endangered species: moving from stalemate to solutions. In: Racke, K.D., McGaughey, B.D. (Eds.), Symposium on the Endangered Species Act and Pesticide Regulation: Scientific and Process Improvements/242nd Meeting of the American-Chemical-Society, Denver, CO, pp. 3–27. August 30-September 01, 2011. Pesticide regulation and the endangered species act. ACS Symposium Series Volume: 111.

Ragnarsdottir, K.V., 2000. Environmental fate and toxicology of organophosphate pesticides. J. Geol. Soc. 157, 859–876.

Relyea, R.A., 2005. The lethal impact of Roundup on aquatic and terrestrial amphibians. Ecol. Appl. 15, 1118–1124.

Relyea, R.A., Jones, D.K., 2009. The toxicity of Roundup Original Max (r) to 13 species of larval amphibians. Environ. Toxicol. Chem. 28, 2004–2008.

Roh, J.Y., Choi, J.Y., Li, M.S., Jin, B.R., Je, Y.H., 2007. *Bacillus thuringiensis* as a specific, safe, and effective tool for insect pest control. J. Microbiol. Biotechnol. 17, 547–559.

Saghir, S.A., Charles, G.D., Bartels, M.J., Kan, L.H.L., Dryzga, M.D., Brzak, K.A., et al., 2008. Mechanism of trifluralin-induced thyroid tumors in rats. Toxicol. Lett. 180, 38–45.

Sandrock, C., Tanadini, M., Tanadini, L.G., Fauser-Misslin, A., Potts, S.G., Neumann, P., 2014. Impact of chronic neonicotinoid exposure on honeybee colony performance and queen supersedure. PLoS One. http://dx.doi.org/10.1371/journal.pone.0103592

Schlatter, J., Lutz, W.K., 1990. The carcinogenic potential of ethyl carbamate (urethane): risk assessment at human dietary exposure levels. Food Chem. Toxicol. 28, 205–211.

Scholar, J., Krischik, V., 2014. Chronic exposure of imidacloprid and clothianidin reduce queen survival, foraging, and nectar storing in colonies of *Bombus impatiens*. PLoS One Article Number: e91573.

Scorecard, 2011. Federal regulatory program lists. Available from: <http://scorecard.goodguide.com/chemical-groups/one-list.tcl?short_list_name=pest#top> (accessed 17.07.15.).

Soderlund, D.M., Bloomquist, J.R., 1989. Neurotoxic actions of pyrethroid insecticides. Ann. Rev. Entomol. 34, 77–96.

Sparling, D.W., Fellers, G., 2007. Comparative toxicity of chlorpyrifos, diazinon, malathion and their oxon derivatives to larval *Rana boylii*. Environ. Poll. 147, 535–539.

Sparling, D.W., Fellers, G.M., 2009. Toxicity of two insecticides to California, USA, anurans and its relevance to declining amphibian populations. Environ. Toxicol. Chem. 28, 1696–1703.

Sparling, D.W., Bickham, J., Cowman, D., Fellers, G.M., Lacher, T., Matxon, C.W., et al., 2015. In situ effects of pesticides on amphibians in the Sierra Nevada. Ecotoxicology 24, 262–278.

Stokstad, E., 2005. Field research on bees raises concern about low-dose pesticides. Science 335, 1555.

Suchail, S., Guez, D., Belzunces, L.P., 2000. Characteristics of imidacloprid toxicity in two *Apis mellifera* subspecies. Environ. Toxicol. Chem. 19, 1901–1905.

Tavera-Mendoza, L., Ruby, S., Brousseau, P., Fournier, M., Cyr, D., Marecogliese, D., 2002. Response of the amphibian tadpole *Xenopus laevis* to atrazine during sexual differentiation of the ovary. Environ. Toxicol. Chem. 21, 1264–1267.

Thompson, H.M., Fryday, S.L., Harkin, S., Milner, S., 2014. Potential impacts of synergism in honeybees (*Apis mellifera*) of exposure to neonicotinoids and sprayed fungicides in crops. Apidologie 45, 545–553.

Tisler, T., Jemec, A., Mozetic, B., Trebse, P., 2009. Hazard identification of imidacloprid to aquatic environment. Chemosphere 76, 907–914.

United States Department of Agriculture [USDA], 2015. Organic market overview. Available from: <http://www.ers.usda.gov/topics/natural-resources-environment/organic-agriculture/organic-market-overview.aspx> (accessed 08.04.15.).

U.S. Environmental Protection Agency [US EPA], 2000. Bt Plant-Pesticides Risk and Benefit Assessments. SAP Report No. 2000–07, March 12, 2001. Available from: <http://www.epa.gov/scipoly/sap/meetings/2000/october/octoberfinal.pdf> (accessed 20.07.15.).

U.S. Environmental Protection Agency [US EPA], 2003. *Metarhizium anisopliae* strain F52 (029056). Biopesticide Fact Sheet. Available from: <http://www.epa.gov/pesticides/chem_search/reg_actions/registration/fs_PC-029056_01-Jun-03.pdf> (accessed 09.04.15.).

U.S. Environmental Protection Agency [US EPA], 2011. Pyrethrins/Pyrethroids Cumulative Risk Assessment. U.S. EPA Office of Pesticide Programs, Washington DC, Available from: <http://www.regulations.gov/#!documentDetail;D=EPA-HQ-OPP-2011-0746-0003>.

U.S. Environmental Protection Agency [US EPA], 2013. Pyrethroids and pyrethrins. Available from: <http://www.epa.gov/oppsrrd1/reevaluation/pyrethroids-pyrethrins.html> (accessed 08.04.15.).

U.S. Environmental Protection Agency [US EPA], 2014. Chemicals that have completed product registration, October 2014. Available from: <http://www2.epa.gov/pesticide-reevaluation/pesticides-have-completed-product-registration> (accessed 10.04.15.).

U.S. Environmental Protection Agency [US EPA], 2015a. Pesticide industry sales and usage. 2006 and 2007 market estimates. Available from: <http://www.epa.gov/opp00001/pestsales/> (accessed 09.04.15.).

U.S. Environmental Protection Agency [US EPA], 2015b. Pentachlorophenol. Hazard Summary-Created in April 1992; Revised in January 2000. Available from: <http://www.epa.gov/airtoxics/hlthef/pentachl.html>.

U.S. Environmental Protection Agency [US EPA], 2015c. FY2016 Budget. Available from: <http://www2.epa.gov/planandbudget/fy2016> (accessed 20.07.15.).

U.S. Environmental Protection Agency [US EPA], 2015d. 2006–2007 Pesticide market estimates: sales. Available from: <http://www.epa.gov/pesticides/pestsales/07pestsales/sales2007.htm> (accessed 20.07.15.).

Velisek, J., Stara, A., Svobodova, Z., 2011. The effects of pyrethroid and triazine pesticides on fish physiology. In: Stoytcheva, M. (Ed.), Pesticides in the Modern World—Pest Control and Pesticide Exposure and Toxicity Assessment. Available from: <http://www.intechopen.com/books/pesticides-in-the-modern-world-pests-control-and-pesticides-exposure-andtoxicity-assessment/the-effects-of-pyrethroid-and-triazine-pesticides-on-fish-physiology>.

Velisek, J., Kouba, A., Stara, A., 2013. Acute toxicity of triazine pesticides to juvenile signal crayfish (*Pacifastacus leniusculus*). Neuroendrocrinol Lett. 34 (Suppl. 2), 31–36.

World Health Organization[WHO], 2004. Toxicological Evaluations: Imidacloprid; International Programme on Chemical Safety. World Health Organization., Available from: <http://www.inchem.org/jmpr/jmprmono/2001pr07.htm>.

Chapter 6

Halogenated Aromatic Hydrocarbons

Terms to Know

Halogenated
Dioxin
Furan
Polychlorinated Biphenyls
Phenyl
Congener
PCDD
PCDF
Biphenyl
Coplanar
Nonplanar
Ortho-substitution
Aroclor
TCDD
Toxic Equivalent Factor
Toxic Equivalent Quotient
Aryl Hydrogen Receptor
Cytochrome P450
CYP
CYP
EROD
PBB
PBDE
Substituted

INTRODUCTION

In Chapter 4, we discussed the organochlorine pesticides which are one of the highly persistent organic pollutants (POPs). This chapter continues our examination of POPs by focusing on a few groups of halogenated chemicals structurally based on two benzene or phenyl rings. These include polychlorinated biphenyls (PCBs), dioxins, furans, polybrominated diphenyl ethers (PBDEs), and polybrominated biphenyls (PBBs). *Halogenated* means that all of these

Ecotoxicology Essentials.

chemicals have one or more halogen atoms attached to their benzene rings. Halogens are the five elements that fall in group 17 of the periodic chart of elements and require one electron to complete their outer valence shells. All are toxic, form acids when combined with hydrogen, and form a bond with carbon that is difficult to break. The halogens we are most interested in include chlorine (Cl), bromine (Br), and fluorine (F). PCBs, dioxins, and furans have had a long and often intertwined history in manufacturing, research, and concern; thus, we will examine them first and then discuss their brominated counterparts.

INTRODUCTION TO POLYCHLORINATED BIPHENYLS, DIOXINS AND FURANS

PCBs are organic molecules composed of two phenyl (hence biphenyl) or benzene rings with 1–10 chlorine atoms attached. Their general molecular formula is $C_{12}H_{10-x}Cl_x$ where x is the number of chlorines (Fig. 6.1). For each chlorine substitution, a hydrogen atom is removed. There is a maximum of 209 different individual PCBs or *congeners* because the number of sites where chlorine can attach is limited. PCBs are entirely synthetic. Up until the 1970s, they were used for several industrial purposes, mostly associated with cooling transformers and similar devices. Like chlorinated pesticides, PCBs are lipophilic and can both bioaccumulate and bioconcentrate. The lipophilicity and persistence are

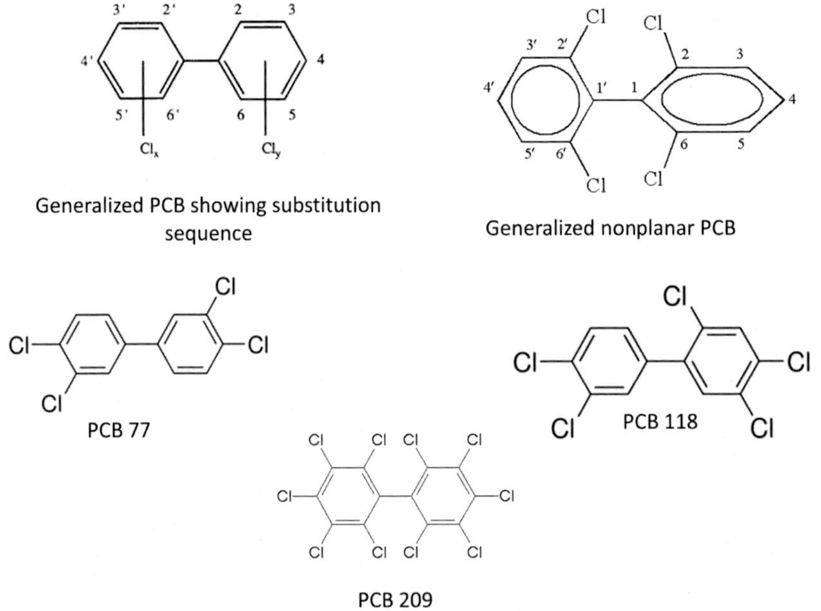

Generalized PCB showing substitution sequence

Generalized nonplanar PCB

PCB 77

PCB 118

PCB 209

FIGURE 6.1 Structural formulas for PCBs including nonplanar, planar, and different degrees of chlorination.

closely related to the number of chlorines in the molecule. Some PCBs cause cancer, endocrine disruption, neurotoxicity, and immunotoxicity in animals and are proven carcinogens in humans.

A small group of PCBs are termed dioxin-like because they behave like dioxins in producing harmful physiological effects. Because of their environmental persistence and suspected human carcinogenicity, PCB production in the United States and several other countries was banned in 1972 and usage was halted in 1979. PCBs were among the POPs recommended to be banned throughout most of the world by the Stockholm Convention in 2001.

Dioxins and furans are structurally very similar to PCBs. Dioxins, more formally known as polychlorinated dibenzo-*p*-dioxins (PCDDs), are chlorinated organic molecules composed of two benzene rings connected to each other via two intermediary bonds with oxygen instead of directly bonded to each other as in PCBs (Fig. 6.2). Due to chlorine substitutions, there are up to 75 PCDDs with seven of these being highly toxic. Furans, also called polychlorinated dibenzofurans (PCDFs) are similar to PCDDs and PCBs because they have two benzene rings, but these are connected by a single oxygen (Fig. 6.2). There are 135 congeners of furans, 10 of which have highly toxic dioxin-like properties.

Both dioxins and furans come from several sources that typically involve high heat conditions. Volcanoes, forest fires, brush fires, the normal processing of sugar cane through burning prior to harvesting, and burning of other biomass are major natural sources of these compounds. Industrial combustion processes also produce substantial dioxins and furans.

Chemical Characteristics of PCBs, Dioxins, and Furans

On each ring of a biphenyl there are potentially five sequentially numbered reactive sites (site # 1 is occupied by the bond connecting the rings in PCBs, Fig. 6.1). If no chlorination occurs, the molecule is simply called biphenyl and the free bonds are occupied by hydrogen atoms. In PCBs, up to 10 chlorine atoms can be "substituted" onto the biphenyl, one at each reactive site. The

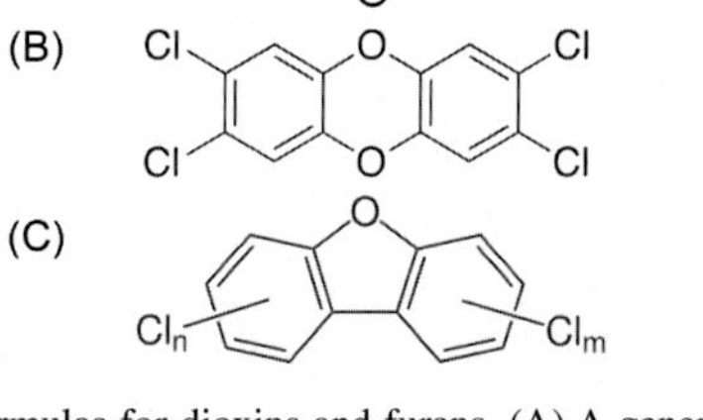

FIGURE 6.2 Structural formulas for dioxins and furans. (A) A generalized dioxin with Cl_n and Cl_m representing the number of chlorine atoms from 1 to 4 on either phenyl. (B) 2,4,7,8-TCDD, the most toxic of dioxins. (C) A generalized furan.

specific congener is given a chemical name based on the number and location of its chlorine atoms. For example, 2,3′,4,4′,5-pentachlorobiphenyl is a PCB with five chlorines (hence the prefix *penta-*) located at sites 2,4, and 5 on the one benzene ring and at sites 3 and 4 on the other ring. Each congener can be placed in a class based on the number of chlorine ions and given a unique number so 2,3′,4,4′,5-pentachlorobiphenyl is a pentachlorobiphenyl identified as PCB 118 (Table 6.1). PCB 209 is the only PCB that is fully substituted by chlorine atoms. The extra bonding between the phenyl groups on dioxins and furans reduces the number of potential congeners, but the naming scheme remains the same as for PCBs.

PCBs can be either *coplanar* or *nonplanar*. Imagine the two benzene rings in space. If the surfaces of the two rings are in the same plane, that is, a 3D representation would have both lie on a sheet of paper, they are coplanar (sometimes simply called planar). Substitutions at the 2 (or 2′) or 6 (6′) positions are called *ortho* substitutions. Ortho substitutions can produce torsion which rotates the benzene rings out of being coplanar and they become nonplanar. Using our previous explanation, one ring could lie flat on a piece of paper, but the other would pierce the axis of the paper. PCBs that do not have any chlorine atoms attached at these sites are referred to as nonortho substituted PCBs. If only one chlorine is attached at any of these sites the PCB is described as being mono-ortho substituted. Nonortho and mono-ortho substitutions give the congeners a specific structure that allows them to react with an enzyme and mimic certain dioxins. Note that, due to the extra bonding between dioxins and furans, these molecules can only be coplanar. Coplanar PCB congeners are dioxin-like because their structure allows them to interact with a specific intracellular protein, the *aryl hydrocarbon* (AH) receptor. The receptor enhances DNA/RNA transcription and affects several regulatory proteins. In simplistic terms, unregulated enhancement of these proteins can really complicate cellular physiology. Thus, dioxins, furans, and dioxin-like PCBs produce many toxic effects. We'll discuss this later in the chapter. There are four nonortho substituted PCBs and eight mono-ortho substituted PCBs that are dioxin-like and of special interest to toxicologists.

The level of chlorination has substantial effects on the chemical nature of all three groups (Table 6.1). PCBs with low chlorination are odorless, tasteless, have moderate viscosity, and a clear to pale yellow color. Highly chlorinated PCBs are more viscous and have a deep yellow color. In their pure form PCDDs and PCDFs are colorless solids. Water solubility of any of these chemicals is very low. While some variation within a group with the same number of chlorine atoms exists, there is a trend for the melting points to increase with degree of chlorination. At room temperatures, PCBs tend to be solids, but some mono- and dichlorophenyls are liquid at room temperature. Dioxins and furans have higher melting points and are solids up to or more than 185°C. PCBs often served as industrial coolants at temperatures higher than their melting points, so in use they may become liquids. Chlorination also decreases water solubility, as

TABLE 6.1 Some Characteristics of Selected PCBs, Dioxins, and Furans

PCB #	Structure	CAS	Log Kow[a]	MW[b]	Solubility ug/L[c]	MP °C	Vapor Pressure (mm Hg)	Air	Half-Life (Years) Water	Sediment
PCBs										
Monochlorobiphenyls (N = 3)										
1	2	2051-61-8	4.60	188	5.5	35	1.11	0.01	0.17	0.6
3	4	2051-62-9	4.40	188	1.2	79	4.96 E-1	0.02	0.33	0.6
Dichlorobiphenyls (N = 12)										
7	2,4	33284-50-3	5.15	223	1.25	25	1.61 E-1	0.02	0.25	0.80
10	2,6	33146-45-1	5.31	223	1.4	36	3.25 E-1	0.01	0.17	0.80
15	4,4'	2050-68-2	5.33	223	0.06	150	8.19 E-2	0.02	0.34	1.45
Trichlorobiphenyls (N = 24)										
18	2,2',5	37680-65-2	5.55	257	0.4	45	9.14 E-2	0.06	0.89	2.13
28	2,4,4'	7012-37-5	5.69	257	0.16	58	2.27 E-2	006	0.89	2.62
37	3,4,4'	38444-90-5	4.94	257	0.015	88	1.81 E-2	0.09	1.27	2.62
Tetrachlorobiphenyls (N = 42)										
47	2,2',4,4'	2437-79-8	6.29	292	0.09	84	9.59 E-3	0.17	2.49	5.71
66	2,3',4,4'	32598-10-0	5.45	292	0.04	125	6.58 E-3	0.29	3.29	5.71
77	3,3',4,4'	32598-13-3	6.52	292	0.001	181	4.92 E-3	0.29	4.01	5.71

(Continued)

TABLE 6.1 Some Characteristics of Selected PCBs, Dioxins, and Furans *Continued*

PCB #	Structure	CAS	Log Kow[a]	MW[b]	Solubility ug/L[c]	MP °C	Vapor Pressure (mm Hg)	Half-Life (Years)			
								Air	Water		Sediment
Pentachlorobiphenyls (N = 46)											
87	2,2′,3,4,5′	38380-02-8	6.37	326	0.004	115	2.72 E-3	0.22	3.09		4.17
101	2,2′,4,5,5′	37680-73-2	7.07	326	0.01	77	3.28 E-3	0.22	3.12		5.11
116	2,3,4,5,6	18259-05-7	6.30	326	0.008	125	4.08 E-3	0.22	2.97		2.28
Hexachlorobiphenyls (N = 42)											
134	2,2′,3,3′,5,6	52704-70-8	7.30	360	0.0004	101	1.32 E-3	0.66	9.03		6.65
153	2,2′,4,4′,5,5′	35065-27-1	7.75	360	0.001	104	5.66 E-4	0.66	9.21		8.56
155	2,2′,4,4′,6,6′	33979-03-2	7.12	360	0.0007	115	1.57 E-3	0.34	4.93		8.56
Heptachlorobiphenyls (N = 24)											
171	2,2′,3,3′,4,4′,6	52663-71-5	6.70	395	0.002	123	1.92 E-4	0.54	7.65		8.56
185	2,2′,3,4,5,5′,6	52712-05-7	7.93	395	0.00045	150	3.47 E-4	0.44	6.31		8.56
Octachlorobiphenyls (N = 12)											
194	2,2′,3,3′,4,4′,5,5′	35694-08-7	8.68	429	0.0002	160	4.74 E-5	2.97	3.83		18.3
202	2,2′,3,3′,5,5′,6,6′	2136-99-4	8.42	429	0.0003	163	1.88 E-4	1.60	22.0		18.3

Nonachlorobiphenyls (N = 3)

206	2,2'3,3',4,4',5,5'6	40186-72-9	9.14	464	0.00011	207	1.83 E-5	3.25	41.5	20.5
208	2,2',3,3',4,5,5'6,6'	52663-77-1	8.16	464	1 E-5	184	3.76 E-5	1.25	17.7	20.5

Decachlorobiphenyls (N = 1)

209	2,2'3,3',4,4',5,5',6,6'	2051-24-3	9.60	498	1.2 E-6	226	9.77 E-6	5.7	68.5	22.8

Dioxins (xxx-dibenzo-p-dioxin)[d]

1-chloro	39227-53-7	4.75	218	0.417	123	9 E-5
2,3-dichloro	29446-15-9	5.60	253	0.0149	164	2.9 E-6
1,2,3,4-tetra	30746-58-8	6.6	322	0.00047	190	4.8 E-8
1,2,3,4,7-penta	39227-28-6	8.64	356	0.00012	196	6.6 E-10
1,2,3,4,7,8-hexa	29227-25-6	9.19	390	0.000004	273	3.8 E-11
1,2,3,4,6,7,8-hepta	35822-46-9	9.69	425	0.000002	265	5.6 E-12
octachloro	3268-87-9	10.07	460	0.00000007	332	8.2 E-13

(Continued)

TABLE 6.1 Some Characteristics of Selected PCBs, Dioxins, and Furans *Continued*

PCB #	Structure	CAS	Log Kow[a]	MW[b]	Solubility ug/L[c]	MP °C	Vapor Pressure (mm Hg)	Half-Life (Years)		
								Air	Water	Sediment
Furans (xxx-chlorodibenzofuran)										
	2,3,7,8-tetra	51201-31-9	5.82	306	0.00042	219	9.21 E-7			
	2,3,4,7,8-penta	57117-31-4	6.92	340	0.00024	225	1.63 E-7			
	1,2,3,6,7,8-hexa	57117-44-9	NA	375	0.000008	225	6.07 E-8			
	1,2,3,4,6,7,8-hepta	67562-39-4	7.92	409	0.000014	236	1.68 E-8			
	1,2,3,4,6,7,8,9-octo	39001-02-0	8.20	444	0.0000012	259	NA			

[a]Eisler (2000).
[b]Shiu and Mackay (1986).
[c]Paasivirta and Sinkkonen (2009).
[d]ATSDR (2000).

seen in both the log K_{ow} values and actual solubilities. In all cases, molecules with few chlorines are moderately soluble compared with many other POPs. However, those with four or more chlorine atoms have very low solubility in water. Chlorination also affects the persistence of the congeners with greater chlorination increasing persistence.

Sources and Uses of PCBs, Dioxins, and Furans

The first PCB-like chemicals were synthesized in 1865 as byproducts from coal tar and were used as sealants in place of tar. In 1881, German chemists produced the first PCBs. Commercial production increased in 1929 when Monsanto Chemical Company, the same manufacturer that produced DDT and Agent Orange during the Vietnam War (along with Dow Chemical Company) and now specializes in genetically modified crops, purchased the rights to produce the chemicals. For more than 40 years, PCBs were widely used in industry as viscous liquids that were effective in keeping electrical transformers cool without being flammable at normal temperatures. They were also used to stabilize various additives in polyvinyl chloride (PVC) plastics and reduce fire hazards. Other uses included casting agents, treating fabrics to be fire resistant, adhesives, paints, coating railroad ties, and components of vacuum pump and hydraulic fluids.

Monsanto blended PCB congeners into commercially sold Aroclors that were distinguished by the degree of chlorination. There were 12 different Aroclors marketed between 1957 and 1971 with chlorine contents ranging from 21 to 68%. Each Aroclor was identified by a four-numbered suffix indicating the number of carbon atoms and the percent of chlorination by mass. For example, Aroclor-1250 had 12 carbons (the usual number for PCBs) and was 50% chlorine. In 1974, the Monsanto Chemical Company produced slightly more than 40 million pounds (18,000 tonnes, or metric tons) of Aroclor mixtures. From 1930 to 1977, an estimated 272,700 tonnes were produced in the United States with an additional 4,204,500 tonnes in Europe (ATSDR, 2000). Under commercial production, PCBs and Aroclors have had several brand names including Asbestol, Askarel, Pyranol, Chlorinol, Saf-T-Kuhl, and Therminol just in the United States and several other names in other countries.

Neither dioxins nor furans are intentionally manufactured except in small quantities for research purposes. Instead, they may be naturally produced through incomplete combustion of organic material during forest fires or volcanic activity. Both are unintentional byproducts of industrial, municipal, and domestic incineration and combustion processes. It is currently believed that dioxin emissions associated with anthropogenic incineration and combustion activities are the predominant environmental source. In addition, dioxins, mainly 2,3,7,8-tetrachlorodibenzo-*p*-dioxin (TCDD), may be formed during the chlorine bleaching process used by pulp and paper mills and they occur during manufacturing certain chlorinated organic chemicals, such as chlorinated phenols. 2,3,7,8-TCDD is a byproduct formed during the manufacture of

2,4,5-trichlorophenol (2,4,5-TCP) which is used to produce hexachlorophene (used to kill bacteria, such as in mouth washes).

The herbicide, 2,4,5-trichlorophenoxyacetic acid (2,4,5-T) contaminated by 2,3,7,8-TCDD was a major component of the herbicide Agent Orange that was widely used to clear jungle growth during the Vietnam war. Thousands of soldiers and Vietnamese came in contact with Agent Orange. Great controversy arose over the effects of this herbicide. It has been repeatedly shown that children in the areas where Agent Orange was used had multiple health problems, including cleft palates, mental disabilities, hernias, extra fingers and toes, and high levels of cancers. In the 1970s, high levels of dioxin were found in the breast milk of South Vietnamese women, and in the blood of US military personnel who had served in Vietnam. The most affected zones were in mountainous areas. That physical abnormalities and diseases including cancer occurred in sprayed areas was not questioned so much as who had the responsibility. The United States argued about the causes of the maladies and how much it was to blame for them. Over the years and hundreds of court cases, the government finally took responsibility, at least for American veterans.

Small amounts of furans enter the environment from several sources. Accidental fires or breakdowns involving capacitors, transformers, and other electrical equipment (eg, fluorescent light fixtures) that contain PCBs can release furans formed by thermal degradation. PCDFs also enter into the environment from burning municipal and industrial waste. Old cars that used leaded gasoline released small amounts of PCDFs in the environment. Small amounts of PCDFs may also enter into the environment from burning coal, wood, or oil for home heating.

Persistence

Table 6.1 lists some estimated half-lives of PCBs in air, water, and sediment or soil. A few words of explanation concerning the data in the table are necessary. Direct measurement of degradation is impractical because of the long half-lives of the highly chlorinated PCBs. Paasivirta and Sinkkonen (2009) modeled hourly degradation rates for all 209 PCBs using available data and structural relationships. We converted their data to years which we felt was more meaningful but, by doing so, we had to use a fairly large correction factor (8760, the number of hours in a year) and this magnified any error in estimated values. Thus, the values are rough estimates to provide some concept of just how persistent PCBs are. There are many environmental factors, both biotic and abiotic, that affect actual half-lives in natural conditions, just like there are with any halogenated organic molecule.

As with other chemical characteristics, degree of chlorination is a very important factor when considering persistence. PCBs in air degrade most rapidly and those with only one to three chlorine atoms have half-lives measured in a few days. In contrast, the half-life of PCB 209 in air is 5–6 years. PCBs are more stable in water than in air and even more stable in soils and sediments than in water. Ultraviolet radiation effectively degrades PCBs and exposure to UV is greatest in air. Low-chlorinated PCBs in sediments have half-lives measured in months, but those with five or more chlorines have half-lives of several years.

Of these environmental sources, water and soil or sediments serve as the primary reservoirs of PCBs (Rice et al., 2003). Although PCBs are not very soluble in water, the vast amounts of salt and freshwater on the planet allow for substantial storage. Even though vapor pressure of PCBs tends to be low, volatilization from sediments and soils is the main way PCBs enter the atmosphere.

The half-life of 2,3,7,8-TCDD ranges from 9 to 15 years in surface soils, 25-100 years in subsurface soils, seven years in humans, and only 8.3 days in air (ATSDR, 1998). Furans are probably very similar, although there is less known about their persistence. As with all the other groups, various environmental factors can greatly alter the persistence. For example, while the whole body half-life in humans is around 8 days, furans can persist in water for 0.4-255 days, depending on the congener and environment (ATSDR, 1994).

Breakdown of PCBs

Despite their persistence, these molecules will break down or dechlorinate eventually. They are slowly metabolized by many living organisms, including mammals, plants, fungi, and bacteria (Tehrani and Van Aken, 2014). One method of natural reduction of PCBs is to convert them to their hydroxylated forms by adding an –OH group. However, hydroxylated PCBs are often more toxic than their parent forms and this step may actually make it more difficult to breakdown the PCB further. OH-PCBs formed by metabolic activity in living tissues can be released into the environment and enter the food chain by excretion, predation, and natural cycling of vegetation (Tehrani and Van Aken, 2014). Hydroxylation can also occur in the atmosphere through chemical binding with free –OH radicals.

The most common methods to remediate heavily contaminated soils and sediments is to mechanically remove the soils and dispose of them in hazard landfills or to cap an onsite landfill, thus containing the PCBs at the contaminated site. The technology to reduce the chlorination of the PCBs in these landfills is improving and often relies on enhancing microbial decay of the molecules. In addition to UV light, under natural conditions a variety of microorganisms will slowly metabolize PCBs, dioxins, and furans in soils or sediments. In fact, many different additives or activities to soil that promote bacterial decay can enhance the dechlorination, even with highly chlorinated congeners. Many of the PCBs in water will eventually deposit into sediments, where they will be buried and removed from contact with organisms. By turning soil over, helping to aerate it, and mixing it with organic matter, even earthworms can accelerate the degradation of PCBs.

As with chlorinated pesticides, plants can take up these contaminants from soils or sediments and then the plants can be disposed of or burnt. Plants can also be used to enhance bacterial decomposition.

PCBs concentrations are diminishing in many environmental sources because they are no longer manufactured or used in the United States or other countries. For example, Buehler et al. (2002) reported on 23 years of monitoring atmospheric PCBs in Lake Michigan and Lake Superior. They observed a general decline in PCB concentration (Fig. 6.3) so that total PCBs dropped by approximately 91% between 1978 and 2000 along Lake Superior.

Examples of PCB Concentrations in Environmental Sources

PCBs can bioconcentrate and biomagnify due to their lipophilic and persistent natures. For instance, the average and maximum congener concentration in oysters was 20 and 60 times higher than in sediments, respectively (Villeneuve et al., 2010, Fig. 6.4). Oysters may obtain some of their contaminant burdens through their food, but they undoubtedly obtain much of it directly from the sediments. Bioconcentration factors (BCF) for specific congeners ranged from 3 for PCB-170 to 60 for PCB-128. Differences in the K_{ow} and K_{oc} values for these PCBs were probably major factors for the differences in BCF.

As an example of biomagnification, Sobek et al. (2010) investigated four PCBs in simplified food chains in the Arctic Barents Sea and in the more temperate environment of the Baltic Sea. In each area, PCB concentrations in water, zooplankton, fish, and seals were measured (Fig. 6.5). BCFs ranged from 18 to 288

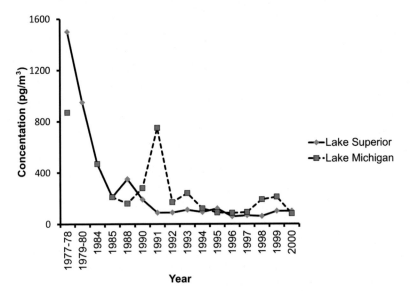

FIGURE 6.3 Decline in atmospheric PCB concentrations in Lake Michigan and Lake Superior from 1977 to 2000. *Buehler et al. (2002).*

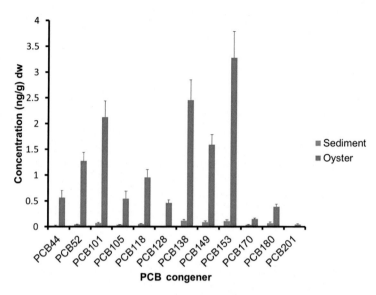

FIGURE 6.4   An example of bioconcentration between sediment and oysters. While the composition by congener is similar between sediments and oysters, oysters were able to bioconcentrate the PCBs by several-fold. *Villeneuve et al. (2010).*

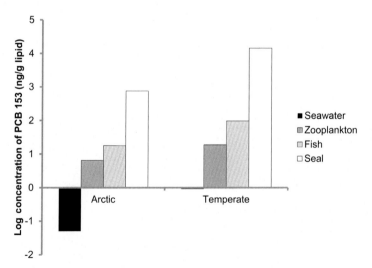

FIGURE 6.5   Bioconcentration dynamics of PCB 153 in the Arctic Sea and the more temperate Baltic Sea. For the Arctic Sea, the dominant invertebrate was the copepod *Calanus glacialis*; the fish was polar cod, *Boreogadus saida*, and the seal was ringed seal, *Pusa hispida*. In the temperate environment, the invertebrates included amphipods, rotifers, and copepods; the predominant fish was the herring (*Clupea harengus*); and seal was gray seal *Halichoerus grupus*. The graph has a log scale so that all trophic levels can be shown. The bioconcentration factors ranged from: (1) 21 and 129 for water to zooplankton; (2) 2.75 to 5.14 for zooplankton to fish; (3) 42 to 147 for fish to seal. *Sobek et al. (2010).*

from saltwater to zooplankton, 2.4 to 7.6 from zooplankton to fish, and 2.3 to 148 from fish to seal. The total BCFs for the Barents Sea food chain from water to seal ranged from 4022 to 15,180, whereas those Baltic Sea ranged from 218 to 21,860.

PCB concentrations are least in air, intermediate in water, and highest in sediments and soil. Hence, atmospheric concentrations of PCBs are usually expressed in pg/m^3 or parts per quintillion; water concentrations are in ng/L or parts per trillion; and soil or sediment concentrations are usually in ng/g or ug/kg—both of which are parts per billion (Table 6.2). Some very contaminated sites, such as landfills, may have soil concentrations in the low mg/kg or parts per million.

TABLE 6.2 Some Representative Concentrations of Total PCBs in Air, Water, and Sediments or Soils

Location	Conditions	Range	Units	Year	Source
Atmosphere					
Argentina	Urban	200 ± 130	pg/m^3	2006–2007	Tombesi et al. (2014)
Argentina	Rural	20 ± 20	pg/m^3		Tombesi et al. (2014)
China	Electronic waste site	7825–76330	pg/m^3	2007–2008	Chen et al. (2014)
China	Rural	24.2–3552	pg/m^3		Chen et al. (2014)
Chile	Rural	40	pg/m^3	2007	Pozo et al. (2012)
Chile	Urban	160	pg/m^3		Pozo et al. (2012)
Chile	Industrial	100–350	pg/m^3		Pozo et al. (2012)
Water					
Pakistan	River Chenab	0.21–27.5	ng/L	2013	Mahmood et al. (2014)
China	Yangtze River	2.02–15.67	pg/L	2010	Zhang et al. (2014)
France	Sewage sludge	0.28–0.70	ng/g	1999–2000	Blanchard et al. (2004)

(Continued)

TABLE 6.2 Some Representative Concentrations of Total PCBs in Air, Water, and Sediments or Soils *Continued*

Location	Conditions	Range	Units	Year	Source
Sediments					
Pakistan	River Chenab	0.83–59.4	ng/g	2013	Mahmood et al. (2014)
Wisconsin	Fox River	0.2–17.6	ng/g	1999	Imamoglu and Christensen (2004)
Sweden	Boat yards	3.0 ± 5.7	ng/g dw	2000–2011	Eklund and Eklund (2014)
United States	Indiana Harbor	12,570	ng/g dw		Liang et al. (2014)
United States	Lake Erie	1.9–245	ng/g	1997–2000	Marvin et al. (2004)
	Lake Ontario	2.6–255	ng/g	1997–2000	Marvin et al. (2004)
	Lake Michigan	220	ng/g	1997–2000	Marvin et al. (2004)
Soil					
China	Solid waste incinerator	413 ± 298	pg/g	NA	Zhang et al. (2014)
	Chemical plant	229 ± 299	pg/g	NA	Zhang et al. (2014)
	Power plant	314 ± 298	pg/g	NA	Zhang et al. (2014)
Tibet	Rural	0.22–2.71	ng/g	2011	

The composition and concentration of PCBs in an environmental source will depend on the source of the PCBs and the Aroclors used in that source, the amount of dechlorination that has occurred, atmospheric deposition of other PCBs, and the type of environmental source. Of the 209 possible polychlorinated biphenyls, only 130 were manufactured commercially (UNEP, 1999). Although the PCB-containing Aroclors that were sold on the market varied in their composition from low to high chlorinated PCBs, most contained at least some of the moderately chlorinated PCBs, from PCB-70 to PCB-140, and these are among the most common congeners found. PCBs with fewer chlorines tend to be more volatile than those with six or more chlorines and tend to be

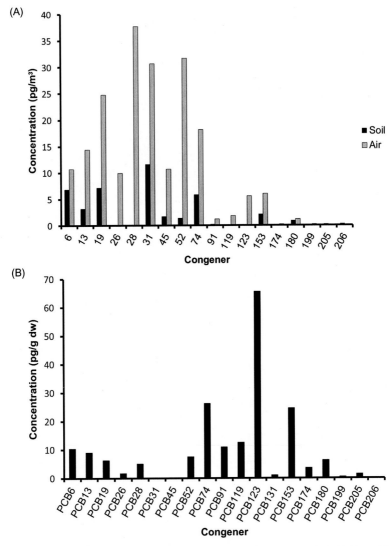

FIGURE 6.6 Concentrations of selected PCB congeners from Mudanya, Turkey. (A) Air, and Soil with Soil concentrations in µg/kg (B) Soils. *Yolsal et al. (2014).*

more common in the atmosphere. Soils and sediments tend to reflect what was used in the region, but older sediments or soils may show some bias towards highly chlorinated PCBs which are less volatile and more persistent. Compare Fig. 6.6B which is a congener profile from Turkish soils to Fig. 6.6A from air above those soils and you'll see that the patterns are not the same. The air profile has predominately low to moderately chlorinated congeners while the soils present a more complete profile of low, moderate, and highly chlorinated congeners.

Concentrations of PCBs in Some Animal Tissues

Eisler (2000) conducted an extensive review of PCB concentrations and effects in organisms up to that time. Changes in PCB concentrations are slow and, if anything, may be lower because PCBs are no longer used—although they are very persistent. We'll use his tables for reference. Concentrations of specific congeners vary predictably by characteristics of the congener, species of concern, and location. As with environmental samples, very low molecular weight congeners will volatize, which may make them unavailable to most organisms. The highest molecular weight congeners are mostly in the soil and sediments and not very mobile so they are less common in organisms. However, the heaviest congeners, if they become assimilated, are also the most persistent and tend to remain in organisms for many years. Generally, middle-weight congeners will most commonly be found in animals and plants. Plants may assimilate PCBs, but the greater concern is animals. Animals with relatively high concentrations of body lipids will tend to have more PCBs stored in their tissues than lean animals. Body concentrations are frequently lipid-adjusted or based on percent or total lipids because of the affinity of PCBs (and the other contaminants described in this chapter) to lipids. It is common sense to suspect that animals collected from contaminated sites will have higher concentrations than those from reference areas.

Marine mammals are the most vulnerable and most likely target of PCBs (Tanabe, 1988; Eisler, 2000). They contain a lot of fat in the form of insulating blubber. In addition, the metabolic potential to degrade organochlorine contaminants is lower in cetaceans than in many terrestrial animals (Tanabe et al., 1987; Kannan et al., 1993). Therefore, we often see lipid-corrected PCB concentrations in these animals at the mg/kg range rather than in the µg/kg range. For example, concentrations in the striped dolphin (*Stenella coeruleoalba*) from the Mediterranean Sea in the 1990s were among the highest reported in the literature (Eisler, 2000). Total PCB concentrations in lipids have been recorded as high as 855 mg/kg and other studies have reported 480 mg/kg fresh weight. Specific congeners ranged from 0.043 to 47 mg/kg lipid weight. Compounding the high overall concentration, approximately 53% of the total PCBs were composed of coplanar congeners (Kannan et al., 1993) with dioxin-like properties that have substantially higher toxicities than other PCBs. The dolphins in that study were experiencing a viral epizootic and it was suspected that the high concentrations of PCBs may have compromised their immune system. In contrast to striped dolphins, studies on other cetaceans and pinnipeds usually report total PCB concentrations in the µg/kg range to 50 mg/kg lipid weight, but killer whales or orca (*Orcinus orca*), at the top of the marine food chain, have been reported with more than 370 mg/kg (Tanabe et al., 1987).

Some fatty fish have fairly high concentrations of PCBs, depending on location. Flounder collected from contaminated waters had 124 to 333 mg/kg total PCBs dry weight whereas those from a reference area had only 4 to 13.3 mg/kg. Lake trout (*Savelinus namayacush*), a top freshwater predator, were reported to

FIGURE 6.7   Fish advisories in the Great Lakes. Note that PCBs are common to all of the lakes. *Courtesy U.S. Environmental Protection Agency. http://www.epa.gov/greatlakes/glindicators/ fishtoxics/sportfishb.html*

have up to 8 mg/kg lipid weight total PCBs in 1977, but this gradually declined to 2.5 mg/kg by 1988 (Borgmann and Whittle, 1991). Fish advisories warning people to limit or cease consuming fish from particular lakes due to high levels of contaminants including PCBs exist for several states in the East and Midwest (Fig. 6.7).

There are some examples that illustrate certain aspects of PCB contamination. Adult male burbots (*Lota lota*) in Lake Erie (Stapanian et al., 2013) had higher PCB concentrations than adult females or juveniles. Lower concentrations in females is often attributed in part to maternal transfer from mother to eggs. Likewise, juveniles have less PCBs than adults due to the accumulation of the contaminant through time. In this case, however, the authors suggested that the higher burdens in males was due to habitat partitioning—males tended to congregate in back bays of Lake Erie that had higher sediment concentrations of PCBs and other organics than the sites used by females and juveniles.

Gender and age differences in PCB concentrations seem to be common among animals. It stands to reason that as a healthy animal ages it will accumulate greater concentrations of PCBs, especially the highly chlorinated, fat soluble, and persistent congeners. Other examples of gender differences may be seen in Fig. 6.8A–C. Adult California sea lion males (*Zalophus californianus*) are around 3.5 times larger than adult females and feed on larger fish, which contributes to their higher body burdens of PCBs (Nino-Torres, Fig. 6.8A); this is in addition to the loss females can experience when pregnant and at parturition. This figure also presents some idea of the difference in PCB concentration

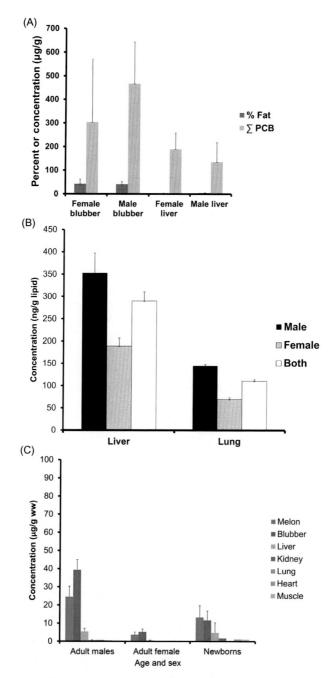

FIGURE 6.8   Distribution of PCB congeners by sex and organ. (A) Liver and blubber in California sea lions in Mexico by percent lipid. (B) Liver and lung in red fox in Poland. (C) Several tissues in striped dolphins from the Mediterranean Sea. *(A) Nino-Torres et al. (2009). (B) Tomaz-Marciniak et al. (2012). (C) Storelli et al. (2012).*

between organs and tissues. Blubber has substantially more fat than liver does and, although both tissues are prime areas for PCB concentration, blubber has a higher PCB concentration than liver. Predictable gender and tissue or organ differences also exist in European red fox (*Vulpes vulpes*, Fig. 6.8B, Tomza-Marciniak et al., 2011) and among striped dolphins in the Mediterranean Sea (Fig. 6.8C, Storelli et al., 2012). Fig. 6.8C also shows variation among organs; both the liver and blubber have high concentrations of lipids.

A study on the little brown bat (*Myotis lucifugus*) (Kannan et al., 2010) focused on a bat colony where the fungal disease white-nose syndrome was prevalent and one where it was not present. White-nose syndrome has devastated many colonies of bats, particularly in the eastern and southern United States. An objective of the study was to determine if organic contaminants were associated with white-nose syndrome, perhaps through compromising bat immune systems. The authors found high concentrations of several halogenated contaminants in both sites and could not conclude that the contaminants were related to the presence of white-nose syndrome. In New York, in a diseased colony lipid weight concentrations of total PCBs ranged from 1900 to 35,000 µg/kg while in a Kentucky reference colony, the concentrations ranged from 17,100 to 18,500 µg/kg lipid weight. On the other hand, total PCBs ranged from 520 to 900 µg/kg lipid weight in New York and 4300 to 13,000 µg/kg in Kentucky.

Recall that there are 209 possible PCB congeners. Even with only 103 manufactured, there is remarkable potential for variation among species and among conditions in the actual congener composition that makes the total PCB burden in an animal. This variation is due in large part to location—what if any PCB Aroclors were used in the region and whatever contribution precipitation of PCB made to this mix. Variation is also due to the specific congeners themselves. Low-chlorinated PCBs that tend to be more water soluble are lost comparatively soon after acquisition, both in the environment and in the organism. Middle-weight congeners were the most widely manufactured and released PCBs when they were used and some may have formed after initial release due to dechlorination of higher weight congeners. Another factor is that many studies that examine congener-specific body burdens do not attempt to quantify the entire range of congeners.

In general, the concentrations of PCBs in abiotic and biotic compartments of the environment are decreasing, but not as a universal phenomenon. Polar bears (*Ursus maritimus*) in the Hudson Bay region actually experienced a 149% increase in $\sum$ PCBs between 1991 and 2003, although this decreased to a wash by 2007 (McKinney et al., 2010). Part of the difference may have been due to sampling issues in that males have much higher PCB concentrations than females and a shift in sex ratios in sampling could make a difference, but it was obvious that the bears are still being exposed to PCBs. Among ivory-billed gulls (*Pagophila eburnea*) living in the Canadian Arctic, PCB concentrations dropped 60% from 1976 to 2004 and dioxins and furans decreased by approximately the same amount although these compounds occur naturally or are still used to a limited extent (Braune et al., 2007, Fig. 6.9). In contrast, however,

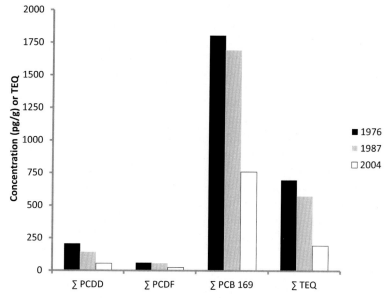

FIGURE 6.9   Concentrations of PCDDs, PCDF, PCBs, and the total toxic equivalents in ivory-billed gull eggs over time. *Braune et al. (2007).*

PCB concentrations did not change between 1991 and 2007 in another polar bear population (Fig. 6.10, McKinney et al., 2010), while DDT concentrations decreased by 60% and PBDEs actually increased by 570% over the same time. Concentrations in human milk seem to be declining in parts of the world. For example, Ryan and Rawn (2014) demonstrated that PCBs in human milk in Canada have declined by 11% between 2002 and 2010. While organochlorine concentrations have decreased in parts of Spain since the 1980s, PCB concentrations are still of concern for some wildlife (Mateo et al., 2012). The air over the Great Lakes is definitely cleaner than it was in the 1970s. Total PCBs dropped 93%, and 91% over Lake Superior and Lake Michigan, respectively (Buehler et al., 2002). The Environmental Protection Agency (2015) has determined that, at least a few years ago, PCBs in Great Lakes commercial fisheries have been declining by 3–7% per year. A few studies have suggested that global climate change involving the melting of Arctic ice may release the PCBs that have been locked there for decades and result in an increase in global concentrations (eg, Bogdal et al., 2013). I guess we'll have to wait and see.

Biological Effects of PCBs

Review the discussion on organochlorine pesticides and plants from Chapter 5 because similar relationships pertain to PCBs and plants. Generally speaking, PCBs are not thought to be toxic to plants because the more toxic, highly chlorinated PCBs including those that have dioxin-like properties are not

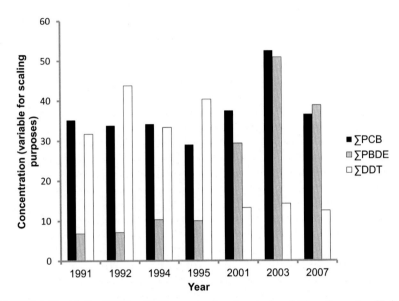

FIGURE 6.10 Median PBDE and PCB concentrations in polar bear fat from the western Hudson Bay, Canada, 1991–2007. PCB concentrations are in ug/g *10 and PBDEs in ng/g in order to put both on the same graph. *McKinney et al. (2010).*

absorbed very well by most plants. Gichner et al. (2007) found that tobacco plants (*Nicotiana tabacum*) grown in soils from a disposal site with heavily contaminated soil (165–265 mg/kg PCBs) had reduced growth, smaller leaves, and increased DNA damage compared with control plants. The authors did not see any increase in actual mutations, however. These effects were considered mild to moderate, especially when the elevated PCB concentrations are considered. I cannot help but think that using this tobacco for smoking would definitely increase one's risk of health problems.

Plants have been used to help remediate contaminated soils because they can assimilate some PCBs. For instance, Viktorova et al. (2014) genetically modified tobacco plants by adding a bacterial gene to produce an enzyme that breaks the diphenyl ring. Genetically modified plants were more tolerant to PCB-contaminated soils than nonmodified plants and they were more efficient in reducing PCB concentrations in a dumpsite.

One way to remove PCBs from aquatic sediments is to dredge the sediments and place them in spoil banks on land; then phytoremediation can be used to remove PCBs. This process is being studied and questions such as ideal plant species to use under various conditions need to be more clearly understood. Plants vary widely in their ability to take up PCBs, suggesting that "weeding" out less efficient species could be beneficial. The common pumpkin (*Curcurbita pepo*) and other members of the squash and gourd family *Curcubitaceace* offer promise in efficient removal of PCBs (Zeeb et al., 2006). *Cucurbits* also have been demonstrated to take up and transport other soil-bound Pops, including DDE, chlordane, dioxins, and furans. Common wisdom suggested that only the more water-soluble PCBs would be taken in by plants, but pumpkin roots, sap, and shoots are unusual in that they can assimilate virtually all PCBs, even those that are highly lipophilic (Greenwood et al., 2011). BCF have been around 5.2 for roots and 1.6 for shoots. Assimilation of PCBs by plants, however, poses some risk if the PCBs are ingested. Obviously, one would not want to eat pumpkins grown on a contaminant landfill (they could still be used in jack-o-lanterns, however), but PCBs can be found in plants used for herbal teas (Amakura et al., 2009) and potentially some vegetables or forage plants affecting livestock.

In Chapter 5, we discussed the benefits of organochlorine pesticides that grizzly bears (*Ursusarctos horribilis*) get from eating plants (Christensen et al., 2013). PCB elimination is also enhanced by herbivory. The authors estimated that inland bears consume 341–1120 µg of PCBs daily when salmon are available. When salmon stop spawning and only plant material is available to inland grizzlies, their herbivorous diets contain a mean of 8.42 µg PCBs per day. During that time, PCB concentrations in grizzly feces may be as high as 4500 times that in their diet and it appears that grizzlies get to reduce substantial body concentrations of PCBs.

Toxicity of Dioxins, Furans, and Dioxin-Like Compounds

Dioxins are highly toxic chemicals that are linked or known to produce a range of effects in humans and other species including cancer, endocrine disruption, neurological disorders, liver toxicity, cardiovascular disorders, genotoxicity, and several other pathological conditions (ATSDR, 1998). The most toxic of the dioxins is 2,3,7,8-tetrachlorodibenzo-*p*-dioxin or 2,3,7,8-TCDD (TCDD for short). When people speak of dioxins, they often mean this particular congener. Animals are usually exposed through the food chain and consuming those animals as part of that food chain is how most humans become exposed to dioxins. Of great concern is that dioxins and other halogenated contaminants can be passed to young mammals and children from their mother through the placenta or from milk. In a study of 418 mothers and their children born between 1997 and 2001 in Denmark, a combination of PCDD, PCDF, and PCBs were sampled and evaluated for their combined toxic equivalency or toxic equivalency quotient (TEQ)—see next section (Wohlfahrt-Veje et al., 2014). Toxic equivalence in milk increased significantly with maternal age and fish consumption. Environmental exposure to dioxin-like chemicals was associated with babies being underweight at birth and with accelerated early childhood growth (rapid catch-up growth).

While we will cover PBDEs in more detail below, it is appropriate to include them in this discussion of maternal transfer. According to Zota et al. (2013), pregnant women at San Francisco General Hospital in 2008–2009 had the highest concentrations of PBDEs and their metabolites ever reported from pregnant women around the world. Another sample of women collected in 2011–2012 showed that concentrations of PBDEs had decreased significantly, especially in the penta- and octapBDEs that were banned by California in 2004. Similarly, there was a modest, insignificant decline in PCB concentrations between 2008–2009 and 2011–2012. The authors concluded that PBDE exposures were likely declining due to regulatory action, but the relative stability in PCB exposures suggests that exposures may eventually plateau and persist for decades.

Other mammals show similar conditions. As one example, female polar bears demonstrated maternal transfer of PCBs and mercury through milk to their cubs (Knott et al., 2012). Total concentrations of 11 PCBs in milk ranged from 160 to 690 µg/kg ww. Although the daily intake levels for PCBs through milk consumption for cubs of the year exceeded the tolerable thresholds, calculated TEQ in milk were below adverse physiological thresholds for aquatic mammals. The authors suggested that relatively high concentrations between nondioxin-like PCBs in polar bear milk and blood could impact endocrine function of Southern Beaufort-Chukchi Sea polar bears.

Maternal transfer can also occur in egg laying animals due, in part, to the high lipid content of yolk. Transfer has been found in all sorts of animals from birds to alligators and marine turtles. One example of maternal transfer can be seen with the eggs of common snapping turtles (Bishop et al., 1995). For PCBs, the first five eggs (mean = 23.9 mg/kg) had 16% more PCB than the last five eggs (mean = 20.1 mg/kg).

General Mechanisms of Toxicity

Among the PCB congeners, only a few have a toxicity worth noting. These are the nonortho and mono-*ortho* substituted PCBs that have dioxin-like properties. There are 12 PCBs with dioxin-like properties, four nonortho substituted congeners (PCB 77, 81, 126, and 169) and eight mono-ortho substituted congeners (PCB105,114,118, 123,156,157,167, 189). The common characteristic of the highly toxic dioxins, and the furans and PCBs with dioxin-like qualities is that they can bind with the AH receptor in animals and exert their toxic effects through this bond. The relative toxicity of dioxin-like chemicals varies considerably. To make some sense of the risk from dioxin-like chemicals, the World Health Organization (WHO) developed the concept of TEQs and factors (Van den Berg et al., 2006). Each dioxin-like compound has a toxic equivalent factor (TEF) and the mixture of dioxin-like compounds is assigned a TEQ based on the sum of TEFs. The TEF is based on the toxicity of 2,3,7,8-TCDD. TCDD is given a TEF of one and the toxicity of all other chemicals are a factor of this. Thus, PCB126 has a TEF of 0.1, which suggests that its toxicity is one-tenth of TCDD. Dioxins other than TCDD, furans, and nonortho-substituted PCBs have TEFs ranging from 0.1 to 0.0003. Mono-ortho-substituted PCBs have much lower toxicity, approximately 0.00003 as toxic as TCDD.

TEFs and TEQ are not intended to be exact figures. Some of the TEFs were developed in vivo from studies with live animals, usually rats. Others were determined from less accurate studies using cell cultures. In addition, many factors, not the least of which is the subject species, can influence toxicity. The real value of these estimates is that animals and people are seldom exposed to only one dioxin-like contaminant. Several of these compounds are most often found together and a subject is exposed to all of these, each with their own toxicity. To evaluate risk, therefore, we need to place some sort of educated guess value on an exposure to these mixtures. It is fairly safe to conclude that a TEQ of five is going to be more toxic than a TEQ of 0.5, but to suggest that the first is exactly 10 times more toxic than the second is stretching the limits of the system quite a bit. It is also fair to say that the toxicity of most dioxin-like PCBs is low to very low compared to TCDD.

Polychlorinated biphenyls, especially those with dioxin-like properties, are much more likely to produce sublethal effects than acute mortality at environmentally realistic concentrations. Concentrations necessary to produce acute mortality are two to three times or higher than most environmental concentrations. However, the sublethal effects of PCBs are numerous. PCBs can cause endocrine disruption, neurotoxicity, immunotoxicity, embryonic developmental problems, teratogenicity, bronchitis, low birth weights in humans, and some genotoxicity at high concentrations. Dioxins are proven carcinogens (WHO, 2010); dioxin-like PCBs have been linked to cancer in animals and appear to produce precancerous conditions in humans. They are considered to be probable carcinogens although there has yet to be direct cause and effect relationship established. The liver is the primary organ of PCB metabolism and *hepatatoxicity* (liver damage), including *hepatamegaly* (liver enlargement) and *necrosis* (cell death or tissue damage) are often seen following PCB exposure in laboratory animals and humans.

Toxicity from dioxin-like PCBs is due to their interaction with the AH receptor. This receptor enhances gene transcription and is involved with many genes. TCDD and to a lesser extent, other dioxin-like compounds bind to the AH receptor. This induces the P450 1A class of enzymes that then break down toxic compounds. During the breakdown process, PCBs may be converted to their even more toxic hydroxylated forms (OH-PCB). Activation of the P450 system can be used to infer exposure to toxic compounds, but the system operates on a broad range of chemicals so exposure to PCBs specifically is not assured.

More on the Cytochrome P450 System

This is probably a good place to discuss the cytochrome P450 system more fully. *Cytochromes* are *hemeproteins* (sometimes also called hemoproteins) containing heme (iron) groups and are primarily responsible for the generation of adenosine triphosphate (ATP) via electron transport. Cytochromes consist of many families that have specific roles in respiration and photosynthesis. Cytochrome P450 or CYPs is a superfamily of hemeproteins that are typically the last of a set of oxidizing enzymes in electron transfer chains. By convention, specific enzymes in the system are referred to by CYP followed by a number such as CYP1A. The genes that code for these enzymes follow the same convention except that they are italicized. The gene *CYP1A,* therefore, codes for the enzyme CYP1A.

CYPs occur throughout living organisms including animals, plants, fungi, single-celled organisms, bacteria, and viruses. More than 18,000 distinct CYPs have been identified across this spectrum. In vertebrates, CYPs can be found in virtually every organ and tissue but those that are involved in the metabolism of toxic compounds such as organochlorine pesticides, PCBs, PBDEs, polycyclic aromatic hydrocarbons, and others have their highest concentrations in the liver. The primary metabolic function of these and most other CYPs is part of what is called Phase I metabolism which involves insertion of one oxygen atom from the O_2 molecule into an organic substrate (eg, PCB) while the other oxygen atom combines with hydrogen to form water:

$$RH + O_2 + NADPH + H^+ \rightarrow ROH + H_2O + NADP^+$$

where R is the organic substrate.

The actual process is quite complex and we will only touch on the basics here. Interested students might want to consult one of the many books recently written on the topic. During this process, various molecules called radicals such as peroxide (O_2^{-2}) and superoxide (O_2^{-1}) can be generated and these radicals can cause *oxidative stress* in cells (Lewis, 2002). Oxidative stress affects several aspects of cellular metabolism and can cause physical damage to cellular components. Several diseases including cancer, muscular dystrophy, autoimmune diseases, emphysema, Parkinson's disease, multiple sclerosis, atherosclerosis, and cancer have been linked to oxidative stress (Halliwell, 1987) and it is reasonable to suggest, if not already demonstrated, that it may also be a cause or at least linked to analogous diseases in animals.

One enzyme, CYP2E, is particularly linked to producing oxidative stress by accentuating the generalized P450 reaction beyond the ability of the cells to remove the harmful byproducts. Whereas *CYP1A* is the primary gene activated by exposure to organic contaminants, *CYP2* genes may also be induced. Moreover, CYP1A can also produce its own toxic effects and further research on this is important. The bottom line is that organisms have developed a system to detoxify chemicals that they would naturally encounter. However, anthropogenic chemicals such as OCPs, PCBs PBDEs, and PAHs can either overwhelm the system or "trick" it into producing metabolites that are many times more toxic than the parent compounds.

General activation of the P450 system is not very predictive in identifying the types of contaminants an organism may be exposed to. Activation of the system and the occurrence of oxidative stress can reveal that an animal has been exposed to some contaminant and may be experiencing pathological effects, but the range of possibilities is huge. Fortunately, there are other enzymes that can be used to narrow down the field of possible contaminants affecting an individual. These enzymes are responsive to certain families of organic contaminants. They will not identify the specific congener or molecule, but they can be used as bioindicators for a group of contaminants. They have been found in most vertebrates in varying concentrations and sensitivities. Ethyoxyresorufin-O-deethylase (EROD) is induced by PAHs, coplanar PCBs, dioxins, and furans in a dose-dependent manner. The activation of EROD in tissues provides evidence for the induction of cytochrome-dependent P450 monooxygenases in the CYP1A subfamily upon exposure to contaminants. From a practical if not oversimplified perspective, while the CYP enzymes assist in the metabolism of the contaminants, EROD and its associated enzymes can be used to denote exposure. Other enzymes that can be used as biomarkers of exposure include benzyloxyresorufin-O-deethylase (BROD) and pentoxyresorufin-O-deethylase (PROD) (Heinonen et al., 1996). BROD and PROD are less sensitive to the types of contaminants that induce CYP1A enzymes and more responsive to those that induce CYPB2B or CYPB3A enzymes. More specifically, EROD and the others are separable in that EROD can be induced or activated by β-napthaflavone whereas PROD and BROD are responsive to phenobarbital. The activity of each of these enzymes can be determined by their ability to breakdown their specific substrates (eg, 7-ethyoxyresorufin for EROD). From a molecular scientist's perspective, that difference provides a good separation between the enzymes.

Because most of the toxicity of PCBs is due to a handful of dioxin-like coplanars, a congener profile can be misleading. For example, take a look at Fig. 6.9 that shows the congener profile of the walrus (Wolkers et al., 2006), European red fox (Mateo et al., 2012), and the great tit (Dauwe et al., 2005). These species were selected simply to show some of the variation in congener profiles. The walrus profile (Fig. 6.11A) was represented by 28 congeners for a total PCB ($\sum$ PCB) concentration of 2000.8 μg/kg based on lipids. Of these congeners, however, PCB 153 accounted for 56% of $\sum$ PCB and the PCBs that had dioxin-like toxicity accounted for only 20% of the total concentration. The TEQ of this profile is 0.003 μg/kg, assuming that the TEFs are reliable for the walrus. The profile for the European red fox (Fig. 6.11B) had 14 congeners measured, but most were very low in concentration for a $\sum$ PCB of 1262 μg/kg. 51% of the $\sum$ PCB was due to congener 180 and dioxin-like PCBs amounted to 70% of the total. The TEQ was 0.017 μg/kg. The great tit had the most homogenous profile with 21 congeners measured and a $\sum$ PCB of 2890 μg/kg lipid weight (Fig. 6.11C). The most common congener, PCB 153, only accounted for 27% of the $\sum$ PCB; 31% of the $\sum$ PCB was in dioxin-like PCBs. The TEQ for great tits was 0.010 μg/kg. Thus, we would have to conclude that the fox was at the greatest risk of the three species.

As briefly mentioned above, the presence of PCBs, dioxins, furans, and other chlorinated and brominated contaminants in human breast milk is a global concern (Porta et al., 2008). In a sample of first-time Irish mothers, Pratt et al. (2012) found that 65% of the total dioxin-like concentration was due to

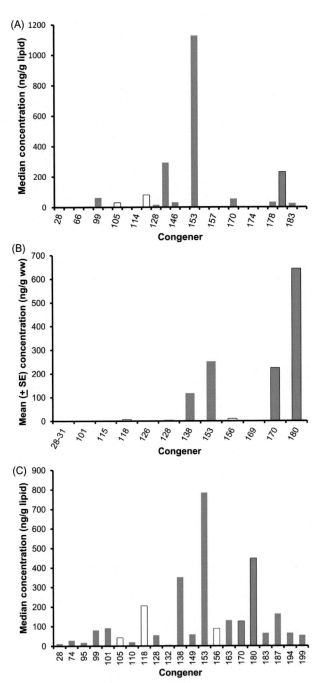

FIGURE 6.11   Congener-specific body burdens of PCBs in animals. (A) Distribution of PCB congers in walrus from Norway. (B) European red fox. (C) Great tit. *(A) Wolkers et al. (2006). (B) Mateo et al. (2012). (C) Dauwe et al. (2005).*

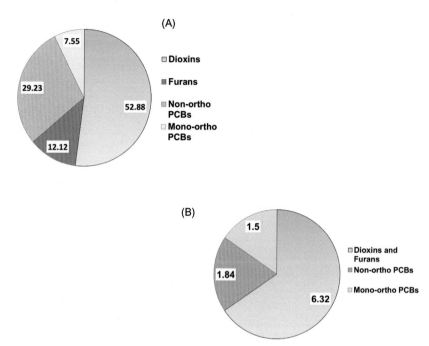

FIGURE 6.12   Concentrations (A) and TEQ (B) of dioxin-like PCBs, dioxins and dioxin-like furans in breast milk of woman from Ireland. Values are pg/g. *Pratt et al. (2012).*

PCDDs and PCDFs (Fig. 6.12A) and that their share of the TEQ (9.66 ng/kg) was a proportional 62% (Fig. 6.12B). As one more example, in a sample of Japanese brown frogs (*Rana japonica*) mono-ortho PCBs accounted for 72% of the dioxin-like substances by concentration, but only 2% of the 5.05 ng/kg ww TEQ (Figs. 6.13A, and B).

POLYBROMINATED DIPHENYL ETHERS AND POLYBROMINATED BIPHENYLS

Instead of chlorine, bromine substitutes for hydrogen on the benzene rings of these chemicals. PBDEs and PBBs (Fig. 6.14) were used as flame retardants in a wide variety of industries including textiles, automotives, building materials, electronics, furnishings, aeronautics, and plastics. They share similar characteristics of persistence, bioconcentration and magnification, and lipophilicity with the other chemicals in this chapter. Both PBDEs and PBBs have 209 congeners because they have the same ring structure as PCBs. Commercially marketed PBDEs were sold as mixtures of several congeners. Penta- and octa-congeners of these brominated compounds are especially persistent and toxic.

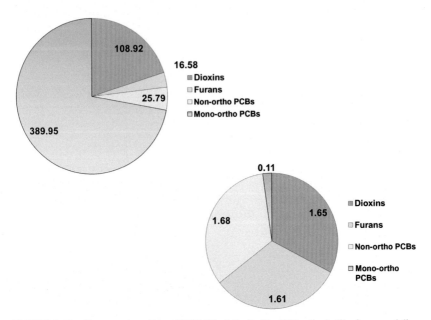

FIGURE 6.13 Concentrations (A) and TEQ (B) of dioxin-like PCBs, dioxin-like furans and dioxins in the Japanese brown frog adults. Values are in pg/g ww. *Kadokami et al. (2002).*

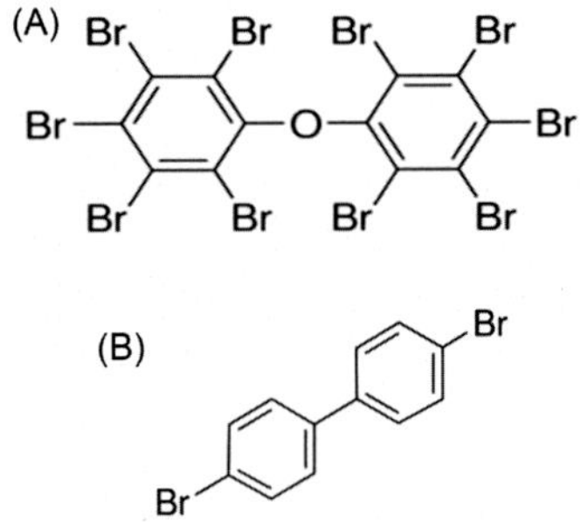

FIGURE 6.14 Structural formulas for (A) Decabromodiphenyl ether, a PBDE and (B) 4,4′-Dibromobiphenyl, a PBB.

PBBs were first voluntarily removed from the market by their manufacturer in the 1970s due to a major accident involving thousands of livestock. In early 1973, both PBB (sold under the trade name FireMaster®) and magnesium oxide (a cattle feed supplement sold under the trade name NutriMaster®) were produced at the same St Louis, Michigan plant by the Michigan Chemical Company. A shortage of preprinted paper bag containers led to 10–20 fifty pound bags of PBB accidentally being sent to the Michigan Farm Bureau Services in place of

NutriMaster®. This accident was not recognized until long after the bags had been shipped to feed mills and used in the production of feed for dairy cattle. By the time the mistake was discovered in April 1974, PBB had entered the food chain through milk and other dairy products, beef products, and contaminated swine, sheep, chickens, and eggs. As a result of this incident, more than 500 contaminated Michigan farms were quarantined, and approximately 30,000 cattle, 4500 swine, 1500 sheep, and 1.5 million chickens were killed, along with more than 800 tons of animal feed, 18,000 pounds of cheese, 2500 pounds of butter, 5 million eggs, and 34,000 pounds of dried milk products (Michigan Dept. Public Health, 2011). While this may seem like ancient history, it is not beyond imagination that a similar accident could occur again with similar compounds. The manufacture of PBB was banned in the United States in 1976 after this catastrophe (US EPA, 2014).

In 2009, both PBDEs and PBBs were listed as POPs by the Stockholm Convention which has effectively reduced their production throughout much of the world. In the United States, bans on PBDEs are being applied on a state-by-state basis. Penta- and octaPBDEs, which are thought to be carcinogenic and are mobile in the environment, were the chief targets in these rules and banned from production in 2004. DecaPBDE was banned from production in 2010. As of 2015, 13 states had implemented regulations ranging from setting upper allowable limits of PBDEs in manufactured goods to total bans (NCSL, 2014). DecaPBDE has also been banned by the European Union. Manufacturing of PBDEs does not occur in Canada, but the use of penta- and oxyPBDEs and decaPBDEs are restricted. The purpose of PBBs and PBDEs—to provide protection against fire—is still of concern and other brominated and organophosphate compounds are being used instead.

FOCUS—IN SITU TESTING WITH TREE SWALLOWS

Problem—Determine Whether the Contaminants in Wetlands Bioaccumulate and If They Have Any Effect on Wildlife Using Them

How would you go about addressing the problem described above? Undoubtedly there are a lot of ways, including doing expensive sampling of sediments, water, invertebrates, and wildlife near or in the wetlands. Unless you have your own analytical laboratory, extensive sampling is costly and the mere presence of contaminants, even in the wildlife species of interest, does not necessarily denote that they are affecting the animals. Organochlorines, for example, could be sequestered in fatty tissues where they have minimal consequences for healthy animals.

One way biologists have addressed this problem is to use avian species as sentinels. The choice of a sentinel species is complex. If you are testing the toxicity of a group of wetlands, the sentinel must extensively, preferably exclusively, use a specific wetland or small group of wetlands. It wouldn't be desirable to have animals using a wide variety of ponds because you couldn't be sure where the contaminants came from. Thus, the sentinel species should be rather sedentary. It might be highly desirable to sample the animals repeatedly throughout the course of a year and, if you could rely on at least some of the animals coming back in subsequent years, that might even be better. It would be ideal if the sentinel species provided some way for sampling without causing the death of the animal itself—to do minimally invasive procedures that would reduce stress and not hamper the survival of the animal or affect the population. You might also want to be reasonably sure of seeing some effects, at least in the wetlands you know to be most highly contaminated. The sentinel species should also have a widespread distribution so that similar studies can be conducted in many different localities.

There are some species that meet all of these criteria. Several researchers have studied freshwater turtles by collecting blood or sometimes euthanatizing them to collect organ samples (eg, Bishop et al., 1995). For the most part, turtles are sedentary and can be representative of a wetland. However, they also seem to be very hardy. While they bioaccumulate and even bioconcentrate some contaminants, there have been few studies that report adverse effects from these contaminants. Other species that may serve as sentinels include crocodilians and colonial species of birds. In these species, however, the investigator is pretty much restricted to studying where the animals are and has limited ability to study wetlands of his or her choice. Birds also migrate and contaminant residues obtained from their different residences may confound interpretations.

Another group of scientists have been using tree swallows (*Tachycineta bicolor*, Fig. 6.15) as sentinels. Tree swallow breeding distribution extends from central Alaska to central California and east to the Atlantic States, covering

FIGURE 6.15 Tree swallow fledgling poking its head out of its nesting box, almost ready to fly. *Courtesy Steve Byland. www.stevebyland.com.*

virtually all of Canada and the northern half of the United States. They readily use nesting boxes for breeding and can be attracted to them with relative ease. Tree swallows are almost exclusively insectivores and during the breeding season, the birds capture insects as they emerge from surface waters. They will usually feed within 500 m of their nest (Quinney and Ankney, 1985) and are opportunistic, so that if the wetland nearest their nest provides sufficient invertebrate food, they will feed extensively from that wetland. Tree swallows are gregarious and will nest in boxes that are closely spaced, which helps in ensuring an adequate statistical replication although this has been a problem in some studies. Finally, tree swallows lay four to seven eggs—this allows two eggs, the mass necessary to obtain reliable analyses, to be sacrificed for residue analysis while allowing the others to develop and hatch so that developmental morphology, hatching, and fledgling success can be compared to residue concentrations.

We mentioned that egg order is sometimes important when measuring contaminants—for some species the first laid eggs may have higher or lower concentrations than subsequent eggs in a clutch. Tree swallows show some of that depending on the specific contaminant (Custer et al., 2010). The authors did not find any difference in egg volume, weight, percent moisture, or percent lipids by laying order. Among the contaminants measured, PCB concentrations varied by as much as 60% from one egg to another within a clutch. The importance of egg sequence varied between years and even between clutches that were highly and lightly contaminated. Ultimately, it appears that the importance of egg sequence cannot be predicted ahead of time so the authors recommended randomly collecting two eggs per clutch and pooling their contents for analysis.

There have been many studies that used tree swallows in this type of experimental presentation and we cannot review them all. Instead, we will examine a few to provide enough information to develop an understanding of the process.

Pulp mills, factories that take trees and make them into paper, historically used many different chemicals in the process and contaminated many water bodies when they dumped their effluents into streams and rivers. In the 1980s and 1990s, many of these mills were either closed down or upgraded to reduce their levels of pollution. Tree swallows were used to evaluate the effects of these upgrades in British Columbia (Harris and Elliott, 2000). The investigators set up nesting boxes upstream and downstream of pulp mills along the Fraser and South Thompson Rivers in central Canada in 1993 and 1994. They found low concentrations of PCBs, PCDDs, and PCDFs. While there were no significant differences between upstream and downstream tissue concentrations, nestlings downstream on the Fraser River had the highest concentrations of organochlorines and those downstream on the south Thompson River had the highest TEQs. Nest success was lower downstream than upstream on both rivers but the authors found little direct evidence of specific contaminant effects. Downstream parents were less attentive than upstream birds, which might have been due to organochlorine toxicity in adults. Reduced attentiveness could also occur if there were fewer insects in downstream sites and this may or may not have been due to the contaminants. However, the authors could not make strong inferences on the effects of these contaminants on breeding success with only one year of study.

Also in Canada, Point Pelee National Park in Ontario has some hotspots of organochlorine contamination (Papp et al., 2005). Tree swallow nesting studies were used to determine that those birds nesting in the more PCB-contaminated sites had higher EROD activities than those that nested at less contaminated sites. However, reproductive success was not related to the level of contamination.

Synthesis of retinoids, including vitamin A, is disrupted by chlorinated hydrocarbon contaminants which can cause a series of problems related to vitamin A-mediated enzyme reactions. Corticosterone is a hormone whose elevated levels indicate stress. Using the tree swallow method, Martinovic et al. (2003) found that PCDD concentrations ranged from 5.4 pg/g to 79.5 pg/g in fledgling swallows along the St. Lawrence River in Canada and the United States. Various measures of retinoid correlated with PCDDs, causing the authors to conclude that PCDD was interfering with vitamin A pathways. Corticosterone was also elevated in highly contaminated sites, indicating that PCDD affected stress and the glucocorticoid axis that produces corticosterone. Both sublethal effects may result in numerous metabolic pathways that could compromise survival or reproduction.

The tree swallow egg removal procedure is flexible. For example, it has been used to compare physiological health among commercial apple orchards and reference sites (Bishop et al., 1998). In this case, pesticides such as pyrethroids and fungicides were the primary contaminants. Swallows in the orchards experienced slight anemia, somewhat impaired immune systems, and altered thymus and thyroid responses compared with birds from reference areas.

In addition to tree swallows, the egg removal for analysis accompanied by examination of the reproductive effects has been adapted to spotted sandpipers (*Actitis macularia*), belted kingfishers (*Megaceryle alcyon*, Custer et al., 2010), night herons (*Nycticorax ncyticorax*, Rattner et al., 2000), wood ducks (*Aix sponsa*, Augspurger et al., 2008), great blue herons (*Ardea herodias*, Baker and Sepulveda, 2009), and other species.

STUDY QUESTIONS

1. What halogens are most often associated with halogenated hydrocarbons?
2. Why are the number of congeners for PCBs, PBBs, and PBDEs limited to exactly 209? Do investigators have to look for all 209 PCB congeners when studying PCB contamination? Why or why not?
3. Of the molecular structures shown below, which is a PCB, PCDD, and PCDF?

(A) 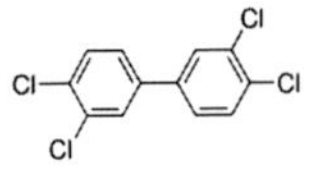   (B) 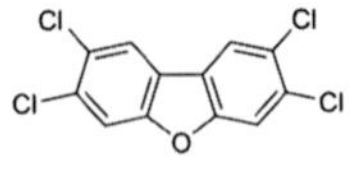   (C)

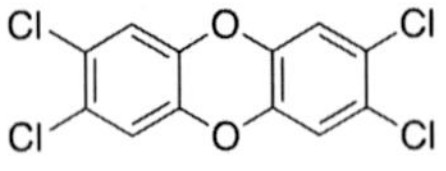

4. List the specific chemical formula names for the three molecules in question 3.

5. What factor on the PCB molecule determines if it is coplanar or not? Why does the molecular structure of coplanar PCBs make them function like dioxins?

6. Trace the changes in behavioral characteristics assumed by PCBs as the degree of chlorination increases.

7. Do dioxins occur naturally in the environment?

8. Most soil that is contaminated with PCBs or dioxins is physically removed and stored in landfills. Discuss some environmental problems associated with this type of disposal.

9. Why is TCDD the standard for toxicity comparison among dioxins and dioxin-like PCBs and furans?

10. Provide some reasons why male mammals tend to have higher PCB concentrations in their tissues than female mammals.

11. Explain how the P450 cytochromes can increase the toxicity of PCBs and related molecules.

12. What features would you look for in a sentinel species?

REFERENCES

Amakura, Y., Tsutsumi, T., Tanno, K., Nomura, K., Yanagi, T., Kono, Y., et al., 2009. Dioxin concentrations in commercial health tea materials in Japan. J. Health Sci. 55, 290–293.

ATSDR Agency for Toxic Substances and Disease Registry, 1994. Toxicological Profile for Chlorodibenzofurans. U.S. Department of Health and Human Services, Public Health Dept., Atlanta, GA.

ATSDR Agency for Toxic Substances and Disease Registry, 1998. Toxicological Profile for Chlorinated Dibenzo-*p*-dioxins. U.S. Department of Health and Human Services, Public Health Department, Atlanta, GA.

ATSDR Agency for Toxic Substances and Disease Registry, 2000. Toxicological Profile for Polychlorinated Biphenyls (PCBs). U.S. Department of Health and Human Services, Public Health Dept., Atlanta, GA.

Augspurger, T.P., Tillitt, D.E., Bursian, S.J., Fitzgerald, S.D., Hinton, D.E., Di Giulio, R.T., 2008. Embryo toxicity of 2,3,7,8-tetrachlorodibenzo-*p*-dioxin to the wood duck (*Aix sponsa*). Arch. Environ. Contam. Toxicol. 55, 659–669.

Baker, S.D., Sepulveda, M.S., 2009. An evaluation of the effects of persistent environmental contaminants on the reproductive success of Great Blue Herons (*Ardea herodias*) in Indiana. Ecotoxicology 18, 271–280.

Bishop, C.A., Lean, D.R.S., Brooks, R.J., Carey, J.H., Ng, P., 1995. Chlorinated hydrocarbons in early life stages of the common snapping turtle (*Chelydra serpentina serpentina*) from a coastal wetland on Lake Ontario, Canada. Environ. Toxicol. Chem. 14, 421–426.

Bishop, C.A., Ng, P., Pettit, K.E., Kennedy, S.W., Stegeman, J.J., Norstrom, R.J., et al., 1998. Environmental contamination and developmental abnormalities in eggs and hatchlings of the common snapping turtle (*Chelydra serpentina* serpentina) from the Great Lakes St Lawrence River basin (1989-91). Environ. Pollut. 101, 143–156.

Blanchard, M., Teil, M.J., Ollivon, D., Legenti, L., Chevreuill, M., 2004. Polycyclic aromatic hydrocarbons and polychlorinated biphenyls in wastewater and sewage sludges from the Paris area (France). Environ. Res. 95, 184–197.

Bogdal, C., Abad, E., Abalos, M., van Bavel, B., Hagberg, J., Scheringer, M., et al., 2013. Worldwide distribution of persistent organic pollutants in air, including results of air monitoring by passive air sampling in five continents. Trac-trends Anal. Chem. 46, 150–161.

Borgmann, U., Whittle, D.M., 1991. Contaminant concentration trends in Lake Ontario Lake trout (Salvelinus namaycush)—1977 to 1988. J. Great Lakes Res. 17, 368–381.

Braune, B.M., Mallory, M.L., Gilchrist, H.G., Letcher, R.J., Drouillard, K.G., 2007. Levels and trends of organochlorines and brominated flame retardants in ivory gull eggs from the Canadian Arctic, 1976 to 2004. Sci. Total Environ. 378, 403–417.

Buehler, S.S., Basu, I., Hites, R.A., 2002. Gas-phase polychlorinated biphenyl and hexachlorocyclohexane concentrations near the Great Lakes: a historical perspective. Environ. Sci. Technol. 36, 5051–5056.

Chen, S.J., Tian, M., Zheng, J., Zhu, Z.C., Luo, Y., Luo, X.J., et al., 2014. Elevated levels of polychlorinated biphenyls in plants, air and soils at an E-waste site in southern China and enantioselective biotransformation of chiral PCBs in plants. Environ. Sci. Technol. 48, 3847–3855.

Christensen, J.R., Yunker, M.B., MacDuffee, M., Ross, P.S., 2013. Plant consumption by grizzly bears reduces biomagnification of salmon-derived polychlorinated biphenyls, polybrominated diphenyl ethers, and organochlorine pesticides. Environ. Toxicol. Chem. 32, 995–1005.

Custer, C.M., Custer, T.W., Dummer, P.M., 2010. Patterns of organic contaminants in eggs of an insectivorous, an omnivorous, and a piscivorous bird nesting on the Hudson River, New York, USA. Environ. Toxicol. Chem. 29, 2286–2296.

Dauwe, T., Jaspers, V., Covaci, A., Scheppens, P., Eens, M., 2005. Feathers as a nondestructive biomonitor for persistent organic pollutants. Environ. Chem. Toxicol. 24, 442–449.

Eisler, R., 2000. Polychlorinated biphenyls. In: Eisler, R. (Ed.), Handbook of Chemical Risk Assessment: Health Hazards to Humans, Plants and Animals. Vol. 2. Organics Lewis Publishers, Boca Raton, pp. 1237–1341.

Eklund, B., Eklund, D., 2014. Pleasure boatyard soils are often highly contaminated. Environ. Manage. 53, 930–946.

Gichner, T., Lovecka, P., Kochankova, L., Mackova, M., Demerova, K., 2007. Monitoring toxicity, DNA damage and somatic mutations in tobacco plants growing in soil heavily polluted with polychlorinated biphenyls. Mutat. Res. Gen. Toxciol. Environ. Mut. 629, 1–6.

Greenwood, S.J., Rutter, A., Zeeb, B.A., 2011. The absorption and translocation of polychlorinated biphenyl congeners by Cucurbita pepo spp. Pepo. Environ. Sci. Technol. 45, 6511–6516.

Halliwell, B., 1987. Oxidants and human diseases: some new concepts. FASEB J. 1, 358–364.

Harris, M.L., Elliott, J.E., 2000. Reproductive success and chlorinated hydrocarbon contamination in tree swallows (Tachycineta bicolor) nesting along rivers receiving pulp and paper mill effluent discharges. Environ. Pollut. 110, 307–320.

Heinonen, J.T., Sidhu, J.S., Reilly, M.T., Farin, F.M., Omiecinski, C.J., Eaton, D.L., et al., 1996. Assessment of regional cytochrome P450 activities in rat liver slices using resorufin substrates and fluorescence confocal laser cytometry. Environ. Health Perspect. 104, 536–543.

Imamoglu, I., Christensen, E.R., 2004. PCB sources, transformations, and contributions in recent Fox River, Wisconsin sediments determined from receptor modeling. Water Res. 36, 3449–3462.

Imamoglu, I., Li, K., Christensen, E.R., Mcmullin, J.K., 2004. Sources and dechlorination of polychlorinated biphenyl congeners in the sediments of the Fox River, Wisconsin. Environ. Sci. Technol. 38, 2574–2583.

Kadokami, K., Takeishi, M., Kuramato, M., Ono, Y., 2002. Congener-specific analysis of polychlorinated dibenzo-*p*-dioxins, dibenzofurans, and coplanar polychlorinated biphenyls in frogs and their habitats, Kitakyushu, Japan. Environ. Toxicol. Chem. 21, 129–137.

Kannan, N., Tanabe, S., Borrell, A., Aguilar, A., Focardi, S., Tatsukawa, R., 1993. Isomer-specific analysis and toxic evaluation of polychlorinated biphenyls in striped dolphins affected by an epizootic in the western Mediterranean Sea. Arch. Environ. Contam. Toxicol. 25, 227–233.

Kannan, K., Yun, S.H., Rudd, R.J., Behr, M., 2010. High concentrations of persistent organic pollutants including PCBs, DDT, PBDEs and PFOs in little brown bats with white-nosed syndrome in New York, USA. Chemosphere 80, 613–618.

Knott, K.K., Boyd, D., Ylitalo, G.M., O'Hara, T.M., 2012. Lactational transfer of mercury and polychlorinated biphenyls in polar bears. Chemosphere 88, 395–402.

Lewis, D.F.V., 2002. Oxidative stress: the role of cytochrome P450 in oxygen activation. J Chem. Technol. Biotechnol. 77, 1095–1100.

Liang, Y., Martinez, A., Hornbuckle, K.C., Mattes, T.E., 2014. Potential for polychlorinated biphenyl biodegradation in sediments from Indian Harbor and Ship Canal. Int. Biodeterior. Biodegradation 89, 50–57.

Mahmood, A., Malik, R.N., Li, J., Zhang, G., 2014. Levels, distribution profile, and risk assessment of polychlorinated biphenyls (PCBs) in water and sediment from two tributaries of the River Chenab, Pakistan. Environ. Sci. Pollut. Res. 21, 7847–7855.

Marvin, C., Painter, S., Williams, D., Richardson, V., Rossmann, R., Van Hoof, P., 2004. Spatial and temporal trends in surface water and sediment contamination in the Laurentian Great Lakes. Environ. Pollut. 129, 131–144.

Martinovic, B., Lean, D.R.S., Bishop, C.A., Birmingham, E., Secord, A., Jock, K., 2003. Health of tree swallow (*Tachycineta bicolor*) nestlings exposed to chlorinated hydrocarbons in the St. Lawrence River Basin. Part 1. Renal and hepatic vitamin A concentrations. J. Toxicol. Environ. Health A 66, 1053–1072.

Mateo, R., Millàn, J., Rodríguez-Estial, J., Camarero, P.R., Palomares, F., Ortiz-Santaliestra, M.E., 2012. Levels of organochlorine pesticides and polychlorinated biphenyls in the critically endangered Iberian lynx and other sympatric carnivores in Spain. Chemosphere 86, 691–700.

McKinney, M.A., Stirling, I., Lunn, N.J., Peacock, E., Letcher, R.J., 2010. The role of diet on long-term concentrations and pattern trends of brominated and chlorinated contaminants in western Hudson Bay polar bears, 1991-2007. Sci. Total Environ. 408, 6210–6222.

Michigan Dept. Public Health, 2011. PBBs (Polybrominated Biphenyls) in Michigan: Frequently Asked Questions—2011 update. Avaialble from: <https://www.michigan.gov/documents/mdch_PBB_FAQ_92051_7.pdf> (accessed 12.08.15.).

NCSL National Conference of State Legislators, 2014. State Regulation of Flame Retardants in Consumer Products. Available from: <http://www.ncsl.org/research/environment-and-natural-resources/flame-retardants-in-consumer-products.aspx> (accessed 12.08.15.).

Nino-Torres, C.A., Gardner, S.C., Zenteno-Savin, T., Ylitalo, G.M., 2009. Organochlorine pesticides and polychlorinated biphenyls in California Sea Lions (*Zalophus californianus californianus*) from the Gulf of California, Mexico. Arch. Environ. Contam. Toxicol. 56, 350–359.

Paasivirta, J., Sinkkonen, S.I., 2009. Environmentally relevant properties of all 209 polychlorinated biphenyl congeners for modeling their fate in different natural and climatic conditions. J. Chem. Eng. Data 54, 1189–1213.

Papp, Z., Bortolotti, G.R., Smits, J.E.G., 2005. Organochlorine contamination and physiological responses in nestling tree swallows in Point Pelee National Park, Canada. Arch. Environ. Contam. Toxicol. 49, 563–568.

Porta, M., Puigdomenech, E., Ballester, F., Selva, J., Ribas-Fito, N., Dominguez-Boadas, L., et al., 2008. Studies conducted in Spain on concentrations of persistent toxic compounds. Gaceta Sanit. 22, 248–266.

Pozo, K., Harner, T., Rudolph, A., Oyola, G., Esellano, V.H., Ahumada-Rudolph, R., et al., 2012. Survey of persistent organic pollutants (POPs) and polycyclic aromatic hydrocarbons (PAHs) in the atmosphere of rural, urban and industrial areas of Concepción, Chile, using passive air samplers. Atmos. Pollut. Res. 3, 426–434.

Pratt, I.S., Anderson, W.A., Crowley, D., Daly, S.F., Evans, R.I., Fernandes, A.R., et al., 2012. Polychlorinated dibenzo-*p*-dioxins (PCDDs), polychlorinated dibenzofurans (PCDFs) and polychlorinated biphenyls (PCBs) in breast milk of first-time Irish mothers: impact of the 2008 dioxin incident in Ireland. Chemosphere 88, 865–872.

Quinney, T.E., Ankney, C.D., 1985. Prey size selection by tree swallows. Auk 102, 245–250.

Rattner, B.A., Hoffman, D.J., Melancon, M.J., Olsen, G.H., Schmidt, S.R., Parsons, K.C., 2000. Organochlorine and metal contaminant exposure and effects in hatching black-crowned night herons (*Nycticorax nycticorax*) in Delaware Bay. Arch. Environ. Contam. Toxicol. 39, 38–45.

Rice, C.P., O'Keefe, P.W., Kubiak, T.J., 2003. Sources, pathways, and effects of PCBs, dioxins, and dibenzofurans. In: Hoffman, D.J., Rattner, B.A., Burton Jr., G.A., Cairns Jr., J. (Eds.), Handbook of Ecotoxicology. Lewis Publishers, Boca Raton, FL, pp. 501–573.

Ryan, J.J., Rawn, D.F.K., 2014. The brominated flame retardants, PBDEs and HBCD, in Canadian human milk samples collected from 1992 to 2005; concentrations and trends. Environ. Int. 70, 1–8.

Shiu, W.Y., Mackay, D., 1986. A critical review of aqueous solubilities, vapor pressure, Henry's law constants, and octanol-water partition coefficients of the polychlorinated biphenyls. J. Phys. Chem. Ref. Data 15, 911–929.

Sobek, A., McLachlan, M.S., Borgå, K., Asplund, L., Lundstekdt-Enkel, K., Polder, A., et al., 2010. A comparison of PCB accumulation factors between an arctic and a temperate marine food web. Sci. Total Environ. 408, 2753–2760.

Stapanian, M.A., Nadenjian, C.P., Rediske, R.R., O'Keefe, J.P., 2013. Sexual difference in PCB congener distributions of burbot (*Lota lota*) from Lake Erie. Chemosphere 93f, 1615–1623.

Storelli, M.M., Barone, G., Giacominelli-Stuffler, R., Marcotrigiano, G., 2012. Contamination by polychlorinated biphenyls (PCBs) in striped dolphins (*Stenella coeruleoalba*) from the south-eastern Mediterranean Sea. Environ. Monit. Assess. 184, 5797–5805.

Tanabe, S., 1988. PCB problems in the future: foresight from current knowledge. Environ. Pollut. 50, 5–28.

Tanabe, S., Kannan, N., Subramanian, A., Watanabe, S., Tatsukawa, T., 1987. Highly toxic coplanar PCBs: occurrence, source, persistency and toxic implications to wildlife and humans. Environ. Pollut. 47, 147–163.

Tehrani, R., Van Aken, B., 2014. Hydroxylated polychlorinated biphenyls in the environment: sources, fate and toxicities. Environ. Sci. Pollut. Res. 21, 6334–6345.

Tombesi, N., Pozzo, K., Harner, T., 2014. Persistent organic pollutants (POPs) in the atmosphere of agricultural and urban areas in the province of Buenos Aires in Argentina using PUF disk passive air samplers. Atmos. Pollut. Res. 5, 170–178.

Tomza-Marciniak, A., Pilarczyk, B., Wiczorek-Dabrowska, M., Bakowska, M., Witczak, A., Hendzel, D., 2011. Polychlorinated biphenyl (PCB) residues in European roe deer (*Capreolus capreolus*) and red deer (*Cervus elaphus*) from north-western Poland. Chem. Ecol. 27, 493–501.

UNEP United Nations Environmental Programme, 1999. Guidelines for the Identification of PCBs and Materials Containing PCBs. UNEP Chemicals, Switzerland.

US EPA U.S. Environmental Protection Agency, 2014. Technical Fact Sheet—Polybrominated Diphenyl Ethers (PBDEs) and Polybrominated Biphenyls (PBBs). Available from: <http://www2.epa.gov/sites/production/files/2014-03/documents/ffrrofactsheet_contaminant_perchlorate_january2014_final_0.pdf> (accessed 12.08.15.).

US EPA U.S. Environmental Protection Agency, 2015. Great Lakes monitoring. Available from: <http://www3.epa.gov/greatlakes/glindicators/fishtoxics/sportfishb.html>.

Van den Berg, M., Birnbaum, L.S., Denison, M., De Vito, M., Farland, W., Feeley, F., et al., 2006. The 2005 World Health Organization reevaluation of human and mammalian toxic equivalency factors for dioxins and dioxin-like compounds. Toxicol. Sci. 93, 223–241.

Villeneuve, J.-P., Cattini, C., Bajet, C.M., Navarro-Calingacion, M.F., Carvalho, F.P., 2010. PCBs in sediments and oysters of Manila Bay, the Philippines. Int. J. Environ. Health Res. 20, 259–269.

Viktorova, J., Novakova, M., Trbolova, L., Vrchotova, B., Lowecka, P., Makova, M., et al., 2014. Characterization of transgenic tobacco plants containing bacterial bphc gene and study of their phytoremediation ability. Int. J. Phytorem. 16, 937–946.

WHO World Health Organization, 2010. Dioxins and their effects on human health. Fact Sheet #225. Available from: <http://www.who.int/mediacentre/factsheets/fs225/en/> (accessed 12.08.15.).

Wohlfahrt-Veje, C., Audouze, K., Brunak, S., Antignac, J.P., Bizec, B.I., Juul, A., et al., 2014. Polychlorinated dibenzo-*p*-dioxins, furans, and biphenyls (PCDDs/PCDFs and PCBs) in breast milk and early childhood growth and IGF1. Reproduction 147, 391–399.

Wolkers, H., van Bavel, B., Ericson, I., Skoglund, E., Kovacs, K.M., Lydersen, C., 2006. Congener-specfic accumulation and patterns of chlorinated and brominated contaminants in adult male walruses from Svalbard, Norway: indications for individual-specific prey selection. Sci. Total Environ. 370, 70–79.

Yolsal, D., Salihoglu, G., Tademir, Y., 2014. Air-soil exchange of PCBs: levels and temporal variations at two sites in Turkey. Environ. Sci. Pollut. Res. 21, 3920–3935.

Zeeb, B.A., Amphlett, J.S., Rutter, A., Reimer, K.J., 2006. Potential for phytoremediation of polychlorinated biphenyl-(PCB-) contaminated soil. Int. J. Phytorem. 8, 199–221.

Zhang, Y.-F., Fu, S., Dong, Y., Nie, H.-F., Li, Z., Liu, X.-C., 2014. Distribution of polychlorinated biphenyls in soil around three typical industrial sites in Beijing, China. Bull. Environ. Contam. Toxicol. 92, 466–471.

Zota, A.R., Linderholm, L., Park, J.S., Petreas, M., Guo, T., Priyalsky, M.L., et al., 2013. Temporal comparison of PBDEs, OH-PBDEs, PCBs, and OH-PCBs in the serum of second trimester pregnant women recruited from San Francisco General Hospital, California. Environ. Sci. Technol. 47, 11776–11784.

Chapter 7

Polycyclic Aromatic Hydrocarbons

Terms to Know

Pyrolysis
Substituted PAH
Saturated PAH
Polyaromatic hydrocarbon
Photo-oxidation
Sorption Coefficient
Adsorption
Absorption
Maximum Contaminant Level (MCL)
Procarcinogen
Priority List of Contaminants
Bay or Fjord Region

INTRODUCTION

Polycyclic aromatic hydrocarbons (PAH), also known as *polyaromatic hydrocarbons,* are a group of more than 100 different chemicals characterized by two or more aromatic or benzene rings connected to each other. Unlike polychlorinated biphenyls (PCBs) and organochlorine pesticides, many PAHs are naturally occurring. Natural PAHs come primarily from combustion of organic materials. Forest, brush, and grassland fires are by far the most productive natural sources of PAHs. In second place are volcanoes, other sources of combustion, and aerobic bacteria. Human-derived or anthropogenic sources of PAHs include industrial combustion, fossil fuel burning, oil spills, vehicular exhausts, asphalts, tars, refuse incineration, and other industrial processes.

PAHs are among the most carcinogenic contaminants in the environment (ATSDR, 2009). Single microgram levels of exposure to certain PAHs can produce carcinogenic tumors in some organisms. All vertebrates including fish, birds, and mammals are prone to developing cancer due to PAH exposure and several PAHs have been declared possible to probable human carcinogens by the US EPA and other agencies. Several other toxic effects can also occur in humans and animals with PAH exposure (ATSDR, 2009).

The U.S. EPA has listed 17 PAHs in their list of Priority Pollutants (US EPA, 2013) based on widespread distribution, potential environmental concentrations, and toxicity including:

Acenapthene	Acenapthylene
Fluoranthene	Anthracene
Napthalene	Benzo[g,h,i] perylene
Benzo[a]anthracene	Fluorene
Benzo[a]pyrene	Phenanthrene
Benzo[b]fluoranthene	Dibenzo[a,h]anthracene
Benzo[k]fluoranthene	Indeno[1,2,3-cd] pyrene
Chrysene	Pyrene

The Agency for Toxic Substances and Diseases Registry (ATSDR, 2014) also lists benzo[e]pyrene, benzo[j]fluorine, and coronene on its priority list of hazardous substances.

CHEMICAL CHARACTERISTISTICS OF PAHs

PAHs can be composed of relatively simple hydrocarbons or they can be more complex with other organic adjuncts. Most discussions on PAHs focus on molecules that are saturated with hydrogen atoms, which means that all the available binding sites on the carbon atoms are filled with hydrogen atoms (Fig. 7.1).

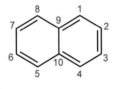

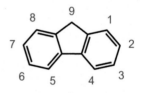

Naphthalene, with two benzene rings

Fluorene, a PAH with three rings

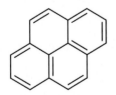

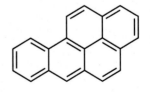

Pyrene with four rings

Benzo [a] pyrene with five rings

FIGURE 7.1   Some examples of saturated PAHs. In saturated forms, all of the available carbons have hydrogen atoms attached.

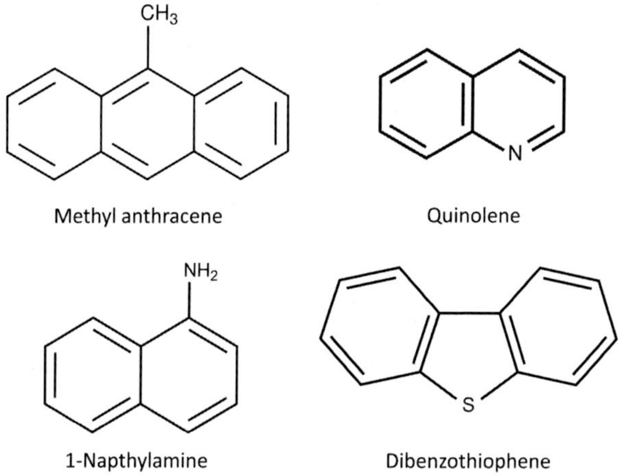

Methyl anthracene   Quinolene

1-Napthylamine   Dibenzothiophene

FIGURE 7.2 A few examples of PAHs substituted with other molecules.

Some PAHs may have other molecules attached to the carbons. These are called substituted PAHs because the molecules substitute for the hydrogen atoms. The substitutes may include alkanes (straight chain hydrocarbons with single bonds), alkenes (hydrocarbon chains with one or more double bond), hydroxyl radicals, amines, and oxygen (Fig. 7.2). Unless otherwise stated, our discussions will focus on saturated PAHs. Environmentally important PAHs range from light molecular weight (128 g), two-ring naphthalene to heavier (300 g), seven-ring coronene.

We've seen in PCBs and polychlorinated dibenzofurans (PCDFs) that as molecular weight and size increases, the molecules become less soluble in water, less volatile, have higher melting points, and are more stable. The same trends occur with PAHs (Table 7.1). Lightweight PAHs, such as naphthalene, volatilize in air and are relatively water soluble compared with many of the organic molecules we have examined. By the time we get to four or more rings, however, water solubility is practically nonexistent and the PAH has very low volatility. PAHs with six and seven rings are very stable.

SOURCES AND USES OF PAHs

Zhang and Tao (2009) conducted a mass budget analysis of PAH deposition and concluded that approximately 520,000 tonnes (metric tons) of PAHs are dumped into the atmosphere each year on a global basis (recall that a tonne or metric ton is equivalent to 1000 kg or 2240 lbs). Of this, combustion of biofuels accounted for 56.9%, wildfires 17%, and consumer product use 6.9%. China was the primary contributor to atmospheric PAH (22%), followed by India (17%), and the United States (6.5%). Eisler (2000) suggested that an additional 209,000 tonnes

TABLE 7.1 Characteristics of Representative PAHs[a]

PAH	CAS	Rings	MW	Melting Point (°C)	Solubility in water (mg/L) (20–25°C)	Vapor Pressure (mm) (25°C)	K_{ow}
Naphthalene	91-20-3	2	128.2	78.2	31.8	0.05	3.3
Acenapthylene	83-32-9	3	154.2	93.4	4	NA	3.7
Anthracene	120-2-7	3	178.2	215.8	0.04	1.9 E-4	4.6
Fluorene	86-73-7	3	166.2	116	2	3.2 E-4	4.2
Phenanthrene	85-01-8	3	178.2	101	1	6.8 E-4	4.6
Chrysene	218-01-9	4	228.3	254	1 E-8	6.2 E-9	5.9
Pyrene	129-00-0	4	202.2	146	0.135	4.5 E-6	4.9
Benzo[a]pyrene	56-55-3	5	252.3	179	2.0 E-4	5.5 E-9	6.0
Benzo(k) fluoranthene	207-08-9	5	252.3	524	5.5 E-4	9.6 E-11	6.1

[a]Taken from various sources.

are released into aquatic environments each year. Advances in vehicle efficiency and in reducing pollution from incinerators and coal-powered generators have gradually reduced the amount of PAH pollution in recent years. However, extensive brush and forest fires in the western United States over the past few years may significantly add to atmospheric pollution. Nevertheless, PAHs are ubiquitous in the environment due to natural synthesis of the compounds and it is safe to say that all humans are exposed to PAHs in many ways.

The US EPA (2007) estimated that in the United States, 10,350 tonnes of PAH are released into the atmosphere each year. Of this amount, residential heating and cooking accounted for just over half (51.7%) of the total and open burning and forest fires constituted approximately 20% of the total. Other sources of PAH included automobile exhausts (3.8%) and, in decreasing amounts: open burning, coke ovens for making steel, other vehicles, commercial cooking, and industrial boilers and heaters.

The production of PAHs occurs during incomplete combustion of organic materials, termed *pyrolysis*, and typically requires temperatures exceeding 700°C. These high temperatures are easily attained in volcanoes, forest fires, and industrial processes. However, some PAHs can accumulate over a long period of time with combustion at lower temperatures, as evidenced by the concentration of PAHs in fossil fuels (Eisler, 2000) and by the natural production of PAHs through bacterial action.

Most PAHs are highly hydrophobic. Most of the PAHs reside in water either as surface films or attached to particulates that can eventually sink and enter sediments where they may persist for a few years. Approximately one-third of the PAHs in water are in the dissolved phase, the other 67% are attached to particles (Lee and Grant, 1981). Stormwater runoff that eventually entered the Anacostia River in Washington, DC, for example, had 1510–12,500 ng/L PAH with 68–97% of that as particulates (Hwang and Foster, 2006). Additional water sources include discharge from industrial sources or runoff from surface spills, particularly oil spills. Petroleum is 0.2–7% PAH (Albers, 2003) and major oil spills can really ramp up concentrations of PAH in water.

Due to their high affinity for soil particles, land-based PAHs that are not part of runoff quickly become absorbed by soil where they are subject to bacterial decomposition. In both sediments and soil, sorption greatly reduces the availability of PAHs to organisms and thus reduces their risk of toxicity.

PERSISTENCE

Many factors affect the persistence of PAHs in the environment. Bacteria, plants, and other organisms can metabolize PAHs and microbial degradation is a major method of removing PAHs from soil and sediment. In comparison to other organic pollutants, loss of PAH from the environment can be rather rapid. Factors that influence the rate of decay include the presence or absence of oxygen, pH, and particle size of the soil or sediment. Oxygen supports aerobic bacteria that are more efficient at metabolizing PAHs than anaerobic microorganisms; low pH inhibits the breakdown of PAHs, therefore removal occurs most rapidly at neutral or slightly alkaline conditions; and large particle size, such as in sand, expedites loss of PAH compared with clays and silts. In addition to biological degradation, PAHs can volatize into the atmosphere or precipitate into sediments where they are less available to organisms.

In the atmosphere, a majority of PAHs are found attached to or in particulate form. Lighter molecular weight PAHs are volatile or semivolatile and are in a gaseous state. Volatile PAHs are particularly sensitive to ultraviolet light and may last only a few hours before they decompose or *photodegrade*; since the process results in the loss of electrons from the PAH or oxidation, the term *photo-oxidation* is also appropriate. Particulate PAHs may also degrade through exposure to UV light eventually, but are more resistant than those in the gaseous phase. Ozone can also be effective in degradation but has less of an effect than sunlight. For example, the half-life of anthracene under simulated sunlight alone was 0.20 hours and dropped to 0.15 hours when ozone was present. Under dark conditions, the half-life increased to more than an hour, even with ozone present (Ravindra et al., 2008).

The persistence of different PAHs on land depends in part on their molecular weight. For example, both coal tar and asphalt are used extensively as sealants on parking lots, roads, and driveways throughout the country. Newly applied coal tar contains 93,000 mg/kg PAH or 9.3% by weight (Mahler et al., 2014). Approximately half of this volatilizes or runs off within a month or so. 80–99% of the lighter, two- to three-ring PAHs such as naphthalene, acenaphthene, and fluorene disappear during that time. The authors attributed the decrease of these lighter compounds to volatilization. Runoff from coal tar sealants were dominated by medium-weight PAHs. The heaviest PAHs tended to be the most sedentary. In contrast, asphalt sealants initially have only approximately 6% of the PAH found in coal tar. Mahler et al. (2014) noted that other asphalt products they had tested averaged only 50 mg/L of PAH, far below that of coal tar. Coal tar is still used throughout the world for paving and patching roads, parking lots, and other surfaces. The results of this study mirror those of several other studies—low molecular weight PAHs tend to volatize to a greater extent, middle-weight PAHs generally contribute the most to runoff and effluent discharges, and the heaviest PAHs are the most sedentary (ATSDR, 1995). On the surface of water, PAH films are exposed to ultraviolet light and decompose rapidly. PAHs that mix with sediments may persist if they are in shaded areas and if the sediments are anaerobic.

Unlike dioxins, furans and organochlorines, PAHs are not included in the list of persistent organic pollutants by the Stockholm Convention, the international agreement to curtail the use and spread of the more persistent pollutants, because PAHs are not nearly as resistant to breakdown as PCBs (www.unido.org). Chlorinated naphthalenes have been proposed for listing.

A good example to illustrate just some of the complexity of PAH degradation in water was demonstrated by Jacobs et al. (2008). They examined photodegradation rates of PAHs in water collected in Gary, Indiana and Wilmington, North Carolina and determined that the rates were affected by the amount of dissolved organic carbon (DOC) and by nitrates in the water. DOC is important in predicting aquatic toxicity and we will discuss it more when we get to metals in Chapter 8. In theory, DOC can adsorb to contaminants and decrease their bioavailability. They can also increase OH^- ions in water, which aids in the oxidation of PAH, thus increasing the rate of degradation. However, DOCs can also reduce light penetration and inhibit photodegradation. Nitrates, which also release OH^-, do not attain high concentrations in water except in urban runoff and in agricultural areas due to the use of fertilizers. In Gary, Indiana water, nitrate levels were higher but DOC was lower than in water from Wilmington, North Carolina. PAHs degraded more quickly in Gary waters than in Wilmington waters and the authors attributed this to the increased OH^- in Gary, which enhanced photolysis and reduced light penetration in Wilmington inhibiting the same.

There are so many factors that affect the stability of PAHs that assigning precise half-life times could be misleading. Rather, it is safe to say that two-ring PAHs have half-lives of days or less in the atmosphere, weeks in water, months in soils, and years in sediments. Those with three to four rings have half-lives that are approximately double those of two-ring structures. PAHs with five or more rings persist for weeks in the atmosphere, months in water, and years in sediments and soil (Wick et al., 2011).

REMEDIATION OF PAH-CONTAMINATED SOILS AND SEDIMENTS

Many PAHs are readily assimilated or adsorbed by green plants and rapidly growing plants can be used to remediate or clean up areas that have been contaminated. One way of evaluating the effectiveness of a material in cleaning up contaminants is determining the *sorption coefficient* of the material. Sorption is the ability of a material to hold another substance like a PAH. It includes *adsorption*, when the contaminant adheres or sticks to the surface of the sorptive material, and *absorption,* when the contaminant is drawn up into the pores or cells of the material. Think of absorption as a kind of sponge and adsorption as masking tape. The higher the coefficient, the more efficient is the material. The idea, therefore, is to maximize efficiency. In addition, scientists want to find a material that is relatively inexpensive. Often contaminants have to be cleaned from a large body of water and highly efficient sorptive materials might not be very useful if they are expensive to use. Xi and Chen (2014) examined the ability of a variety of ground-up agriculture wastes ranging from bamboo, pine bark, pine needles, and several others to determine which ones might be preferred over others. They found that treating the materials with an acid bath removed much of the sugars inherent in the materials and increased the absorptive capacity from eightfold to tenfold over raw materials by opening pores in the materials and increasing surface area. Synthetic sorbents were approximately 10 times better than natural materials, even when given the acid bath, but were more expensive.

Biological degradation or biodegradation of PAHs can be facilitated by microorganisms. The rate of biodegradation tends to be increased by aerobic conditions compared with anaerobic conditions. Animals and plants can also assimilate and metabolize PAHs. For example, earthworms (*Eisenia foetia*) can be very effective in consuming and metabolizing PAHs from contaminated soils.

Yet another way of removing PAHs from water and soil is through various oxidizing processes. An underlying goal is to generate hydroxyl ions (OH^-), which are extremely efficient in oxidizing PAHs, leading to breakup of the ring structures. A variety of chemicals including chlorine, potassium permanganate, ozone, and hydrogen peroxide have been used to accomplish this oxidation. Most of these methods work best under low pH conditions. Advanced methods can reduce 100% of PAHs to their basic elemental constituents within a relatively short period of time. These methods use combinations of light, chemical oxidants, and sound to break up PAHs. To reduce the cost of these methods, they can be combined with forms of biodegradation. Fungi can also detoxify and metabolize PAHs. Cerniglia et al. (1985) found that the fungus *Cunninghamella elegans* inhibited the mutagenic activity of benzo[a]pyrene, benzo[a]anthracene, and the methylated 7,12-diemethylbenz[a]anthracene in cultures.

Although PAHs may be found in a wide variety of products including coal tar, creosote, petroleum, fire wood, asphalt, and some dyes, their deliberate manufacture is limited. A few PAHs such as naphthalene, fluorene, anthracene, phenanthrene, fluoranthene, and pyrene are manufactured for pharmaceuticals or plastics (Ravindra et al., 2008).

ENVIRONMENTAL CONCENTRATIONS

Because PAHs are naturally produced compounds, they are found just about everywhere in the world. PAHs typically come in mixtures. Seldom, if ever, do we see an area with only one or two PAHs. The multitude of PAHs in an area, each with its own toxicity, makes it difficult to compare risk from one area to another. Two sites may have comparable concentrations of total PAH but differ extensively in the specific PAHs that make up those totals and therefore demonstrate very different impacts due to PAHs.

Atmospheric concentrations are often in the low ng/m^3 range. Lightweight PAHs are often the most common forms in the atmosphere due to their higher volatility. These lightweight molecules form a major portion of the gaseous state of PAHs, whereas the middle-weight molecules are usually associated with particulate matter in the atmosphere; some studies differentiate the two forms of PAHs in their risk assessments. Some long-term monitoring studies such as that reported in US EPA (2014) suggest that PAH concentrations in air are declining slowly. However, natural renewal is preventing them from declining as rapidly as they might if they were totally produced by humans and could be controlled or regulated. PAH concentrations in air tend to be higher in winter than in summer, which makes sense if you consider that winter leads to increased burning of fuels for heat. Following similar logic, the atmosphere above cities tends to have higher concentrations than that of rural areas (Eisler, 2000). Eisler also suggested that the concentration of the carcinogen benzo[a]pyrene (BaP) provides a rough estimate of total PAHs in the air with the total being approximately 10 times greater than the BaP concentration. Mean concentrations of BaP over the Great Lakes range from very low or nondetectable around Lake Superior to approximately $0.6\,ng/m^3$ over Chicago (US EPA, 2014). As an example of high values, total PAHs ranged from 13.7 to $95.1\,ng/m^3$ in several cities in 1959 (US EPA, 1980) and a maximum BaP concentration was reported as $80\,ng/m^3$ downwind from a coal gasification plant in Yugoslavia (Edwards, 1983). The Occupational Safety and Health Administration, which sets the standards for air quality in work places, has established that the Permissible Environmental Limit or PEL for BaP concentrations should not exceed $0.045\,\mu g/m^3$ (4 ppb).

Worldwide, total and specific PAH concentrations in water tend to be at the low $\mu g/L$ levels, but near sources of contamination such as oil spills, concentrations may be 1000 times greater. Whereas the low solubility and tendency to volatize reduces the concern about PAH contamination in water, the petroleum that often serves as the source of aquatic PAHs is a different matter. The Maximum Contaminant Level Goal (MCLG) or target values for pollution reduction that are set by the US EPA for BaP in drinking water is $0\,ng/L$.

However, the more practical Maximum Contaminant Level (MCL) that is actually enforced is allowed at 0.2 μg/kg.

Sediment and soil concentrations of total PAH range broadly from below detection limits in reference areas to several μg/kg or even mg/kg in contaminated sites. Similarly, specific PAHs are at higher concentrations where they are used for industrial purposes. Creosote and other preservatives used to maintain wood products such as railroad ties and telephone poles are high in naphthalene, phenanthrene, and pyrene, and soil concentrations near wood-preserving plants are similarly high in these PAHs. Ellis et al. (1991) determined that subsoil concentrations of phenanthrene were as high as 3400 mg/kg (3.4%) near creosote plants and anthracene peaked at 693 mg/kg. In comparison, BaP, chrysene, and some other common PAHs which are not used in creosote very much were absent altogether.

SOME EXAMPLES OF BIOLOGICAL CONCENTRATIONS

PAHs have a great tendency to bioconcentrate from the environment to organisms. As with other organic compounds, bioconcentration is facilitated by the lipophilicity of PAHs and we often find them concentrated in the fat of animals and in plant oils (Eisler, 2000). The biological concentration factor (BCF) for a given contaminant depends on the specific PAH, the length of exposure, and the organism of concern and their tissue. Eisler (2000) summarized BCFs from several different studies and the values range widely. For example, BCFs for anthracene ranged from 200 in the water flea, *Daphnia magna*, exposed for 1 hour to 9200 in rainbow trout (*Oncorhynchus mykiss)* exposed for 72 hours. For BaP, BCFs ranged from 9 in clams (*Rangia cuneata*) exposed for 24 hours to more 134,000 in another species of water flea (*Daphnia pulex*) in a laboratory solution for 3 days.

Fig. 7.3 illustrates some of the relationships between time of exposure and organ in uptake of BaP by northern pike (*Esox lucius*; Balk et al., 1984). The authors used radioactive BaP and determined concentration through scintillation. Unfortunately, this method does not distinguish between the parent compound and radioactive metabolites, but it does indicate areas of storage and activity. The authors stated that most of the PAH was taken up by tissues within a day and then dispersed into organs. Here, we see that the BCF for BaP increases early but tapers off after several days. The decrease after is commonly seen as organisms adjusting to the presence of PAH and activating mechanisms, including the P450 system, to increase metabolism of the contaminants. Note also that bile had the highest BCFs and therefore the highest concentrations of the PAHs. Concentration in bile offers a major way to depurate PAHs through feces.

BCFs may occur in the tens of thousands. For example, snails had a BCF for BaP of 82,000 after 3 days of exposure and water fleas, had a BCF of 134,000 over the same time period (Lu et al., 1977). Some species, such as algae and mollusks, do not have the ability to metabolize PAHs and particularly prone to bioaccumulation (Eisler, 2000). For example, under controlled laboratory

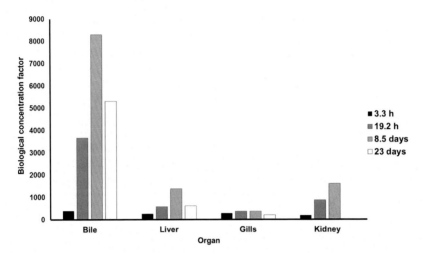

FIGURE 7.3   Biological concentration factors (BCFs) for benzo[a]pyrene in northern pike tissues based on length of exposure. BCFs have been reduced to 1/10 of actual in bile for presentation. *Balk et al. (1984).*

conditions, the BCFs for the blue mussel, *Mytilus edulis,* varied from 13,300 for lightweight PAHs, such as naphthalene, to 376,600 for the heavier dibenzothiophene (Baussant et al., 2001). In contrast, turbots (*Scophthalmus maximus*), a species of marine fish, had its highest BCF of 17,800 for naphthalene but only 2600 for dibenzothiophene. These species are also likely to have a higher tolerance to PAHs because they do not oxidize them efficiently, thus reducing the presence of DNA adducts (see the following discussion).

For those species that do digest PAHs, metabolism can be fairly rapid. In a mixture of PAHs injected into chicken eggs, anywhere from 52% to 99% of the PAHs were metabolized after 2 weeks (Näf et al., 1992). The single PAH that remained above 11% of the injected amount was the largest PAH, sevenring coronene. Similarly, concentrations of a variety of PAHs in common eider (*Somateria mollissima*) eggs went from 1100 ng/egg to below detection limits 19 days later; total PAH concentrations dropped 99.2%. Since eggs cannot lose PAHs except through some minor amounts volatizing through the shell, they had to be lost through metabolism (Näf et al., 1992).

Depuration rates in animals, which may include metabolism as well as elimination through feces or urine, are widely variable. Many species can remove PAHs within minutes as, for example, the water flea that can depurate 50% of their assimilated PAH within an hour when placed in clean water (Southworth et al., 1978). However, 50% depuration rates took several hours in rainbow trout (Kennedy and Law, 1990), 36 hours for marine copepods and naphthalene (Neff, 1982), and weeks for some marine mollusks (Jackim and Lake, 1978). Although PAHs can bioconcentrate, the ability of organisms to metabolize the chemicals nearly eliminates their potential to biomagnify.

BIOLOGICAL EFFECTS OF PAHs

Effects on Plants

Terrestrial plants can absorb PAHs from the soil through their roots and move them to other parts of the plant including leaves and stems (US EPA, 1980), which tend to have higher concentrations than roots or rhizomes. Low molecular weight PAHs are more readily absorbed than higher molecular weight chemicals. Under laboratory conditions, plants can bioconcentrate PAHs (Eisler, 2000) but the BCFs tend to be low except for some species that can be hyperaccumulators. Under natural conditions, plant tissues usually have lower concentrations than their immediate environment. However, leafy surfaces can accumulate higher concentrations of medium- and heavyweight PAHs to make the total concentration of PAH higher than the surrounding soils. These superficial deposits are readily removed through washing and peeling, thus reducing risk to humans (Kipopoulou et al., 1999). At least some plants have the ability to breakdown BaP into simpler molecules.

Toxic effects on green plants is rare at environmentally realistic concentrations However, Dubrovskaya et al. (2014) determined that phenanthrene interfered with several physiological processes in sorghum (*Sorghum bicolor*) including photosynthetic pigment quantities and ratios, germination, survival of seedlings, and growth. They used 10 and 100 mg/kg concentration in soil which could be expected in contaminated soils.

Some plants contain chemicals that protect them against the toxic effects of PAHs. Some of these chemicals include anticarcinogenic compounds found in kale, cabbage, Brussel sprouts, broccoli, and cauliflower and their protective effects can be transferred to animals, including humans, that eat them (ATSDR, 2009).

Effects in Animals

In contrast to plants, many animals have a high capacity to bioconcentrate PAHs. Specific values vary by species and chemical but BCFs range from 10 to 10,000 or more (ATSDR, 2009).

The primary concern for health in humans and other animals from PAHs is cancer. Studies have shown that PAHs can cause or at least are associated with tumors and cancers in humans, other mammals, reptiles, and fish. PAHs can cause a wide variety of other disorders including depressed growth and survival, genotoxicity, metabolic effects, reproductive suppression, and endocrine disruption. In most instances, concentrations that cause acute lethality in the laboratory are at least one to two orders of magnitude higher than those routinely found in the environment, so acute lethal effects are less of a concern than chronic disorders. The lighter-weight, more–water soluble, two- and three-ring PAHs tend to have higher acute toxicity than larger molecules; four- and five-ring PAHs tend to be the most carcinogenic; and the heaviest PAHs tend to

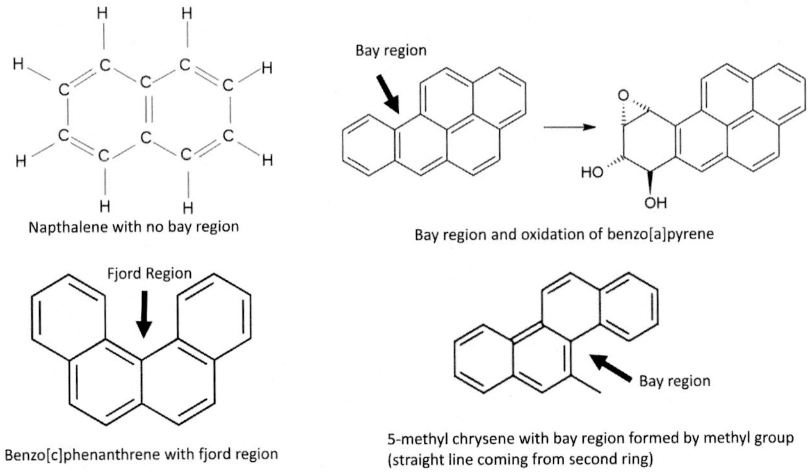

Napthalene with no bay region

Bay region and oxidation of benzo[a]pyrene

Benzo[c]phenanthrene with fjord region

5-methyl chrysene with bay region formed by methyl group (straight line coming from second ring)

FIGURE 7.4   The bay and fjord regions of PAHs are where bonds with DNA or RNA nucleotides are most likely to occur.

be relatively biologically inert, perhaps because they have such low solubility (Eisler, 2000). In vertebrates, BaP is one of the most, if not *the* most, carcinogenic contaminants in the environment although some of the methylated PAHs also have a very pronounced carcinogenic activity. Thus BaP is often used as the standard to which other PAHs are compared. Efforts have been made to develop toxic equivalent factors (TEFs) based on BaP or other factors (Okparanma et al., 2014; Nguyen et al., 2014), but none have been as widely accepted as 2,3,7,8 TCDD as the standard for PCBs, dioxins, and furans.

Cancer is most often of concern with PAHs and the presence of a *bay region* or fjord region in the molecular structure of the PAH (Fig. 7.4) is a useful predictor of PAH carcinogenicity. We know, for instance, that the bay region of BaP allows it to bind with guanine of DNA and RNA with the linkage taking place between the C-10 carbon of BaP and the C-2 carbon of guanine. Once attached, BaP can enter the cell and disrupt the proper synthesis of protein (Fernandez and L'Haridon, 1994).

In the body, PAH molecules are subject to treatment by the cytochrome P450 system which we discussed in the preceding chapter. Through a set of complex reactions, PAH can enter different metabolic pathways, all of which can produce carcinogenic derivatives. Parent PAH molecules are not toxic, but they must be activated by the P450 system to become toxic. Thus, they are rightly called *procarcinogens*. One metabolic route (Fig. 7.5) is for the PAH to become oxidized by the enzymes CYP1A1 and CYP1B1. The terminal ring associated with the bay or fjord region is very reactive to this oxidation. Oxidation may lead to epoxides, which are then hydrolyzed to diols. These diols may be quickly reconverted to epoxides through further action of P450. Epoxides gain entrance

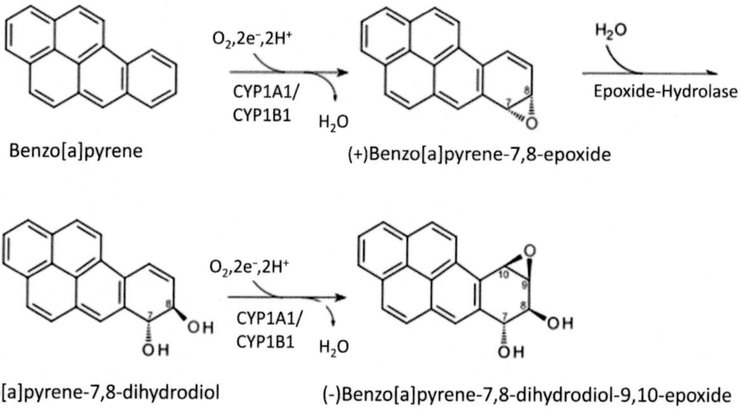

FIGURE 7.5 Oxidation and hydrolysis of benzo[a] pyrene to a form that facilitates the production of DNA adducts.

into the cell and its nucleus where they bind with DNA to form adducts. These adducts interfere with normal translation. If they occur in rapidly growing or replicating tissues such as bone marrow, skin, or lungs, they can develop tumors and eventually cancers.

Another metabolic route is for the PAH to be oxidized by aldo-keto reductases, a type of enzyme in the P450 system (Fig. 7.6). Oxidation leads to the formation of ketols which spontaneously form catechols. Catechols are highly reactive and bind with DNA, also causing adducts and interfering with normal DNA activity. In addition to cancers, adducts can cause genotoxicity, malformations, and many other maladies. The radical cation pathway involves excitation of PAHs with low ionization potential directly creating radicals or radical ions that can bind with the nucleus of the cell and activate genes. The top part of Fig. 7.6 shows what is called Phase I P450. Phase II involves other enzymes and eventually leads to the elimination of the PAHs.

A great many studies have been conducted on the effects of PAH on fish. Eisler (2000) summarized much of this body of information. He related how brown bullheads (*Ameiurus nebulosus*) collected from a highly contaminated area in the Detroit River had numerous external abnormalities including truncated barbels, skin lesions, and liver lesions and that the frequency of occurrence correlated with the concentration of PAHs in sediments (Leadley et al., 1998). As further support (but not proof) of the relationship between PAH and abnormalities in bullheads, Baumann and Harshbarger (1995) related how between 1980 and 1982 PAH concentrations in the Black River of Ohio declined by 65% in the sediments and 93% in the livers of brown bullheads; between 1982 and 1986 the sediment concentrations had declined another 99% due to the closure of a coke plant. During this 6-year period, the rate of cancer in the bullheads declined to 10% from a high of 39%.

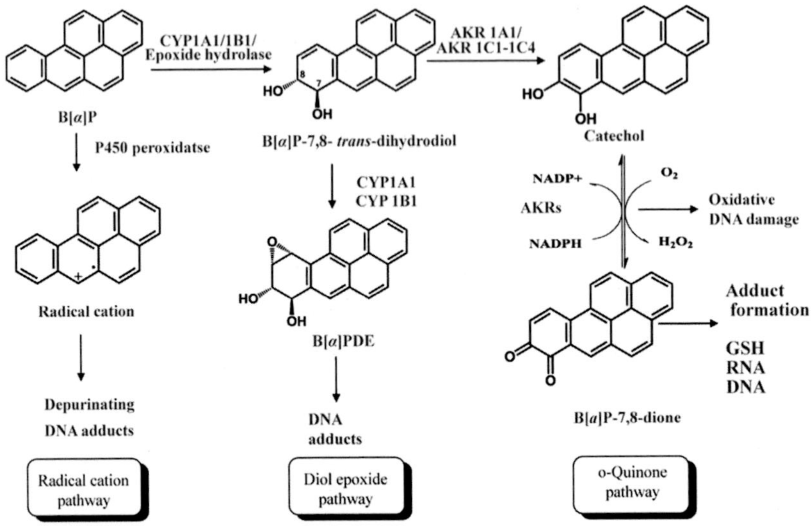

FIGURE 7.6 Pathways for PAH metabolism. Phase I involves CYP1A and the aryl receptor. Phase II leads to elimination. *Zhang et al. (2012).*

Studies have shown that PAHs, like PCBs, PBDEs, and other organic contaminants can act through the aryl hydrogen receptor. Brummell et al. (2010) found that BaP was more effective in activating CYP1A in sunfish than was PCB77 and claimed that the presence of CYP1A may be useful in identifying exposure of fish to PAHs. Wang et al. (2010) exposed killifish (*Fundulus heteroclitius*) to 5 mg/L BaP and the same concentration of the methylated PAH dimethylbenzanthracene (DMBA). After only 6 hours of exposure, both chemicals induced CYP1A and another P450 enzyme CYP1C1. CYP1C1 had not been well studied before this but it did seem to be tissue specific. After 8 months postexposure, 20% of the killifish exposed to BaP and 35% exposed to DBMA had benign and cancerous tumors in the liver, bile ducts, and elsewhere.

Further studies have been conducted on skin and liver tumors in brown bullheads because the species lives on the bottom of rivers in fresh and brackish water and is susceptible to contaminants (Fig. 7.7). Pinkney et al. (2011) reported on brown bullhead populations near the capitol of the United States, Washington, DC. His studies on the Anacostia River, long recognized as being one of the most polluted rivers in the United States and a poster child for neglected rivers (Fig. 7.8, NRDC, 2014), showed that the highest liver tumor record for bullhead samples was 50–68%, while 53% had skin tumors. The South River, closer to Annapolis, MD, had liver tumor rates in brown bullheads ranging from 19 to 58% and the Severn River, also near Annapolis, had rates of 2–52%. Other rivers whose watersheds were primarily in agriculture had lower

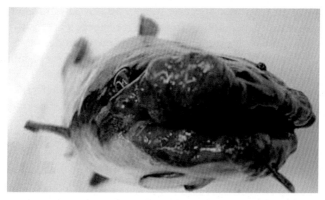

FIGURE 7.7   Tumors in brown bullhead. *Courtesy U.S. Fish and Wildlife Service http://www.fws. gov/chesapeakebay/newsletter/Spring12/Bullheads/Bullheads.html. Photo by Fred Pickney.*

FIGURE 7.8   The Anacostia River, Washington, DC. *Photo by the Anacostia Watershed Society.*

rates of 2–5%. Liver and skin cancer rates increased with the age of the fish, so that in the Anacostia the largest fish were nearly 100% affected. Chemical analyses of sediments and bullhead bile showed a clear correlation between the rate of tumors and the concentrations of PAHs. Also, using radiolabeling, the researchers found that liver ^{32}P-PAH DNA adducts in bullheads from the nearby South and Choptank Rivers in Maryland averaged 5–12 nmol aberrant/ mole normal nucleotides, whereas the rate in the Anacostia was 122 nmol/mole normal nucleotides. Adduct levels greater than 1 nmol/mole of normal nucleo- tides confirmed exposure to genotoxic chemicals (Aas et al., 2003).

Just one briefer example of cancer in fish. Atlantic killifish are resistant to PAH toxicity because their aryl hydrocarbon receptor is less sensitive than that of some other fish and therefore they are less likely to produce P450 enzymes that breakdown PAHs (Wills et al., 2010). Two populations of killifish were tested for the sensitivity to PAHs. One came from a Superfund site that had been exposed to creosote for 66 years and the other came from a reference site. Fish from both populations were exposed in the laboratory to a range of BaP from 0 to 400 µg/L. After 24 hours, reference fish exposed to the higher concentrations of BaP began to synthesize CYP1A, but those from the Superfund site did not. After 9 months in the laboratory, the reference fish had developed liver cancer while those from the Superfund site did not. Based on this study and related ones, the authors concluded that the Superfund fish, after living in contaminated waters for so long, had developed a genetic resistance to PAHs.

With regards to other, noncancerous effects, PAHs may negatively impact immune systems. Rainbow trout exposed to 17 µl/L creosote solution (=611.6 ng/L PAH) showed reduced immune response in the pronephros (equivalent to the kidney) and decreased plasma lysozyme concentrations (Karrow et al., 1999). Lysozymes are enzymes that break down bacterial cells and aid immunity.

PAH may also have reproductive effects. Killifish seem to be a favorite species for these types of tests. Booc et al. (2014) exposed adult killifish to BaP concentrations of 0, 1, or 10 µg/L for 28 days. In the first 14 days, males and females were kept separately, but in the last 2 weeks they were combined and their reproductive output was examined. The killifish kept at 10 µg/L had reduced reproductive output—egg fertilization and successful hatching were impaired. In males, gonad weight and plasma testosterone were reduced at 10 µg/L compared with controls. In females, estradiol concentrations were significantly reduced after BaP exposure, but egg production and gonad weight were not altered. No other effects were observed and the test was too short to develop tumors.

Guppies (*Poecilia vivipara*) were exposed to phenanthrene at 0, 10, 20, and 200 µg/L for 96 hours (de Souza Machado et al., 2014) to determine the effects of the PAH on oxidative stress. Lipid peroxidation occurred in muscles at 200 µg/L phenanthrene. Genotoxicity increased at 20 µg/L, while 200 µg/L caused a relative decrease in erythrocyte concentration in blood. According to the authors, these findings indicated that phenanthrene is genotoxic and can induce oxidative stress. However, the solvent used to carry the PAHs actually produced some greater responses than the PAHs themselves. Thus, researchers should always be very aware of all aspects of their experiments.

Some research has been done on the effects of PAHs on amphibians. In one study, tiger salamanders (*Ambystoma tigrinum*) were captured from a contaminated pond on a Texas Air Force base (Anderson et al., 1982). Several PAH-containing pollutants had either been dumped or allowed to runoff into the pond over many years and the salamanders had a 24–42% rate of skin tumors some of which were benign, others cancerous. Under in vitro laboratory conditions, the captive salamanders were exposed to BaP and perylene to measure the rate of oxidation or hydrolysis of these PAHs. In short, neither PAH was activated by the salamander preparations and the authors concluded that the species did not have a P450 system that was sensitive to PAHs. In light of the Wills et al. (2010) study cited previously, one has to wonder, however, if salamanders taken from a clean pond would have greater sensitivity.

We have already mentioned that street and parking lot sealants, especially coal tar, release a considerable amount of PAHs through runoff. Bommarito et al. (2010a,b) studied the effects of coal tar and asphalt on two species of salamanders. In one study, spotted salamanders (*Ambystoma maculatum*) were exposed to 0, 60, 280, or 1500 mg coal tar sealant/kg sediment for 28 days. Half of the dosed animals were exposed to fluorescent lighting only and the other half were exposed to fluorescent lighting plus ultraviolet radiation (UV). There has been some data which show that PAHs and UV light can interact to increase toxicity (Fernandez and L'Haridon, 1994; Steevens et al., 1999). There was no difference in the rate of mortality between treated animals and controls. However, exposure to 1500 mg/kg sealants resulted in slower rates of growth and diminished tendency to swim in a dose-dependent fashion. Exposure to UV affected the frequencies of leukocytes which suggested an immune response and increased the incidence of micronucleated erythrocytes (see Chapter 3). These effects in blood were not seen with coal tar and fluorescent lighting, but there was an interactive effect of sealant and UV on impairing swimming behavior.

In a companion study, eastern newts (*Notophthalmus viridescens*) were exposed to asphalt and coal tar sealants. Newts were again exposed to sediments containing dried sealants ranging from 0 to 1500 mg/kg under UV radiation and visible light to determine concentration/response relationships. Total PAH in the asphalt treatment ranged from 0.07 to 20.58 mg/kg in sediment and 30 to 281 μg/L in water. For the coal tar sediment, concentrations ran from 1.51 to 1149 mg/kg and water from 30 to 1464 μg/L. As in the other study, the concentrations were not lethal. Significant effects due to sealants included decreased righting ability and diminished liver enzyme activities. Coal tar sealant was more effective in inducing these changes than asphalt sealant. In neither case was serious damage observed, but it would have been interesting if the animals could have been kept for several months to determine if any tumors developed.

Among reptiles, Bell et al. (2005) found high prevalence of malformations in snapping turtles (*Chelydra serpentina*) and painted turtles (*Chrysemys picta*) at the John Heinz National Wildlife Refuge in Philadelphia, PA. For snappers, annual rates of malformations ranged from 13% to 19% of the sampled population and for painted turtles it was 45–71%. Several different types of malformations were found but the greatest frequencies involved bone or shell. There were too many different kinds of contaminants for all of them to be measured, but the authors determined that among those that were quantified, total PAHs and alkanes (straight chain, saturated hydrocarbons) were possible causes for the malformations. Alkane concentrations were as high as 837 mg/kg, but maximum total PAHs was 4.5 mg/kg.

Van Meter et al. (2005) confirmed that PAHs and alkanes caused malformations. They exposed snapping turtle eggs to petroleum crude oil ranging from 1 to 10 μL crude applied to the shell; 0.12% by mass crude mixed into sand; 10 μL of 0.02 to 1.0% BaP; and the same range of concentrations for DMBA, a methylated PAH. The crude oil produced several malformations ranging from minor to lethal with frequency of lethal malformations increasing with concentration and having a maximum of approximately 50% at 10 μL/egg. Survival decreased with crude oil concentration. With BaP and DMBA, maximum malformations of approximately 50% were among controls taken from the refuge; those from other locations were far lower.

Many studies on the effects of PAHs on birds have been associated with oil spills because shorebirds are among the fauna that are often heavily affected by such spills. We will cover oil spills in more detail in the FOCUS section at the end of this chapter, but for now we will mention a couple of examples. During the early 1990s, Brunström et al. (1991) tested the toxicity of PAHs to chick embryos by injecting small quantities dissolved in peanut oil through the shell into the airspace above the developing chick. They found that the LD50s were as low as 14 mg/kg of egg or in the low ppm range. The astute student might well ask "why test chicken eggs? They can't be exposed can they?" The answer is, "well, yes they can be exposed through seepage past the eggshell," but the important part of this study was that the scientists discovered that PAHs elicit EROD activity. As you remember, EROD is an enzyme of the P450 system and its presence demonstrated that PAHs, in addition to PCBs, also acted through the aryl hydrocarbon receptor which was not known before then.

Bustanes (2013) related a tale where common eiders living along a fjord in Norway suffered population declines during the 1980s when PAH pollution was entering the waters of the fjord. The population declined from 800 breeding pairs to less than 500 pairs in 1991. Hatching success was around 90% compared to nearly 100% in another reference area. The issue seemed to be that blue mussels (*Mytilus edulis*), a primary food for the eiders, had accumulated very high concentrations of PAH and had BCFs of approximately 500. The eiders were consuming this PAH and possibly passing it on to their eggs. After PAH disposal was stopped in the mid-1990s, the population began to return and by 2000 there were again 700 breeding pairs of eiders. Duckling mortality remained high, however. Bustanes admitted that the evidence for PAH being the cause of the decline was circumstantial, but that's often what happens in biological field studies—it is generally difficult to identify a specific contaminant for population declines. In this study, however, the only group of contaminants that was apparent after extensive sampling was PAH.

Studies on laboratory mice and rats have contributed extensively to our understanding of the mechanisms of PAH metabolism, depuration, and toxicity. A recent use of a common search engine on the terms "PAH" and "mice" yielded more than 6000 hits on separate studies, and a similar search with "rat" produced nearly 7000 hits. Researchers interested in human toxicology often rely on information derived from rodent studies to further understand risks to PAHs but, in general, laboratory rats and mice are less widely used by wildlife toxicologists because those mice and rats have existed for a great many generations in captivity and may not provide reliable data for studies conducted on wild species.

Among those things that laboratory studies have identified is that PAHs can be endocrine disruptors in mammals. In regards to hormones associated with sex and reproduction, there are several ways that PAHs may influence their activity. One way is by binding with estrogen receptors on plasma membranes of cells that are responsive to estrogen, such as those in the uterus, cervix, mammary glands, pituitaries, or ovaries. Many activated PAHs act as estrogen mimics. Compared to natural estrogens, such as 17-β estradiol, these PAHs are weak estrogens but they can produce enough activity to have effect and estrogenic activity varies among specific PAHs. For example, BaP has comparatively high activity while benzo[a]anthrene and fluorene have low estrogenic activity. However, the possible estrogen effects do not stop there. There can be "crosstalk," or interaction, between the membrane receptors and AhR (Kummer et al., 2008). Binding between the PAH, receptors, and AhR may inhibit estrogenic activity by: (1) increasing the metabolism of estrogens; (2) through direct interaction of the AhR on critical promotor regions in the genes necessary for estradiol production; (3) through competition for regulators of transcription; or (4) through a degradation of the membrane receptors themselves. Moreover, binding with AhR can elicit CY1A1, CYP1A2, and CYPB1 enzymes that can metabolize estrogen and act as an estrogen blocker. In addition to the organs that are stimulated by estrogen, PAH activation might also work on the pituitary and hypothalamus, organs that regulate estrogen activity. As one example of PAHs' negative potential, benzo[a]fluorene has been proposed as an antiestrogen (Arcaro et al., 1999a,b), at least in vitro. All of this is complicated and we have tried to simplify the possible effects of PAHs on estrogen. The bottom line is that PAHs, either through the membrane receptors or in the interaction between the receptors and AhR at the cellular level, can result in smaller reproductive organs, embryonic damage, damaged ovarian follicles, or decreased fertility. Similar interactions with the processes that lead to testosterone production can cause developmental defects in testes and diminish spermatogenesis.

Mammals are not the only group to have their reproduction negatively impacted by PAH-caused endocrine disruption. The study by Booc et al. (2014) described previously in killifish is another example of endocrine disruption and its effects on reproduction.

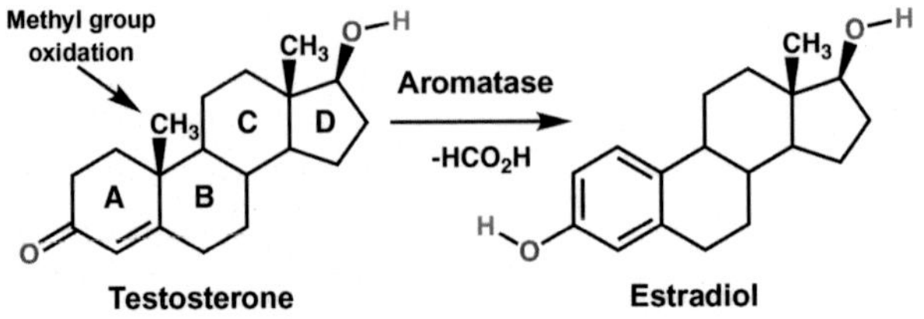

FIGURE 7.9   Structural formulae showing the conversion of testosterone to estradiol through the enzyme aromatase. This process can be hampered by PAH adducts.

One more interesting tidbit. Among vertebrates, organs that produce estrogen including the ovaries and adrenals make androgens first, even in females. The enzyme aromatase converts the androgens to estrogens (Fig. 7.9). Obviously this activity is substantially higher in females than in males although it can occur in both genders. Dong et al. (2008) found that in vitro exposure of BaP in killifish decreased brain and ovarian aromatase mRNA activity. In vivo this could result in the masculinizing of genetic females.

FOCUS—OIL SPILLS AND WILDLIFE

How many oil spills would you guess occur each year in the United States? If you guessed 1000 or even 10,000, you would be much too low. The US EPA estimates that upwards of 14,000 spills occur each year in this country alone. Add to that the number of spills in other countries such as the Middle East and you have a whole lot of potentially toxic oil dumped into the environment. Fortunately, we seem to be getting better over time in conserving our oil, at least in terms of tanker spills (Fig. 7.10). Events like the BP Deepwater Horizon oil spill do still occur, but the total number of major oil spills has declined over the past couple of decades. The top 10 oil spill disasters (Table 7.2) resulted in conservative estimates of 30.32–33.72 million barrels of oil (5.74–6.38 billion liters).

You might ask, "where is the infamous *Exxon Valdez* in this?" Well, the *Exxon Valdez* at 257,000–750,000 barrels ranks below number 50 on the list of greatest oil spills in the world (Alaska Oil Spill Commission, 1990). The BP Deepwater Horizon spillage is the worst to occur in the United States. The exact amount of oil spilled for this disaster varies among sources. BP claimed that 3.26mbbl of oil were spilled (BP, 2014), but the Smithsonian Institute Ocean Portal (2014) and NOAA (2014) cited 4.9–5mbbl; and the Natural Resources Defense Council (NRDC, 2014) likes to phrase it as "170 million gallons," which sounds worse than the equivalent 3.4mbbl.

Tanker oil spills that occur far from land, such as the ABT Summer Oil Spill and the *Amoco Caldiz,* although large by total amount of oil, have less potential for doing damage to wildlife or their habitats because much of the oil will sink into deep sediments, be broken down by microbial action, or be dispersed by tides and wave action before ever coming to land; and except for sedentary

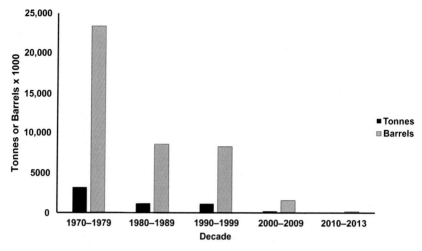

FIGURE 7.10   The amount of oil spilled from tankers by decade. *International Tanker Owners Pollution Federation 2014.*

TABLE 7.2 Ten Worst Oil Spills in Global History

Name	Location	Year	Amount	Environment
Gulf War	Kuwait, Persian Gulf	1991	7.6–10.4 mbbl[a]	Marine
Lakeview Gusher	Kern Co., California	1910–1911	7.56 mbbl	Land
BP Deepwater Horizon	Gulf of Mexico	2010	3.26 mbbl	Marine
Ixtoc Oil Spill	Bay of Campeche, Mexico	1979–1980	2.8 mbbl	Marine
Atlantic Empress/ Aegean Captain Tanker Collision	Off coast of Trinidad and Tobago		1.8 mbbl	Marine
Kolva River Oil Spill	Kolva River, Russia	1994	1.7 mbbl	Inland
Nowruz Oil Field Oil Spray	Persian Gulf, Iran	1983	1.6 mbbl	Docked at offshore pipeline
Castillo de Bellver Oil Spill	Saldanha Bay, South Africa		1.6 mbbl	Marine
Amoco Caldiz Oil Spill	Portsall France	1978	1.4 mbbl	69 nautical miles out at sea
ABT Sumner Oil Spill	Genoa, Italy	1991	1.0–1.6 mbbl	700 nautical miles at sea

[a]*Million barrels of oil.*

species that tend to reside in shallower water, most marine animals can avoid contaminated areas. Effects on marine invertebrates, fish, and marine mammals in shallow water are likely to be more devastating, however. Spills close to shore can wreak havoc on vegetation, sediments, shorelines, and wildlife. Petroleum can negatively affect fish, wildlife, and their habitats through physical covering and smothering sediment, soils, and vegetation; reducing light penetration in water; altering water chemistry, including pH and dissolved oxygen concentration; and through direct toxic effects on the organisms themselves. Shoreline wildlife can become covered with goo while foraging for organisms debilitated or killed by previous contact with oil. In addition to physical effects of coating, ingestion of oil can be toxic. Under aerobic conditions oil can be broken down by microorganisms, but it can take years to finally reduce a major oil spill to its elemental constituents.

Crude oil or petroleum is a complex of thousands of hydrocarbon and non-hydrocarbon substances (Albers, 2003). By weight, most crude oil and refined oils are composed of 75% hydrocarbons, but heavy crude can contain more than 50% of nonhydrocarbons (Albers, 2003). The hydrocarbon component consists of straight chain, saturated alkanes, branched alkanes, cycloalkanes, and aromatics including PAHs. Crude oil contains from 0.2 to 7% PAH, with the percent PAH increasing with the specific gravity of the oil (Albers, 2003).

The transport and fate of oil products are similar to that of PAHs. While they originate primarily from spills instead of pyrogenesis, oils can spread on the surface of waters where they are exposed to sunlight, wave action, and microbial degradation. Approximately 40–80% of crude oil can be decomposed by microorganisms (Atlas and Bartha, 1973). Heavier oils sink and form globules in sediments or on beaches (Fig. 7.11) and these may mix with soils and become more persistent.

Petroleum damages vegetation in many ways. Mangroves have been frequently hit hard by oil pollution in tropical and semitropical regions because

FIGURE 7.11 Oil globules on Løngsturp Beach, Denmark. *Credit; Stefan Thiesen, Wikimedia Commons.*

FIGURE 7.12   Mangrove swamp in Malaysia coated with oil up to high tide line. *Image: Wikimedia Commons.*

the mangrove trees grow very thickly and trap oil amongst their trunks and root systems (Fig. 7.12). Other plant communities that are often affected include salt marshes, large intertidal algae, sea grass communities, and freshwater vegetation affected by inland spills (Albers, 2003). Harmful effects of petroleum to plants often involves coating of leaves, thereby reducing photosynthesis and growth.

Oil spills have effects at the individual, population, and community levels of organization. At the individual level, environmentally realistic concentrations of petroleum can cause mortality and sublethal changes in physiology, increased cancer rates, and genotoxicity. Smothering is often the most common cause of death among small and less mobile invertebrates. Population effects include changes in abundance, age structure, reproductive output, and recruitment. Community effects include disruptions in predator and prey interactions, inter-specific competition, and symbiotic relationships because some species may be more vulnerable than others. Perhaps the factor that causes the greatest amount of upheaval is altered trophic relationships through the disproportionate loss of smaller organisms (Albers, 2003).

While freshwater and marine fish are usually very mobile and can escape major effects of oil spills, if spills occur in small, disconnected bodies of water or rapidly move through the water they can affect fish populations. Eggs, however, are not mobile and larvae may be less able to escape than adults so the smothering effect of petroleum can quickly kill embryos and very young fish (Albers, 2003). Thus, timing of oil spills, especially in freshwater where fish have definite breeding seasons, can be very important in determining the overall effect of spills.

About 3 years after the *Exxon Valdez* spill in 1989, the herring (*Clupea pallasii*) population in Prince William Sound collapsed and never recovered.

The decline is both ecologically and economically problematic with harvests of herring being greatly curtailed. Many factors were probably involved in this decline and the oil spill might not even be involved with the initial or continued depression of herring populations (Pearson et al., 2012), but a leading hypothesis is that the oil spill somehow disrupted the trophic web involving the herring (Malakoff, 2014). We do know that billions of salmon and herring eggs were destroyed at the time. At this time, the damages concerning the BP Deepwater Horizon spill are still being debated and reliable estimates on its effects on fish are continually being updated—it will be decades before final evaluation of ecosystem damage is assessed.

Very little is known about the effects of oil spills on amphibians or reptiles. Sea snakes may be smothered and sea turtles may consume toxic levels of oil (Albers, 2003). Following the BP oil spill, 457 sea turtles were found with visible oiling, of these 18 were dead and 439 were alive. An additional 517 turtles were found dead by search crews with no visible oiling and 80 live turtles that were collected had no oil (US FWS, 2011). These results are inconclusive about the lethal effects of oiling. Other studies on sea turtles either were hampered by small sample sizes or also inconclusive.

Seabirds and mammals are among the hardest hit of animals from spills. The *Exxon Valdez* spill created a 750 km slick of crude oil and killed an estimated 250,000 seabirds, 2800 sea otters (*Enhydra lutris*), 22 orca (*Orcinus orca*), as well as the salmon and herring eggs mentioned previously (Malakoff, 2014). The embryos of birds are particularly sensitive to oil pollution. Oil contaminated hens return to the nest and deposit on eggs. As little as 1–20 µL of light oil that penetrates the eggshell can cause death (Parenell et al., 1984; Hoffman, 1990). Since 1 mL of fluid is roughly equivalent to 10 drops and 1 µL is 1/1000 of a milliliter, much less than a drop on the shell can result in dead embryos.

While there is a lot of attention and media coverage of petroleum disasters such as BP Deepwater Horizon and *Exxon Valdez,* there is not a lot of coverage on a more chronic problem associated with oil; that of chronic pollution produced by natural seeps, leaking shipwrecks, vessels legally and illegally cleaning bilges in marine waters, and other nonpoint sources (Henkel et al., 2014). Fortunately, dumping waste oil and bilge water in US waters was declared illegal in 1990 so that one source has been reduced, but enforcement can sometimes be problematic. Chronic pollution can have serious effects on seabird populations without causing a lot of hype because only a few birds are killed or affected at a time. However, when all of the individual incidents are accumulated, nearly 300,000 seabirds may be killed annually by chronic oiling in waters surrounding Newfoundland (Wiese and Roberston, 2004) and major die-offs have been reported in Europe, the southwest Atlantic coast, and in California (Henkel et al., 2014).

One of the very important areas in this regard is the coast of Santa Barbara Channel in California which contains some of the most productive marine oil seeps in the world, and may contribute 20,000 tonnes of crude oil to the marine environment per year (Kvenolden and Cooper, 2003). The channel was also the

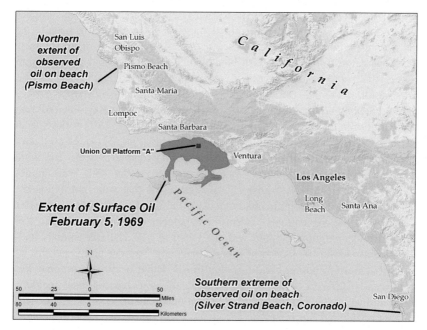

FIGURE 7.13   Location of the Santa Barbara oil spill, 1969. *Image: Wikimedia Commons.*

site of the huge Santa Barbara oil spill in 1969 (Fig. 7.13). Beach watches for oiled wildlife have been ongoing along the shores of the Santa Barbara Channel since 1970, the year after the famed oil spill. While sporadic mortality is observed to the present day, an unexpected oiling event occurred in 2005 when almost 1500 birds, mostly western (*Aechmophorus occidentalis*) and Clark's (*Amphiprion clarkii*) Grebes (Humple et al., 2011) died. More than 50,000 seabirds have died in central California between 1990 and 2003 from bunker fuel oil leaking from the *S.S. Jacob Luckenbach,* a cargo ship that sunk more than 50 years ago off of San Francisco (Luckenbach Trustee Council, 2006).

Henkel et al. (2014) reported on the number of oiled birds collected from 2005 to 2010 along central and southern California shorelines. To determine the source of the oil, they used "oil fingerprinting," which relies on telltale ratios of specific petroleum components that vary among sources. During these 6 years, 1475 live oiled birds were collected and, when possible, cleaned and released. Santa Barbara County had the greatest frequency of live birds, with 64.6% of the total number captured; numbers decreased with distance north and south of that county. The beach surveillance identified 43 different oiled species as either live or dead with western/Clark's Grebe (41%), common murre (*Uria aalge*; 29%), loons (*Gavia* spp. 12%), and brown pelican (*Pelecanus occidentalis*; 7%) as the most common. Other species that were commonly found offshore included rhinoceros auklet (*Cerorhinca monocerata*), northern fulmar

(*Fulmarus glacialis*), and sooty shearwater (*Puffinus griseus*); these were usually found dead. The authors determined that 89% of the oiled birds had oil from the Monterey Formation which was a natural source coming from oil seeps. Only 6% came from Luckenbach bunker fuel or an additional anthropogenic source; the rest (5%) could not be identified to source.

Among mammals, 32 common bottlenose dolphins (*Tursiops truncatus*) were captured, examined and released from Barataria Bay, LA (BB) and compared with 27 dolphins captured in Sarasota Bay, FL (SB) following the BP Deepwater Horizon oil spill in 2011 (Schwake et al., 2014). Many factors may affect the health of dolphins, but the animals captured from BB which received prolonged and heavy contamination from the spill were noticeably in poorer health than those captured from SB that were not exposed to oil. Three of the BB dolphins had extensive tooth loss compared to no extensive loss in the SB animals. Twenty-five percent of the BB dolphins were deemed underweight compared with 1 (3.4%) in SB. The BB animals had a significantly higher incidence of a particular lung disorder and generally had lungs that were in poorer condition that in SB. Other health factors that were significantly different between the two populations included greater incidence of inflammation, hypoglycemia, altered iron metabolism, lower serum concentrations of adrenal hormones, lower serum cholesterol, and higher incidence of liver disorders in BB dolphins than in SB dolphins. Fourteen of 29 dolphins fully examined in BB were classified as being in guarded or worse condition whereas only one of 15 was considered to be in guarded condition in SB with the rest being in fair to good health status. Somewhat surprisingly, the authors determined the concentrations of several organochlorine pesticides, PCBs and PBDEs in blubber biopsies and animals in SB actually had higher concentrations than those in BB, ruling out effects from these contaminants. They did not report on PAH concentrations, but since these contaminants can be rapidly metabolized and were probably dissipated by the time animals were caught, their analysis might not be very meaningful. In summary, there was a "smoking gun" hypothesis; namely, that the oil in BB negatively affected the health of dolphins living there.

STUDY QUESTIONS

1. Fill in the blank: PAHs are among the most_____ contaminants in the environment.
2. Is this structure of pyrene saturated or substituted? Why?

3. Name four chemical characteristics of PAHs that change in accordance with the number of rings (or molecular weight) in the molecule.

4. True or False. PAHs can be naturally occurring molecules in the environment.
5. List several anthropogenic sources of PAHs.
6. True or False. Compared with PCBs and organochlorine pesticides, most saturated PAHs are relatively short lived but some substituted PAHs can persist for decades.
7. Discuss three ways of remediating soil, water, or sediment that has become contaminated by PAHs.
8. What molecule is often considered to be the benchmark of cancer-causing PAHs?
9. Do PAHs typically bioaccumulate? Bioconcentrate? Biomagnify?
10. Discuss some of the effects of PAHs on organisms.
11. What is the role of a bay region in a PAH molecule and DNA adducts?
12. Speculate with your class on why fish like the brown bullhead may be more seriously affected by PAH than other species of fish.
13. What is the role of sunlight on PAH persistence?
14. True or False. Because PAHs are already organic, their methylation would not affect their toxicity.
15. Describe how PAHs are thought to interact with molecular receptors on plasma membranes to produce endocrine disrupting effects.
16. What is the relationship between oil spills and PAH pollution?
17. True or False. Major spills from tankers or drilling platforms are the only form of oil pollution worthy of mention with regards to wildlife mortality.

REFERENCES

Aas, E., Liewenborg, B., Grosvik, B.E., Camus, L., Jonsson, G., Borseth, J.F., et al., 2003. DNA adduct levels in fish from pristine areas are not detectable or low when analysed using the nuclease P1 version of the 32P-postlabelling technique. Biomarkers 8, 445–460.

Alaska Oil Spill Commission, 1990. Final report. State of Alaska, Juneau. Available through Exxon Valdez Oil Spill Trustee Council. Available from: <http://www.evostc.state.ak.us/index.cfm?FA=facts.details>.

Albers, P.H., 2003. Petroleum and individual polycyclic aromatic hydrocarbons. In: Hoffman, D.J., Rattner, B.A., Burton Jr., G.A., Cairns Jr., J. (Eds.), Handbook of Ecotoxicology, second ed. Lewis Publishers, Boca Raton, pp. 341–371.

Anderson, R.S., Doos, J.E., Rose, F.L., 1982. Differential ability of *Ambystoma tigrinum* hepatic microsomes to produce mutagenic metabolites from polycyclic aromatic hydrocarbons and aromatic amines. Cancer Lett. 16, 33–414.

Arcaro, K.F., O'Keefe, P.W., Yang, Y., Clayton, W., Gierthy, J.F., 1999a. Antiestrogenicity of environmental polycyclic aromatic hydrocarbons in human breast cancer cells. Toxicology 133, 115–127.

Arcaro, K.F., O'Keefe, P.W., Yang, Y., Clayton, W., Gierthy, J.F., 1999b. Benzo[k]fluoranthene enhancement and suppression of 17-β-estradiol catabolism in MCF-7 breast cancer cells. J. Toxicol. Environ. Health A 58, 413–426.

Atlas, R.M., Bartha, R., 1973. Fate and effects of polluting petroleum in the marine environment. Residue Rev. 49, 49–85.

ATSDR [Agency for Toxic Substances and Disease Registry], 1995. Toxicological Profile for Polycyclic Aromatic Hydrocarbons. U.S. Health and Human Services, Public Health Administration, Agency for Toxic Substances and Disease Registry, Atlanta, GA.

ATSDR [Agency for Toxic Substances and Disease Registry], 2009. Studies in Environmental Medicine: Toxicity of Polycyclic Aromatic Hydrocarbons (PAHs). Available from: <http://www.atsdr.cdc.gov/csem/pah/docs/pah.pdf> (accessed 08.08.15.).

ATSDR [Agency for Toxic Substances and Disease Registry], 2014. Priority List of Hazardous Substances. Available from: <http://www.atsdr.cdc.gov/spl/> (accessed 08.08.15.).

Balk, L., Meijer, J., DePierre, J.W., Appelgren, L., 1984. The uptake and distribution of [3H] benzo[a]pyrene in the Northern pike (*Esox lucius*). Examination by whole-body autoradiography and scintillation counting. Toxicol. Appl. Pharmacol. 74, 430–449.

Baumann, P.C., Harshbargar, J.C., 1995. Decline in liver neoplasms in wild brown bullhead catfish after coking plant closes and environmental PAHs plummet. Environ. Health Perspect. 103, 168–170.

Baussant, T., Sanni, S., Jonsson, G., Arnfinn, S., Børseth, J.F., 2001. Bioaccumulation of polycyclic aromatic compounds: 1. Bioconcentration in two marine species and in semipermeable membrane devices during chronic exposure to dispersed crude oil. Environ. Toxicol. Chem. 20, 1175–1184.

Bell, B., Spotila, J.R., Congdon, J., 2005. High incidence of deformity in aquatic turtles in the John Heinz National Wildlife Refuge. Environ. Poll. 142, 457–465.

Bommarito, T., Sparling, D.W., Halbrook, R.S., 2010a. Toxicity of coal-tar pavement sealants and ultraviolet radiation to *Ambystoma maculatum*. Ecotoxicology 19, 1147–1156.

Bommarito, T., Sparling, D.W., Halbrook, R.S., 2010b. Toxicity of coal–tar and asphalt sealants to eastern newts, *Notophthalmus viridescens*. Chemosphere 81, 187–193.

Booc, F., Thornton, C., Lister, A., MacLatchy, D., Willett, K.L., 2014. Benzo[a]pyrene effects on reproductive endpoints in *Fundulus heteroclitus*. Toxicol. Sci. 140, 173–182.

BP, 2014. Containing the Leak. Available from: <http://www.bp.com/en/global/corporate/gulf-of-mexico-restoration/deepwater-horizon-accident-and-response/containing-the-leak.html> (accessed 08.08.15.).

Brummell, B.F., Price, D.J., Birge, W.J., Harmel-Laws, E.M., Hitrond, J.A., Elskus, A.A., 2010. Differential sensitivity of CYP1A to 3,3',4',4-tetrachlorobiphenyl and benzo(a)pyrene in two *Lepomis* species. Comp. Biochem. Physiol. C 152, 42–50.

Brunström, B., Bröman, D., Näf, C., 1991. Toxicity and erod-inducing potency of 24 polycyclic aromatic-hydrocarbons (PAHS) in chick-embryos. Arch. Toxicol. 65, 485–489.

Bustanes, J.O., 2013. Reproductive recovery of a common eider *Somateria mollissima* population following reductions in discharges of polycyclic aromatic hydrocarbons (PAHs). Bull. Environ. Contam. Toxicol. 91, 202–207.

Cerniglia, C.E., White, G.L., Heflich, R.H., 1985. Fungal metabolism and detoxification of polycyclic aromatic hydrocarbons. Arch. Microbiol. 143, 105–110.

De Souza Machado, A.A., Müller Hoff, M.L., Klein, R.D., Cordeiro, G.J., Avila, J.M.L., Costa, P.G., et al., 2014. Oxidative stress and DNA damage responses to phenanthrene exposure in the estuarine guppy *Poecilia vivipara*. Mar. Sci. Res. 98, 96–105.

Dong, W., Wang, L., Thornton, C., Scheffler, B.E., Willett, K.L., 2008. Benzo(a)pyrene decreases brain and ovarian aromatase mRNA expression in *Fundulus heteroclitus*. Aquat. Toxicol. 88, 289–300.

Dubrovskaya, E.V., Polikarpova, I.O., Muratova, A.Y., Pozdnyakova, N.N., Chernyshova, M.P., Turkovskaya, O.V., 2014. Changes in physiological, biochemical, and growth parameters of sorghum in the presence of phenanthrene. Russ. J. Plant Physiol. 61, 529–536.

Edwards, N.T., 1983. Polycyclic aromatic hydrocarbons (PAH's) in the terrestrial environment—a review. J. Environ. Qual. 12, 427–441.

Eisler, R., 2000. Handbook of Chemical Risk Assessment: Health Hazards to Humans, Plants and Animals. Vol 2. Organics. Lewis Publishers, Boca Raton.

Ellis, B., Harold, P., Kronberg, H., 1991. Bioremediation of a creosote contaminated site. Environ. Technol. 12, 447–459.

Fernandez, M., L'Haridon, J., 1994. Effects of light on the cytotoxicity and genotoxicity of benzo[a] pyrene and an oil refinery effluent in the newt. Environ. Mol. Mutagen. 24, 124–136.

Henkel, L.A., Nevins, H., Martin, M., Sugarman, S., Harvey, J.T., Ziccardi, M.D., 2014. Chronic oiling of marine birds in California by natural petroleum seeps, shipwrecks, and other sources. Mar. Poll. Bull. 79, 155–163.

Hoffman, D.J., 1990. Embryotoxicity and teratogenicity of environmental contaminants to bird eggs. Rev. Environ. Contam. Toxicol. 115, 39–45.

Humple, D.L., Nevins, H.M., Phillips, E.M., Gibble, C., Henkel, L.A., Boylan, K., et al., 2011. Demographics of *Aechmophorus* grebes killed in three mortality events in California. Mar. Ornithol. 39, 235–242.

Hwang, H., Foster, G.D., 2006. Characterization of polycyclic aromatic hydrocarbons in urban stormwater runoff flowing into the tidal Anacostia River, Washington DC. Environ. Poll. 140, 416–426.

International Tanker Owners Pollution Federation Limited, 2014. Oil Tanker Spill Statistics 2013. International Tanker Owners Pollution Federation Limited, London.

Jackim, E., Lake, C., 1978. Polynuclear aromatic hydrocarbons in estuarine and near shore environments. In: Wiley, M.L. (Ed.), Proceedings of 10th National Shellfish Sanitation Workshop, Hunt Valley, NY, pp. 415–428.

Jacobs, L.E., Weavers, L.K., Chin, Y., 2008. Direct and indirect photolysis of polycyclic aromatic hydrocarbons in nitrate-rich surface waters. Environ. Toxicol. Chem. 27, 1643–1648.

Karrow, N.A., Boermans, H.G., Dixon, D.G., Hontella, A., Solomon, K.R., Whyte, J.J., et al., 1999. Characterizing the immunotoxicity of creosote to rainbow trout (*Oncorhynchus mykiss*): a microcosm study. Aquat. Toxicol. 45, 223–239.

Kennedy, C.J., Law, F.C.P., 1990. Toxicokinetics of selected polycyclic aromatic hydrocarbons in rainbow trout following different routes of exposure. Environ. Toxicol. Chem. 98, 133–139.

Kipopoulou, A.M., Manoli, E., Samara, C., 1999. Bioconcentration of polycyclic aromatic hydrocarbons in vegetables grown in an industrial area. Environ. Poll. 106, 369–380.

Kummer, V., Mašková, J., Zralý, Z., Neča, J., Šimečková, P., Vondráček, J., et al., 2008. Estrogenic activity of environmental polycyclic aromatic hydrocarbons in uterus of immature Wistar rats. Toxicol. Lett. 180, 212–221.

Kvenolden, K.A., Cooper, C.K., 2003. Natural seepage of crude oil into the marine environment. Geo-Mar. Lett. 23, 140–146.

Leadley, T.A., Balch, G., Metcalfe, C.D., Lazar, R., Mazak, E., Habowsky, J., et al., 1998. Chemical accumulation and toxicological stress in three brown bullhead (*Ameirurus nebulosus*) populations of the Detroit River, Michigan, USA. Environ. Toxicol. Chem. 17, 1756–1766.

Lee, S.D., Grant, L. (Eds.), 1981. Health and Ecological Assessment of Polynuclear Aromatic Hydrocarbons Pathotex Publ., Park Forest South IL.

Lu, P.-Y., Metcalf, R.L., Plummer, N., Mandrel, D., 1977. The environmental fate of three carcinogens: benzo-(a)-pyrene, benzidine, and vinyl chloride evaluated in laboratory model ecosystems. Arch. Environ. Contam. Toxicol. 6, 129–142.

Luckenbach Trustee Council, 2006. S.S. Jacob Luckenbach and Associated Mystery Oil Spills Final Damage Assessment and Restoration Plan/Environmental Assessment. Prepared by California Department of Fish and Game, National Oceanic and Atmospheric Administration, US Fish and Wildlife Service, National Park Service. Available from: <https://www.wildlife.ca.gov/OSPR/NRDA/Jacob-Luckenbach> (accessed 08.08.15.).

Mahler, B.J., Van Metre, P.C., Foreman, W.T., 2014. Concentrations of polycyclic aromatic hydrocarbons (PAHs) and azaarenes in runoff from coal-tar- and asphalt-seal coated pavement. Environ. Poll. 188, 81–87.

Malakoff, D., 2014. 25 Years after the Exxon Valdez, where are the herring? Science 343, 1416.

Näf, C., Broman, D., Brunström, B., 1992. Distribution and metabolism of polycyclic aromatic hydrocarbons (PAHs) injected into eggs of chicken (*Gallus domesticus*) and common eider duck (*Somateria mollissima*). Environ. Toxicol. Chem. 11, 1653–1660.

Neff, J.M., 1982. Accumulation and release of polycyclic aromatic hydrocarbons from water, food, and sediment by marine animals. In: Richards, N.L., Jackson, B.L. (Eds.), Symposium: Carcinogenic Polynuclear Aromatic Hydrocarbons in the Marine Environment. pp. 282–320. US. EPA Rep 600/9-82-013.

Nguyen, T.C., Loganathan, P., Nguyen, T.V., Vigneswaran, S., Kandasamy, J., Slee, D., et al., 2014. Polycyclic aromatic hydrocarbons in road-deposited sediments, water sediments, and soils in Sydney, Australia: comparisons of concentration distribution, sources and potential toxicity. Ecotox. Environ. Safe. 104, 339–348.

NOAA [National Oceanic and Atmospheric Administration], 2014. Gulf Spill Restoration. Available from: <http://www.gulfspillrestoration.noaa.gov/oil-spill/> (accessed 08.08.15.).

NRDC [Natural Resources Defense Council], 2014. Cleaning up the Anacostia River. Available from: <http://www.nrdc.org/water/pollution/fanacost.asp> (accessed 08.08.15.).

Okparanma, R.N., Coulon, F., Mayr, T., Mouazen, A.M., 2014. Mapping polycyclic aromatic hydrocarbon and total toxicity equivalent soil concentrations by visible and near-infrared spectroscopy. Environ. Poll. 192, 162–170.

Parnell, J.F., Shield, M.A., Frierson, D., 1984. Hatching success of brown pelican eggs after contamination with oil. Colony Waterbirds 7, 22–24.

Pearson, W.H., Deriso, R.B., Elston, R.A., Hook, S.E., Parker, K.R., Anderson, J.W., 2012. Hypotheses concerning the decline and poor recovery of Pacific herring in Prince William Sound, Alaska. Rev. Fish Biol. Fish. 22, 95–135.

Pinkney, A.E., Harshbarger, J.C., Karouna-Renier, N.K., Jenko, K., Balk, L., Skarphéðinsdóttir, H., et al., 2011. Tumor prevalence and biomarkers of genotoxicity in brown bullhead (*Ameiurus nebulosus*) in Chesapeake Bay tributaries. Sci. Total Environ. 410–411, 248–257.

Ravindra, K., Sokhia, R., Van Grieken, R., 2008. Atmospheric polycyclic aromatic hydrocarbons: source attribution, emission factors and regulation. Atmos. Environ. 42, 2895–2921.

Schwacke, L.H., Smith, C.R., Townsend, F.I., Wells, R.S., Hart, L.B., Balmer, B.C., et al., 2014. Health of common bottlenose dolphins (*Tursiops truncatus*) in Barataria Bay, Louisiana, following the Deepwater Horizon oil spill. Environ. Sci. Toxicol. 48, 93–103.

Smithsonian Institute Ocean Portal, 2014. Gulf oil Spill. Available from: <http://ocean.si.edu/gulf-oil-spill>.

Southworth, G.R., Beauchamp, J.J., Schmeider, P.K., 1978. Bioaccumulation potential of polycyclic aromatic hydrocarbons in *Daphnia pulex*. Water Res. 12, 973–977.

Steevens, J.A., Slattery, M., Schlenk, D., Aryl, A., Benson, W.H., 1999. Effects of ultraviolet-B light and polyaromatic hydrocarbon exposure on sea urchin development and bacterial bioluminescence. Mar. Environ. Res. 48, 439–457.

US EPA [U.S. Environmental Protection Agency], 1980. Ambient water quality criteria for polynuclear aromatic hydrocarbons, Washington, DC. U.S. EPA Rep. 440/5-80-069.

US EPA [U.S. Environmental Protection Agency], 2007. Survey of new findings in scientific literature related to atmospheric deposition to the great waters: polycyclic aromatic hydrocarbons (PAH). U.S. Environmental Protection Agency Office of Air Quality Planning and Standards. Research Triangle Park, North Carolina. 31 pp.

US EPA [U.S. Environmental Protection Agency], 2013. Priority Pollutants. Available from: <http://water.epa.gov/scitech/methods/cwa/pollutants.cfm> (accessed 08.08.15.).

US EPA [U.S. Environmental Protection Agency], 2014. Great Lakes Monitoring: Air Indicators, Atmospheric Deposition of Toxic Pollutants. Available from: <www.epa.gov/greatlakes/glindicators/air/airb.html> (accessed 08.08.15.).

Van Meter, R.J., Spotila, J.R., Avery, H.W., 2005. Polycyclic aromatic hydrocarbons affect survival and development of common snapping turtle (*Chelydra serpentina*) embryos and hatchlings. Environ. Poll. 142, 466–475.

U.S. Fish and Wildlife Service, 2011. Deepwater Horizon Response Consolidated Fish and Wildlife Collection Report. Available from: <http://www.fws.gov/home/dhoilspill/pdfs/ConsolidatedWildlifeTable042011.pdf> (accessed 08.08.15.).

Wang, L., Camus, A.C., Dong, W., Thorton, C., Willett, K.L., 2010. Expression of CYP1C1 and CYP1A in *Fundulus heteroclitus* during PAH-induced cancer. Aquat. Toxicol. 99, 439–447.

Wick, A.F., Haus, N.W., Sukkariyah, B.F., Haering, K.C., Daniels, W.L., 2011. Remediation of PAH Contaminants in Soils and Sediments: A Literature Review. Virginia Polytech Inst State Univ. Dept Crops Soil Sci, Blacksburg VA.

Wiese, F.K., Roberston, G.J., 2004. Assessing seabird mortality from chronic oil discharges at sea. J. Wildl. Manage. 68, 627–638.

Wills, L.P., Jung, D., Koehrn, K., Zhu, S., Willett, K.L., Hinton, D.E., et al., 2010. Comparative chronic liver toxicity of benzo[a]pyrene in two populations of the Atlantic killifish (*Fundulus heteroclitus*) with different exposure histories. Environ. Health Perspect. 118, 1376–1381.

Xi, Z., Chen, B., 2014. Removal of polycyclic aromatic hydrocarbons from aqueous solution by raw and modified plant residue materials as biosorbents. J. Environ. Sci. 26, 737–748.

Zhang, L., Jin, Y., Huang, M., Penning, T.M., 2012. The role of human aldo-keto reductases in the metabolic activation and detoxication of polycyclic aromatic hydrocarbons: interconversion of PAH catechols and PAH quinones. Front. Pharmacol. http://dx.doi.org/10.3369/fpharm.2012.00193.

Zhang, Y., Tao, S., 2009. Global atmospheric emission inventory of polycyclic aromatic hydrocarbons (PAHs) for 2004. Atmos. Environ. 43, 812–819.

Chapter 8

Metals

Terms to Know

Metal
Metalloid
Heavy Metals
Ductile
Malleable
Paracelsus
Minimum daily requirement
Recommended daily allowance
Tolerable upper limits
Permissible exposure limits
Hyperaccumulator
Competitive binding
Metallothionein
Depurated
Minamata disease
Organometals
Methylmercury

INTRODUCTION

In Chapters 4 and 5, we studied organochlorine and current-use pesticides. These compounds are intentionally designed to be lethal towards unwanted organisms including plants and animals. In Chapters 6 and 7, we looked at polychlorinated biphenyls (PCBs) and polycyclic aromatic hydrocarbons (PAHs) that are not designed to kill, but are persistent and can be very toxic. All of these are organic molecules because they have carbon and hydrogen as principal constituents, however none are required for living organisms. Other than PAHs, which can be naturally occurring, the vast majority of the chemicals in those chapters were also synthetic. In this chapter, we discuss a very different type of contaminant that is both inorganic and, in some cases, required for life; they all are naturally occurring in their native form. Metals and metalloids are elements that occupy a large portion of the Periodic Chart of Elements (Fig. 8.1).

Given the breadth of elements that belong to the metals group, there is a wide diversity in their chemical behavior. In general, with the exception of

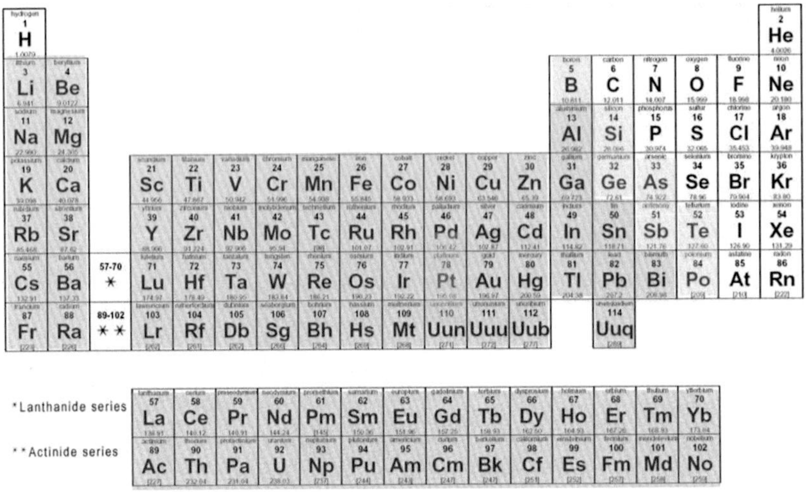

FIGURE 8.1 Periodic chart of elements. Elements in darker gray are metals, those in light gray are metalloids, and those in white are nonmetals.

mercury, **metals** share common properties of being **malleable** (can be shaped permanently); **ductile** (capable of being drawn into thin strands); generally good electrical and thermal conductors; **fusible** (capable of being melted and blended); opaque; have a metallic luster; solid at room temperature; typically contain two to three electrons in outermost valence shell; and easily ionized. Solubility of metals in water varies from those elements at the far left of the periodic chart—that need to be protected from water due to high reactivity—to those that, as we progress from left to right, are soluble especially in acidic solutions, and those that are reactive only with strong acids to those that are virtually inert. Mercury is unique in that it is a liquid at room temperature and hence is neither malleable nor ductile except at low temperatures. **Metalloids** share some but not all of the properties of metals and include boron (B), selenium (Se), germanium (Ge), arsenic (As), actinium (Ac), tellurium (Te), and polonium (Po).

From a toxicological perspective, the metals that have received the greatest attention are cadmium (Cd), chromium (Cr), lead (Pb), mercury (Hg), copper (Cu), nickel (Ni), and zinc (Zn). These are called **heavy metals** because most have higher molecular weights and densities than iron (Fe). However, that source of infinite wisdom, Wikipedia, considers any metal of environmental concern a heavy metal. Thus, the term seems to be in the eye of the beholder, so to speak. Suffice to say that the technicalities of what constitutes a heavy metal or even what distinguishes a metal from a metalloid are not all that pertinent for our study. If you are really interested in these matters, many universities offer courses in inorganic chemistry, perhaps you have already taken one. For our purposes, we will focus on the metals that are of greatest ecotoxicological significance.

Important characteristics of heavy metals include their solubility in water and their concomitant lipophobicity. Metals may bioconcentrate for short durations but do not biomagnify because they have these characteristics. However, some heavy metals complex with organic molecules and these can bioconcentrate along food chains.

SOURCES OF METALS IN THE ENVIRONMENT

The natural sources of metals are in the lithosphere or terrestrial environment. Volcanoes are a natural source of metals from deep within the earth to the crust. Sometimes they occur in their elemental or native forms, but are often combined with other metals, sulfides, carbamates, oxides, and other chemical groups. Metals with economic value are mined and smelted in various ways to produce purer forms of the raw substance. The mining and processing of metal ores are among the principal anthropogenic sources of metals in the environment. Other sources of environmental contamination occur when metals are used in industrial applications. Historically, mining and industrial processes have been far from environmentally friendly and often allowed effluents to contaminate rivers and streams. This still goes on despite regulations to the contrary, but the amount of pollution has decreased dramatically since the 1960s. Another way that some metals enter the atmosphere is through combustion of fossil fuels, especially coal. Technological advancements can remove many of these contaminants from the coal and from combustion products—the so-called smokestack scrubbers, leading to cleaner exhausts. The key to clean burning, however, is installing these scrubbers which easily cost many thousands of dollars. Prior to catalytic converters, automobile exhausts were also significant sources of lead, but the converters now require unleaded gasoline.

Slag, or the residue that is left behind after minerals are extracted from ore, can still contain high concentrations of metal mixtures, even with more efficient extraction procedures. Modern treatment of this slag usually involves depositing it into landfills. Older and poorly managed landfills can be inadequately isolated from the underlying soil, allowing the metals and other toxins to leach into groundwater.

BIOLOGICAL EFFECTS OF METALS

Paracelsus (1493–1541) who some call the "father of modern toxicology," stated that the "dose makes the poison," meaning that all things can be poisonous if given in sufficient amounts. He may well have been thinking of metals. Among animals and plants some metals are necessary in comparatively large amounts and are macronutrients. Most essential metals, however are needed in only small quantities and are called micronutrients. What distinguishes large versus small quantities? One way is to consider lipids, carbohydrates, and proteins as macronutrients. However, the amount of iron, copper, or zinc needed by animals is high compared to other metals and these can be considered as macronutrients as well. The amount of these macronutrients and micronutrients that humans require each day is called the **Minimum Daily Requirements (MDRs)** that are established by the US Food and Drug Administration. These MDRs may vary somewhat by sex and age. Other standards for nutrients including these essential minerals include the **Recommended Daily Allowances (RDAs)** that are set somewhat higher and indicate what people in certain age and sex brackets should be ingesting; these are set by the Food and Nutrition Board of the National Institute of Medicine. **Tolerable Upper Levels (TUL)**, also defined by the Food and Nutrition Board, indicate the maximum amount of intake before toxicity occurs. Maximum allowable amounts usually exceed MDRs by at least an order of magnitude but eventually metal toxicity can occur at persistently high dietary concentrations.

There is a balance between sufficient intake of metals and toxic intake (Fig. 8.2). At very low intake, physiological problems can occur because the processes dependent on the metals cannot take place. At somewhat higher levels, minor physiological problems that might be difficult to diagnose can occur. At some range, there is an optimum amount of metal being assimilated. At higher doses than optimum, toxicity may begin to set in and become a serious problem, possibly leading to death at even higher intake concentrations.

This means that even biologically essential metals can exert a wide variety of toxic effects given sufficient dosage. Elemental metals can chelate with enzymes, thereby interfering with the processes that these enzymes synergize. Depending on the metal and exposure concentration, serious neurological, metabolic, reproductive, teratogenic, and immunological damage can occur through

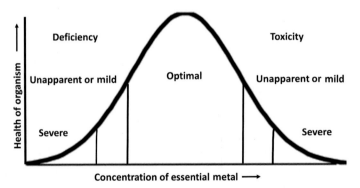

FIGURE 8.2 The relationship between concentration of essential metals, deficiency, and toxicity. At severely low concentrations, the concentration of the metal is insufficient to provide for physiological needs. The severity of the deficiency decreases as the metal concentration increases, giving way to ideal conditions. At even higher concentrations toxicity begins to set in until it becomes lethal. *Modified from Alloway, B.J. 1995. Metals in Soils, 2nd ed. New York: Blackie Academic & Professional, Chapman & Hall.*

exposure. For instance, both lead and mercury, two of the most toxic metals, can produce very serious neurological effects in humans and wildlife.

Plants tend to be more tolerant to metal exposure than animals, but they too can experience problems with photosynthesis, discoloration of leaves, reduced growth, and root death. **Hyper accumulator** plants can tolerate, uptake, and translocate high levels of certain heavy metals that would be toxic to most organisms. These plants can have leaves that contain >100 mg/kg of Cd, >1000 mg/kg of Ni and Cu, or >10,000 mg/kg of Zn and Mn (dry weight) when grown in metal-rich environments (Kamal et al., 2004).

Many factors affect the toxicity of metals. Some of these factors include valence state of the metal, whether the metal is combined with an organic molecule, environmental pH, concentration of organic matter, and calcium concentrations in soil or water, the relative concentration of other metals, and a class of metal-transporting proteins call metallothioneins. Many metals can assume multiple valence states and toxicity may vary with state. Chromium, for example, is mostly found in its elemental state, as a trivalent Cr^{3+} ion, or as a hexavalent Cr^{6+} ion. Of these, elemental chromium is essentially nontoxic, Cr^{3+} has a low toxicity, but Cr^{6+} has high toxicity.

Metals can occur in their inorganic forms either as ions or as nonionic elements. They can also combine with organic molecules such as ethyl or methyl groups. **Organometals** can be intentionally produced for a variety of industrial or agricultural purposes, but they can also occur naturally. Most often the organic portions augment the toxicity of the metal itself because they facilitate biological assimilation and ease the passage into cells. Inside the cells, the organometals can become DNA adducts and cause cancers, malformations, and other toxic effects. Some of the metals used to form organometals include mercury, boron, silicon, selenium, germanium, tin, lead, arsenic, and platinum. Serious environmental issues have been caused by methylmercury and selenomethionine, both of which form through natural processes.

A major factor in determining the valence state of a metal is the pH of the environment. Metals are oxidized at low pH, which means they move from a low ionization state to a higher ionization state (eg, elemental Zn or Zn^0 to Zn^{2+}). In a dry environment, these ionic forms may combine with oxygen to form oxides or sulfur to form sulfides. In aqueous or moist environments such as lakes, streams, or after heavy rains the ionic forms are more water soluble than the elemental metal. Soluble metallic molecules are also more readily assimilated by organisms than elemental forms.

In the 1980s, a great deal of concern was given to acid precipitation which is produced when hydrogen sulfide, nitrites, or nitrates are released through fossil fuel combustion, then transported through the atmosphere, and come down in wet (eg, rain) or dry (particulate) deposition. Upon entering lakes or streams, these pollutants reduce the pH of the water and increase the solubility of metals. Aluminum (Al) has been closely associated with the toxic effects of acid precipitation. Aluminum is the most common metal and the third most common mineral in the Earth's crust. The valence state of Al varies from 2^- to 3^+ with anions occurring in alkaline waters and cations in acidic systems. Under acidic conditions, dissolved Al results in reduced reproduction among fish and amphibians. Acidification could also affect the growth and development of young waterbirds because aluminum forms insoluble complexes with phosphorus, preventing the normal development of bony tissue (Sparling, 1991). Studies determined that for much of the United States, natural buffering or acid-neutralizing capacity of soils reduced the risks of acid precipitation but regions of low buffering in the United States and Canada remain at greater threat of damage (more on this in Chapter 9).

Metals tend to bind with organic matter either in water or soil. This is a physical binding, not chemical bonding. Bound metals are less bioavailable than free, dissolved metals in the water column. Numerous studies have shown that the concentration of dissolved organic matter (DOM) in water significantly affects the availability and hence toxicity of metals in aquatic organisms. Similarly, soils and sediments with high organic content are generally less toxic than those with low content because of binding.

Calcium and other metals affect toxicity in another way. Calcium in soil or water often exists as calcium carbonate ($CaCO_3$). Calcium carbonate buffers acidity, thus reducing the solubility of metals. In addition, the calcium ion (Ca^{2+}) is preferentially taken up by the digestive systems of animals or by plant cells compared with metals. In this way, environments with moderate to high calcium-soluble concentrations of metals may be reduced and the solutions impeded from entering the blood or cells, thus further reducing toxicity. Metals with biological functions such as zinc may also preferentially bond to cells and inhibit the uptake of other metals. This is called **competitive binding** and it can ameliorate the effects of more toxic metals.

Metallthionein (MT) is a family of cysteine-rich, low molecular weight proteins that have the capacity to bind to metals, whether they are essential or not. Metallothioneins are found across the fungi, plant, and animal kingdoms and are even found in prokaryotes. In higher animals, they are produced by the liver and kidneys and are localized in the membranes of the Golgi apparatus. Metallothioneins provide protection against metal toxicity and oxidative stress and are involved in regulation of essential metals. Their production is induced by the presence of metals and other minerals in the bloodstream. Metallthionein has been documented to bind with a wide range of metals including cadmium, zinc, mercury, copper, arsenic, and silver. Metallothioneins specific for copper and zinc occur in many organisms. Thionein, the organic basis for metallothionein, picks up a metal when it enters a cell and carries the metal to another part of the cell where it is released or secreted. Cysteine residues from MTs can capture harmful oxidant radicals like the superoxide and hydroxyl radicals. In this reaction, cysteine is oxidized to cystine, and the metal ions that were bound to cysteine are released to the bloodstream and away from cells. This mechanism is important in the control of oxidative stress by metallothioneins.

CHARACTERISTICS OF SELECTED METALS

In this section, we describe several metals in greater detail, some of these are essential nutrients, some are very lethal, so as to provide a deeper understanding of the ecotoxicology of metals.

The US EPA (2014) lists several metals as priority pollutants; these are chemical pollutants that are regulated and have analytical methods. This list includes antimony, arsenic, beryllium, cadmium, chromium, copper, lead, mercury, nickel, selenium, silver, thallium, and zinc. Here we will focus on arsenic (As), cadmium (Cd), chromium (Cr), copper (Cu), lead (Pb), mercury (Hg), and zinc (Zn). As we have previously suggested, other metals can be toxic at high concentrations and may have environmental relevance in highly contaminated sites such as around mining or certain industries. However, many of the details presented in this select list can be applied to other metals as well. Very basic characteristics of these metals can be found in Table 8.1. The US EPA's **Maximum Contaminant Level (MCL)** is the maximum concentration allowable in drinking water. The Occupational Safety and Health Administration (OSHA) publishes **Permissible Exposure Limits (PELs)** that define the maximum concentration of a contaminant in workplace air. In contrast to RDA and TUL, these standards are for most contaminants, not just nutrients.

ARSENIC

General Characteristics

Arsenic is a metalloid, which means that it has several characteristics in common with metals but some features that differ. Arsenic, unlike other metallic elements, does not have a true melting point. Rather, it sublimates (goes directly from solid to gas) at what would be its boiling point under standard conditions. It is rarely found naturally in elemental form, being combined with several other elements in the Earth's crust, and it occurs in ionization states of 5^+, 3^+, 2^+, 1^+, and 3^-. Arsenic makes up about 1.5 mg/kg (0.00015%) of the Earth's crust, making it the 53rd most abundant element and moderately rare. In 2012, 45,000 tonnes (metric tons) of AsO_3 were produced, with Chile and China as the major producers (USGS, 2014a).

Environmental Concentrations and Uses of Arsenic

Soil generally contains 1–40 mg/kg of arsenic with an average of 3–4 mg/kg (ATSDR, 2007b). Contaminated soils, of course, can have higher concentrations of As. These values were higher than the US EPA's recommended maximum concentration of 7.2 mg/kg As, but generally lower than the Probable Effects Concentration (PEC) of 42.0 mg/kg. The PEC is the level of a contaminant in the media (ie, surface water, sediment, soil) that is likely to cause adverse effects. Arsenic in soil is predominantly either arsenite (As^{5+}) or arsenate (As^{3+}); of

TABLE 8.1 Basic Characteristics of Selected Metals

Metal	Symbol	Atomic Number	Atomic Mass	Melting Point	Boiling Point	Density (g/cm³)	EPA MCL[a] (mg/L)	OSHA PEL[b] (mg/m³)
Arsenic	As	33	74.9	Undefined	615°C, 1137°F	74.9	0.010	0.2
Cadmium	Cd	48	112.4	321°C, 610°F	767°C, 1413°F	8.65	0.005	0.005
Chromium	Cr	24	52.0	1907°C, 3465°F	2671°C, 4840°F	7.19	0.1	0.5 (Cr⁶⁺)
Copper	Cu	29	63.5	1084°C, 1984°F	2502°C, 4643°F	8.96	1.3	1
Lead	Pb	82	207.2	327°C, 621°F	1749°C, 3180°F	11.34	0	0.05
Mercury	Hg	80	200.6	−38.8°C, −37.9°F	356.7°C, 674.1°F	3.53	0.002	0.1
Zinc	Zn	30	65.4	419.5°C, 787.1°F	907°C, 1665°F	7.14	5	Not defined

[a]Maximum concentration level in drinking water.
[b]Permissible environmental level in workplace air.

these two, arsenite is more soluble, mobile, and toxic. However, arsenate affects cellular phosphorylation more strongly than arsenite. Seawater and freshwater have only 1.6 µg/L As on average, but contaminated sites can have up to 1000 µg/L. Air can have 1 to 2000 ng As/m^3 (ATSDR, 2007a).

In addition to being the literary poison of choice for little old ladies with blue hair, As was and is used in a variety of industries. In agriculture, As was used as a wood preservative between the 1950s and 2004 when concern about its toxicity led to its ban in the European Union and the United States; other nations still use As for this purpose, however. Until 2013, As was used in insecticides. Surprisingly, As has been used in poultry and swine food to increase food efficiency and weight gain—increased growth led to the idea that As may be an essential nutrient but, if it is, no specific function has ever been identified. The practice has been almost totally abandoned, but some producers still use it for turkey in a product labeled nitrasone. Most authorities discount the possibility that As may be a nutrient.

The primary industrial use of As is in alloys with other metals, especially lead and copper. An arsenic/lead alloy can extend the life of car batteries by reducing the loss of zinc from the electrode plates. Arsenic is also used in alloys as a semiconductor. Arsenic has been used as a pigment in Paris Green and, in the 1800s, was used to make candies green. Imagine passing out As-laced candies for Halloween today!

One more historical use of As was important to biologists. Taxidermists and naturalists would use As as a preservative for study skins and mounts. The taxidermist would have a small cup of As powder nearby, and every so often he would lick his finger, pick up the powder on the moistened finger, and apply it to the skin. Unfortunately, there would be residues of As on his finger the next time he licked it, so a lot of taxidermists were poisoned in that way. Old taxidermist mounts may still have As residues and should be treated with care. Hat makers who used natural furs and early human embalmers had the same occupational risk.

Biological Effects of Arsenic

Arsenic and phosphorus are similar in molecular size and reactivity and, consequently, As can compete with phosphorus in biological reactions. Arsenic interferes with ATP, RNA, and DNA production through several mechanisms. It can interfere with the citric acid cycle of respiration by inhibiting cofactors for pyruvate dehydrogenase. It can uncouple oxidative phosphorylation, thus inhibiting the energy-linked reduction of NAD$^+$ and mitochondrial respiration, and it increases hydrogen peroxide production in cells which can increase oxidative stress. These metabolic disturbances can lead to death from multisystem organ failure.

Arsenic can function as an endocrine disruptor by affecting gene regulation through receptors on thyroid cells. The metal binds with the receptor, gains access to a cell, and disrupts gene activity associated with those receptors. As a result, As has been implicated for interfering with thyroid activity in rats, fish, and perhaps humans (ATSDR, 2007a).

Indian cricket frogs (*Rana limnocharis*) were exposed to sodium arsenite at concentrations ranging from 0 to 400 µg/L through metamorphosis (Singha et al., 2014). Neither increased lethality nor reduced body mass was noted. However, sodium arsenite accelerated the rate of metamorphosis at 100 and 400 µg/L, reduced body size, and induced developmental deformities such as loss of limbs. Significant genotoxicity occurred at both concentrations of sodium arsenite/L. Naïve earthworms also showed genotoxicity that was positively related to the concentration of As in soil. However, there was evidence of adaptation in that worms that had been collected from As-contaminated soil did not show any adverse DNA effects (Button et al., 2010).

Arsenic may be toxic to plants, but there are species-specific differences. White clover (*Trifolium repens*) was exposed to soil concentrations of 5–20 mg/kg As, 20–60 mg/kg Cd, or a combination of both metals (Ghiani et al., 2014). The As-only treated plants assimilated As proportionally to the amount in soil. However when As and Cd were mixed, As reduced the uptake of Cd while Cd facilitated the uptake of As. All As and Cd treatments resulted in increased genotoxicity and As was more effective in inducing genotoxicity than Cd. The fern *Pteris vittata* is an As hyper accumulator and is used to remove As during remediation or clean-up operations. Arsenic can actually increase plant growth in hyper accumulators (Tu and Ma, 2002).

CADMIUM

General Characteristics

Cadmium is insoluble in water, has a high sheen, and is resistant to corrosion. The average concentration of Cd in the Earth's crust is between 0.1 and 0.5 mg/kg (0.0001 to 0.0005%) although, it can be much higher in contaminated sites. This means that Cd is moderately rare compared with many other metals. As a salt, Cd binds with sulfates and chlorides and is considerably more soluble than elemental Cd. Typical concentrations in fresh or salt water are around 0.05 µg/kg and 0.003 pg/m^3 in air (Eisler, 2000). Most of the world's Cd is in soil but because of the vast amount of water, its other principal location is oceans. Cadmium is usually found with other minerals such as zinc and is extracted during the mining of those metals. Global production of Cd is approaching 23,000 tonnes per year with most from eastern Asia.

Environmental Concentrations and Uses of Cadmium

Approximately 86% of mined Cd is used in the production of Ni-Cad for use in batteries. Another 6% of available Cd is used in electroplating, particularly in aircraft due to the metal's resistance to corrosion. Other industrial uses include a protectant in nuclear fission studies; yellow, orange, or red pigments in paint;

solders; or in polyvinyl chloride (PVC) pipe as a stabilizer. Many instruments use Cd in semiconductors.

Some Cd may enter the environment through natural weathering of soils but most comes from human-related activities associated with mining, industrial effluents, smelting, fuel combustion, improper disposal of metal containing materials, through the application of phosphate fertilizers, or in sewage sludge. Wet and dry deposition of Cd from the atmosphere may also contribute sizable amounts of Cd to soil in the areas surrounding sources of atmospheric emissions. The implementation of the Resource Conservation and Recovery Act (RCRA) in 1976 affected disposal of Cd and many other materials through increased recycling; nearly 100% of Cd can be recycled. Nevertheless, Cd has been identified in at least 61% of the 1669 hazardous waste sites that have been proposed for inclusion on the EPA National Priorities List (ATSDR, 2012).

Stormwater detention ponds are designed to temporarily hold runoff to allow solids to settle and thereby reduce pollution in urban streams. Stephensen et al. (2014) found that concentrations of Cd in sediments of detention ponds ranged from approximately 0.08–0.9 mg/kg dosing weight (dw) compared with 0.1–3.5 mg/kg dw in nearby lakes. In fauna of these water bodies, largely consisting of mollusks and dragonfly nymphs, average Cd concentrations ranged from 0.05–0.53 mg/kg in ponds and 0.02–0.18 mg/kg in lakes. Overall the authors concluded that there really were not many differences in Cd concentrations between lakes and stormwater ponds; both sources seemed to be equally polluted.

Unless there is a strong source of Cd, the metal is often lower in concentration than many other metals. For example, in boat yards servicing pleasure boats, sediment concentrations of Cd averaged 0.52 mg/kg dw compared with 540, 440, 400, and 2.50 mg/kg for copper, lead, zinc, and mercury, respectively (Eisler, 2000).

In comparison with other metals, Cd concentrations in natural organisms also tend to be low. In essence, Cd concentrations in biota are not very interesting. With the exception of a few values in the hundreds of mg/kg, the vast majority of aquatic organisms, regardless of taxa, have concentrations from below detection level to approximately 20 mg/kg (Eisler, 2000). Most have values less than 5 mg/kg, which is not considered to be of concern. In addition, there were few to no patterns among groups of organisms with similar habits such as freshwater plants or aquatic birds.

Biological Effects of Cadmium

For humans, acute inhalation exposure to cadmium at concentrations more than approximately 5 mg/m^3 may cause destruction of lung epithelial cells, resulting in pulmonary edema, tracheobronchitis (inflammation of the trachea or bronchi), and pneumonia. Target organs for Cd storage include the kidney and liver and these organs are most likely to be affected by Cd toxicity (ATSDR, 2012).

Aquatic organisms tend to be more sensitive to Cd than terrestrial animals, especially birds and mammals. The 96-h LC50 concentrations for the water flea *Daphnia magna* are approximately 10 μg/L (US EPA, 1980). For fish, the LC50 values are similar: for example, 1–2 μg/L in striped bass (*Morone saxatilis*, Hughes, 1973); and 1–6 μg/L in rainbow trout (*Onchorhynchus mykiss*, Chapman, 1978). In contrast, mallard (*Anas platyrhynchus*) drakes fed up to 200 mg/kg in their diets for 90 days experienced no mortality or any loss of body weight. Laying in mallard hens that were fed that amount of Cd decreased, but did not stop altogether (White and Finley, 1978). Similarly, it takes at least 250 mg Cd/kg body weight in rats and 150 mg Cd/kg body weight for guinea pigs (*Cavia* sp.) before lethality is attained (US EPA, 1980).

Sublethal effects in invertebrates ranged from decreased growth rates to population declines over a period of several days. Earthworms (*Eisenia fetida*) exposed to Cd concentrations greater than 100 mg/kg in soil experienced higher mortality than controls (Žatauskaitė and Sodienė, 2014). Cadmium reduced the weight of juveniles, retarded growth, and delayed sexual maturation with worms at the highest concentrations (250 and 500 mg/kg) failing to mature. In addition, there was evidence that Cd increased lipid peroxidation, a form of oxidative stress.

Cadmium has been linked to cancer in several organisms. Lerebours et al. (2014) conducted a field study on European flatfish (*Limanda limanda*) in the North Sea and English Channel. While these sites have many contaminants, variations in the concentration of Cd highly correlated with the occurrence of retinoblastoma, an eye cancer. Cadmium was also associated with other cancerous or precancerous tumors in flatfish. The frequency of malignant tumors ranged from 0–20% of sampled animals and precancerous tumors ranged from 19–43%. Mean Cd concentrations in fish tissues ranged from 48–406 μg/kg.

Cancer and other effects of Cd toxicity have been observed in terrestrial animals. The US Department of Health and Human Services and the International Agency for Research on Cancer both list Cd as a known human carcinogen and the US EPA lists it as a probable carcinogen. Although studies on humans have sometimes been equivocal, laboratory studies on other animals strongly support the carcinogenic activity of Cd; lung cancer from inhalation seems to be a major risk in humans (ATSDR, 2012). Cadmium combined with a methyl or ethyl group is more serious than elemental Cd. Recall from previous chapters that adducts such as methyl groups to DNA can interfere with normal gene activation which can result in cancers.

In waterfowl, sublethal effects of Cd include reduced growth, kidney damage, and testicular damage (Blus et al., 1993). These effects, however, occurred at mg/kg concentrations, many times greater than effects seen in aquatic organisms and at concentrations that are higher than most environmental circumstances. Cadmium can also affect the endocrine system, influencing hormone production. The metal is associated with reduced gonadal function, altered secretions of adrenal corticotropic hormone (ACTH), growth hormone (GH), corticosterone, and thyroid stimulating hormone (TSH) by the pituitary. Some other effects in birds include bone marrow loss, anemia, liver hypertrophy, kidney damage, and testicular damage.

CHROMIUM

General Characteristics

Chromium (Cr) is described as a hard, brittle, lustrous metal that resists tarnishing and takes a high polish. These attributes have made Cr a highly desired metal for automobile enthusiasts for generations. The metal is the 22nd most common element in the Earth's crust and has an average concentration in uncontaminated soil of 100 mg/kg (with a range of 1–300 mg/kg). Concentrations range from 5–800 μg/L in seawater and 26 μg to 5.2 mg/L in freshwater. Approximately 44% of the 8.7 million tons of chromium ore mined each year comes from South Africa, but Eastern Europe and Turkey are also major mining areas. In nature, Cr occurs mostly as Cr^0 (elemental), Cr^{3+}, and Cr^{6+}, with Cr^{3+} being the most common ion and hexavalent or Cr^{6+} the most toxic.

Environmental Concentrations and Uses of Chromium

Chromium adds considerable strength to metal alloys and is used with iron in high-speed tool steels to reduce wear. Nickel-based alloys often contain Cr because of its strength and to enhance resistance to high temperatures. Nickel/chromium alloys are especially useful in jet engines and gas turbines. Chromates are also used as protective oxide layers on metals like aluminum, zinc, and cadmium.

Crocoite or lead chromate ($PbCrO_4$) was used as a yellow pigment in paint, resulting in the color Chrome Yellow (as on school buses in the United States). Due to the toxicity of both the Pb and Cr, the pigment was discontinued. By varying the associated elements and pH, Cr can also produce red and green pigments. Trace amounts of Cr^{3+} give natural and synthetic rubies their red color. The first laser, built in 1960, used a synthetic ruby with Cr^{3+}. Cr^{6+} is both highly toxic and magnetic. It is used in wood preservatives, especially in the chemical chromated copper arsenate which contains 35–65% Cr. Believe it or not, back in the "old days," recordings were made on tape and the magnetic coating of the higher-quality tapes was chromium dioxide or CrO_2. These tapes can still be purchased by audiophiles although other media such as compact discs and MP3 have taken over much of the popular market.

Environmental concentrations of Cr are more variable and can be much higher than cadmium. In terrestrial environments, contaminated sites can have Cr concentrations in excess of 4700 mg/kg. Pfeiffer et al. (1980) found that sediments just outside of the discharge pipe of a Brazilian electroplating plant had concentrations ranging from 1420–54,300 mg/kg. The concentrations decreased with distance from the plant. Fortunately, Cr tends to firmly bind with soil and sediment particles so the amounts that would be biologically available to fish and organisms living in the water column are probably far less than the measured concentrations (Eisler, 2000). Exposure would be greater for organisms that live in sediments such as worms and various insects because sediment metals can mix with pore water; the water that fills the pores among sediment and soil particles.

Chromium, like other metals has an affinity to adhere to soil particles. The same Brazilian study discussed previously (Pfeiffer et al., 1980) found that filtered, suspended particles had 2210–61,070 mg/kg of Cr attached to them. Chromium attached to these particles can easily be transported through the body of water and eventually deposited. For instance, at 50 m from a discharge point, the Cr concentration was 15,260 mg/kg, but at 600 m, it was 22% greater. Water can have high Cr concentrations in contaminated areas. Unpolluted waters have concentrations in the low-to-medium-µg/L range. However, untreated industrial effluents can have concentrations exceeding 5000 mg/L. While background atmospheric concentrations are approximately $0.001 \, \mu g/m^3$, urban concentrations in North America are 10–30 times higher and up to 500 times higher in cities that have steel mills (ATSDR, 2012).

Biological Effects of Chromium

Chromium is readily taken up by organisms and some bioconcentration in plants and in aquatic animals occurs because it is an essential element. However, the element is also **depurated** or expelled fairly rapidly and even in highly contaminated sites does not appear to biomagnify through the food chain. In general, Cr concentrations range from below detection level to 100 mg/kg in marine algae, 25 mg/kg in marine or freshwater mollusks, and 3 mg/kg in crustaceans (Eisler, 2000). Of course, animals living in the immediate vicinity of industrial effluents can have elevated concentrations of the metal. For example, snails living approximately 9 km below a tannery had 450 mg Cr/kg dw body weight in their tissues (Eisler, 2000). Mammals generally have Cd concentrations below 20 mg/kg dw (Eisler, 2000).

The value of Cr as a nutrient in mammals, including humans, has been debated for many years. Currently, the National Institutes of Health cites Cr as "a mineral that humans require in trace amounts, although its mechanisms of action in the body and the amounts needed for optimal health are not well defined." The institute describes Cr as having a role in counteracting diabetes, especially diabetes type II; regulating the metabolism of fats, carbohydrates, and proteins; and enhancing body weight and condition. No specific RDA has been set, but less stringent "Adequate Intake" levels are 35 µg/day for adult men; 24 to 45 µg/day for women, depending on pregnancy status; and 0.2 to 15 µg/day for children, depending on age. The typical normal intake for adult women and men is 23 to 29 μg/day and 39 to 54 μg/day, respectively, so the recommended amount is easily obtained from diets for all except pregnant or lactating women. Based on an extensive review on the effects of Cr to aquatic organisms, Eisler (2000) concluded the following: (1) Cr^{6+} is more toxic than Cr^{3+} to freshwater organisms in soft or acidic waters; (2) organisms at younger life stages are more sensitive than those at older life stages; (3) the 96-h LC50 assays are inadequate to explain the effects of Cr on population mortality patterns; (4) in saltwater environments, algae tend to be more resistant to Cr toxicity as salinity increases; and (5) pH seriously affects the toxicity of Cr^{6+}. Item 3 is a very common concern when trying to estimate risk. Long-term exposures to most contaminants induce harm that short-term, acute exposures often do not reveal and, under natural conditions, other factors that we have mentioned can ameliorate or intensify toxicity. Van der Putte et al. (1981) supported the findings of many other studies in showing that the toxicity of Cr was inversely related to pH in water (Fig. 8.3). Chromium in soft waters is more toxic than in hard water. In bluegill (*Lepomis macrochirus*), for instance, the 96-h LC50 was 118 mg/L in soft water and 213 mg/L in hard water (US EPA, 1980).

With regard to sublethal effects, growth of several freshwater algae species is inhibited by Cr^{6+} concentrations between 10 and 45 μg/L and effects were most observable at low pHs. The US EPA (1980) reported that 1900 μg Cr^{6+}/L significantly inhibited root growth in water milfoil *Myriophyllum spicatum*, but that it took five times that amount of Cr^{3+} to have the same effect. Chromium can also reduce growth of terrestrial plant shoots and roots. For instance, Cr^{6+} was 10 times more powerful in stunting growth of barley than Cr^{3+}.

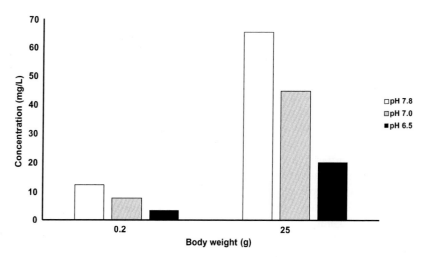

FIGURE 8.3 Toxicity of chromium in rainbow trout as a function of water pH and the size of fish. Values are the LC50s.

Among freshwater fish, hexavalent Cr decreased growth in rainbow trout and Chinook salmon (*Onchorhynchus tshwytscha*) fingerlings at $21\,\mu g/L$ (US EPA, 1980). Coho salmon (*O. kisutch*) that were migrating towards the sea experienced problems in osmoregulation when subjected to $230\,\mu g/L$ and they showed decreased immune functions at $500\,\mu g\ Cr^{6+}/L$ (Sugatt, 1980a).

Birds can accumulate Cr in their tissues including feathers and eggshells which offer noninvasive means of evaluating exposure and uptake of this and other metals. However, as for lethal toxicity, few studies have examined the effects of Cr in wild birds or mammals. Dietary supplements with $2\,mg/kg$ chromium chloride ($CrCl_3$) increased weight gain, food conversion efficiency, and immune responses while they decreased respiration rate and body temperature in broiler chicks (Norain et al., 2013). Most current research seems to be focused on how Cr supplements can increase growth and immune functions rather than on adverse effects of Cr on birds, especially domestic fowl. Concentrations of $CrCl_3$ of approximately $1.5\,mg/kg$ in the diet apparently have many beneficial results. However, teratogenicity or improper embryonic development of broiler chicks was observed when Cr^{6+} was injected into eggs (Gilani and Marano, 1979). These abnormalities included twisted and shortened limbs, small eyes, everted viscera, exposed brains, parrot-like beaks, and stunted growth.

According to Eisler (2000), under appropriate conditions hexavalent Cr is a human and animal carcinogen and mutagen. Trivalent Cr can also produce tumors if it is in solution at concentrations far exceeding those that would be environmentally relevant. Hexavalent Cr can also cause skin ulcerations, dermatitis, mucous membrane ulcerations, and bronchial cancer in humans. Hexavalent Cr is also a spermicide, causes birth defects, and causes spontaneous abortions in rodents.

COPPER

General Characteristics

Next to iron, copper is arguably the most important metal to living organisms. All living organisms require it for normal growth and physiology. Copper has a high electroconductivity and high thermoconductivity and is easily malleable. Copper will oxidize, but unlike rust on iron, the oxidized coating protects the copper from further oxidation. If exposed to the atmosphere, a green patina or verdigris (copper carbonate) forms. The Statue of Liberty is a good example of this (Fig. 8.4). Copper occurs naturally as elemental, Cu^{1+} (cuprous), or Cu^{2+} (cupric) but it can also be found as Cu^{3+} and Cu^{4+} ionic states. Typical concentrations of Cu in the earth's crust range from 2–$250\,mg/kg$. Freshwater and marine concentrations range from 0.5–$1000\,\mu g/L$ with an average of approximately $10\,\mu g/L$. The amount of Cu in uncontaminated air varies from a few nanograms per cubic meter to $200\,ng/m^3$ (ATSDR, 2005). Copper is a valuable metal with many different purposes and, for that reason, some $197,000,000$

FIGURE 8.4   The Statue of Liberty is made of copper and has a green patina called *verdigris* which is caused by oxidation of the metal.

tonnes of Cu are extracted per year (USGS, 2014b). Chile, followed by the United States, Indonesia, and Peru are the top producers of Cu globally. Copper recycling is extensive and, excluding Cu wire which requires new metal, nearly 75% of copper use in the United States comes from recycling.

Environmental Concentrations and Uses of Copper

Copper is often blended into brass (copper mixed with zinc) or bronze (typically copper with tin, but other metals can also be used with copper). Cupronickel is a blend of copper and nickel and used in low-valued coins such as—you guessed it—nickels. Today, the "copper" penny is actually only 2.5 % copper and 97.5% plated zinc. The major applications of Cu are in electrical wires (60%), roofing and plumbing (20%), and industrial machinery (15%). Other uses include cookware, architecture and building supplies, and antifouling paint on boats. Copper sulfate is used in agriculture as fungicides, insecticides, repellents, or algicides; medicines for humans and livestock; and in nutritional supplements. Unfortunately, the amount of Cu sulfate needed to adequately control algae blooms in lakes and ponds can also be toxic to other organisms.

The largest source of Cu pollution through human activities is to land. Major sources of this contamination include mining and milling operations. Other sources include agriculture, sludge from publicly owned treatment works, and municipal and industrial solid wastes. Fertilizers made from livestock feces often contain high concentrations of heavy metals from animal feed that is poorly digested. Spread over a long period of time, heavy metal concentrations can build up until they become toxic to crops.

Copper is released into water as a result of natural weathering of soil and discharges from industries and sewage treatment plants. According to the ATSDR (2005), domestic wastewater is the major anthropogenic source of Cu in waterways with concentrations of Cu discharged into wastewater treatment plants, averaging approximately 0.5 mg/L but can be several times higher. Acid mine drainages usually have low (<4.0) pH, which leads to high solubility and bioavailability of metals. The US EPA found copper in 96% of 86 samples of runoff from 19 cities throughout the United States with concentrations ranging from 1–100 μg/L and a mean of 18.7 μg/L (Cole et al., 1984). Copper enters the atmosphere naturally from windblown dust and volcanoes, but the principal anthropogenic atmospheric sources of Cu are smelters where concentrations can range from 7–138 ng/m^3 (ATSDR, 2005).

Concentrations of Copper in Organisms

Copper and other metals can be taken up by plants following at least one of three pathways: (1) metals may not be taken up at all; (2) the metal is picked up by the roots, but the plant may have ways of inhibiting the ability of metals to move from roots to shoots; or (3) metals may move more or less freely through the plant and may accumulate in certain plant parts. Since Cu is required by living organisms, it is more readily taken up than some other metals. Some plants are hyper accumulators of Cu. Kamal et al. (2004) tested the ability of three aquatic plant species; water mint (*Mentha aquatic*), parrot feather (*Myriophyllum aquaticum*), and creeping primrose (*Ludwigia palustris*); to accumulate Cu in a laboratory situation. Following a starting concentration of 5.56 mg Cu/L in solution, all three species partially depleted Cu in the water column over the course of 21 days. At this time, the final Cu concentrations in water were reduced by approximately 40% for all three plant species. The final plant concentrations ranged from 304–840 mg/kg dw. In contrast, controls had 11–25 mg/kg dw at the end of the study. This study shows that bioconcentration of Cu does occur and it varies among species. In a review of several terrestrial species of plants, Eisler (2000) determined that such species usually have concentrations below 50 mg/kg dw. However, those near contaminated sites can have concentrations that exceed 10,000 mg/kg dw.

De Jonge et al. (2014) conducted a study in which they measured water chemistry and body burdens of Cu and other metals in four groups of insects: two stoneflies (*Lectura* sp. and Perlodidae), black flies (Simulidae), and mayflies (*Rhithrogena* sp). Water concentrations of Cu ranged from 0.19–9.52 μg/L. In the insects, the concentrations ranged from 11.4–876 μg/L so bioconcentration was evident in all four groups with biological concentration factors of 60–92. Several water chemistry factors influenced body concentrations including concentrations of free-ion Cu, pH, dissolved organic matter, and water hardness. Body burdens of Cu increased with Cu water concentrations, and as pH, dissolved organic matter, and hardness decreased. Terrestrial invertebrates such

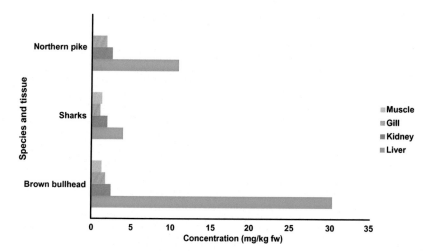

FIGURE 8.5   Copper concentration in fish as a function of organ. Note that liver is the primary reservoir for copper in the body followed by kidneys. *From Jenkins (1980).*

as insects have Cu concentrations in the range from less than 5–140 mg/kg dw (Eisler, 2000).

Typical concentrations of Cu in fish range from 1.5 to 25 mg/kg dw (Eisler, 2000). In studies that examined organ-specific concentrations, kidneys and livers tended to have higher concentrations than other organs (Fig. 8.5). Among the few amphibians examined, concentrations were similar to those found in fish.

In birds, low-background whole body concentrations are approximately 5 mg/kg dw and highs are near 50 mg/kg (Eisler, 2000). Concentrations in mammals are in the same region as birds.

Biological Effects of Copper

Too much or too little Cu can be harmful to any organism including plants where deficiency of Cu is characterized by reduced growth, dark coloration of roots, and chlorotic (blotchy or yellowish) leaves, reduced fertility, and withering. Most plants have the ability to efficiently eliminate excess Cu. It is only when levels of the metal get very high that serious harm can occur. "Very high," of course, depends on the species. For instance, in cucumbers soil concentrations <2 mg/kg dw cause Cu deficiency, cucumbers grown with concentrations between 2 and 10 mg/kg dw are healthy, and when concentrations exceed 10 mg/kg dw the plants show signs of Cu toxicity (Eisler, 2000). In soybeans, the concentrations for deficiency, optimal, and toxicity are <4, between 10 and 30, and >50 mg/kg, respectively, with some margin of leeway between 4 and 10 and between 30 and 50, depending on other soil factors. Copper toxicity inhibits root elongation and branching, which reduces the ability of the plant to absorb water and nutrients from the soil.

In vertebrates, Cu is involved with many enzyme functions including cytochrome C oxidase which is part of the electron transport system in mitochondria; Cu/Zn superoxide dismutase; and in many enzymes involved with protein synthesis, oxidative phosphorylation, iron transport, and synthesis of neurotransmitters. Daily dietary ingestion in humans may provide 1–5 mg/day (ATSDR, 2005), of which only 20–50% is absorbed. Cu deficiency is very rare unless there is some genetic or physiological issue that inhibits Cu uptake because it is generally available in most foods. In addition, Cu deficiency has been tied to neurological conditions, including sensory ataxia (loss of coordination), spasticity, muscle weakness, and loss of vision, or damage to the peripheral nerves or spine.

Copper deficiencies have not been reported in wild birds or mammals, but have occurred in domestic livestock and laboratory animals. Sudden death can occur in chickens, swine, and cattle that are deprived of Cu. Other effects center on the multitude of enzyme reactions that Cu is involved in and result in increased EROD activity, anemia, acute inflammation, lesions in the central nervous system, and reduced phospholipid synthesis. In other words, having too little Cu is not a good thing.

At high Cu concentrations, acute (96-h) lethal toxicity concentrations of Cu were 260 µg/L in nematodes, 1700 µg/L in snails, 560 µg/L in oysters, 26 µg/L in a species of clam, and 50 µg/L in black abalone (*Haliotis cracherodil*, reviewed by Eisler, 2000). Among fish, the US EPA (1980) reported that low LC50s ranged from 13.8 µg/L for rainbow trout, and for goldfish (*Crassius auratus*) it was 36 µg/L when calcium carbonate was low and 300 µg/L at higher levels of $CaCO_3$/L. For a few less sensitive species, the LC50s included 8000 µg/L for mummichog (*Fundulus heteroclitus*), 937 µg/L in green sunfish (*Lepomis cyanellus*), and 1100 µg/L for bluegill larvae (summarized by Eisler, 2000).

In amphibians, a few LC50 values included 2696 µg/L for the tolerant Fowler's toad (*Bufo [Anaxyrus] fowleri*) embryos, 1120 µg/L in two-lined salamander (*Eurycea bislineata*) juveniles over 48 h, and 50 µg/L in the more sensitive northern leopard frog (*Rana [Lithobates] pipens*) embryos over eight days (Eisler, 2000). Marbled salamander embryos (*Ambystoma opacum*) did not die from Cu concentrations as high as 1000 µg/L, but they did hatch earlier than controls and those exposed to lower concentrations of Cu (Soteropoulos et al., 2014). Larvae experienced high mortality at both 500 and 1000 µg/L. The lack of mortality among embryos is most likely due to protection provided by the jelly coating around eggs.

In general, mammals and birds are 100–1000 times more resistant to Cu than aquatic animals, but some ruminant animals (ie, sheep, goats, and cattle) are significantly more sensitive to Cu toxicity than nonruminants (Eisler, 2000). Among ruminants, sheep seem to be particularly sensitive. Copper deficiency can be one cause of swayback or lordosis in ungulates (Jaiser and Winston, 2010).

The liver is a major storage organ for copper and often has the highest concentrations of all organs. Effects in mammals include kidney damage, increased mortality, gastric ulcers, and liver pathology. Toxic reference values, which are concentrations that might produce some effects, for nine species of mammals including six wildlife species and three domesticated animals ranged from 0.9 to 4.5 mg/kg (Ford and Beyer, 2014). Herbivores ingest considerable amounts of soil with their foods, so accounting for food consumption, soil concentrations of copper that are considered safe ranged from 109 mg/kg in bighorn sheep (*Ovis candadensis*) to 2013 mg/kg in horses.

Sublethal effects in birds include Cu accumulation in the liver and kidneys, heart and skeletal lesions, anorexia, and listlessness. It does not appear that mortality would occur at environmentally realistic concentrations except perhaps in highly contaminated sites. Ford and Beyer (2014) estimated that mourning doves (*Zenidura macroura)*, mallards, and Canada geese (*Branta canadensis*) could ingest 3.3 mg Cu/kg/day safely. They further calculated that soils could have from 689 mg/kg for mourning doves to 1008 mg/kg for the waterfowl species and still be considered safe.

When it became apparent that waterfowl and some endangered species were dying due to lead shot exposure (see the FOCUS section), several studies evaluated the suitability of other metals as substitutes. In one study, copper shotgun pellets were fed to American kestrels (*Falco sparverius*). Birds were fed 5 mg Cu/g body weight nine times during a 38-day exposure trial (Franson et al., 2012). Essentially all birds retained the pellets for at least 1 h, but most regurgitated them within 12 h. Copper concentrations were higher in the livers of dosed birds compared to controls, but there was no apparent difference in blood concentrations between the two groups. Copper exposure elicited metallothionein in dosed male birds, but not females. No clinical signs were observed, and there was no treatment effect on body mass; hemoglobin or methemoglobin (a form of hemoglobin, in which the iron in the heme group is in the Fe^{3+} (ferric) state, not the Fe^{2+} (ferrous) of normal hemoglobin) in the blood; or on Cu concentrations in kidney, plasma biochemistries, or hematocrit. The authors concluded that the copper pellets posed little threat to the kestrels or (presumably) related species although longer exposures might produce negative effects.

LEAD

General Characteristics

Finally, we get to a metal that has high risk of environmental toxicity. Along with mercury, lead is a very toxic metal whose distribution has increased tremendously due to human activities. Lead is soft, malleable, and heavy. Metallic lead has a bluish-white color after being freshly cut, but it soon tarnishes to a dull grayish color when exposed to air. Lead has a shiny chrome-silver luster when it is melted into a liquid. It is also the heaviest nonradioactive element. Its most common valence states are either Pb^0 or Pb^{2+}. Lead is ranked 37th in abundance among elements with an average concentration in soil and sediments of 10 mg/kg (0.001%); actual concentrations of course will vary from site to site. Lead seldom occurs as a pure metal, rather it is found in various oxides and sulfides, such as the ores galena, cerussite, and anglesite. Other metals often found with Pb include zinc, silver, and copper. Annual global extraction of lead ore is around 8.8 million tons with Australia, China, and the United States as the principal mining countries.

Uses of Lead

Contrary to public opinion, lead is not found in pencils—it is actually a form of carbon called graphite. Approximately half the amount of Pb mined goes to the automobile industry as leaded batteries. The other major uses of lead include small arms ammunition, shotgun pellets, fishing sinkers and weights for fishing nets, and tire balancing. Less common uses include electrodes, radiation shielding, building industry, sculptures, leaded glass, ceramic glazes, and semiconductors. Lead is still used in pigments in industrial paints but its use in home paints was banned in the United States in 1978 due to its high toxicity. We still find lead paint in slums and houses that have not been well maintained. Somewhat appalling, lead carbonate was used as a white pigment on the faces of Geisha girls (Fig. 8.6) and for other "white face" uses. The lead chromate was indeed toxic to those who used it. Lead found a use in plumbing joints until 1998 when the EPA banned its use for that purpose. Organic lead, especially tetraethyl lead, was used in gasoline for automobiles from the 1920s until the early 1970s as an octane booster. However, after the organic portion was combusted, the lead was emitted through tail pipes into the atmosphere or onto road surfaces where it could be part of runoff into rivers and streams. Following the ban, the output of Pb from vehicles declined by 95% and atmospheric Pb has declined by 94% (US EPA, 2014, Fig. 8.7). The EPA promulgated a maximum atmospheric discharge of Pb from any single source at $1.5 \, \mu g/m^3$ in 1977, which still remains the standard (US EPA, 2009).

Environmental Concentrations of Lead

In 1978, Nriagu (1978) estimated that only the lithosphere—soil and sediments—retained an appreciable amount of lead (approximately 99.9% of the total amount in the world). This estimate has probably not changed very much. If anything, as stricter controls were placed on Pb emissions, the atmosphere has even less lead than at that time.

Eisler (2000) reported that air typically has a Pb concentration of around $0.1 \, \mu g/m^3$ to maybe 100 times that in metropolitan areas. In the smelting and refining of lead, mean concentrations of lead in air can reach $4400 \, \mu g/m^3$; in the

FIGURE 8.6   In the 19th century and earlier, geishas in Japan wore a "white face" that had lead as a base.

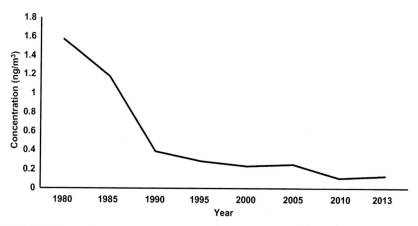

FIGURE 8.7   Lead in the atmosphere has declined since the cessation of leaded gasoline. *From US EPA, 2014.*

manufacture of storage batteries, mean airborne concentrations of lead ranging from 50–5400 μg/m^3 have been recorded; and in the breathing zone of welders of structural steel, an average lead concentration of 1200 μg/m^3 has been found (ATSDR, 2007b).

The amount of lead released by industry to water sources is less than that in the atmosphere or on land. The EPA Toxic Release Inventory in 2006 reported that 118,700 kg (118.7 tonnes) of Pb had been dumped into bodies of water, mostly streams and rivers. Of the known aquatic releases of lead, the largest ones are from the steel and iron industries and Pb production and processing operations (US EPA, 1982). Urban runoff and atmospheric deposition are significant, indirect sources of lead found in the aquatic environment (ATSDR, 2007b). For example, in a brief review of urban stormwater ponds in the Netherlands, Langeveld et al. (2012) reported that Pb concentrations ranged from 2–239 μg/L.

Water pH and dissolved salt concentrations heavily influence the amount of soluble Pb in surface waters (ATSDR, 2007b). At pH >5.4, the total solubility of lead is approximately 30 μg/L in hard water (high calcium or magnesium concentrations) and approximately 500 μg/L in soft water. Sulfate ions, if present in soft water, limit the lead concentration in solution through the formation of lead sulfate. In general, relatively clean bodies of water in the United States have up to 50 μg/L of dissolved lead. However, most of the lead in the water column is bound to particulates, either finely suspended particles or coarse particles that eventually precipitate to the bottom. Demayo et al. (1982) found that Tennessee streams had concentrations of dissolved Pb that ranged from 0.01–0.02 μg/L, the amount bound to dissolved organic carbon ranged from 30–84 μg/L, colloidal or suspended lead on fine particles from 62–2820 μg/L, and that coarse particle Pb from 124–653 μg/L.

The Toxic Release Inventory (US EPA, 2015) reported that in 2004, 6221 tonnes of lead were released to the land, both onsite and offsite, despite that, approximately 80% of current lead usage is in batteries and 95% of that is recycled (International Lead Association, 2015). In addition, 83 and 3977 tons of lead and lead compounds, respectively, were injected underground (ATSDR, 2007b). As the ATSDR pointed out, however, while the majority of lead releases are to land, they constitute much lower exposure risks to humans or wildlife than releases to air and water. Terrestrial releases are required by law to be isolated from environmental areas such as aquifers and water bodies. Of course, these landfills can leak so they need to be monitored regularly. Risk is further reduced because most of the lead released to land is tightly bound to organic material and becomes immobile (ATSDR, 2007b). Concentrations in soil or sediment can range from 10 to 11,000 mg/kg (Eisler, 2000). In the United States, sediments along a 195-mile (295-km) stretch of the Mississippi River in Missouri had 8.06–13.25 mg Pb/kg dw, but one location near the effluent of a smelter had 1710 mg/kg (Missouri DNR, 2010). Baseline measurements of sediment Pb in the Coeur d' Alene River Valley, an historically important mining area, ranged from 1900 to more than 5000 mg/kg dw; waterfowl sampled from that area had blood Pb levels that indicated clinical or severely critical clinical toxicity concentrations (Spears et al., 2007).

Concentrations of Lead in Organisms

Lead that is present on the surface of plants reflects deposition levels but not what is actually assimilated; for that, one needs Pb concentrations to be incorporated into tissues. Although the bioavailability of lead to plants is limited because of the strong adsorption of lead to soil organic matter, it increases as pH and organic matter content of the soil drop. Most plants, if they assimilate Pb to any degree, seem to sequester it in their roots, allowing little to enter shoots or leaves. While there are hundreds of plant species that are known hyper accumulators of metals, there are only a few that have been identified as hyper accumulators of Pb, meaning that they can have more than 1000 mg Pb/kg in their tissues (Auguy et al., 2013). Among these include the grasses *Agrostis tenuis* and *Festuca ovina*; some penny cresses; a sorrel *Rumex acetosa*; and the aptly named leadworts, family Plumbaginaceae. Auguy et al. (2013) investigated the mustard *Hirschfeldia incana*, another suspected hyper accumulator. They found the plant growing wild on lead-mined land in Morocco with soils having 26–9479 mg Pb/kg dw. Leaves from these plants had 0.53–1.43 g Pb/kg dw (yes, grams), with an average of 0.79 g/kg. The authors cultivated seeds from these plants in both hydroponic media and soil. In soil spiked with 7531 mg Pb/kg dw, leaves contained 3.58 g/kg after a few weeks. In hydroponic solutions of 62.1 mg/L, roots had 121 g Pb/kg dw, but shoots had only 3% (3.6 g/kg dw) of that.

In uncontaminated sites, plants and invertebrates tend to have higher concentrations of Pb than birds or mammals. Aquatic invertebrates had 5–58 mg Pb/kg dw in their tissues unless they lived near a contaminated site where they could have >11,000 mg/kg dw (Eisler, 2000). High values for invertebrates included 14,233 mg/kg dw in black fly larvae (*Simulium* sp.) near a Missouri tailings, 981 mg/kg in worms (*Eisenia rosea*) in an Illinois area receiving lead-laden sludge, and 931 mg/kg dw in the limpet (*Acmaea digitalis*) near bridges in California.

Fish tend to have relatively low concentrations of Pb in their tissues. In a couple of recent papers a food fish common to Pakistan had 0.12–1.74 mg/kg Pb in its muscle which was the lowest concentration among six heavy metals (Ahmed et al., 2015). In India, Pb concentrations in the muscles of five food fish ranged from 0.073–0.386 mg/kg fresh weight; Pb concentrations ranked second lowest among 10 metals (Giri and Singh, 2015). Aquatic organisms will pick up lead through their gills, ingestion of food and sediment, or through their skin.

There are not enough data on Pb levels in amphibians to make meaningful generalizations. Sparling and Lowe (1996) found that northern cricket frogs (*Acris crepitans*) collected from experimental ponds had whole body concentrations of Pb ranging from 6.7–19.7 mg/kg dw. These concentrations were significantly and positively correlated with sediment concentrations. More recently,

Ilizaliturri-Hernández et al. (2012) found that blood lead levels in giant toads (*Rhinella marina*) of Mexico ranged from 10.8–70.6 μg/dL (deciliters—multiply those values by 10 to determine concentrations per liter) with significant differences existing among giant toads that came from rural areas (mean 4.7 μg/dL), urban/industrial areas, (8.46 μg/dL), and industrial areas (22.0 μg/dL).

Grillitsch and Schiesari (2010) produced an extensive list of metal concentrations in reptiles. Concentrations in unpolluted sites were similar or lower than those observed in birds and mammals while contaminated sites had higher concentrations. Among hundreds of data points some high concentrations included 115 mg/kg ww in bones of snapping turtles (*Chelydra serpentine*) located within the lead-mining area of Missouri, 136 mg/kg dw in the femur of box turtles (*Terrepene carolina*) collected from an area near a lead smelter in Missouri, and 105–386 mg/kg ww in captive alligators (*Alligator mississippensis*) in Louisiana.

Terrestrial vertebrates will assimilate Pb through inhalation or ingestion, but ingestion is by far the more common way. Inhaled Pb will enter the lungs and almost immediately enter the bloodstream. Ingested lead may be reduced by the pH of the gut and absorbed more slowly into the blood; much of the ingested lead might be depurated with feces. The amount of lead actually absorbed by the digestive system depends on the pH of the gut and certain dietary factors. Anything that facilitates elimination, such as fiber, will facilitate depuration of lead. Diets high in calcium and iron will inhibit lead absorption. Elemental Pb seems to be absorbed less readily than organically bound lead, but elemental Pb is more readily absorbed than some inorganic lead complexes.

We focused on the liver, which was the most consistently used organ in Eisler's (2000) summary and an organ for Pb storage in animals, because lead concentrations can vary significantly among tissue types (Table 8.2). Actually, bone tends to have the highest tissue concentrations of inorganic Pb among organs (Fig. 8.8), but Pb is accumulated and stored in bone with little bioavailability to the rest of the organism while concentrations in liver can become freely available. Organic forms of Pb are usually highest in kidney. In birds, blood, liver, and kidney were the most frequently analyzed tissues, but samples also included gut contents, various bones, brain, pancreas, spleen, feathers, carcass, eggs, lungs, and muscle. Tissues analyzed in mammals were similarly variable but also included hair, antlers, feces, blubber, skin, and teeth. Basically, over all these studies, if an organ could be analyzed it was. Concentrations in birds and mammals, even in contaminated sites, did not reach the highs observed in plants or invertebrates but were many times greater than those seen in supposedly clean sites.

Biological Effects of Lead

Lead can negatively affect every organism and virtually every biological system within organisms; it is a highly toxic, cumulative, metabolic poison. According to Eisler (2000), environmental pollution from Pb is so high that its body burdens

TABLE 8.2 Summary of Lead Concentrations Found as Either Dry Weight (dw) or Fresh Weight (fw) in Various Biological Groups

Group	Conditions	N	Mean Concentration + SE (mg/kg)	Maximum (mg/kg)
Plants	Noncontaminated	41	14.3 ± 3.86 dw	123 dw
Plants	Contaminated sites	30	1228 ± 557 dw	11300 dw
Terrestrial invertebrates	Noncontaminated Sites	23	15.1 ± 3.7 dw	52 dw
Aquatic invertebrates	Noncontaminated sites	15	10.6 ± 3.6 dw	58 dw
All invertebrates	Contaminated sites	28	680 ± 503 dw	14233 dw
Fish liver	Noncontaminated sites	7	0.43 ± 0.06	0.6
Fish liver	Contaminated sites	3	9.1 ± 3.0	6.5
Bird livers	Noncontaminated sites	19	0.69 ± 0.37 fw	6 fw
Bird livers	Contaminated sites	25	37.5 ± 12.8 fw	315 fw
Mammal livers	Noncontaminated sites	20	3.02 ± 0.6 dw	8.7 dw
Mammal livers	Noncontaminated sites	15	1.05 ± 0.34 fw	5.1 fw
Mammal livers	Contaminated sites	12	13.9 ± 3.2 dw	30 dw
Mammal livers	Contaminated sites	10	22.9 ± 11.4 dw	120 fw

From Eisler (2000).

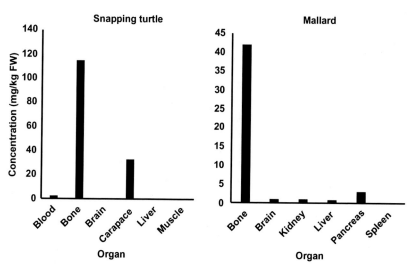

FIGURE 8.8 Organ-specific concentrations of lead in snapping turtles and mallards. Bone is a primary reservoir for lead. *Overmann and Krajicek (1995) for turtles, Guitart et al. (1994) for mallards.*

in many human populations are closer than any other contaminant to producing clinical toxicity. This can often be said about wildlife as well. Lead can be a mutagen, teratogen, carcinogen (in animals, but not conclusively in humans) or cocarcinogen (demonstrated in humans). Lead can disrupt reproduction, impair liver and thyroid functions, and attack the immune system. Its primary target is the nervous system especially in children (ATSDR, 2007b) and, by extension, young animals. A primary mechanism of Pb toxicity is that it binds or deactivates many proteins and enzymes in organisms (ATSDR, 2007b).

Plant responses to lead vary widely. The bioavailability of Pb in circumneutral to alkaline soils is very low. Regardless of pH, organic matter can bind Pb and reduce its uptake; thus, soils with low pH and low organic matter would make Pb most available. When Pb is assimilated, it reduces photosynthesis, reduces mitosis, inhibits growth, reduces pollen germination and seed viability, and impairs water absorption. Inhibition of photosynthesis occurs through blocking of sulfhydryl groups and diminishing phosphate concentrations in cells.

Acute toxicity in aquatic invertebrates can occur at ppb concentrations under laboratory conditions. In the water flea, *Daphnia magna*, 96-h LC50s varied with water hardness with increased softness resulting in greater toxicity (US EPA, 1985). Tolerance over multiple generations can also occur—isopods (*Asellus meridianus*) raised in clean conditions had a 48-h LC50 of 280 µg/L while those collected from a lead-contaminated river needed 3500 µg/L to attain their LC50 (Demayo et al., 1982).

Hariharan et al. (2014) conducted a multiphase study on the effects of environmentally relevant concentrations of Pb on the green mussel (*Perna viridis*). Under laboratory conditions they conducted an acute toxicity test with Pb concentrations ranging from 0–11.55 mg/L and for a chronic, 30-day test, they reduced the concentrations to range from 0–0.232 mg/L. They found that the acute 96-h LC50 was 2.62 mg/L. Mortality increased with the duration of the study, so that at the end of the 30-day trial, only 45% of the mussels survived at 0.11 mg Pb/L. Lead exposure increased oxidative stress in the animals, as determined by several bioindicators including reduced glutathione and glutathione-S-transferase. Histopathology was observed in the chronic test with gill filament and lamellar structures being damaged at 0.054 mg/L and higher concentrations; the same levels of Pb resulted in damage to the adductor muscle—the muscle that closes the shell—with 0.11 mg/kg, resulting in separation of muscle fibers.

In a routine investigation of ponds at the Prime Hook National Wildlife Refuge, US Fish and Wildlife Service contaminants biologist Sherry Krest located a wetland on the refuge border that had numerous adult frogs but no evidence of any larval forms or tadpoles. Further investigation revealed that the wetland was located behind a berm used by a hunting club to stop ammunition from entering the refuge. Unfortunately, overshooting gradually deposited lead shot into the wetland over a long period of time and the sediments had accumulated a mean of 5700 mg Pb/kg dw. Sparling et al. (2006) conducted a lab study to assess the effects of these high concentrations on larval southern leopard frogs (*Rana* [*Lithobates*] *sphenocephala*). They exposed recently hatched lar-vae to sediments that were mixed with up to 7580 mg/kg dw, which equated to 24.4 mg/L in soil pore water. Although reduced growth and development were noted at 2360 mg/kg dw, the long term LC50 was 3228 mg/kg (12.5 mg/L) and all animals died within 5 days at concentrations ≥3940 mg/kg. Those that died, however, were probably the fortunate ones because severe skeletal and spinal problems, including twisted spines, shortened long bones, and deformed digits were common among survivors, reaching a 100% occurrence rate at concentra-tions ≥2360 mg/kg (Fig. 8.9). As a side note, those concentrations may seem high, and they are, but in this case they were environmentally relevant.

For most aquatic biota, Eisler (2000) concluded that: (1) dissolved water-borne Pb was the most toxic form; (2) organic lead compounds were more toxic than inorganic compounds; (3) adverse effects in some species were observed at 1 μg Pb/L; and (4) effects were more apparent at elevated water temperatures, reduced pH, comparatively soft waters, and in younger life stages.

A lot of research has been conducted on the effects of Pb on birds and mam-mals. In the case of birds, much of this interest has been generated by poi-soning due to lead shotgun pellets and fishing sinkers that enter the aquatic environment and are picked up by waterfowl or result in secondary toxicity

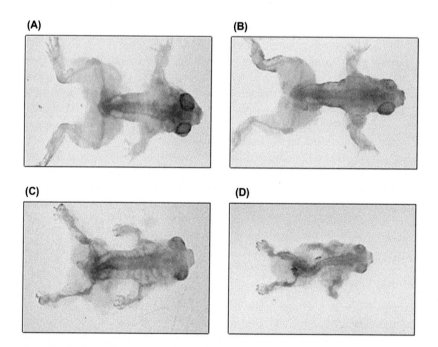

(A) **(B)**

(C) **(D)**

FIGURE 8.9 Increasing degree of skeletal malformations in a *Rana sphenocephala* juvenile exposed to varying levels of lead in water. (A) 45 ppm, (B) 75 ppm, (C) 540 ppm, and (D) 2360 ppm.

when scavengers consume contaminated carcasses. These findings will be summarized in the FOCUS section later in this chapter.

Signs of Pb toxicosis in birds include loss of appetite, lethargy, weakness, emaciation, tremors, drooped wings, green-stained feces, impaired locomotion, unsteadiness, and poor depth perception. Internally, birds will have microscopic lesions in the proventriculus (an upper portion of the digestive tract), pectoral muscles, and proximal tubules of the kidney. They may display enlarged, bile-filled gall bladder; anemia; reduced brain weight; fluid or edema around the brain; abnormal skeletal development; and esophageal impaction. Other signs include fluid-filled lungs; abnormal gizzard linings; an unusually pale, emaciated, and dehydrated carcass; and elevated Pb concentrations in liver, kidneys, and blood. Organic lead is about 10 to 100 times more toxic than inorganic lead, depending on species or sex. Altricial chicks—those that spend some time in the nest before becoming independent—seem to be more sensitive to lead than precocial chicks and more sensitive than adults of the same species.

As we discussed in Chapter 3, one of the enzymes necessary for heme or blood formation in vertebrates and many invertebrates that can be useful in diagnosing if an animal has been exposed to Pb is the enzyme delta-aminolevulinate dehydratase or ALAD. Lead inhibits the gene *Aminoleuvulinic dehydratase* from producing the enzyme and even small exposures to lead can reduce ALAD activity by 90% or more. In the study on giant toads reported previously (Ilizaliturri-Hernández et al., 2012), the authors found a significant drop in ALAD with increasing concentrations of blood Pb.

Several generalizations on mammalian toxicity have been developed from extensive research on the effects of lead in laboratory and wild animals: (1) lead toxicosis can occur in real environments with actual exposure concentrations; (2) organic lead is usually more toxic than inorganic lead; (3) there is considerable variation among species in sensitivity to Pb; and (4) as with assimilation of Pb, many environmental factors (including calcium, magnesium, pH, and organic matter) can affect lead toxicity; diets deficient in some basic nutrients such as calcium, minerals, and fats can contribute to lethal and sublethal expression (Eisler, 2000). Signs of Pb poisoning in mammals include those seen in birds, but also involve spontaneous abortions, blindness, peripheral nerve disease, poor performance in tests involving learning or memory, and various blood disorders.

MERCURY

General Characteristics

Mercury (Hg) is a contaminant of global concern. Several international conferences, agreements, and conventions have been established to regulate and reduce Hg concentrations in the environment, particularly as they relate to human health. Mercury is unique among metals in several ways. Along with lead, Hg is extremely toxic to living organisms and has no known biological function. However, Hg is the only metal that is liquid under standard temperature and pressure conditions. It can form organic complexes such as methylmercury (MeHg), which are fat soluble and can both bioconcentrate and biomagnify through food chains. Mercury is resistant to most acids, although concentrated sulfuric acid can dissolve the metal. In contrast, Hg itself dissolves several other metals to form amalgams. Gold and silver are two commercially useful amalgams. Dental amalgams using Hg are used in dentistry although their popularity is fading. Mercury is also infamous for forming an amalgam with aluminum by dissolving the lighter metal. For this reason, the transport of Hg by aircraft is largely banned—imagine having your plane dissolve while you are flying it!

Mercury is an exceptionally rare metal in the Earth's crust, it is the 66th most common element in the crust and has an average concentration of 0.08 mg/kg in soils and sediments. That means that 0.000008% of the earth's soil is Hg. When found, however, Hg tends to pool and form relatively rich but widely scattered pockets. Mercury commonly occurs in one of three valence conditions: 0, 1^+, and 2^+; higher valence states can occur but are very rare. All forms of inorganic Hg are toxic. As we will see, Hg has a tendency to become organic by becoming methylated or ethylated under certain environmental conditions. Methylmercury, the most common form of organomercury, is considerably more toxic than any of the inorganic ionic states. The 1^+ oxidation state often takes the form of Hg^{2+}_2 with two Hg atoms forming a dimer with a 2^+ charge. When combined with Cl or some other anion, the Hg atoms remain united. The most common Hg-containing ore is cinnabar (HgS). Approximately 1810 tons of Hg are extracted each year with China accounting for 75% of the total production (USGS, 2014a). The last Hg mine in the United States closed in 1992 (USGS, 2014c).

Environmental Concentrations of Mercury

Mercury occurs in the lithosphere, atmosphere, hydrosphere, and biosphere. Since it occurs in both organic and inorganic forms in all of these media and because bacteria are involved in the conversion of Hg from one state to another, a mercury cycle has been recognized (Fig. 8.10). Starting with the atmosphere, either inorganic or organic Hg can be transported long distances until it eventually precipitates on land or water. The primary sources for the 5500–8900 tons of Hg emitted into the atmosphere each year are natural processes such as volcanoes and combustion in coal-fired plants. Together these account for 82% of atmospheric Hg (Pacyna et al., 2006). Other sources include gold mining and smelters.

On land or water Hg can volatilize as elemental Hg back into the atmosphere, enter the food chain, or runoff into water bodies as ionic or elemental Hg. Mercury in runoff or effluents amasses to approximately 1000 tons per year globally (UNEP, 2013). In water, inorganic Hg can be combined with organic molecules, usually a methyl group, and become organomercury; similarly, microorganisms can convert organomercury into inorganic Hg. Organomercury,

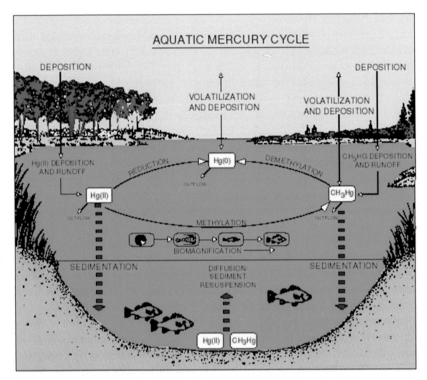

FIGURE 8.10   The mercury cycle. *Courtesy of US Geological Survey. http://wi.water.usgs.gov/ mercury/mercury-cycling.html.*

because it is fat soluble due the organic portion, readily bioaccumulates at low trophic levels and even bioconcentrates among predators such as fish-eating birds and mammals. Much of the Hg that enters water bodies settles into sediments and can be locked in place for millions of years until some natural event such as a volcano eruption releases it back into the atmosphere.

Over the past few decades, scientists have become increasingly concerned about Hg in the South Water District of Florida which includes the Everglades. Fish consumption advisories cover more than a million acres in the district; sale of captured alligators for food has been curtailed because of Hg contamination and records show that the concentration of Hg jumped sixfold between 1900 and 1992 due to changes in hydrology, agriculture, and urban development (USGS, 2013).

The exact proportions may have changed somewhat, but in 1994, Sundolf et al. (1994) estimated that 61% of the Hg was due to atmospheric deposition from human sources, especially solid waste combustion facilities (15%); medical waste incinerators (14%); paint manufacturing and applications (11%); electric utility industries (11%); private residences (2%); and combustion of fossil fuels, electrical apparatuses including lighting (6%). All other anthropogenic sources, including burning of sugar cane (also a major source of PAHs), open burning, sewage sludge disposal, and dental industries accounted for approximately 3% of the total emitted to the environment. Virtually all natural sources (39%) were attributed to release from the soils due to natural processes including microbiological transformations between organic and inorganic forms. In general, organomercury represents only 1% or less of total Hg, even in contaminated sites but, as we will see, this small percent accounts for considerable harm to the environment.

Mercury is persistent. While its mean retention time in the atmosphere ranges from 6–90 days, it can last in soils for 1000 years; ocean water for 2000 years and ocean sediments for more than 1 million years (Eisler, 2000). These retention times are correlated with the importance of each reservoir in regards to total Hg contained. Ocean sediments account for 330×10^9 tons, ocean water for 4.15×10^9 tons; soils for 21×10^6 tons, the atmosphere for 850 tons, and freshwater for only 4 tons. Although the atmosphere is not a primary reservoir for holding Hg, it serves as the major conduit from one reservoir to another.

Uses for Mercury

Mercury has been used in industry, medicine, and agriculture but its demand has been halved since 1980, primarily due to concern regarding its toxicity. In industry, Hg has found use in measuring devices such as thermometers, barometers, manometers, and sphygmomanometers; in float valves; in electrical purposes such as mercury switches, mercury relays, fluorescent lamps, and lighting; and in the production of some types of telescopes. Today, Hg in thermometers and other instruments has been mostly replaced with colored alcohol or digital devices. If you have a thermometer with red fluid used to indicate the temperature, it uses alcohol; if it is silver, it is likely to be Hg. In the 1880s, Hg was used as a bath in making felt hats. The metal neatly separated the fur from the skin to facilitate obtaining the felt. This was in addition to the arsenic that they also used. Unfortunately, hat manufacturers did not recognize the neurotoxicity produced by Hg, and many hat makers developed various neurological disorders. Combined with arsenic toxicity, it is no wonder that the phrase "mad as a hatter" became widespread and perhaps was the inspiration for the Mad Hatter in Lewis Carroll's novel *Alice Adventures in Wonderland* (Fig. 8.11). Although hat makers were affected by both arsenic and Hg, they didn't even have hazard pay at that time.

In agriculture, Hg has been used as a seed coating to protect against fungal and other infections. Unfortunately, some people ate seed corn and other seeds protected with the organomercury coating and became poisoned. This practice of coating seeds with Hg was stopped in the United States in 1995, but still occurs in some other nations.

In medicine, Hg has been used for hundreds of years—going all the way back to ancient Egypt and China. It is still used in Chinese medicine, but its use in western medicine has almost disappeared. As mentioned above, Hg is

FIGURE 8.11 The Mad Hatter, as played by Johnny Depp. *From Walt Disney Picture, 2010.*

used with other metals in dental amalgams. It is also found in thimerosal as a preservative for vaccines and related fluids.

Concentrations of Mercury in Organisms

Organomercury and elemental Hg are very different with regards to bioaccumulation. In particular, organomercury can easily bioconcentrate. This is due to a few salient features of organomercury. Organomercury can biomagnify in aquatic food chains so that almost all Hg accumulated in tertiary levels consists of methylmercury because it can be generated within lower trophic levels and is lipid soluble; inorganic mercury has far less propensity to biomagnify. While inorganic Hg is tightly bound to metallothioneins that facilitate Hg loss, relations between organomercury and metallothioneins are weaker and organisms are less capable of regulating organomercury so there is a greater tendency to retain it. Organomercury is more completely absorbed or assimilated from foods than inorganic mercury, it is also more soluble in lipids, passes more readily through biological membranes; and is slower to be excreted (Eisler, 2000). All of these factors add up to considerably more risk to organisms from organomercury than from inorganic forms of the metal.

Among plants in clean freshwater, the average concentration of Hg was 0.11 mg/kg fw; in contaminated sites, the mean was 10.6 mg/kg fw (Eisler, 2000). Freshwater invertebrates had mean concentrations of 0.21 mg/kg in clean water and 3.66 mg/kg in contaminated sites. Marine invertebrates averaged 0.40 mg/kg dw in reference sites and 6.97 mg/kg dw in polluted sites.

In general, mean Hg concentrations in muscle or whole bodies of fish seldom exceed 0.5 mg/kg fw although maxima can exceed 1.1 mg/kg fw. Various species of tuna collected from highly polluted areas in the Indian Ocean had muscle concentrations exceeding 24 mg/kg (NAS, 1978). Older fish, as measured by age, length, or weight, tend to have higher concentrations than young ones in the same waters because organisms are quick to take up Hg and slow to lose it.

Among amphibians and reptiles, Hg concentrations generally ranged from below detection limits to less than 0.5 mg/kg fw (Eisler, 2000). Amphibians collected below an acetate fiber mill with a history of Hg use in Virginia streams had Hg concentrations that were 3.5 to 22 times those of the same species collected from reference streams; mean values were 3.45 mg/kg for adults and 2.48 mg/kg for larvae (Bergeron et al., 2010).

Considerable research has been conducted on Hg levels in birds and mammals. Many of the tissues and whole body residue levels were below 1.0 mg/kg but the frequency of higher concentrations was greater than seen in aquatic organisms, suggesting that terrestrial organisms—or perhaps homeothermic animals—are more prone to Hg assimilation than aquatic organisms. High concentrations included great blue heron (*Areda herodias*) livers from Lake St. Clair, Ontario (175 mg/kg fw); livers in male (187 mg/kg fw) and female (100 mg/kg fw) common loons (*Gavia immer*) collected as dead birds from New England; and 295 mg/kg dw in livers of several species of albatrosses collected in New Zealand (reviewed by Eisler, 2000). Carnivorous birds generally have higher concentrations of Hg than herbivores.

Feathers have been a reliable source for nondestructive Hg residue analyses. Feathers can be used to compare recently collected animals to museum specimens and get a time sequence perspective on changes in Hg concentrations. They can also be used to compare different ages, sites, or other factors among rare or endangered species of birds without fear of affecting populations. Residues in feathers can reflect differences in conditions between the two sites because molting patterns vary among species with some species molting and regrowing feathers on the breeding and on the wintering grounds. Unlike with some other metals, Hg seems to have a particular affinity with feathers and concentrations in these structures are often higher than in soft tissues. Hair offers a similar benefit for mammals.

If you recall, marine mammals were among those that had the highest concentrations of organic contaminants such as PCBs and organochlorine pesticides. Such is the case with Hg, especially organomercury, because it shares the lipophilicity of other organic molecules and can biomagnify. Mercury concentrations in marine mammals typically measure less than 15 mg/kg (Eisler,

2000). However, among the highest concentrations reported were in livers of harbor seals (*Phoca vitulina*) in California (81–700 mg/kg fw); livers of adult harbor porpoises from the North Sea (504 mg/kg dw); and 1026 mg/kg in the livers of female California sea lions (*Zalophus californianus*). In contrast, elevated concentrations in nonmarine mammals seldom exceeded 50 mg/kg, although hair from dead Florida panthers (*Felis concolor coryi*) had Hg concentrations of 130 mg/kg fw (Roelke et al., 1991) and domestic cats that ate fish from below a chlor-alkali plant in Ontario had up to 392 mg/kg fw in their fur (Jensen, 1980).

Effects of Mercury on Animals

The 96-h LC50 for *Dapnia* sp. is 5 μg/L (US EPA, 1980), and this species is a model for several other species of aquatic invertebrates. Lethal concentrations of total Hg generally range from 0.1–2.0 μg/L for sensitive fish, but the high end may be around 150 μg/L to produce a 96-h LC50. Among sensitive species of fish, adverse sublethal effects can occur at water concentrations of 0.03–0.1 μg Hg/L.

Lethal concentrations in amphibians also range widely from 1.3 μg/L in embryos and larvae of narrow-mouthed toads (*Gastrophyrne carolinensis*, Birge et al., 1979) to 4400 μg/L in adult female river frogs (*Rana heckscheri*, Punzo, 1993).

Birds and mammals are substantially more resistant to Hg toxicosis than most aquatic animals. Median lethal doses to organomercury typically range from 2.2–50 mg/kg body weight or 4–40 mg/kg dietary among sensitive birds. Toxicity of inorganic mercury is much, much lower. For instance, in Japanese quail (*Coturnix japonica*), the LC50 using inorganic Hg was between 2956 and 5086 mg/kg in the diet over 5 days of exposure followed by 7 days of observation. However, only 31–47 mg/kg dietary methylmercury was needed to produce 50% mortality. Given as an acute oral dose, 26–54 mg/kg inorganic Hg and 11–33.7 methylmercury produced LC50s over a 14-day observation period in the same species (Hill and Soares, 1987).

Among mammals, a 96-h LD50 of 17.9 mg/kg oral dose was obtained for mule deer (*Odocoileus hemionus*, Hudson et al., 1984). Mink are particularly sensitive to Hg, and it only took 1.0 mg/kg to produce 100% mortality of the test animals over a 2-month period (Sheffy and Stamant, 1982).

In birds and mammals, sublethal effects of Hg include teratogenesis, mutagenesis, and cancer. It can negatively affect reproduction and growth, behavior, and blood chemistry. It is also infamous for causing neurological disorders involving motor coordination, hearing, vision, and (in humans) thought processes. Histopathology in kidneys, livers, pancreas, and heart are common. As we have mentioned many times, organic mercury is substantially more potent in producing these effects that inorganic Hg, but both can produce significant harm at environmentally realistic concentrations. For sensitive birds, sublethal toxicity can occur at 640 μg Hg/kg body weight or 50–500 μg/kg in diet. Sensitive mammals may be affected at 250 μg Hg daily/kg body weight or 1100 μg/kg in diet.

A Major Human Catastrophe

One of the great human environmental tragedies occurred due to methylmercury. The Japanese-owned Chisso Corporation dumped untreated effluents into Minamata Bay, Japan, from 1932 to 1968; in fact, dumping occurred well past 1968, but in that year some minor treatment of effluents was installed. The operation started as a fertilizer plant but eventually moved into chemical production, making acetylaldehyde and related products with no prevention of dumping their wastes into nearby rivers and ultimately into Minamata Bay. The resulting effluents were also heavily contaminated with several metals and other pollutants. Through this period, shellfish in the bay bioaccumulated exceedingly high concentrations of methylmercury. The local populace fed extensively on these shellfish, not having any idea of the consequences. The first human victim was a young girl in April 1956. The physicians employed by the Chisso Corporation had no concept of what was causing her difficulty in walking and talking or convulsions. By May of that year, an epidemic of neurological conditions had been declared and through that summer the number of cases continued to increase. Patients with the "Minamata disease" complained of a loss of sensation and numbness in their hands and feet. The disease affected their ability to grasp small objects; hindered their ability to run or walk; altered the pitch of their voices; and compromised their ability to see, hear, or swallow. At the same time, domestic cats and fish-eating wildlife were observed having spasms and convulsions. Cats were moving erratically, crows were falling from the sky, and fish were dying along the seashore (Withrow and Vail, 2007). By November 1956, investigators began to put the pieces of the puzzle together and identified that the one common factor among humans and animals was a diet high in fish and shellfish. They concluded that a heavy metal was involved in the disease and Chisso became the chief suspect. Rather that cooperating with the investigation, Chisso refused to provide information or offer any other assistance. They even conducted some of their own experiments that led to confirmation of Hg as the toxic agent but did not release their findings to the public or to the authorities. As a result, progress in finding the actual cause and solving it was slow, and it was not until 1959 that organic mercury was identified as the toxic agent. At this point, Chisso did "help" by agreeing to dredge the contaminated sediments around its effluent areas, but whether they had a change of conscience or only wanted to reclaim the Hg in the discharge is unclear.

Between 1953 and 1957, the fishing industry in Minamata Bay declined by 91%. The fishermen demanded reimbursement and, after more stonewalling, Chisso finally set up funds for reparations amounting to an astounding sum equivalent to $55,600 directly to the fishers and another $41,700 to reestablish fishing. In 1959, the government required that Chisso provide a "sympathy fund" for those affected by the pollution. If you were officially certified as having Minamata disease, the company would compensate you the equivalent of $278 per year (the equivalent of $2455 today after inflation), $83/year if you were a child, and your family could claim a one-time benefit of $889 if it was proven that you had died due to the disease. Many other examples of corruption occurred in this situation and the full extent of reparations and responsibility have still not been met. As of March 2001, 2265 victims had been officially certified (1784 of whom have died), and more than 10,000 people have received financial compensation from Chisso, although they are not recognized as official victims. As a result of this disaster, the United Nations established the Minamata Convention on Mercury in 2013. The principal articles of this international convention include a ban on new Hg mines, the phase-out of existing ones, control measures on air emissions, and regulation of small-scale gold mining (UNEP, 2015). Wikipedia does a good job of detailing the entire situation (http://en.wikipedia.org/wiki/Minamata_disease), but an official summary may be seen at the website on the topic established by the Japan Ministry of the Environment (2002).

As a closing note on Hg, it has been known since the 1960s that selenium (Se) can reduce methylmercury toxicity in vertebrates (eg, Ralston and Raymond, 2010). Selenium, especially at high concentrations, substantially reduces the neurotoxic and lethal effects of methylmercury. Recent studies have shown that Se-enriched diets can even reverse some of the more severe symptoms of methylmercury toxicity. Methylmercury is a highly specific, irreversible inhibitor of Se-dependent enzymes (selenoenzymes) that are necessary to combat oxidative damage, particularly in the brain and neuroendocrine tissues. Inhibition of selenoenzyme activities in these vulnerable tissues appears to be a direct cause of the pathology produced by methylmercury. Mercury has a very high binding affinity for selenium so that under conditions of low dietary selenium, most or all of it is bound to Hg.

ZINC
General Characteristics

The last metal that we will discuss is one, like copper, that is required by all living organisms. As a result, we'll often see relatively high concentrations of this metal in tissues, even from plants and animals living in reference sites. As with copper, Zn deficiencies are more common than Zn toxicity. Approximately 31% of the global human population lives at risk of Zn deficiency (ATSDR, 2005). Severe deficiencies contribute to 176,000 diarrheal deaths, 406,000 pneumonia deaths, and 207,000 deaths due to malaria each year (Caulfield and Black, 2004). On the other hand, there are natural populations of plants and animals that are at risk from Zn toxicosis near smelters and industries that utilize Zn.

Mining of Zn ore dates back to 10,000 BC when the metal, along with copper, was blended into bronze, establishing the Bronze Age. It is the 24th most abundant element and the 4th most abundant metal in the earth's crust. If the distribution of Zn were homogeneous, it would amount to 75 mg Zn/kg of soil or sediment (0.0075%) but the actual distribution can range from 5 to 770 mg/kg with an average of 64 mg/kg in the crust and 30 µg/L in seawater. Zinc naturally occurs as Zn^{1+} and Zn^{2+}, although Zn^{2+} is by far the most common oxidation state. India, followed by Australia and Peru are currently the largest Zn producers of the 12,250 tons of elemental Zn extracted per year (USGS, 2014a).

Environmental Concentrations and Uses of Zinc

In the atmosphere, Zn typically ranges from 0.1 to 4 µg/m^3 (USGS, 2014a). The behavior of zinc and its cycling is similar to the other metals we have discussed. Zinc in the atmosphere generally adheres to particulates which precipitate onto land and water. Additional Zn comes from volcanoes, smelters, and industrial processes. Approximately 21,450 tons of Zn are released by the United States each year (ATSDR, 2005). Of that, 414 tons (1.9%) enters the air, 26 tons (0.12%) is released into water, 16,900 tons (78%) consists of contaminants on land, and 4100 tons (19%) is stored on land. Zinc is found in soils and surficial materials of the contiguous United States at concentrations between <5 and 2900 mg/kg, with a mean of 60 mg/kg. The zinc background concentrations in surface waters are usually <0.05 mg/L, but can range from 0.002–50 mg/L (ATSDR, 2005). Similar concentrations occur in seawater. Canadian sediments from reference sites ranged from 50–180 mg/kg dw. Such sediments in the United States were also 10–75 mg/kg dw.

Zinc is one of the most commonly used metals. The primary use of Zn is in galvanizing steel, which accounts for 55% of the world's production. Other uses include mixing into different alloys for industrial purposes (21%), brass and bronze production (16%), and miscellaneous uses (8%). Zinc is used in a host of metal and other products including solder, pipe organs, machine bearings, pigments, fire retardants, and wood preservatives. Zinc is also present in virtually every multivitamin on the market and is found in many pharmaceutical products. If you've ever used a white cream on your nose to prevent sunburn, you've probably used zinc.

Biological Effects of Zinc

Red and brown algae in marine environments are very efficient accumulators of Zn and can have concentrations greater than 1 g/kg (Eisler, 2000). Zinc concentrations in terrestrial and aquatic invertebrates often run into the high mg/kg or even g/kg ranges. Earthworm concentrations run from 120 to 1600 mg/kg dw among several sampling sites. Oysters from contaminated sites in England had as much as 12.6 g/kg in their soft tissues (Bryan et al., 1987).

Fish, birds, and mammals typically have lower concentrations than seen in invertebrates. Whereas concentrations can exceed 500 mg/kg in fish, the majority of reported values ranged from less than 10 mg/kg to approximately 200 mg/kg. Birds collected from reference sites generally had concentrations that were less than

60 mg Zn/kg dw. The highest Zn concentrations in birds tend to be in liver and kidney and the lowest in muscle (Eisler, 2000). Feathers appear to have similar concentrations as other body tissues and can be used to monitor Zn levels without being invasive.

Ungulates have often been reported with elevated Zn concentrations exceeding 100 mg/kg fw in liver or kidneys. These include white-tailed deer (*Odocoileus virginianus*) and red deer (*Cervus elaphus*) living near a Zn smelter (Sileo and Beyer, 1985). Typical concentrations for small mammals, mice, voles, and shrews range in relatively clean sites ranged from 16–204 mg Zn/kg dw (Eisler, 2000). In contaminated sites, concentrations could be higher at 370 mg Zn/kg dw. Marine mammals do not have unusually high concentrations of Zn as they do with some other contaminants because it is not fat soluble and does not bioconcentrate. Among several seals and whales, concentrations ranged from 47–406 mg/kg dw in liver, 14–140 mg/kg dw in muscle, and 37–353 mg/kg in kidneys (Eisler, 2000).

From a health perspective, Zn is involved in 100–300 enzymes in plants and animals. In addition, it provides structural support for DNA transcription, has numerous neurological benefits, and is very active with metallothioneins. Zinc can bolster the immune system and there are all sorts of products that one can take to avoid or reduce colds. The USDA established the recommended daily allowance for women as 8 mg/day, during breastfeeding and pregnancy, the recommended intake goes up to 13 mg/day. For men, the dose is 11 mg/day. Vitamin supplements can provide assurance of adequate intake, but most people in developed countries obtain sufficient Zn from a well-balanced diet.

As mentioned, Zn deficiency poses a greater risk than too much Zn for humans and for wildlife. Signs of deficiencies include depression; diarrhea; lack of appetite; depressed growth; reproductive and maturation delays; teratogenesis; alopecia (hair loss); eye and skin lesions; altered behavior including cognition, reduced activity, and less play among animals; impaired immunity; and defects in carbohydrate utilization (ATSDR, 2005). Mild zinc deficiency depresses immunity, although excessive zinc does as well. Animals with a diet deficient in zinc require twice as much food in order to attain the same weight gain as animals given sufficient zinc (ATSDR, 2005).

However, Zn toxicosis can affect all groups of organisms around highly contaminated sites. Sensitive plants can die when soil concentrations exceed 100 mg/kg dw. Sublethal effects in plants include poor growth and inhibited reproduction. Among invertebrates, reduced growth may be seen at 300–1000 mg Zn/kg diet dw (Eisler, 2000) and reduced survival may be experienced in the range of 470–6400 mg/kg in soil. Sensitive invertebrates include the earthworm *Eisenia foetida*, a common lab animal, and the slug *Arion ater*. Effects on invertebrate communities including loss of species richness and lower numbers of survivors occurred at 1600 mg/kg dw in the soil (Beyer and Anderson, 1985).

For fish, a few generalizations can be made (Eisler, 2000). Freshwater species tend to be more sensitive than marine species. Embryos and larvae are more sensitive than adults. For brown trout (*Salmo trutta*), fry, a very sensitive species and life stage, death occurred within the range of 4.9–9.8 µg/L (Sayer et al., 1989). For other species, lethal and sublethal effects begin at 50–230 µg/L in sensitive species. Zinc toxicosis in fish is indicated by hyperactivity followed by lethargy, loss of coordination, gill hemorrhaging, and extensive production of mucus. In extreme cases, death is often a result of apoxia (loss of blood oxygen) due to gill deterioration.

Less sensitive birds demonstrated reduced growth with diets having more than 2000 mg/kg in food and reduced survival at concentrations exceeding 3000 mg/kg diet. Ducks (*Anas* sp.) demonstrated decreased survival with 2500–3000 mg/kg dietary Zn and, when force-fed Zn, died at 746 mg/kg body weight (Eisler, 2000). Signs of Zn poisoning in birds include ataxia, flaccid paralysis, and histopathology and necrosis of pancreas. Chickens fed 15,000 mg/kg in their food as zinc oxide showed reproductive problems but no mortality after 7 days.

Mammals tend to be quite tolerant to Zn toxicity. Lab animals such as rats suffer adverse effects at 0.8 mg Zn/m^3 in air, 90 to 300 mg/kg in their diets, 300 mg/L in drinking water, and 350 mg/kg body weight in oral doses (Eisler, 2000). Zinc targets the pancreas of mammals, resulting in loss of function. It also has serious neurological, hematological, immunological, and cardiovascular effects. Zinc toxicosis adversely affects development, growth, pancreatic fibrosis, acute diarrhea, copper deficiencies, and impaired reproduction (ATSDR, 2005).

FOCUS—AVIAN MORTALITY DUE TO LEAD SHOT, BULLETS, AND WEIGHTS

This is a story of how sportsmen unwittingly contributed to the death of millions of birds over the course of several decades.

Since the invention of firearms, the material of choice for ammunition has been lead. Lead shot, musket balls, and bullets have the density to fly well and accurately and have the compressibility so that when they strike their intended target, they flatten out to some extent, increasing the likelihood of inflicting maximum damage and making a clean kill.

However, this section focuses on a problem that have little to do with the actual killing power of firearm-propelled ammunition and more to do with the leftover lead following a hunt.

There are three situations dealing with sportsmen-derived environmental lead that contribute to avian mortality in a major way. These include concentrations of shotgun pellets in aquatic and upland situations; lead bullets in the tissues of animals that were shot but escaped, later to die; and lead fishing sinkers and weights.

With regard to shotgun pellets and their effect on birds, initial attention was paid to waterfowl and Pb contamination in wetland areas. Now that regulatory action has curtailed the use of lead ammunition in wetlands (see the discussion of this topic later in this chapter), greater focus is being placed on upland birds such as doves and their habitats. As early as 1880, hunters and biologists began to be aware that lead shot inflicted an insidious mortality factor. If a gamebird, such as a mallard duck is struck by lead shot but not killed or crippled, it has a good likelihood of surviving because after the shot enters muscle tissue, a capsule of scar tissue forms around the pellets and isolates them from the rest of the body with little or no adverse effects. However, shotgun pellets span the range of size occupied by the seeds that serve as natural foods for these birds. As a result, birds may also ingest pellets when feeding in contaminated areas. Ingested lead is by far the principal mortality factor in all three scenarios mentioned above. Once in the gizzard, the grinding effect of the muscles combined with acidic secretions dissolve the lead pellets and facilitates movement of the soluble lead into the circulatory system where it can have a host of adverse effects. Waterfowl suffering from lead toxicosis, for example, are weak and often unable to fly. They may demonstrate edema in the head, atrophy of breast muscles, lack of appetite, impaction of the proventriculus, ulcers, green feces, and diarrhea. Death often occurs through gradual starvation.

Public hunting areas operated privately and by states and federal governments are highly attractive to birds and to hunters and, through the course of time, substantial concentrations of lead pellets can build up. In areas managed for dove hunting, for example, estimates of the number of shot present range from 3228–167,593/ha before hunting to 17,628 to 860,185 after hunting (Lewis and Legler, 1985). An earlier study conducted in a public area for hunting Canada geese measured only 7512 pellets/ha, but the pellets were larger than those used for doves and in one year, the authors found more than 900 dead geese in the area (Szymczak and Adrian, 1978). Canada geese are field feeders, so this concentration of lead pellets was an upland estimate. In what has to be an extreme case, Perroy et al. (2014) estimated that the concentration of pellets from a wetland on an old trap shooting range was greater than 50,000 pellets/m^2 or 5 million pellets/ha (Fig. 8.12). Perhaps the highest density of pellets found in a National Wildlife Refuge was 2 million pellets/ha (Thomas et al., 2001). The authors reported that conventional tilling practices can reduce the amount of pellets available to birds by almost 50% and these practices are being used on public lands. Since it often takes one or two #4 pellets to be lethal (Puckett and Slota, 2014), the abundance of shot is very important in determining how many birds could die. In semicontrolled studies, it was estimated that 1–4% of the mourning doves that visit public hunting areas could die from lead ingestion (Plautz et al., 2011). While the percentage values seem small, they translate into many thousands of birds.

It took decades from the first scientific studies on the effects of spent shot in wetland environments to the execution of federal and state laws regulating lead shot. The first solid evidence that lead shot was a mortality factor for waterfowl occurred in 1919. Friend (1989) and others estimated that 1.8–2.4 million

FIGURE 8.12 Photograph showing lead shot in a shooting preserve. *Courtesy of <Nebraskalandmagazine.com>, <http://outdoornebraska.ne.gov/blogs/2010/08/toxic-game-part-ii>.*

waterfowl were dying each year due to ingested shot and federal legislation was enacted in 1991. At that time, Congress banned the use of lead shot on National Wildlife Refuges and waterfowl production areas. Shortly thereafter, some states enacted bans on their properties and by 2011, 35 states had banned the use of lead shot on designated hunting areas. One of the barriers to legislation was finding an adequate substitute for lead, something that would have similar flight patterns and killing effectiveness. Ultimately, this led to such substitutes as steel, bismuth, and various alloys of tungsten shot.

Another method of lead toxicity that is of concern for wildlife is secondary toxicity. In this scenario, an animal is shot with bullets or shotgun pellets that lodge in its tissues and is not recovered by the hunter for whatever reason. When the carcass is scavenged by other animals, the scavenger consumes the contaminated tissues and becomes poisoned. Secondary poisoning can be serious for a variety of scavenging species but arguably, the most vulnerable of these is the California condor (*Gymnogyps californianus*) (Fig. 8.13) and the bald eagle (*Haliaeetus leucocephalus*). The trouble begins when the lead bullet or buckshot sometimes fragments into scores of small pieces in the body.

Warner et al. (2014) examined the bodies of 58 bald eagles found dead in 2012 in the Midwest. Sixty percent of the birds had detectable lead concentrations with 38% of them having lethal concentrations. Raptors show many of the same signs of lead toxicity as waterfowl (Fig. 8.14). The authors went on to identify the source of this lead on the Fish and Wildlife Service's Upper

FIGURE 8.13 California condors have been victims of secondary poisoning of animals, dying from lead bullets or shot. The steer here was not necessarily shot, but the picture does represent the scavenging behavior of condors.

FIGURE 8.14 Ferruginous hawk (*Buteo regalis*) showing drooped wings and head characteristic of lead poisoning. *Courtesy Gregory Stempien, US Fish and Wildlife Service.*

Mississippi River National Wildlife and Fish Refuge which allows deer hunting. Thirty-six percent of offal piles contained lead fragments ranging from 1–107 particles per pile. They concluded that offal piles were likely the source of lead for bald eagles.

Overhunting large carnivores, intentional killing of condors in a mistaken belief that they preyed on lambs and calves, and habitat destruction coupled with a low reproductive rate are factors that have led to the decline and near

FIGURE 8.15   Lead shot (arrows) in the gizzard of a Canada goose. *Courtesy Milton Friend, US Geological Survey.*

extinction of condors, but lead ingestion has also been an important factor. In the 1980s, all wild condors were captured and brought to zoos where they encouraged to breed. Offspring of these birds were released back into the wild in areas that had been closed to hunting by federal law. Approximately 62% of 150 free-ranging condors sampled 15 years after the release program began had blood lead levels, and the percent of animals with lead increased with age and independence away from sanctuaries where clean food was provided (Kelly et al., 2014). Maximum lead concentrations in blood also remained high over those years. It does not seem that condors are free from the risk of lead at this time. However, steps continue to be taken to protect the endangered bird. In 2008, eight counties in California within the heart of Condor country were closed to all forms of lead shot and bullets.

The last topic of lead toxicity in this section is on lead weights and sinkers used in fishing. Similar to lead shot used in waterfowl hunting, lead weights can be ingested and cause toxicosis. Many of the split shot and other forms of weights are also within the size range of seeds (Fig. 8.15) and are mistakenly eaten. Concern about lead weights began in England in the 1970s and 1980s and centered on swans that were found dead or dying. An estimated 4000 mute swans (*Cygnus olor*) died from lead poisoning each year. In 1987, Britain banned use of leaded weights ranging in size from 0.06–28.36 g. The swan population has increased in size since then, but it is not clear how much of the increase was due to the lead ban. Kelly and Kelly (2004) reported that of 1421 swans brought into a wildlife hospital in England between 2000 and 2002, between 15% and 18% had fishing tackle issues including getting caught in lines, getting hooked, and showing signs of possible lead toxicity. Of 921 birds tested for the presence of lead in their blood, 74% proved positive with concentrations ranging from 0.2 µg/dL to 2.3 mg/dL. Lead in sediment continues to be a problem in Britain, both as lead weights and gunshot.

STUDY QUESTIONS

1. What are some of the common characteristics of metals?
2. How many elements are listed as metals on the periodic chart of elements? How about metalloids? Nonmetals?
3. How would you define a heavy metal (please ignore the music style)?
4. What did Paracelsus mean when he said that the dose makes the poison?
5. Describe the relationship between deficiency and toxicity for metals that are essential for diets.
6. Discuss some reasons why organometals are usually more toxic than elemental or inorganic metals.
7. What role does pH have in the solubility and bioavailability of metals?
8. True or False. Hard water with a high concentration of calcium reduces the toxicity of many metals compared with soft water.
9. What group of proteins is highly involved with metal transport and excretion?
10. True or False. Arsenic does not have a true melting point.
11. List four or five major uses of the following metals: arsenic, copper, cadmium, lead, mercury, zinc.
12. Why was the occupation of making hats in the 19th century a risky business?
13. Which of the metals that were individually described are suspected or confirmed endocrine disruptors? Which are suspected or confirmed carcinogens? Which are essential to living organisms? Include both elemental and organic forms of the metals in your answer.
14. As a general rule, is the toxicity of metals to fish more or less than to birds or mammals?
15. What is the effect of particulate matter such as dissolved organic matter in water concerning bioavailability of metals?
16. Rank the individually described metals from low to high in the amount mined each year globally.
17. Why are lead pellets that become imbedded in muscle tissue not very toxic whereas those that are consumed can be lethal?
18. What is secondary toxicity and how does it affect California condors?

REFERENCES

Ahmed, Q., Benzer, S., Elahi, N., Ali, Q.M., 2015. Concentrations of heavy metals in *Sardinella gibbosa* (Bleeker, 1849) from the coast of Balochistan, Pakistan. Bull. Environ. Contam. Toxicol. 95, 221–225.

ATSDR, Agency for Toxic Substances and Disease Registry, 2005. Toxicological Profile for Zinc. Public Health Service Agency for/Toxic Substances and Disease Registry, Atlanta, GA.

ATSDR, Agency for Toxic Substances and Disease Registry, 2007a. Toxicological Profiles for Arsenic. U.S. Public Health Service, Atlanta, GA.

ATSDR, Agency for Toxic Substances and Disease Registry, 2007b. Toxicological Profiles for Lead. U.S. Public Health Service, Atlanta, GA.

ATSDR, Agency for Toxic Substances and Disease Registry, 2012. Toxicological Profiles for Cadmium. U.S. Public Health Services, Atlanta, GA.

Auguy, F., Fahr, M., Moulin, P., Brugel, A., Laplaze, L., El Mzibri, M., et al., 2013. Lead tolerance and accumulation in *Hirschfeldia incana*, a Mediterranean Brassicaceae from metalliferous mine spoils. PLoS One. http://dx.doi.org/101371/journal.pone.0061932.

Bergeron, C.M., Bodinoff, C.M., Unrine, J.M., Hopkins, W.A., 2010. Mercury accumulation along a contaminant gradient and nondestructive indices of bioaccumulation in amphibians. Environ. Contam. Toxicol. 29, 980–988.

Beyer, W.N., Anderson, A., 1985. Toxicity to woodlice of zinc and lead-oxides added to soil litter. Ambio 14, 173–174.

Birge, W.J., Black, J.A., Westerman, A.G., Hudson, J.E., 1979. The effect of mercury on reproduction of fish and amphibians. In: Niragu, J.O. (Ed.), The Biogeochemistry of Mercury in the Environment. Elsevier, New York, NY, pp. 629–655.

Blus, L.J., Henny, C.J., Hoffman, D.J., Grove, R.A., 1993. Accumulation and effects of lead and cadmium on wood ducks near a mining and smelting complex in Idaho. Ecotoxicology 2, 139–154.

Bryan, G.W., Gibbs, P.E., Hummerstone, L.G., Burt, G.R., 1987. Copper, zinc and organotin as long-term factors governing the distribution of organisms in the Fal Estuary in southwest England. Estuaries 10, 208–219.

Button, M., Jenkin, G.R.T., Bowman, K.J., Harrington, C.F., Brewer, T.S., Jones, G.D.D., et al., 2010. DNA damage in earthworms from highly contaminated soils: assessing resistance to arsenic toxicity by use of the Comet assay. Mut. Res/Gen. Toxicol. Environ. Mutagen. 696, 95–100.

Caulfield, L.E., Black, R.E., 2004. Zinc deficiency. In: Ezzati, M., Lopez, A.D., Rodgers, A., Murray, C.J.L. (Eds.), Comparative Quantification of Health Risks. Vol. 1. World Health Organization (WHO), Geneva, pp. 157–279.

Chapman, G.A., 1978. Toxicity of cadmium, copper and zinc to four juvenile stages of Chinook salmon and steelhead. Trans. Am. Fish. Soc. 107, 841–847.

Cole, R.H., Frederick, R.E., Healy, R.P., Rolan, R.G., 1984. Preliminary findings of the priority pollutant monitoring project of the National Urban Runoff Program. J. Water. Poll. Control Fed. 56, 898–908.

De Jonge, M., Lofts, S., Bervoets, L., Blust, R., 2014. Relating metal exposure and chemical speciation to trace metal accumulation in aquatic insects under natural field conditions. Sci. Tot. Environ. 496, 11–21.

Demayo, A., Taylor, M.C., Taylor, K.W., Hodson, P.V., 1982. Toxic effects of lead and lead compounds on human health, aquatic life, wildlife plants, and livestock. CRC Crit. Rev. Environ. Contam. 12, 257–305.

Eisler, R., 2000. Handbook of chemical risk assessment. Health Hazards to Humans, Plants, and animals Metals, vol. 1. Lewis Publ, Boca Raton, FL.

Ford, K.L., Beyer, W.N., 2014. Soil criteria to protect terrestrial wildlife and open-rang livestock from metal toxicity at mining sites. Environ. Monit. Assess. 186, 1899–1905.

Franson, J.C., Lahner, L.L., Meteyer, C.U., Rattner, B.A., 2012. Copper pellets simulating oral exposure to copper ammunition: absence of toxicity in American kestrels (*Falco sparverius*). Arch. Environ. Contam. Toxicol. 62, 145–153.

Friend, M., 1989. Lead poisoning: the invisible disease. Fish and Wildlife Leaflet 13.2.6.

Ghiani, A., Fumagalli, P., Nguyen Van, T., Gentili, R., Citterio, S., 2014. The combined toxic and genotoxic effects of Cd and As to plant bioindicator *Trifolium repens* L. PLoS One 9, e99239. http://dx.doi.org/10.1371/journal.pone.0099239

Gilani, S.H., Marano, M., 1979. Chromium poisoning and chick embryogenesis. Environ. Res. 19, 427–431.

Giri, S., Singh, A.K., 2015. Metals in some edible fish and shrimp species collected in dry season from Subarnarekha River, India. Bull. Environ. Contam. Toxicol. 95, 226–233.

Grillitsch, B., Schiesari, L., 2010. Appendix: Metal concentrations in reptiles. An appendix of data compiled from the existing literature. In: Sparling, D.W., Linder, G., Bishop, C.A., Krest, S.K. (Eds.), Ecotoxicology of Amphibians and Reptiles, second ed. SETAC/CRC Press, Boca Raton, FL, pp. 553–901.

Guitart, R., Tofigueras, J., Mateo, R., Bertolero, A., Cerradelo, S., Martinezilata, A., 1994. Lead-poisoning in waterfowl from the Ebro Delta, Spain—calculation of lead-exposure thresholds for mallards. Arch. Environ. Contam. Toxicol. 27, 289–293.

Hariharan, G., Purvaja, R., Ramesh, R., 2014. Toxic effects of lead on biochemical and histological alterations in green mussels (*Perna viridis*). J. Toxicol. Environ. Health A 77, 246–260.

Hill, E.F., Soares, J.H., 1987. Oral and intramuscular toxicity of inorganic and organic mercury-chloride to growing quail. J. Toxicol. Environ. Health 20, 105–116.

Hudson, R.H., Tucker, R.K., Haegle, M.A.,1984. Handbook of toxicity of pesticides to wildlife. U.S. Fish Wildl. Serv. Resour. Publ. p. 153.

Hughes, J.S., 1973. Acute toxicity of thirty chemical to striped bass (Morone saxatilis). Proc. West Assoc. State Game Fish Commun. 53, 399–413.

International Lead Association, 2015. Lead recycling. <http://www.ila-lead.org/lead-facts/lead-recycling> (accessed 25.04.15.).

Ilizaliturri-Hernández, C.A., González-Mille, D.J., Mejía-Saavedra, J., Espinosa-Reyes, G., Torres-Dosal, A., Pérez-Maldonado, I., 2012. Blood lead levels, δ-ALAD inhibition, and hemoglobin content in blood of giant toad (*Rhinella marina*) to asses lead exposure in three areas surrounding an industrial complex in Coatzacoalcos, Veracruz, Mexico. Environ. Monit. Assess. 185, 1685–1698.

Jaiser, S.R., Winston, G.P., 2010. Copper deficiency myelopathy. J. Neurol. 257, 869–881.

Japan Ministry of the Environment, 2002. Minamata Disease, History and Measures. Environ Health Dept., Tokyo.

Jenkins, D.W., 1980. Biological monitoring of toxic trace metals. Vol. 2. Toxic trace metals in plants and animals of the world. Part II. U.S. Environ. Protect. Ag. Rep. 600/3-80-091-505-618.

Jensen, D.W., 1980. Biological monitoring of toxic trace metals. Vol. 2. Toxic trace metals in plants and animals of the world. Part II. U.S. Environmental Protection Agency Rept 600/3-80-091:505–618.

Kamal, M., Ghaly, A.E., Mahmoud, N., Cote, R., 2004. Phytoaccumulation of heavy metals by aquatic plants. Environ. Inter. 29, 1027–1039.

Kelly, A., Kelly, S., 2004. Fishing tackle injury and blood lead levels in mute swans. Waterbirds 27, 60–68.

Kelly, T.R., Grantham, J., George, D., Welch, A., Brandt, J., Burnett, L.J., et al., 2014. Spatiotemporal patterns and risk factors for lead exposure in endangered California condors during 15 years of reintroduction. Cons. Biol. 28, 1721–1730.

Langeveld, L.G., Liefting, H.J., Boogaard, F.C., 2012. Uncertainties of stormwater characteristics and removal rates of stormwater treatment facilities: Implications for stormwater handling. Water Res. 46, 6868–6880.

Lerebours, A., Stentiford, G.D., Lyons, B.P., Bignell, J.P., Derocles, S.A.P., Rotchell, J.M., 2014. Genetic alterations and cancer formation in a European flatfish at sites of different contaminant burdens. Environ. Sci. Technol. 45, 10448–10455.

Lewis, J.C., Legler, E., 1985. Lead shot ingestion by mourning doves and incidence in soil. J. Wildl. Manage. 32, 476–482.

Missouri Department of Natural Resources, 2010. Total maximum daily load information sheet. Mississippi River. Missouri Department of Natural Resources Water Protection Program. Jefferson City, MO. <http://www.dnr.mo.gov/env/wpp/tmdl/info/1707-miss-r-info.pdf> (accessed 12.10.14.).

NAS, National Academy of Sciences, 1978. An Assessment of Mercury in the Environment. Natl Acad Sci, Washington, DC.

Norain, T.M., Ismail, I.B., Abdoun, K.A., Al-Haidary, A.A., 2013. Dietary inclusion of chromium to improve growth performance and immune-competence of broilers under heat stress. Ital. J. Anim. Sci. 12, e92. http://dx.doi.org/10.4081/ijas.2013.e92.

Nriagu, J.O. (Ed.), 1978. The Biogeochemistry of Lead in the Environment. Part A. Ecological Cycles. Elsevier/Holland Biomedical Press, Amsterdam.

Overmann, S.R., Krajicek, J.J., 1995. Snapping turtles (*Chelydra-serpentina*) as biomonitors of lead contamination of the Big River in Missouri Old Lead Belt. Environ. Toxicol. Chem. 14, 689–695.

Pacyna, E.G., Pacyna, J.M., Steenhausen, F., Wilson, S., 2006. Global anthropogenic mercury emission inventory for 2000. Atmos. Environ. 40, 4048–4063.

Perroy, R.L., Belby, C.S., Mertens, C.J., 2014. Mapping and modeling three dimensional lead contamination in the wetland sediments of a former trap-shooting range. Sci. Total Environ. 487, 72–81.

Pfeiffer, W.C., Fiszman, M., Carbonell, N., 1980. Fate of chromium in a tributary of the Iraja River, Rio de Janerio. Environ. Poll. 1B, 117–126.

Plautz, S.C., Halbrook, R.S., Sparling, D.W., 2011. Lead shot ingestion by mourning doves on a disked field. J. Wildl. Manage. 75, 779–785.

Puckett, C., Slota, P., 2014. Lead shot and sinkers: weighty implications for fish and wildlife health. U.S. Geological Survey Office of Communication, Reston VA. <http://www.usgs.gov/newsroom/article.asp?ID=1972&from=rss_home#.VMbrTP7F_T8> (accessed 5.02.15.).

Punzo, F., 1993. Ovarian effects of a sublethal concentration of mercuric chloride in the river frog, *Rana heckscheri* (Anura: Ranidae). Bull. Environ. Contam. Toxicol. 50, 385–391.

Ralston, N.V.C., Raymond, L.J., 2010. Dietary selenium's protective effects against methylmercury toxicity. Toxicology 278, 112–123.

Roelke, M.E., Schultz, D.P., Facemire, C.F., Sundlof, S.F., 1991. Mercury contamination in the free-ranging endangered Florida panther (*Felis-concolor-coryi*). In: Junge RE (ed.), Proc Am Assoc Zoo Vet Ann Meeting. pp. 277–283.

Sayer, M.D.J., Reader, J.P., Morris, R., 1989. The effect of calcium concentration on the toxicity of copper, lead and zinc in yolk-sac fry of brown trout (*Salmo trutta* L.) in soft, acid water. J. Fish. Biol. 35, 323–332.

Sheffy, T.B., Stamant, J.R., 1982. Mercury burdens in furbearers in Wisconsin. J.Wildl. Manage. 46, 1117–1120.

Sileo, L., Beyer, W.N., 1985. Heavy metals in white-tailed deer living near a zinc smelter in Pennsylvania. J. Wildl. Dis. 21, 289–296.

Singha, U., Pandey, N., Boroa, F., Girib, S., Giria, A., Biswas, S., 2014. Sodium arsenite induced changes in survival, growth, metamorphosis and genotoxicity in the Indian cricket frog (*Rana limnocharis*). Chemosphere 112, 333–339.

Soteropoulos, D.L., Lance, S.L., Flynn, R.W., Scott, D.E., 2014. Effects of copper exposure on hatching success and early larval survival in marbled salamanders, *Ambystoma opacum*. Environ. Toxicol. Chem. 33, 1631–1637.

Sparling, D.W., 1991. Acid precipitation and food quality: effects of dietary Al, Ca, and P on bone and liver characteristics in American black ducks and mallards. Arch. Environ. Contam. Toxicol. 21, 281–288.

Sparling, D.W., Lowe, T.P., 1996. Metal concentrations of tadpoles in experimental ponds. Environ. Poll. 91, 149–156.

Sparling, D.W., Krest, S., Ortiz-Santiliestra, M., 2006. Effects of lead-contaminated sediment on *Ranasphenocephala* tadpoles. Arch. Environ. Contam. Toxicol. 51, 458–466.

Spears, B.L., Hanson, J.A., Audet, D.J., 2007. Blood lead levels in waterfowl using Lake Coeur d'Alene, Idaho. Arch. Environ. Chem. Toxicol. 52, 121–128.

Stephensen, D.A., Nielsen, A.H., Hvitved-Jacobsena, T., Arias, C.A., Brix, H., Vollertsen, J., 2014. Distribution of metals in fauna, flora and sediments of wet detention ponds and natural shallow lakes. Ecol. Eng. 66, 43–51.

Sugatt, R.H., 1980a. Effects of sublethal sodium dichromate exposure in freshwater on the salinity tolerance and serum osmolarity of juvenile Coho salmon, *Onchorynchus kisutsch*, in seawater. Arch. Environ. Contam. Toxicol. 9, 41–52.

Sundolf, S.F., Spalding, M.G., Wentworth, J.D., Steible, C.K., 1994. Mercury in livers of wading birds (Ciconiiformes) in southern Florida. Arch. Environ. Contam. Toxicol. 27, 299–305.

Szymczak, M.R., Adrian, W.J., 1978. Lead poisoning in Canada geese in Southeast Colorado. J. Wildl. Manage. 42, 299–306.

Thomas, C.M., Mensik, J.G., Feldheim, C.L., 2001. Effects of tillage on lead shot distribution in wetland sediments. J. Wildl. Manage. 65, 40–46.

Tu, C., Ma, L.Q., 2002. Effects of arsenic concentrations and forms on arsenic uptake by the hyper-accumulator ladder brake. J. Environ. Qual. 31, 641–647.

UNEP, United Nations Environmental Program, 2013. Global Mercury Assessment 2013: Sources, Emissions Releases and Environmental Transport. UNEP Chemical Branch, Geneva, Switzerland. 299–305.

UNEP United Nations Environmental Program Minamata Convention on Mercury, 2015. <http://www.mercuryconvention.org/Convention/tabid/3426/Default.aspx> (accessed 19.1.15.).

US EPA, U.S. Environmental Protection Agency, 1980. Ambient Water Quality Criteria for Cadmium. U.S. Environmental Protection Agency Rep. 440/5-80-025, Washington, DC.

US EPA, U.S. Environmental Protection Agency, 1982. Standards of performance for lead-acid battery manufacturing plants. U.S. Environmental Protection Agency. Code of Federal Regulations. 40 CFR 60. Subpart KK.

US EPA, U.S. Environmental Protection Agency, 1985. Ambient water quality for lead. 1984. US EPA Rept 440/5-84-027. Washington DC.

US EPA, U.S. Environmental Protection Agency, 2009.Air quality criteria for lead (2006) final report. <http://cfpub.epa.gov/ncea/cfm/recordisplay.cfm?deid=158823> (accessed 12.8.14.).

US EPA, U.S. Environmental Protection Agency, 2014. Lead. <http://www.epa.gov/airtrends/lead. html> (accessed 12.8.14.).

US EPA, U.S. Environmental Protection Agency, 2015. The toxic release inventory (TRI) program. <http://www2.epa.gov/toxics-release-inventory-tri-program>.

USGS, U.S. Geological Survey, 2013. The south Florida environment: A region under stress. US Dept Interior, USGS, Circular 1134. <http://sofia.usgs.gov/publications/circular/1134/wes/merc.html> (accessed 20.1.15.).

USGS, U.S. Geological Survey, 2014a. Mineral commodity summaries, February 2014. <http://minerals.usgs.gov/minerals/pubs/commodity/arsenic/mcs-2014-arsen.pdf> (accessed 14.11.14).

U.S. Geological Survey, 2014b. Copper statistics. <http://minerals.usgs.gov/minerals/pubs/historical-statistics/ds140-coppe.pdf> (accessed 26.11.14.).

US Geological Survey, 2014c. Mercury in the environment. <http://www.usgs.gov/mercury/>.

Van der Putte, I., Lubbers, J., Kolar, Z., 1981. Effect of pH on uptake, tissue distribution and retention of hexavalent chromium in rainbow trout (*Salmo gardineri*). Aquat. Toxicol. 1, 6–18.

Warner, S.E., Britton, E.E., Becker, D.N., Coffey, M.J., 2014. Bald eagle lead exposure in the Upper Midwest. J. Fish Wildl. Manage. 5, 208–216.

Westing, A.H., 1972. Herbicides in war: current status and future doubt. Biol. Cons. 4, 322.

White, D.H., Finely, M.T., 1978. Uptake and retention of dietary cadmium in mallard ducks. Environ. Res. 17, 53–59.

Withrow, S.J., Vail, D.M., 2007. Withrow and MacEwen's small animal clinical oncology. Elsevier, Cambridge, MA.

Young, D.R., Jan, T.K., Hershelman, G.P., 1980. Cycling of zinc in the nearshore marine environment. In: Nraigu, J.O. (Ed.), Zinc in the Environment. Part I. Ecological Cycling. John Wiley, New York, NY, pp. 297–335.

Žatauskaitė, J., Sodienė, I., 2014. Effects of cadmium and lead on the life-cycle parameters of juvenile earthworm *Eisenia fetida*. Ecotox. Environ. Safe 103, 9–16.

Chapter 9

Other Contaminants

Terms to Know

Plastic
Macroplastic
Microplastic
Thermoset
Thermoplastic
Phthalate
Bisphenol A
Nitroamines
Nitroaromatics
Acid deposition
Dry deposition
Wet deposition
ANC
NO_x
Circumneutral
Pharmaceuticals
PPCP
Nanoparticles
Liposome
Nanotubules

INTRODUCTION

It would be appropriate to consider this chapter as a "catch-all" because it includes many types of contaminants that have not been discussed in other chapters. It is not that these contaminants are not important, but in the larger picture, they arguably represent less diversity, lower historical prominence, smaller global distribution, or are just simply less well known than the major families of contaminants we have covered up to now. We will cover, in no specific order:

Plastics
Munitions
Acid Deposition
Pharmaceuticals
Nanoparticles

PLASTICS

It is rather dated by now, but in the 1967 movie *The Graduate* the hero of the story, Ben Braddock, played by a much younger Dustin Hoffman, is given the advice upon his graduation from college to "go into plastics." That would have been excellent advice back then and it would not be bad advice today. According to the US Environmental Protection Agency (2015a), 32 million tonnes (29.1 million tons) of plastic waste were generated in 2012, representing 12.7% of total manufactured solid wastes. Approximately 14 million tonnes of plastics were containers and packaging, 11 million tonnes were durable goods such as appliances, and nearly 7 million tonnes were nondurable goods, such as plates and cups. Of that total, only 9% was recovered for recycling despite the fact that 86% of the American public have access to either curbside or recycling center pick up (EPA, 2015a). Some plastics have higher recycling rates than others. For example, the category of plastics which includes bags, sacks, and wraps was recycled at approximately 12%; and the plastics used most in commercial products—polyethylene terephthalate (abbreviated PET or PETE) and high-density polyethylene (HDPE) experienced recycling rates around 22%.

What, exactly, is a *plastic*? Well, it is hard to be precise. Plastic is a material consisting of any of a wide range of synthetic or semisynthetic organic chemicals that can be molded into solid objects of diverse shapes. They are typically organic polymers of high molecular mass, but they often contain other substances. They are mostly derived from petrochemicals, but many are at least partially derived from other materials. *Malleability* is a general property of most plastics that are able to irreversibly deform without breaking, but this occurs in a wide degree of variability. Take, for instance, plastic bottles which can be easily deformed and compare them to polyvinyl chlorine (PVC) plumbing pipes which are far more rigid—both are plastics.

Plastics can be divided in to two major categories: thermosets and thermoplastics. A *thermoset* solidifies or "sets" irreversibly when heated. For example, the bodies of many newer cars are made from thermosets. Polyvinylchloride is another form of thermoset. Thermosets are also used as adhesives, inks, and coatings. A *thermoplastic* material softens when exposed to heat and returns to original condition at room temperature. Plastic water bottles, "Tupperware" type of plastic, milk jugs, floor coverings, credit cards, and carpet fibers are types of thermoplastics.

Another way of classifying plastics is by the type of chemical used to make them. Thus, we get (the respective recycling codes are in brackets after each name):

PETE (Polyethylene Terephthalate) [1]—Most often used for cooking oil bottles, soft drink bottles, and peanut butter jars.
HDPE (High-Density Polyethylene) [2]—Commonly used for milk jugs and detergent bottles.
PVC (Polyvinyl Chloride) [3]—Used in plastic pipes, some water bottles, outdoor furniture, shrink-wrap, liquid detergent containers, and salad dressing containers.

LDPE (Low-Density Polyethylene) [4]—Found in trash can liners, dry-cleaning bags, produce bags, and food storage containers.
PP (Polypropylene) [5]—used for drinking straws and bottle caps.
PS (Polystyrene) [6]—makes up packaging pellets like "Styrofoam peanuts" and some types of plastic drinking cups and plates.
OTHER [7]—Plastics listed in the OTHER category are those not listed in the first six categories. Certain food storage containers fall within this category.

While plastics in the environment do not last forever, they many take a long time, even centuries, to decompose and when they do decompose, they may release toxic chemicals and produce smaller particles that offer different kinds of problems than the original bulk object. On land, polystyrene (ie, foam cups) may take up to 50 years to decompose. In oceans, however, recent studies (Saido et al., 2014) show that exposure to wind, sunlight, and water accelerates decomposition of this plastic to only a few years. During decomposition, several toxic by-products including styrene monomer, styrene dimers such as 2,4-diphenyl-1-butene, 1,3-diphenyl propane, and styrene trimer 2,4,6-triphenyl-1-hexene are produced. Other plastics take even longer to decompose. Polyethylene terephthalate (note the "phthalate"—we will come back to that later) or plastic water bottles may require 400 years, disposable diapers 450 years, and monofilament fishing line 600 years. Often toxic chemicals such as PCBs and PBDEs adhere to polyethylene and polypropylene and when these products decompose the other chemicals are released into the environment.

In recent years, biodegradable plastics have been developed, especially for plastic trash bags that end up in landfills by adding cellulose or starch in the manufacturing process. Supposedly the cellulose or starch break down and only microscopic plastic particles remain. A problem with this method, of course, is that the plastic does remain. Another issue is that "biodegradable" bags and other plastics vary in their rates of degradation. Vaverkova et al. (2014) examined the rates of decomposition of biodegradable shopping bags made of HDPE or polypropylene with an addition of pro-oxidants and materials certified as compostable. At the end of 12 weeks, the HDPE with the additives had not decomposed and even their original color remained. At the same time, polypropylene samples with compostable products had broken down, so maybe we can make an effectively decomposable plastic bag.

Enormous production of plastics and a lack of recycling lead to a massive amount of plastics in our environment. Oceans, in particular, seem to be the final dumping grounds for plastics. It is estimated that 165 million tons (150 tonnes) of plastics consisting of more than 5 trillion pieces contaminate ocean waters (Barnes et al., 2009). This is a very rough estimate because many of the fine particles that enter the ocean or result from the breakdown of larger pieces of plastic are very difficult to quantify. Andrady (2009) stated: "It is likely that nearly all of the plastic that has ever entered the environment still occurs as polymers and very little or any plastic fully degrades in the marine environment." These small

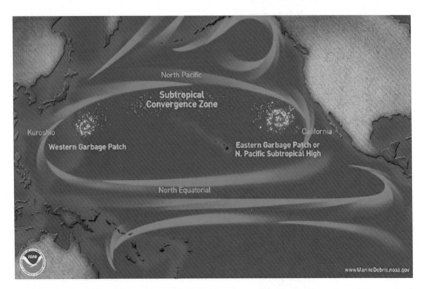

FIGURE 9.1   Pacific garbage patches, 2010. *Courtesy of National Oceanic and Atmospheric Administration. http://marinedebris.noaa.gov/info/patch.html.*

particles are called *microplastics*. One type of which, colloquially called *nurdles*, are produced in a country for transport around the world by ship to where they will be used to manufacture plastic materials. These nurdles can enter the ocean through spills or even through deliberate discharge. Approximately 20% of the ocean-bound plastics comes from oceanic sources such as dumping bilge water, discarded fishing gear, and other sources of waste. Sources on land such as landfills contribute an extensive amount through leaching and breakdown of the landfills, resulting in discharge into rivers and streams and gradual movement downstream. Globally, the vast majority of plastics in the ocean, approximately 80% comes from land sources such as agricultural runoff, discharge of nutrients and pesticides, and untreated sewage (UNESCO, 2015).

Due to currents in the Pacific Ocean there are two very large garbage patches composed of plastics and other debris that are connected by a subtropical convergence zone (Fig. 9.1). Millions of tons of micro- (defined as particles <5 mm in diameter), meso- (5–20 mm), and macroplastics (>200 mm) float on these patches. Andrew Turgeon, a writer for National Geographic, stated that one expedition to the garbage patch collected 750,000 bits of plastic/km^2, which is equivalent to about 1.9 million bits/mile^2 (National Geographic, n.d.). *Macroplastics* that accumulate closer to shore (Fig. 9.2) and eventually breakdown into the microplastics that are found in the oceanic garbage patches. The United Nations Environment Program estimated in 2006 that every square mile of ocean in the garbage patches contains 46,000 pieces of floating plastic (UNESCO, 2015).

FIGURE 9.2 Macroplastics awash in the sea near shore. *From Wikimedia Commons.*

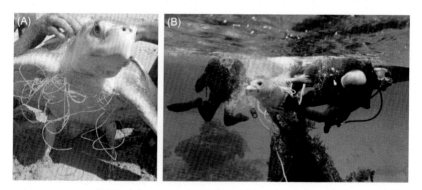

FIGURE 9.3 Examples of wildlife caught in plastic. (A) Sea turtle caught in fishing line. (B) Seal caught in fishing nets. *From Wiki Media Commons.*

There is great concern for wildlife due to the amount of macroplastics, especially fishing lines and nets in the ocean. The United Nations estimates that plastic debris causes the deaths of more than one million seabirds every year, as well as more than 100,000 marine mammals (UNESCO, 2015), not to mention sea turtles, fish, and other marine organisms that get caught in discarded or lost fishing nets (Fig. 9.3). Other animals die because they ingest plastics that block their digestive systems. On Midway Island, which lies approximately halfway between Japan and California, hundreds of carcasses of dead albatross litter the landscape; not all died from ingesting plastics, but almost all of the carcasses have plastic in their digestive systems (Fig. 9.4). Plastic bags, the type used when shopping, last for years to decades and look like jellyfish to hungry sea

FIGURE 9.4   Laysan albatross (*Diomedea immutabilis*) carcass on Midway Island with plastic particles and other debris in its body due to ingestion. *NOAA photo by Claire Fackler.*

FIGURE 9.5   Plastic bags floating in the ocean look a lot like jellyfish to unsuspecting sea turtles. *Courtesy medasset.org.*

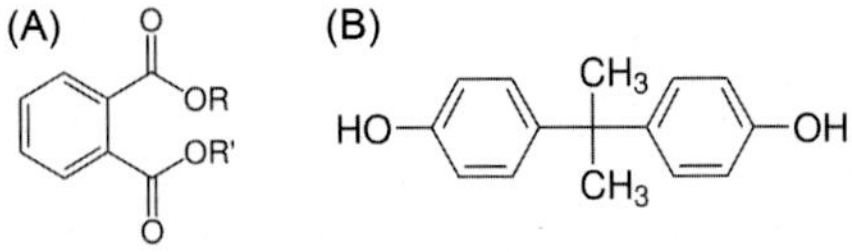

(A)

(B)

FIGURE 9.6   Structural formulas for two endocrine disrupting chemicals associated with plastics. (A) Phthalates (R and R′ are general placeholders); (B) Bisphenol A.

turtles (Fig. 9.5). When they are swallowed by the turtles, they block gastrointestinal passages, resulting in mortality.

In addition to physical difficulties caused by ingesting plastics, partially decomposed plastics, if ingested, can release two toxic types of contaminants. One of these types is *phthalates* (Fig. 9.6A), also called *plasticizers* because they make plastics flexible and more durable. Recall that we said we would come back to this when we were discussing polyethylene terephthalate. Phthalates are used as solvents (dissolving agents) for other materials and in hundreds of products such as vinyl flooring, adhesives, detergents, lubricating oils, automotive plastics, plastic clothes (raincoats), and personal-care products (ie, soaps, shampoos, hairsprays, and nail polishes). They are widely used in polyvinyl chloride plastics, which are used to make products such as plastic packaging film and sheets, pipes, garden hoses, inflatable toys, blood-storage containers, medical tubing, and some children's toys.

Human ingestion of phthalates is low and, as a result, some federal agencies such as the Center for Disease Control and US Food and Drug Administration do not consider humans to be at significant risk. However, the US Environmental Protection Agency (US EPA) stated that studies have associated phthalates with human health concerns, especially as endocrine disruptors and impediments to male fertility; consequently the US EPA is beginning to restrict use of the chemicals (US EPA, 2015b). In laboratory animals, phthalates induce multiorgan damage through oxidative stress, DNA damage, lipid peroxidation, interfering with normal cell functions, and altering the expression-important antioxidant enzymes. Phthalates also interfere with the GSH/GSSG ratios, a sign of oxidative stress. The reproductive system is particularly affected by phthalate-caused oxidative stress and this is expressed as disruption of spermatogenesis, inducing mitochondrial dysfunction in gametocytes, and impairing cellular redox mechanisms in spermatocytes (Asghari et al., 2015).

Bisphenol A (BPA, Fig. 9.6B) is used in some plastics and epoxy resins to make them clear and strong. It is common in water bottles, sports equipment, CDs, DVDs, lining water pipes, and as coatings on the inside of many food and beverage cans (eg, the white inside lining seen in many cans).

BPA is an endocrine disruptor in laboratory animals and as a result, raises concern about its suitability in some consumer products and food containers. In 2012, The US Food and Drug Administration banned its use in baby bottles and toddler cups; Canada and the European Union have also banned BPA in those uses.

An aspect of BPA that adds to the concern is that its effects may span multiple generations. Ziv-Gal et al. (2015) orally dosed pregnant mice with vehicle (serving as control), BPA (0.5, 20, and 50 μg/kg/day) or 0.05 μg/kg/day diethylstilbestrol, a potent estrogen, from gestation day 11 until birth. They examined breeding at the ages of three, six, and nine months in the F1, F2, and F3 generations to evaluate reproductive capacity over time. They also examined age of maturation, litter size, and percentage of dead pups in each generation. From those data, they calculated pregnancy rate, mating, fertility, and gestational indices. The results indicate that BPA exposure significantly delayed the age of maturation in the F3 generation compared to controls. BPA exposure also reduced gestational indices in the F1 and F2 generations, compromised the fertility index in the F3 generation, and reduced the ability of female mice to maintain pregnancies as they aged.

Crain et al. (2007) reviewed the literature on BPA and wildlife. They found extensive evidence that BPA induces feminization during the development of gonads in fishes, reptiles, and birds, but in all cases the amount of BPA necessary to cause such disruption in acute exposures exceeded environmentally realistic concentrations. While that may be the case, it is well known that the longer an animal is exposed to a contaminant, the lower the concentration of that chemical needs to be to have an effect; thus, longer exposures than what have been in laboratory studies or interaction with other chemicals may produce different results. They also found evidence that adult exposure to environmental concentrations of BPA is detrimental to spermatogenesis and stimulates vitellogenin synthesis in fish. Most of the studies they reviewed showed BPA acting as an estrogen-receptor agonist. Note that chronic exposure to BPA causes effects at lower concentrations than acute exposures and the authors concluded that long term exposure to BPA could exert reproductive effects at environmentally realistic concentrations. Further, they concluded that chronic exposures to BPA afforded no significant margin of safety for the protection of aquatic communities.

MUNITIONS

The US Department of Defense (DOD) is the largest landowner in the United States with 264 million ha (1.019 million square miles) under its jurisdiction. Much of this land is remote in the western and southwestern portion of the country. This remoteness makes access to much of this land difficult, but also provides habitat for large numbers of species of concern, especially threatened and endangered species. In fact, DOD lands have the highest density of species of concern of any federal department (Stein et al., 2008). Substantial and sometimes conflicting activity occurs on these lands as the department tries to simultaneously maintain combat readiness and protect the environment.

Much of this activity involves high explosives used in target practice (Fig. 9.7) and military maneuvers. The explosives used are mostly nitroamines or nitroaromatics. *Nitroaromatics* include trinitrotoluene or TNT and have a central phenol group which is attached to nitrates ($-NO_2$, Fig. 9.8). *Nitroamines* lack the central phenol core and include other explosive compounds such as hexahydro-1,3,5-triazine (RDX, also known as C-4 in the movies), octahydro-1,3,5,7-tetrazocine (HMX), and hexanitrohexaazaisowurtzitane (CL-20, Fig. 9.8). Of these, the relative explosive detonation velocities (akin to explosive force) is TNT<RDX<HMX, CL-20. The military uses other highly explosive compounds, but these are the most common (Anderson, 2010). TNT also has considerable commercial applications, such as in mining and fracturing of shale for oil and natural gas or fracking. HMX is used as a solid rocket fuel. These compounds are also commonly found in munitions and in the soil and water of war-torn countries such as Afghanistan, Iraq, Syria, and too many others. In the United States alone, the DOD has more than 12,000 sites that require some form of clean up due to the presence of explosive compounds or their degradates (Anderson, 2010).

FIGURE 9.7   Firing a howitzer and other military activities can release munition chemicals into the environment. *Courtesy US Department of Defense and Wikimedia Commons.*

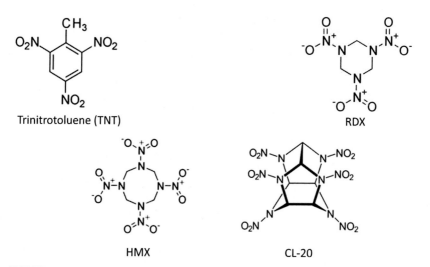

FIGURE 9.8   Structural formulas for commonly used munition compounds.

TNT was first used in artillery shells in 1902 (Anderson, 2010); HMX and RDX were used during World War II; but CL-20 was not created until the later 1980s. Thus, much is known about the chemical fate and transport of TNT, RDX, and HMX, but less is known about CL-20, one of the "second generation" of explosives. Each of these chemicals and the other explosive compounds react somewhat differently in the environment, but certain generalities pertain. Table 9.1 lists some specific physical properties of the four explosives which provide some inference to similar compounds.

TABLE 9.1 Physical Characteristics of Some Explosive Compounds

Characteristic	TNT	RDX	HMX	CL-20
CAS number	118-96-7	121-82-4	2691-41-0	Consists of 8 isomers
Molecular weight (g/mole)	227.1	222.1	296.1	438.2
Chemical formula	$C_7H_5N_3O_6$	$C_3H_6N_6O_6$	$C_4H_8N_8O_8$	$C_6H_6N_{12}O_{12}$
Melting point (°C)	80.35	205	296	112-315
Water solubility (mg/L)	130	38.4	5	5-17
Log K_{ow}	1.6	0.87	0.26	3.72-4.14
Log K_{oc}	2.5	1.87	0.54	3.6-3.8
Vapor pressure (mm Hg)	1.99×10^{-4}	1×10^{-9}	3.3×10^{-14}	3.5×10^{-5} to 4.5×10^{-5}
Detonation velocity (m/s)	6900	8750	9100	9380
Relative effectiveness factor	1	1.6	1.7	1.8

In general, the ratio of carbon: nitrogen provides some information on the explosive power of a compound. TNT, with a ratio of 7:3, has a detonation velocity of 6900 m/s and is the standard for relative effectiveness with a value of 1.0 and CL-20 has a ratio of 6:12, a detonation velocity of 9380 m/s and a relative effectiveness factor of 1.8, meaning that 1 ton of CL-20 has the explosive effectiveness of 1.8 tons of TNT. Water solubility varies from fair for TNT to essentially nonsoluble for HMX and CL-20. K_{ow} values suggest that TNT and RDX are apt to be found in water and not in lipids. Bioconcentration is unlikely to be very high. None of these four are likely to volatilize substantially under standard conditions of temperature and humidity.

Nitroamines and nitroaromatics are subject to photolytic decay, especially in water where half-lives can be measured in hours or days (Rao et al., 2013). These rates are somewhat slower in soil, but even there half-lives are seldom longer that a few weeks. Biodegradation half-life of TNT, for example, was estimated at one to six months, depending on specific conditions (ATSDR, 1995). Over the years, a slew of additives including microbes, yeast, and chemicals have reduced half-lives considerably during studies on remediating nitroamine and nitroaromatic contamination. In addition, like many other compounds, plants that readily take up these compounds have been used in remediation of contaminated sites, but bioconcentration factors are typically less than 10 so we do not see the effectiveness of phytoremediation that we do with some other contaminants.

These commonly used munitions have moderate to low toxicity. Among aquatic species and CL-20, 1.9 mg/L was the EC50 for a seven-day reproductive test with the water flea *Ceriodaphnia dubia* (Robidoux et al., 2004) and 2.7 mg/L was the EC50 for seven-day growth test on the fathead minnow (*Pimephales promelas*, Haley et al., 2007). With HMX, the highest concentration tested, 32 mg/L, had no effect on *Daphnia magna* or the midge *Chironomus tentans* (Bentley et al., 1977). Similarly, 32 mg/L had no effect on adults of four species of fish, but 15 mg/L was the LC50 for seven-day-old fathead minnows. RDX had no effect on tadpoles of *Rana [Lithobates] pipiens* up to 5.9 mg/L in a 10-day test or 28 mg/L on a 28-day exposure. The "insensitive" munition DNAN (called insensitive because it is not prone to blowing up due to shocks, heat, or fire and is therefore safer than the older munitions) had an LC50 on *R. pipiens* of 24.3 mg/L. The LC50 with TNT and this species was 4.4 mg/kg (Stanley et al., 2015). DNX caused spinal malformations in larval zebrafish (*Danio rerio*) at an EC50 of 20.8 mg/kg and had a 96 h LC50 of 23–25 mg/kg (Mukhi et al., 2005).

Among terrestrial species, CL-20 had an LC50 of 53–125 mg/kg in the earthworm *Eisenia andrei*, depending on soil type, and reduced its reproductive output at 0.05–0.09 mg/kg (Robidoux et al., 2004). The EC20 for reproductive effects with HMX and this species was 0.4 mg/kg (Robidoux et al., 2002). The seven-day LC50 for *E. andrei* and RDX ranged from 262 to 390 mg/kg depending on the age of organism and the EC50 on cocoon emergence was 9.5 mg/kg and 27 mg/kg on cocoon production (Zhang et al., 2008).

Among other terrestrial species, LC50 values for the potworm (*Enchytraeus sp.*) ranged from 0.1 mg/kg to over 1 mg/kg, depending on soil type and species (Dodard et al., 2005). EC50s for various reproductive parameters and soil types in this species ranged from less than 0.01 mg/kg to 0.12 mg/kg. In contrast to CL-20, potworms showed no response to HMX in terms of survival or reproduction, even up to concentrations of 918 mg/kg (Dodard et al., 2005).

As seems to be consistent with many contaminants, birds tended to be somewhat less sensitive to munition compounds than aquatic organisms. Concentrations up to 5304 mg/kg HMX in the diets of Japanese quail (*Coturnix japonica*, Bardai et al., 2005) caused no signs of toxicity. In northern bobwhite (*Colinus virginianus*), HMX caused food aversion and consequent weight loss (Brunjes et al., 2007). All northern bobwhite that were orally exposed daily via gavage to 17 mg/kg of RDX in corn oil and 67% of those exposed to 12 mg/kg for 14 days died (Quinn et al., 2009). Sublethal effects included convulsions, weight loss, increased serum globulin, decreased total leukocytes, and degeneration of testicular and splenic tissues. In another study, Quinn et al. (2010) first found that the acute median lethal dose of 2A-DNT, a metabolite of TNT was 1167 mg/kg. They then gavaged the birds daily for either 14 or 60 days at various concentrations. Some mortality occurred in each of the dosage groups, but sublethal effects included tremors, lack of food movement through the gastrointestinal tract, decreased leukocyte counts in females, and increased plasma triglycerides in males. The lowest observed adverse effects level (LOAEL) was determined to be 14 mg/kg/day based on mortality, and the no observed adverse effects level (NOAEL) was determined to be 3 mg/kg/day based on lack of effects at this exposure level.

The responses of domestic pigeons (*Columba livia*) to TNT were examined by Johnson et al. (2005). The authors dosed the birds at 0–200 mg TNT/kg body weight/day) for 60 days. Between two and three weeks of exposure, overt signs of toxicity began to be seen in both sexes. Signs included weight loss, neuromuscular effects (eg, ataxia, tremors, etc.), and scant, red feces. Vomiting following dosing was common and proportional to dose; but it did not appear that the birds were very successful at removing TNT following administration. A total of 8 of 12 males and 2 of 12 females died or were moribund in the 200- and 120-mg/kg-day groups, respectively. Sublethal effects included changes in blood characteristics. Liver and kidney weight relative to body weight increased with dose concentration, but ovary weights decreased. The authors concluded that exposure to TNT can adversely affect the central nervous system and blood parameters in birds.

Most of the studies on the effects of munitions on mammals are on laboratory mice and rats. However, Smith et al. (2009) tested TNX, a degradate of RDX, on deer mice (*Peromyscus maniculatus*) at 1 mg/L in drinking water and found that it impaired reproduction in offspring. Acute mortality of RDX metabolites of RDX were also tested on *P. maniculatus*. Using age classes of 21–200 days, LD50s for MNX varied from 181 to 542 mg/kg body weight and for TNX they were from 338 to 999 mg/kg body weight (Smith et al., 2007).

In summary, we are not getting rid of munitions, at least in the foreseeable future. As a result, they will be of concern, especially on military lands, for the next several decades. Efforts are being made to make them less sensitive to uncontrolled detonation but these new "insensitive" explosives are largely unexplored from an ecotoxicological perspective. Fortunately, the current generation of nitroamines and nitroaromatics have comparatively short half-lives and moderate to low toxicity. With careful stewardship of these compounds, ecological damage can be minimized.

ACID DEPOSITION

Since the start of the Industrial Revolution, which began in the mid-18th century and was highly developed by the mid-19th, many contaminants have bellowed from the smokestacks of every coal-fired utility plant in North America and northern Europe. Among these are carbon dioxide, other gases, mercury, lead, and—for this discussion—sulfur dioxide and nitrous oxides (NO_2 and NO_3, collectively referred to as NO_x). While there continues to be discharge of these pollutants, considerable amounts of mercury, lead, sulfur, and nitrogen are scrubbed out of the smokestacks and recycled or disposed of each year. This proactive cleansing is the result of regulations passed by federal governments and regulated by the US EPA or its counterparts in other nations. Still, roughly two-thirds of all SO_2 and one-fourth of NO_x in the atmosphere comes from fossil fuel combustion for electricity and coal-fired plants are main contributors of these chemicals (US EPA, 2015c).

How does acid form from these gases? Sulfur dioxide forms sulfuric acid in the atmosphere by combining with oxygen to form sulfur trioxide; this combines with water to produce sulfuric acid:

$$SO_2 + \tfrac{1}{2}O_2 \rightarrow SO_3$$
$$SO_3 + H_2O \rightarrow H_2SO_4$$

In water, sulfuric acid dissociates to release hydrogen and sulfate ions:

$$H_2SO_4 \rightarrow 2H^+ + SO_4^{2-}$$

and the H^+ contributes to acidity.

Similarly, nitric oxide from fossil fuel combustion combines with oxygen in the atmosphere to form nitrogen dioxide (NO_2) and this combines with water to form nitric acid. Nitric acid dissociates when mixed with water to add hydrogen ions:

$$NO + \tfrac{1}{2}O_2 \rightarrow NO_2$$
$$3NO_2 + H_2O \rightarrow 2HNO_3 + NO$$
$$HNO_3 \rightarrow H^+ + NO_3^-$$

The process of acidification is sometimes incorrectly called acid rain. However, acid-causing ions can be mixed with any form of moisture such as rain, sleet, fog, or mist; the phenomenon is called *wet deposition* (Fig. 9.9). These chemicals can also adhere to dust particles in the atmosphere and come down as *dry deposition*. A widely accepted term to describe all such events is *acid deposition*.

Starting in the 1960s and continuing through the 1980s, electrical generating plants were provided with tall smokestacks designed to separate generated air pollution from humans, but these stacks simply placed the pollutants into higher strata of the atmosphere where they could be more easily be transported by winds. Although the pollution was largely produced in the Midwest, prevailing winds carried the sulfates and nitrates to the northeast United States and southeast Canada. Automobile exhausts and other sources of fossil fuel combustion added to the air-blown pollution. Soils in the northeast are largely shallow, granitic, and have poor buffering capacity or *acid neutralizing capacity (ANC)*. Wet deposition, often has a pH of 3, the same as vinegar, and dry deposition precipitates onto soils where the acid-forming molecules would flush into lakes, ponds, and streams, reducing their pH.

The Midwest was less affected by its own acid deposition because much of the soil there has a limestone or calcium carbonate base:

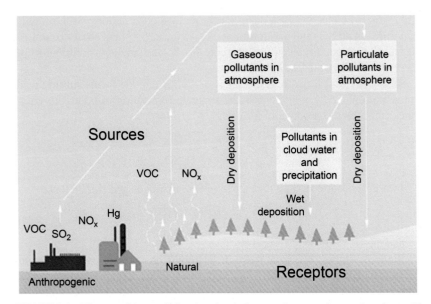

FIGURE 9.9   Diagram of how acid-forming chemicals enter the atmosphere and produce acid deposition. *Courtesy US Environmental Protection Agency. http://www.epa.gov/acidrain/what/.*

$$CaCO_3 + H_2SO_4 \rightarrow CaSO_4 + CO_2 + H_2O$$

In the process, bicarbonate H_2CO_3, can be formed which is a valuable buffering agent and fewer acid-forming molecules can enter the bodies of water. In addition, Midwestern soils have a higher amount of natural, organic acids such as tannic or humic acid that help buffer soils.

Have the regulations that were made 30 years ago to reduce acid deposition been effective? The answer, fortunately, is yes. Compared to preregulatory times, total deposition of nitrates, sulfates (Fig. 9.10), and total acidity (Fig. 9.11) have declined markedly. Similarly, in Europe the deposition of acid-forming molecules has diminished. At the same time, the pH of water bodies has increased, sometimes with the help of adding tons of limestone to streams and lakes.

Acid deposition has had significant harmful effects on plants, animals, and structures. Technically speaking, solutions are acidic if their pH drops below 7.0 and alkaline if the pH is above that mark. In practice, however, the pH of water is considered *circumneutral* and safe for most organisms at a range of 6.5 to 8, with some experts stating as low as pH 6.0. Alkaline waters often occur naturally and are very localized while acidic waters are often due to anthropogenic processes and are much more widespread.

Organisms are often at risk when pH levels drop below 6.0. For example, both aquatic and terrestrial snail shells, which are mostly calcium carbonate, begin to dissolve at pH values of approximately 6. Snails cannot survive without

1989-1991   2013

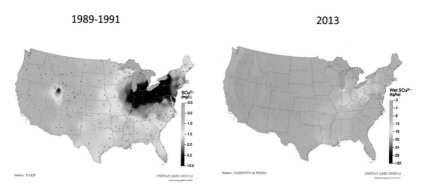

FIGURE 9.10   Annual deposition of sulfates, 2000 and 2013. *From EPA CASTNET. http://epa. gov/castnet/javaweb/precipchem.html.*

1989-1991   2013

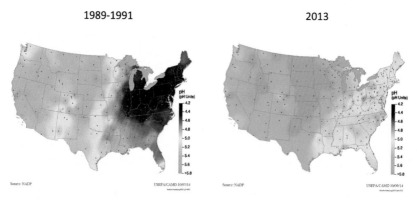

FIGURE 9.11   Acidity measured as pH. *From EPA CASTNET. http://epa.gov/castnet/javaweb/ precipchem.html.*

their shells and begin to disappear at this point. Reduced snail abundance can, in turn, affect their predators. For example, Graveland et al. (1994) documented a decline in great tits, *Parus major*, and other passerines in Netherland forests. The declines were associated with thin eggshells. This time the eggshell thinning was not due to DDT derivatives but to deficient calcium in the soil. Low pH dissolves calcium, which is then leached from soils. Snails require high soil calcium for their shells and contribute to the calcium balance when they die and their shells decompose. The authors found that snail abundance was particularly low in areas with acidified soils and that snail abundance was positively related to calcium.

Understanding of this relationship between snails, bird populations, and soil-calcium concentrations was strengthened when it became clear that augmenting acidic soils with calcium could benefit songbirds. For example, Pabian and Brittingham (2011) experimentally elevated soil calcium by spreading limestone sand on forest floors in central Pennsylvania and observing any responses in ovenbirds (*Seirus aurocapilla*). During the breeding season after spreading the limestone, the authors recorded a 1.8-fold increase in territory density, larger clutch sizes, and more nests, but no effects on egg characteristics such as eggshell thickness. They also witnessed positive relationships between natural soil calcium levels and ovenbird territory density, clutch size, and nest density in forests that were not augmented. The authors conjectured that snails were the link between soils and birds because snail abundance increased with liming and was positively correlated with soil calcium. We should note that, although the degree of acidic deposition has been reduced over the past decades, this study demonstrated that problems with the acidification of soils still exist.

Among aquatic organisms such as clams, mussels, and snails, early signs of ecosystem stress are also observed at pH below 6.0 due to their shells dissolving. At pH 5.5, largemouth bass (*Micropterus salmoides*) begin to suffer and rainbow trout (*Oncorhynchus mykiss*) die due to problems with ion regulation and enzyme functioning (US EPA, 2015d). Amphibians vary in sensitivity due to developmental stage and species. Embryonic amphibians may be unable to hatch at pH from 4.5 to 5.0 because the vitelline membrane that surrounds the egg hardens and prevents the embryo from hatching; continued growth causes the body of the embryo to form a v-shape as it fills the space within the membrane until death occurs (Clark and Hall, 1985). Tadpoles also experience problems with ion regulation around the same pH, should they hatch. Pierce (1985) summarized findings on 16 species of amphibian embryos and found that 96 h median lethal concentrations ranged from pH 3.6 to 3.8 in the carpenter frog (*Rana virgatipes*) to 6.0 in the sensitive *Rana* [*Lithobates*] *pipiens*.

In addition to the direct effects of low pH, aluminum begins to dissolve and form aluminum hydroxide ($Al(OH)_3$) and further dissolves into monomeric aluminum (Al^{3+}) at a pH of approximately 4.5. Aluminum hydroxide is toxic, but monomeric aluminum is very toxic to most aquatic organisms. At pH 4.5 to 5.5, aluminum hydroxide adheres to the gills of fish, causing them to produce a thick mucus and the resulting combination of mucus and aluminum reduces gas exchange, sometimes leading to suffocation. Below pH 4.5, monomeric aluminum interferes with enzyme activities and ion regulation. If there is a sudden increase in pH, such as when acidified waters flow into a circumneutral stream, aluminum drops out of solution and the resulting solids can severely impact gill structure. At a pH of 5.0 to 4.5, aluminum is toxic to fish in unbuffered waters starting at a concentration of 0.1 mg/L. When pH values decrease further, aluminum ions and low pH deplete calcium and sodium and block phosphorus uptake (Driscoll, 1985).

Ocean acidification occurs through a different mechanism (Guinotte and Fabry, 2008; NOAA, n.d.). Rather than being caused by NO_x and SO_2, the acidification of oceans is caused by another product of fossil fuel combustion, carbon dioxide (CO_2). Excessive amounts of CO_2 in waterleads to the production of carbonic acid:

$$CO_{2(aq)} + H_2O \rightarrow H_2CO_3 \rightarrow HCO_3^{3-} + H^+$$

The average pH of ocean waters since roughly 1850 has dropped from 8.2 to 8.1.While this may seem small, keep in mind that the pH scale is logarithmic, which means that there has been a 30% increase in hydrogen ion concentration (NOAA, n.d.). If the output of greenhouse gases continues, we may see several ramifications of increased oceanic acidity. Increased acidity may alter the calcium carbonate balance, making calcium less bioavailable in oceans. If this happens, the numerous species that use calcium for their outer structures such as oysters, clams, corals, echinoderms, and calcareous plankton may be in jeopardy (NOAA, n.d.).

We have seen that forest-dwelling snails may be impacted by acidification. Other terrestrial organisms that live in soil may also be negatively affected, but the risks on land tend to be lower than those in water. However, coniferous forests in the northeast United States, eastern Canada, and Scandinavia have been severely harmed by acid deposition (Fig. 9.12). As also shown in the figure, acid deposition has taken its toll on marble statues. Acidification can leach important minerals such as calcium and magnesium from the soil, making them unavailable to plants. Loss of nutrients weakens trees and other plants, making them more susceptible to other factors such as disease.

Much, but not all, of the pollution resulting in acid deposition is being reduced in the United States, Canada, and European Union. However, not all of the affected lands and waters have been mitigated. This means that environmental damage due to acidification is still occurring. While the northeastern United States, eastern Canada, and Scandinavia have been most impacted by acid deposition, other areas that have largely granitic or quartzite, that is, poorly buffered, soils are also potentially exposed to the effects of acidification. Continued efforts to further reduce acid-causing chemicals in the atmosphere are essential to combat this form of pollution.

Damage to statuary and buildings   **Damage to forests**

FIGURE 9.12   Acid deposition can do severe damage to art and to forests.

PHARMACEUTICALS

Between 2006 and 2010 at any given time 18% of all women aged 15 to 44 use birth control pills or other hormonal treatments to prevent pregnancy. This amounts to somewhere between 11 and 12 million women according to a recent survey conducted by the Center for Disease Control (Jones et al., 2012). These medications contain female hormones, notably progestins and estrogens. Approximately 27 million people have Type II diabetes (CDC, 2014) and many of these use Metformin or other medications to treat their disease. Almost everyone uses antibacterial soaps, hand cleansers, and other products that contain triclosin, triclocarban, or chloroxylenol. People often take as good or better care of their pets, and healthcare products for animals contain many of the same ingredients that may be found in products for humans. These and hundreds of other drugs including antibiotics, antidepressants, antiinflammatories, and analgesics, beta-blockers, oral contraceptives, and hormone replacement therapies are found in increasing concentrations in natural waterways because they are present in urine or feces and are flushed down the toilet, rinsed off into the sink, or disposed of improperly by dumping down the sink or toilet. Most wastewater treatment plants are not equipped to remove these chemicals and therefore discharge them unaltered into rivers or large lakes. The possible ramifications of endocrine disruptors including these pharmaceuticals entering the environmental was made evident to the general public in the 1997 book *Our Stolen Future: Are We Threatening Our Fertility, Intelligence, and Survival?–A Scientific Detective Story,* written by Theo Colborn, Dianne Dumanoski, and John Peter Meyers.

The extent of contamination from these products and the possible effects they can cause are not yet well understood, but are the focus of many ongoing studies. Several studies have documented scores of pharmaceuticals in various sample areas. For example, Yu et al. (2006) sampled only 18 pharmaceuticals and personal care products (PPCPs) in the Baltimore Black River wastewater treatment plant (WWTP) in 2005. Overall, they found 16 PPCPs, which spanned a range of therapeutic classes and some *commonly* used PPCPs in raw sewage. The majority of the target analytes were detected in both the ingoing and outgoing waters of WWTP samples at ng/L levels. The highest concentrations were found for ibuprofen ($1.9\,\mu g/L$), naproxen ($3.2\,\mu g/L$), and ketoprofen ($1.2\,\mu g/L$), all of which are pain-reducing NSAIDs. In total, 10 of the 18 selected PPCPs were detected in the effluent from the sewage treatment plant, indicating that these chemicals were not completely removed from the sewage. Biodegradation was important in removing much of the compounds before being discharged in effluent. Experiments showed that for 13 of the PPCPs, biodegradation was greater than 80% after 50 days of incubation. Diclofenac, another NSAID, had the poorest remediation of only 18%.

Masoner et al. (2014) identified 129 PPCP chemicals in untreated leachate samples from 19 landfills. Untreated leachates occur prior to treatment and environmental release and it is unknown how many of these chemicals would be quantifiable after treatment. Up to 82 chemicals were found in a single landfill, but there was a median of 31 chemicals across the sample sites. The chemicals most frequently found during this study included bisphenol A (from plastics), cotinine (a byproduct of tobacco), N,N-diethyltoluamide (DEET, an insect repellent), lidocaine (local anesthetic), and camphor (mostly a skin treatment), all of which occurred in 90% or more of the landfill leachates. Steroid hormone concentrations were low, generally ranging from 1 to 100 ng/L; prescription and nonprescription pharmaceutical concentrations generally ranged from 100 to several 1000 ng/L; and household and industrial chemical concentrations generally ranged from 1000 to >1,000,000 ng/L.

Blair et al. (2013) found that 32 of 55 sampled PPCPs were detected in Lake Michigan during 2009 to 2010, and 30 were detected in the sediment. The most frequently detected products in Lake Michigan were Metformin, caffeine, sulfamethoxazole (used to treat internal infections), and triclosin. The authors determined the ecological risk of some of these compounds by comparing the maximum measured environmental concentrations to predicted no-effect concentrations; 14 products were medium or high ecological risk. Quantifiable concentrations of some of these products were detected as far as 3 miles into the lake, but the highest concentrations tended to be near shore. The authors concluded that the concentrations found in this study, and their corresponding risk quotients, indicated a significant threat to the health of the Great Lakes, particularly to near shore organisms. As confirmation that Metformin can cause problems, Niemuth et al. (2015) exposed fathead minnows to Metformin for 4 weeks at 40 mu g/L, a level similar to the average found in WWTP effluent in Milwaukee, Wisconsin. Metformin induced significant ribonucleic acid (mRNA) coding for vitellogenin in male fish, an indication of endocrine disruption.

Many of these pharmaceuticals can concentrate in tissues. Of 114 PPCPs and artificial sweeteners, 83 had biological concentration factors (BCFs) greater than 1, and 15 had BCFs greater than 100. Those with the highest BCFs included orlistat (a weight reducing drug) at 123,000, fenofibrate (counteracts high cholesterol and triglycerides)—15,000, and piperonylbutoxide (synergist in pesticides)—2400. Others had BCFs ranging from more than 140 to approximately 700 (Lazarus et al., 2015). The same study collected environmental samples in water and organisms and looked for 24 analytes. The most commonly occurring pharmaceutical in environmental samples was diphenhydramine, an antihistamine found in many cold and allergy medicines. Diltiazem, an antidepressant was also common and occurred in both water and several species of fish.

There is no doubt that pharmaceuticals can injure wildlife, at least under some conditions. Perhaps the worst case of drug-induced mortality in wildlife to date befell the Oriental white-backed vulture (*Gyps bengalensis*) in India (Oaks et al., 2004). A population decline greater than 95% first started in the 1990s and has spread to catastrophic declines involving *Gyps indicus* and *Gyps tenuirostris*. These once common vultures were declared critically endangered by IUCN (2015). Oaks et al. (2004) conducted gross postmortem examinations on 259 adult and subadult *G. bengalensis* of which 219 (85%) had grossly apparent signs of visceral gout on the surface of internal organs. Visceral gout in birds is most commonly the result of renal failure leading to the deposition of uric acid on and within the internal organs. Normally it can be caused by degenerative, metabolic, infectious, or toxic diseases. Extensive necropsies of recently dead vultures with visceral gout strongly suggested acute renal failure due to a toxic cause that accompanied the gout. Additional testing on subsamples of vultures eliminated known causes of avian renal disease including toxic concentrations of cadmium, lead, mercury, copper, iron, manganese, molybdenum, or zinc. Likewise, diseases such as avian influenza, infectious bronchitis, and West Nile viruses were ruled out. There was no evidence of poisoning from carbamate or organophosphate pesticides, organochlorine pesticides, or polychlorinated biphenyls. In addition, there were no viruses isolated from kidney, spleen, lung, or intestine.

The authors surmised that ingested veterinary pharmaceuticals might be responsible for the renal disease in the birds because the primary food source for Oriental vultures in Pakistan was dead domestic livestock. They narrowed the search for a likely cause to diclofenac, a nonsteroidal antiinflammatory drug (NSAID) that was extensively used as an analgesic, antiinflammatory, and anti-pyretic for ailing cattle. Additional analysis of dead birds confirmed with very high probability that diclofenac was associated with renal failure. Diclofenac was confirmed as the cause through experimental dosing of a few birds; three of four developed visceral gout and kidney failure. The authors even fed captive vultures meat from cattle and goats that had been dosed with diclofenac and several of these birds developed gout and kidney failure in a dose-response manner. Subsequently, diclofenac was banned in the Indian subcontinent for veterinary use and the vulture populations seem to be making a slow comeback. It seems that *Gyps* vultures are particularly sensitive to NSAIDS and the search continues for a medication that is effective for cattle but safe to vultures and other scavengers.

Whereas humans take many medicines, we also give our pets and livestock all sorts of drugs. Some regulate the estral cycle or influence other hormonal processes, others combat disease, still others are used as supplements, etc. Virtually all of these have the potential to enter the environment, either in water or in soil, where they can affect living organisms. If animals do not come in direct contact with the compounds, they may be exposed by consuming living prey or scavenging carcasses. Some of the effects on wildlife and fish may be predictable based on what the chemicals have been designed for, but others may have unpredictable results once they get into the environment. An area of research that is particularly lacking is an understanding of how these drugs may interact. Wildlife and fish are often exposed to multiple pharmaceuticals and we really do not know what these cocktails of chemicals might do.

The life cycle of pharmaceuticals is immensely complex and convoluted (Fig. 9.13). We don't presume that you'll be able to read all of the boxes in this diagram, just observe it to gain a perception of the complex nature of this environmental problem.

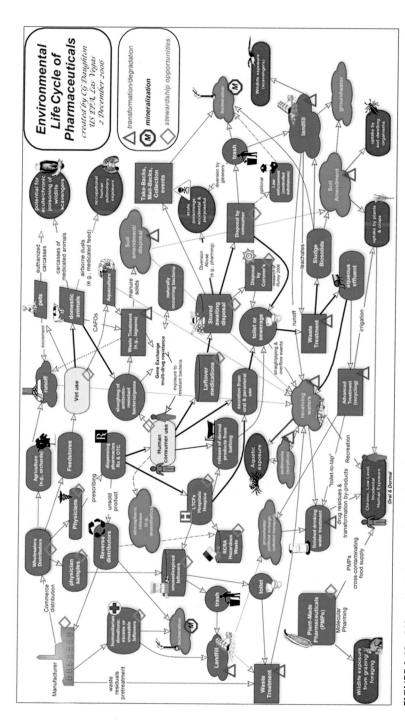

FIGURE 9.13 Life cycle of pharmaceuticals in the environment. *From Daughton (2008). Don't spend too much time studying the box titles, rather focus on the complexity of this situation.*

NANOPARTICLES

By definition, *nanoparticles* are very small, 1 to 100 nanometers (nm) in size. While they are miniscule, they are larger than most molecules. Nanoparticles have characteristics that may differ from the same materials at larger scales. An important aspect of nanoparticles is that they have a very large surface area to volume ratio. This alone may confer special characteristics to these particles. For example, one microparticle of carbon atoms with a diameter of 60 μm has a mass of 0.3 μg and a surface area of 0.01 mm². The same mass of carbon in nanoparticulate form, with each particle having a diameter of 60 nm, has a surface area of 11.3 mm² and consists of 1 billion nanoparticles. This large surface area can increase the rate of chemical reactions by 1000 times (Buzea et al., 2007). Among other characteristics, nanoparticles often possess unexpected optical and magnetic properties because they are small enough to confine their electrons and produce quantum effects. For example, bulk gold particles have the characteristic yellowish color and high sheen, but gold nanoparticles appear deep-red to black in solution. Thermodynamic properties may also differ between bulk and nanoparticles. Gold nanoparticles, for instance, melt at approximately 300°C for a 2.5-nm size, while bulk gold particles melt at a much higher 1064°C (Buffat and Borel, 1976). In addition, nanoparticles absorb solar radiation more efficiently than bulk particles, a property that can be effectively used in sun blockers and building materials. To make accurate predictions of nanoparticle behavior compared with bulk particle behavior, quantum mechanics must often be applied.

Nanoparticles have been used since ancient times and they are environmentally common. In the 9th century BCE, artisans in Mesopotamia used glazes containing nanoparticle materials to produce a glittering effect. Similar glazes were used during the MiddleAges (Khan, 2012). Nanoparticles are found in dust, volcanic emissions, and many other natural processes. Notably, viruses are nanoparticles and cause many diseases from the common cold to cancers. Through evolution, humans and other organisms have developed defense mechanisms including multilayered dermis to reduce the ability of these particles to enter bodies. As we will see, if nanoparticles enter an organism, they are so small that they can readily enter cells. What happens after penetration varies from beneficial to serious health issues.

Whereas naturally occurring nanoparticles have been in existence from the earliest of Earth's history, we are mainly concerned with the much more recent explosion of nanoparticle production in many industrial applications. Keep in mind that it is the size, as well as the chemical composition, that make the difference. Many of the nanoparticles now used in manufacturing, medicine, and other areas are composed of the same materials used in bulk particles. Thus, nanoparticles can occur in a wide variety of manufacturing processes.

We have only experienced the tip of the iceberg in the number of potential uses of nanoparticles. There is great promise for nanoparticles in food preservation and packaging. Engineers are looking at ways of identifying bacteria and monitoring food quality through the use of biosensors in packaging materials. Foods can be encapsulated with specific nanomaterials to keep them fresh. A nanocomposite coating process could improve food packaging by placing antimicrobial agents directly on the surface of the coated film. Foods containing nanoparticles are also being manufactured. The Project on Emerging Nanotechnology (PEN), a think tank on nanotechnology, lists several "nanofoods" including Canola Active Oil, Nanotea, and a chocolate diet shake called Nanoceuticals Slim Shake Chocolate. According to the PEN website, the canola oil contains an additive called "nanodrops" designed to carry vitamins, minerals, and phytochemicals through the digestive system. The shake uses cocoa-infused "NanoClusters" to enhance the taste and health benefits of cocoa without the need for extra sugar (summary courtesy of Consumer Products Inventory (http://www.nanotechproject.org/cpi/products/nanoceuticalstm-slim-shake-chocolate/)).

Nanoparticles are also used in ceramics to provide smooth surfaces in household "self-cleaning" ovens. The textile industry is using nanofibers that enhance stain resistance and wrinkle-free clothing. In cosmetics, nanoparticles of zinc oxide and titanium oxide provide great improvements in sunburn protection. In sports, baseball bats are being made with carbon nanotubes that increase their strength while reducing their weight. Other items, such as sport towels, yoga mats, and exercise mats, which use antimicrobial nanotechnology to prevent users from illnesses caused by bacteria such as Methicillin-resistant *Staphylococcus aureus* (MRSA) are on the market and used by players in the National Football League (NFL).

Nanoparticles have already experienced broad application in medicine (Murthy, 2007). They have been used in medical imagery including improved dyes, fluorescent chips, or quantum dots that can adhere to tissues for emitting light, and as super magnets in MRIs. Another area of rapidly developing success is in drug and chemical delivery to specific tissues. Nanoparticles can be shaped into *lipsosomes* which are vesicles that entrap fluids between phospholipid molecules to form a bilayer (Fig. 9.13). These bilayers consist of oriented lipid molecules that come together with their hydrophobic sides in contact with one another. Minute amounts of drugs can be inserted between these bilayers and then presented to the affected tissues. DNA can also be inserted to genetically modify a seed or ovum. These liposomes and other nanoparticles are being used to treat cancers, HIV/AIDS, neurodegenerative disorders, ocular treatments, and respiratory treatments among other functions (Murthy, 2007).

Nanoparticles organized into carbon *nanotubes* (Fig. 9.14) are very strong and very lightweight. Therefore, they are extremely useful in many aeronautical and military applications. Nanoparticles can be incorporated into the uniforms of soldiers to make them more resistant to penetration. The "invisible camouflage" which reflects background colors and patterns and is (as far as we know) part of science fiction, may someday be possible through highly reflectant nanoparticles. The list of possibilities is awe inspiring.

Liposome for drug delivery

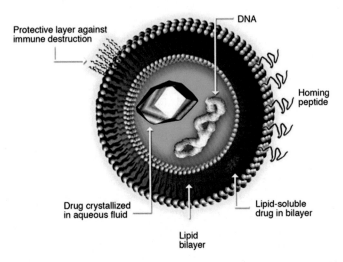

FIGURE 9.14   A liposome composed of nanoparticles for transporting drugs and DNA into cells.

Given this list of potential and actual uses of nanoparticles, it is not surprising that they represent the next generation or two of new materials. As the use of nanoparticles increases environmentalists are beginning to ask, "Should we be concerned about environmental impacts this technology may pose now and in the future?" For several reasons the answer is "yes, we should be concerned." At the current time, understanding of the fate and transport of nanoparticles, not to mention their possible effects in the environment, is not well developed. Clearly, studies in these areas are underway or recently published. However, given the scope of applications and the magnitude of future production of the vast array of materials that can become nanoparticles, our understanding about their environmental consequences is marginal at the very best. At the present time, manufactured nanoparticles represent a small fraction of these tiny particles compared to naturally occurring particles, but that ratio may change in the not too distant future (Buzea et al., 2007).

Some factors to consider with regards to the toxicity of nanoparticles follow (Buzea et al., 2007). The three ways that nanoparticles may gain entrance into an organism such as a human include skin, lungs or other respiratory structures, and gastrointestinal tracts. Due to their small size, nanoparticles can more easily move from these entry points into the circulatory and lymphatic systems, and ultimately to body tissues and organs than bulk particles. For example, the average diameter of a hepatocyte or liver cell is 20 to 30 μm (roughly 1000 times the size of a nanoparticle). In some ways, the movement of nanoparticles through an organism is similar to that of dissolved particles but nanoparticles may even penetrate tissues such as plasma membranes and brains that provide barriers to solutes (Murthy, 2007). Some nanoparticles, depending on their composition and size, can produce irreversible damage to cells by oxidative stress or physical injury to organelles. It is easy to see that because nanoparticles can enter cells rather easily, they have a lot of potential to disrupt the functions of many organelles such as nucleus, mitochondria, and endoplasmic reticulum.

The toxicity of nanoparticles depends on various factors, including: size, aggregation, composition, crystallinity, surface functionalization, etc. (Buzea et al., 2007). The toxicity of any nanoparticle is also influenced by an organism's genetics and other health factors. The most extreme health hazards due to inhalation of nanoparticles (in general, not necessarily manufactured particles) include asthma, bronchitis, emphysema, lung cancer, and Parkinson's. Ingested nanoparticles are associated with Crohn's disease and colon cancer. Nanoparticles present in the circulatory systems are related to occurrence of arteriosclerosis, blood clots, arrhythmia, heart diseases, and ultimately cardiac death. Nanoparticles have also been linked to immunodeficiency. These diseases have been mostly studied in relation to human health and maladies possibly incurred by wild animals are less well known. This clearly is an area of research offering substantial promise.

We can offer a couple of examples of nanoparticle testing in wildlife. Gaiser et al. (2012) compared the effects of silver at nominal diameters of 35 nm, 0.6 μm, and 1.6 μm). By definition, the 35 nm silver particles would be in the nanoparticle range while the other particles are in the microparticle size range. They found that 100% of *Daphnia magna* died at nanosilver concentrations of 1 and 10 mg/L and 56% died at 0.1 mg/L, whereas with microsilver mortality at 10 mg/L was 100% was but only 80% at 1 mg/L. In a 21-day chronic test with carp (*Cyprinus carpio*), nanosilver particles attained higher concentrations in liver, intestines, and gall bladder than micro particles. In an in vitro exposure to rainbow trout hepatocytes, nanosilver was six times more toxic than microsilver.

Powers et al. (2011) also tested silver on fish, but this time they compared nanosilver to dissolved silver (Ag^+) on zebrafish (*Danio rerio*) embryos. Both dissolved silver and nanosilver delayed hatching to a similar extent, but Ag(+) was more effective in slowing swim bladder inflation, and elicited greater malformations and mortality. In behavioral assessments, dissolved silver-exposed fish were hyper-responsive to light changes, whereas nanosilver-exposed fish showed normal responses. Nanosilver did not affect survival or morphology. The authors concluded that nanosilver was less potent than Ag+ with respect to malformations and survival, but did produce neurobehavioral effects. These behavioral responses varied depending on the size of the nanoparticles.

STUDY QUESTIONS

1. Approximately how much plastic garbage is produced in the United States each year? What are the main types of plastics in this garbage?
2. Which is expected to last longer in the environment—microplastics or PCBs?

3. Which types of plastics are most commonly recycled? Why do you suppose other plastics are not recycled to the same extent as these?
4. True or False. Nurdles are a type of microplastic used in the manufacture of plastic objects.
5. What is an oceanic garbage patch? Where are they located? What factors deter us from cleaning these garbage patches and recycling the materials contained in them?
6. List at least four ways that plastics can harm wildlife.
7. Would you consider munitions to be a major global ecotoxicological threat to wildlife? Why or why not?
8. True or False. An ecological advantage of commonly used munitions is that they have short half-lives.
9. What greenhouse gases contribute most to acid deposition?
10. Can acid-forming chemicals in the atmosphere be deposited into water bodies even though it is not foggy, raining, or snowing?
11. True or False. Granitic soils tend to have greater acid-neutralizing capacity than limestone-based soils.
12. What range of pH is generally considered circumneutral and safe for most freshwater, aquatic organisms?
13. Have efforts to remediate acid deposition been effective? Should there be any more concern about acid deposition in the United States? Canada? Europe?
14. How is aluminum toxicity associated with acidification?
15. Globally, the pH in oceans has dropped from 8.2 to 8.1, a decline of only 0.1 pH unit. Why is this apparently slight drop of concern to ecologists?
16. Based on this chapter and what you may have read elsewhere, do you consider pharmaceuticals to be a significant ecological threat? Why or why not?
17. Do some research on the current status of old world or *Gyps* vultures in India. Do they have other threats to their populations than NSAID pharmaceuticals?
18. Based on what is known about nanoparticles, is it correct to assume that they promise greater benefits to humans than damage to the environment?

REFERENCES

Anderson, T.A., 2010. Environmental toxicology of munitions-related compounds: nitroaromatics and nitramines. In: Kendall, R.J., Lacher, T.E., Cobb, G.P., Cox, S.B. (Eds.), Wildlife Toxicology: Emerging contaminant and biodiversity issues. CRC Press, Boca Raton FL, pp. 15–39.

Andrady, A., 2009. Fate of plastics debris in the marine environment. Presentation. In: Arthur, C., Baker, J., Bamford, H. (Eds.), Proceedings of the International Research Workshop on the Occurrence, Effects, and Fate of Microplastic Marine Debris. Sept 9–11, 2008. NOAA Technical Memorandum NOS-OR&R-30.

Asghari, M.H., Saeidnia, S., Abdollah, M., 2015. A review on the biochemical and molecular mechanisms of phthalate-induced toxicity in various organs with a focus on the reproductive system. Int. J. Pharmacol. 11, 95–105.

ATSDR Agency for Toxic Substances and Disease Registry, 1995. Toxicological Profile for 2,4,6-trinitrotoluene. U.S. Public Health Service, Atlanta, GA.

Bardai, G., Sunahara, G.I., Spear, P.A., Martel, M., Gong, P., Hawari, J., 2005. Effects of dietary administration of CL-20 on Japanese quail *Coturnix coturnix japonica*. Arch. Environ. Contam. Toxicol. 49, 215–222.

Barnes, D.K.A., Galgani, F., Thompson, R.C., Barlaz, M., 2009. Accumulation and fragmentation of plastic debris in global environments. Philos. Trans. R Soc. B Biol. Sci. 364, 1985–1998.

Bentley, R.E., LeBlanc, G.A., Hollister, T.A., Sleight, B.H., 1977. Acute Toxicity of 1,3,5,7-Tetranitrooctahydro-1,3,5,7-tetrazocine (HMX) to Aquatic Organisms. U.S. Army Medical Research and Development Command, Washington, DC.

Blair, B.D., Crago, J.P., Hedman, C.J., Klaper, R.D., Barnes, D.K.A., Galgani, 2013. Pharmaceuticals and personal care products found in the Great Lakes above concentrations of environmental concern. Chemosphere 93, 2116–2123.

Brunjes, K.J., Severt, S.A., Liu, J., Pan, X.P., Brausch, J., Cox, S.A., et al., 2007. Effects of octahydro-1,3,5,7-tetranitro-1,3,5,7-tetrazocine (HMX) exposure on reproduction and hatchling development in Northern bobwhite quail. J. Toxicol. Environ. Health A Curr. Issues 70, 682–687.

Buffat, P.H., Borel, J.P., 1976. Size effect on the melting temperature of gold particles. Phys. Rev. A 13, 2287–2298.

Buzea, C., Blandion II, P., Robbie, K., 2007. Nanomaterials and nanoparticles: sources and toxicity. Biointerphases 2, MR17–MR172.

CDC Center for Disease Control, 2014. Diabetes latest. <http://www.cdc.gov/features/diabetesfactsheet/>(accessed 15.06.15.).

Clark, K.L., Hall, R.J., 1985. Effects of elevated hydrogen-ion and aluminum concentrations on the survival of amphibian embryos and larvae. Can. J. Zool. 63, 116–123.

Colborn, T., Dumanoski, D., Meyers, J.P., 1985. Our Stolen Future: Are We Threatening Our Fertility, Intelligence, and Survival?–A Scientific Detective Story. Plume, New York, NY.

Crain, D.A., Eriksen, M., Iguchi, T., Jobling, S., Laufer, H., LeBlanc, G.A., et al., 2007. An ecological assessment of bisphenol-A: evidence from comparative biology. Reprod. Toxicol. 24, 225–239.

Daughton, C.G., 2008. Drug Usage and Disposal: Overview of Environmental Stewardship and Pollution Prevention (with an emphasis on some activities in the federal government). U.S. Environmental Protection Agency, Las Vegas, NV.

Dodard, S.G., Sunahara, G.I., Kuperman, R.G., Sarrazin, M., Gong, P., Ampleman, G., et al., 2005. Survival and reproduction of Enchytraeid worms, Oligochaeta, in different soil types amended with energetic cyclic nitramines. Environ. Toxicol. Chem. 24, 2579–2587.

Driscoll, C.T., 1985. Aluminum in acidic surface waters—chemistry, transport, and effects. Environ. Health Perspect. 63, 93–104.

Gaiser, B.K., Fernandes, T.F., Jepson, M.A., Lead, J.R., Tyler, C.R., Baalousha, M., et al., 2012. Interspecies comparisons on the uptake and toxicity of silver and cerium dioxide nanoparticles. Environ. Toxicol. Chem. 31, 144–154.

Graveland, J., Vanderwal, R., Vanbalen, J.H., Vannoordwijk, A.J., 1994. Poor reproduction in forest passerines from decline of snail abundance on acidified soils. Nature 369, 446–448.

Guinotte, J.M., Fabry, V.J., 2008. Ocean acidification and its potential effects on marine ecosystems. In: Ostfeld, R.S. Schlesinger, W.H. (Eds.), Year in Ecology and Conservation Biology 2008, vol. 1134 Ann. NY Acad. Sci, pp. 320–342.

Haley, M.V., Anthony, J.S., Davis, E.A., Kurenas, C.W., Kuperman, R.G., Checkai, R.T., 2007. Toxicity of the Cyclic Nitramine Energetic Material CL-20 to Aquatic Receptors. Edgewood Chemical Biological Center, Aberdeen Proving Grounds, Aberdeen, MD.

IUCN International Union for the Conservation of Nature, 2015. IUCN Red book of threatened species 2015.1. <IUCNredlist.org>(accessed 20.06.15.).

Johnson, M.S., Michie, M.W., Bazar, M.A., Salice, C.I., Gogal, R.M., 2005. Responses of oral 2,4,6-trinitrotoluene (TNT) exposure to the common pigeon (*Columba livia*): a phylogenic and methodological comparison. Int. J. Toxicol. 24, 221–229.

Jones, J., Mosher, W., Daniels, K., 2012. Current contraceptive use in the United States, 2006–2010, and changes in patterns of use since 1995. Natl. Health Stat. Report. 60, 1–25.

Khan, F.A., 2012. Biotechnology Fundamentals. CRC Press, Boca Raton, FL.

Lazarus, R.S., Rattner, B.A., Brooks, B.W., Du, B., McGowan, P.C., Blazer, V.S., et al., 2015. Exposure and food web transfer of pharmaceuticals in ospreys (*Pandion haliaetus*): predictive model and empirical data. Integr. Environ. Assess. Manage. 11, 118–129.

Masoner, J.R., Kolpin, D.W., Furlong, E.T., Cozzarelli, I.M., Gray, J.L., Schwab, E.A., 2014. Contaminants of emerging concern in fresh leachate from landfills in the conterminous United States. Environ. Sci. Process. Impacts 16, 2335–2354.

Mukhi, S., Pan, X.P., Cobb, G.P., Patino, R., 2005. Toxicity of hexahydro-1,3,5-trinitro-1,3,5-tri-azine to larval zebrafish (*Danio rerio*). Chemosphere 61, 178–185.

Murthy, S.K., 2007. Nanoparticles in modern medicine: state of the art and future challenges. Int. J. Nanomedicine 2, 129–141.

National Geographic. No date. Great Pacific garbage patch. <http://education.nationalgeographic.com/education/encyclopedia/great-pacific-garbage-patch/?ar_a=1> (accessed 12.06.15.).

Niemuth, N.J., Jordan, R., Crago, J., Blanksma, C., Johnson, R., Klaper, R.D., 2015. Metformin exposure at environmentally relevant concentrations causes potential endocrine disruption in adult male fish. Environ. Toxicol. Chem. 34, 291–296.

NOAA National Oceanic and Atmospheric Administration, n.d. PMEL Carbon Program. What is ocean acidification? <http://www.pmel.noaa.gov/co2/story/What+is+Ocean+Acidification%3F>(accessed 13.06.15.).

Oaks, J.L., Gilbert, M., Virani, M.Z., Watson, R.T., Meteyer, C.U., Ridebout, B.A., et al., 2004. Diclofenac residues as the cause of vulture population decline in Pakistan. Nature 427, 630–633.

Pabian, S.E., Brittingham, M.C., 2011. Soil calcium availability limits forest songbird productivity and density. Auk 128, 441–447.

Pierce, B.A., 1985. Acid tolerances in amphibians. Bioscience 35, 239–243.

Powers, C.M., Slotkin, T.A., Seidler, F.J., Badireddy, A.R., Padilla, S., 2011. Silver nanoparticles alter zebrafish development and larval behavior: distinct roles for particle size, coating and composition. Neurotoxicol. Teratol. 33, 708–714.

Quinn, M.J., Bazar, M.A., McFarland, C.A., Perkins, E.J., Gust, K.A., Johnson, M.S., 2009. Sublethal effects of subacute exposure to RDX (1,3,5-trinitro-1,3,5-triazine) in the northern bobwhite (*Colinus virginianus*). Environ. Toxicol. Chem. 28, 1266–1270.

Quinn, M.J., McFarland, C.A., LaFiandra, E.M., Bazar, M.A., Johnson, M.S., 2010. Acute, sub-acute, and subchronic exposure to 2A-DNT (2-amino-4,6-dinitrotoluene) in the northern bobwhite (*Colinus virginianus*). Ecotoxicology 19, 945–952.

Rao, B., Wang, W., Cai, Q.S., Anderson, T., Gu, B.H., 2013. Photochemical transformation of the insensitive munitions compound 2,4-dinitroanisole. Sci. Total Environ. 443, 692–699.

Robidoux, P.Y., Hawari, J., Bardai, G., Paquet, L., Ampleman, G., Thiboutot, S., et al., 2002. TNT, RDX, and HMX decrease earthworm (*Eisenia andrei*) life-cycle responses in a spiked natural forest soil. Arch. Environ. Contam. Toxicol. 43, 379–388.

Robidoux, P.Y., Sunahara, G.I., Savard, K., Berthelot, Y., Dodard, S., Martel, M., et al., 2004. Acute and chronic toxicity of the new explosive CL-20 to the earthworm (*Eisenia andrei*) exposed to amended natural soils. Environ. Toxicol. Chem. 23, 1026–1034.

Saido, K., Koizumi, K., Sato, H., Ogawa, N., Kwon, B.G., Chung, S.Y., et al., 2014. New analytical method for the determination of styrene oligomers formed from polystyrene decomposition and its application at the coastlines of the North-West Pacific Ocean. Sci. Total Environ. 473, 490–495.

Smith, J.N., Lin, J., Espino, M.A., Cobb, G.P., 2007. Age dependent acute oral toxicity of hexa-hydro-1,3,5-triazone (RDX) and two anaerobic N-nitroso metabolites in deer mice (*Peromyscus maniculatus*). Chemosphere 67, 2267–2273.

Smith, J.N., Espino, M.A., Liu, J., Romero, N.A., Cox, S.B., Cobb, G.P., 2009. Multigenerational effects in deer mice (*Peromyscus maniculatus*) exposed to hexahydro-1,3,5-trinitroso-1,3,5-triazine (TNX). Chemosphere 75, 910–914.

Stanley, J.K., Lotufo, G.R., Biedenbach, J.M., Chappell, P., Gust, K.A., 2015. Toxicity of the conventional energetics TNT and RDX relative to new insensitive munitions constituents DNAN and NTO in *Rana pipiens* tadpoles. Environ. Toxicol. Chem. 34, 873–879.

Stein, B.A., Scott, C., Benton, N., 2008. Federal lands and endangered species: the role of military and other federal lands in sustaining biodiversity. Bioscience 58, 339–347.

UNESCO United Nations Educational, Scientific and Cultural Organization, 2015. Facts and figures on marine pollution. <http://www.unesco.org/new/en/natural-sciences/ioc-oceans/priority-areas/rio-20-ocean/blueprint-for-the-future-we-want/marine-pollution/facts-and-figures-on-marine-pollution/> (accessed 07.06.15.).

US EPA U.S. Environmental Protection Agency, 2015a. Plastics. <http://www.epa.gov/epawaste/conserve/materials/plastics.htm> (accessed 07.07.15.).

US EPA U.S. Environmental Protection Agency, 2015b. Phthalates. <http://www.epa.gov/oppt/existingchemicals/pubs/actionplans/phthalates.html> (accessed 12.06.15.).

US EPA U.S. Environmental Protection Agency, 2015c. What is acid rain? <http://www.epa.gov/acidrain/what/> (accessed 12.06.15.).

US EPA U.S. Environmental Protection Agency, 2015d. Effects of acid rain—Surface waters and aquatic organisms. <http://www.epa.gov/acidrain/effects/surface_water.html> (accessed 13.06.15.).

Vaverkova, M., Adamcova, D., Kotovicova, J., Toman, F., 2014. Evaluation of biodegradability of plastics bags in composting conditions. Ecol. Chem. Eng. 21, 45–57.

Yu, J.T., Bouwer, E.J., Coelhan, M., 2006. Occurrence and biodegradability studies of selected pharmaceuticals and personal care products in sewage effluent. Agric. Water Manage. 86, 72–80.

Zhang, B.H., Cox, S.B., McMurry, S.T., Jackson, W.A., Cobb, G.P., Anderson, T.A., 2008. Effect of two major N-nitroso hexahydro-1,3,5-trinitro-1,3,5-triazine (RDX) metabolites on earthworm reproductive success. Environ. Pollut. 153, 658–667.

Ziv-Gal, A., Wang, W., Zhou, C.Q., Flaws, J.A., 2015. The effects of in utero bisphenol A exposure on reproductive capacity in several generations of mice. Toxicol. Appl. Pharmacol. 284, 354–362.

Section III

Identifying and Evaluating Large Scale Contaminant Hazards

Chapter 10

Population Ecotoxicology: Exposure and Effects of Environmental Chemicals

---

[E]ach level (of biological integration) offers unique problems and insights, and ... each level finds its explanations of mechanisms in the levels below, and its significance in the levels above.

George A. Bartholomew (1964)

Terms to Know

Scale (From an Ecological Perspective)
Population
Population Ecotoxicology
Mesocosm
Additive Mortality
Compensatory Mortality
Demographic Stochasticity
Environmental Stochasticity
Genetic Stochasticity
Autecology
Synecology

INTRODUCTION: WORKING BEYOND INDIVIDUAL ORGANISMS

The study of the effects of chemicals on individual organisms is most often encountered in the guise of toxicity assessment, as summarized in Chapter 2. Yet, many ecotoxicologists are primarily interested in answering questions that focus on outcomes linked to effects observed in the field and not necessarily constrained to a single species, such as when working with an endangered species. Thus, an understanding of the principles of population ecology is necessary for ecotoxicologists to pursue knowledge of adverse effects in the field. Describing cause and effect relationships between chemicals released to the environment and their effects on populations—or as we will encounter in the

Ecotoxicology Essentials.
309

next chapter, their effects on communities and ecosystems—is frequently a conundrum for ecotoxicologists to solve.

Although often observed and noted, large or *scale* differences enter our analysis of exposure and effects perhaps because a scale difference related to shifts from organism to population levels of biological organization, or due to a realization that exposures in the field capture spatiotemporal (space and time) scales inadequately considered in laboratory toxicity tests. Indeed, the simple observation of scale differences serves as a source of confounding factors linked to multiple stressors co-occurring with chemicals in the environment. Multiple stressors present even more challenges in characterizing exposure and effects relationships at any level of biological organization, particularly if chemicals occur at low levels for long time periods over large spatial scales. However, when we leave the realm of individual organisms—for example, exposing 10 organisms in each concentration of a laboratory toxicity test, or estimating species-specific effects in resident species based on the outcomes of these tests—we enter a larger range of possibilities in populations or communities and ecosystems where ecotoxicologists use various analytical approaches and deploy a wide range of tools to characterize these scale-dependent interactions. In other words, the larger the scale of investigation, the greater number of confounding factors to deal with.

We collectively discover conceptual frameworks, some of which are presented in published literature, while others are unique to the specific questions for the population, habitats, or landscapes at hand. Biological and ecological systems exist as hierarchical structures, wherein interactions among different levels within a system can be functionally linked and regulated "control systems" similar to the banners of physiology, endocrinology, molecular biology, or biochemistry seen at the individual scale. Such functional or process-oriented interplay also occurs at differing spatiotemporal scales, running the spectrum from events that occur relatively slowly (eg, geological events accompanied by ecological events) to more rapidly occurring events (eg, cellular levels of biological organization). Indeed, as agents responding to these contrasting spatiotemporal scales, organisms operate in a region of overlap between biology and ecology (Fig. 10.1).

For example, scientists may focus on one of two broad classifications in ecological studies—*autecology* or the study of an organism and its environment, and *synecology*, which looks at entire populations or communities. In either case, however, scientists must be aware of the overlap of both the ecological and biological requirements of the organisms under study. Indeed, autecology and synecology may be considered complementary terms distinguished by their interpretive context; for autecology, it is just a matter of scale focusing on the spatiotemporal dynamics of species, whereas synecology encompasses studies focused on distribution, abundance, demography, and interactions between coexisting groups of organisms or species (Cain et al., 2014; Gimme and Hengeveld, 2014). As such, population ecology easily captures the transition between

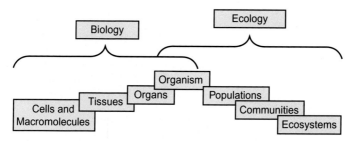

FIGURE 10.1 Interrelationships between levels of biological organization, and biology and ecology. *From http://www.esd.ornl.gov/PGG/HERMES/description.html, last accessed November 10, 2015.*

autecology and synecology frames of reference, a transition that affords ecotoxicologists an opportunity to consider chemical effects at a species-specific level of organization. This often includes the study of biological effects for chemical exposures through single-species toxicity tests in the laboratory, followed by estimating responses in field settings by considering relevant endpoint, such as reproductive effects potentially linked to population-level changes in abundance or population dynamics. However, before diving into our overview of population-level studies and tools that an ecotoxicologist commonly deploys in such studies, a brief sojourn would benefit this chapter and the next. Both this chapter and the next anticipate our integrating of a species-specific population focus, largely within an autecological frame of reference, that builds toward a synecological interpretative context pertinent to moving beyond a strictly species-specific, at best an organism-centric view of ecotoxicology.

SPATIOTEMPORAL SCALES IN ECOLOGY AND ECOTOXICOLOGY

Regardless of the level of biological organization, responses and their relevance within biological and toxicological, or ecological and ecotoxicological contexts cannot be characterized in the absence of spatiotemporal scale (Fig. 10.2). Biological systems are complex, highly interwoven, and, in many respects, entangled systems that are typically characterized by linear rather than nonlinear interactions among their many components. These components can be considered along hierarchical levels of organization. Understanding how changes at one level of biological organization alter emergent patterns or mechanisms at another level of biological organization may be an intractable problem at this time, yet such problems must be considered if we aim to improve the understanding of how contaminants may affect these levels or scales of ecotoxicology. Indeed, ecology—and, to a lesser extent, ecotoxicology as most often practiced today—explicitly or implicitly focuses on landscapes in time and space, landscapes wherein individuals of a species collectively became populations, and

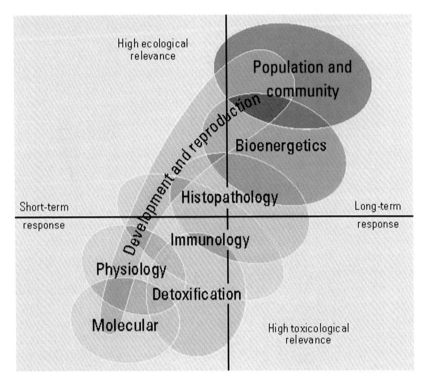

FIGURE 10.2   The relationship between response levels of biological organization and the toxicological relevance and time scale of responses. (*http://www.amap.no/documents/doc/the-relationship-between-response-levels-of-biological-organization-and-the-toxicological-relevance-and-time-scale-of-responses/594, last accessed November 10, 2015*).

populations of various species with all of their various life histories and niches subsequently assembled into communities. All are parts of highly adaptive, dynamic systems of interacting biological and physicochemical components distributed across a wide array of habitats.

This chapter and the next focus on scale-dependent networks, some simple, others much more complex. Some may be characterized by interactions that involve chemicals as one of many stressors affecting organisms in field settings. Within such field settings and populations, the important endpoints are often those that relate to abundance or actual reproduction. In contrast, laboratory studies often emphasize finer molecular or physiological responses that are of less concern to population ecologists. Within the context of ecological or ecotoxicological relevance, some populations may be more highly valued than others. For example, populations of some species may have greater economic, recreational, aesthetic, or biological significance to resource managers. In general, population dynamics are often more well-defined and predictable than community and ecosystem responses to perturbations.

POPULATION ECOTOXICOLOGY

Before we forge onward for an overview of *population ecotoxicology*, it would be helpful to review some elements of population ecology. For our purposes, population ecology focuses on microbial, plant, or animal populations and how these populations change over time and space, and perhaps more importantly, how these changes may be influenced by interactions with their environment. A simple definition of population is 'a collection of individuals of the same species living in the same area at the same time.' As such, populations are generally characterized based on attributes of population size, which is generally measured by counting individuals in a specific area; population density, which refers to how many individuals live in a given area; and population growth, which is most often considered as a change in population size through time. Conditioned on this brief characterization of population ecology, we focus on the responses that populations present when field exposures to environmental chemicals occur. Few monographs have been published strictly on population ecotoxicology (see Albers et al., 2000; Newman, 2001), yet we can find lengthy discussions of the topic in larger compendiums of related articles (Linder et al., 2003; Shugart et al., 2003) or within some monographs that include sections that have populations as their frame of reference (eg, Newman and Clements, 2007).

Population-level effects displayed by biota exposed to environmental chemicals range from dramatic to subtle. The most drastic of population-level effects is extirpation or extinction. In general parlance, *extirpation* refers to the complete loss of a local population whereas *extinction* is the loss of an entire species; extirpation may be temporary but extinction is permanent. Both effects are well defined and potentially have biological significance well beyond the immediate loss of biodiversity consequent to the loss of a population. Within the context of time, however, both might be expressed as a catastrophic outcome to a global calamity. More likely, however, the loss of an entire species will be the outcome of subtle processes, with exposure to environmental chemicals being only one of many stressors that will contribute to a series of local extirpations that cumulatively express themselves as an extinction event. The rate at which these subtler processes occur may reflect combinations of stressors acting jointly or serially. For example, habitat loss coupled with chronic or subchronic exposures to environmental chemicals may be as detrimental to long-term sustainability of a species as a release of chemicals at high concentrations that produce acute toxicity and increased mortality in exposed populations. Rapidly progressing extinction processes, such as those linked to acute exposures, can be monitored without great effort, especially if the species is relatively easy to observe in the field.

In general, local extirpations are relatively common, often directly resulting from toxic effects linked to intentional or accidental releases of chemicals to the environment, but they more commonly result from a combination of factors, such as habitat degradation or loss of competitive advantage when challenged by a species with increased tolerance to the released chemicals (see Chapter 11). In contrast, diminished population numbers or fewer areas of occupation are subtle changes in populations that may be more difficult to monitor than total extirpations.

Yield, abundance, and production are each expressions of a population's capacity to fulfill its biological role within a community or ecosystem. Depending on the species' role in the environment, these and other attributes of populations may characterize a species' biological significance within a community or ecosystem. However, such characterizations require that we understand the species' natural history with sufficient detail to distinguish natural variability in population numbers from negative responses to contaminants over the course of time. Thus, natural fluctuations in population numbers must first be understood before assigning cause and effect relationships between population fluctuations and environmental stressors, including released chemicals. Similarly, the effects of habitat modification on wildlife may adversely affect populations, and habitat evaluations would need to be completed prior to characterizing linkages between chemical exposure and population effects. Indeed, information on the response of organisms to chemicals derived from toxicity tests can be critical for identifying responses to contaminants under field conditions. As such, characterizing adverse effects based on observations of populations in the field can be very important to the evaluation process. Population-level endpoints are frequently used to evaluate chemical exposures and are most appropriate for the assessment process when (1) habitats are similar between the affected and study areas; (2) chemical exposures to individuals are well characterized; and (3) adverse effects exhibited by individuals of a species may be causally linked to significant effects on the population as a whole.

Field Surveys

Conventional population parameters (ie, occurrence, abundance, age structure, birth and death rates, and yield) are poor subjects for laboratory tests, yet should still be considered as key components of field studies, given that they are directly interpretable with respect to potentially exposed populations. Some characteristics of populations such as birth rates, death rates, and yield require long-term studies, sometimes extending over a period of years depending on the longevity of the population's individuals. Age and sex structure can usually be assessed through studies of intermediate length, perhaps a year or less, again depending on the nature of the species. With some exceptions, these population characteristics are more easily measured for annual plants and small animals than for perennials and larger animals, but there are substantial differences among species, regardless of size. The easiest data to obtain typically include estimates of population size at a given time; if only approximations are necessary, often a few statistically designed surveys are sufficient. Spatiotemporal scale will critically influence what we measure to evaluate population responses to chemical exposure. For example, are life-history attributes such as home range of sufficient size to notice a reduction due to chemical influences or do we have data on population density prior to a chemical incident that can be used for comparison? Otherwise, movement of individuals on or off of areas of concern will potentially confound empirical data gathered during field surveys.

While we could not easily find a study that actually proved chemical aversion influences population density or site usage, Kristensen et al. (2000) presented some data suggesting it could be a possibility. Ammonia gas is one of the most abundant aerial pollutants of modern poultry buildings. The authors, therefore, conducted an experiment to test domestic chickens' aversion to ammonia. They set up six groups of six laying hens, where each group was given one of three concentrations of ammonia (0–45 ppm in air) in a preference chamber over a period of 6 days and recorded their location and behavior throughout the day. The authors found significant differences in some key behaviors: hens foraged ($p = 0.018$), preened ($p = 0.009$), and rested ($p = 0.029$) significantly more in fresh air than in ammonia-polluted environments. The differences occurred between the 0 and 25 ppm levels ($p > 0.05$), but not between 25 and 45 ppm, which suggested that ammonia may aggravate hens at levels ≤25 ppm. Other species could express an aversion to other contaminants in the field and reject areas that have concentrations higher than a specified threshold value.

Sometimes areas suspected of receiving chemical releases potentially serve as adversely impacted habitat islands with distinct exposed populations that can be compared with reference populations over time to assess the effects of acute or chronic exposures. Although methods for population surveys are not as standardized as toxicity tests, there are well published and widely used methods applicable to reconnaissance survey needs (eg, Sutherland, 2006; US EPA, 2007, 2011).

The frequency of mass mortalities, and the frequency and nature of overt signs of toxicity correspond to assessment endpoints. Overt signs might be readily measured in the field for most vertebrates through adequate sampling designs. However, mass mortalities are unlikely to occur during a field survey, so local residents or agencies may have to suffice as sources of data. Frequencies of overt signs are quite variable and care must be taken in the diagnosis of lesions and tumors to distinguish effects of toxicants from those of parasites or mechanical injury. These endpoints are not standardized and, with the possible exception of fish kills and a few signs in birds and mammals, are unlikely to be interpreted through the use of existing data. A significant problem in surveying a population of animals for signs or mortality is that animals frequently seek shelter when feeling ill. Thus, sick animals or carcasses may be concealed from investigators, which may severely reduce the observed impact of contaminant exposure relative to the real effect.

Field surveys potentially call for measurements of structural and functional characteristics of populations and communities. Recommended methods for field surveys are many: for aquatic ecosystems—freshwater, estuarine, or marine—(eg, Flotemersch et al., 2006; Peck et al., 2007; US EPA, 2002); terrestrial vegetation (eg, Elzinga et al., 1998; USDA, USFS, 2005; Bonham, 2013); terrestrial vertebrates (eg, Gregory et al., 2004; Silvy, 2012); and terrestrial invertebrates (eg, Hodkinson and Jackson, 2005). Methods are readily available from many different sources, but likely require case-specific implementation depending on the habitats and biota of concern. For example, are reptiles, small mammals, or both of interest in evaluating effects from chemical releases? Methods may differ substantially depending on the type of questions asked and the organisms under investigation. The advantages and limitations of using field surveys in assessing ecological effects are briefly summarized in Table 10.1, yet integrated field and laboratory approaches tend to offset or at least diminish the downsides of either method when used alone and well-designed studies will employ both. On the one hand, without at least some field validation, results of toxicity tests only infer potential population- and community-level effects; thus, field surveys and monitoring activities provide a means for empirical verification of actual larger-scale effects. On the other hand, survey data may be used to identify problem areas, but without knowing what signs denote exposure,

TABLE 10.1 Advantages and Limitations of Field Surveys in Ecological Assessments

Advantages

- Characterizes the basic ecology of the site, identifying important resident species and community types. Based on results from the field survey, relevant species for use in toxicity testing and biomarker analyses can be identified

- Potentially demonstrates definitive ecological effects in the field, delineating zones of effect and no apparent effect

- Field responses integrate temporal and spatial variations in exposure and contaminant concentrations

- Information on the status of terrestrial vegetation can be obtained from aerial photographs, eliminating the need to visit the hazardous waste site to survey terrestrial vegetation

Limitations

- Results from field surveys may be highly variable, requiring extensive sampling to measure ecological status with sufficient precision for detection of effects; as a result, the absence of a measurable effect cannot always be interpreted as no effect

- With survey data alone, causes for observed effects are difficult to determine

- Results represent only a snapshot of the ecological status at the time of the survey

- Procedures for quality assurance/quality control are not well established; difficult to measure precision and accuracy

an investigator may miss or fail to sample for important evidence. Organisms are exposed in real-world settings and measured effects represent an integrated response to the temporal and spatial variations in exposure and contaminant concentrations in the field. As noted in other chapters and in published case studies, causality is best established through a combination of approaches, including chemical sampling, toxicity testing, biomarkers, and field surveys.

Field surveys often result in data that contain a high degree of variability that reflects the high degree of stochasticity (both spatial and temporal) inherent in populations or communities (see Chapter 11). For instance, sampling of animals is influenced by the biota of interest to the study and their responses to trap type, seasonal variation in occurrence, habitat heterogeneity, and other factors. As a result, background variability may be relatively high, and extensive sampling may be needed to measure the ecological characteristics of interest with a sufficient level of precision to detect the effects clearly linked to chemical releases. In addition, particular attention to sampling plans helps optimize survey and monitoring designs. Procedures for quality assurance/quality control exist for field surveys, but are not nearly as well-established or clear-cut as the protocols for toxicity testing.

Integration of Toxicity Tests With Field Surveys

Toxicity assessments are usually derived from laboratory-generated data, but in situ toxicity assessments, while not as well-standardized as laboratory toxicity tests, are becoming more prominent in the ecological assessment process. In situ methods more closely approximate the complexities of real-life situations and reduce the problems associated with extrapolating lab data to field conditions. Ecological effects assessments also rely upon field methods that measure ecological endpoints, either at onsite or at reference sites, and yield survey data relevant to the estimates of adverse ecological effects associated with contamination. Integration of toxicity assessments (be those in situ or laboratory-generated) and field assessments requires a well-designed sample plan to establish linkages among toxicity, site-sample chemistry, and adverse ecological effects if they are apparent.

Instead of conducting tests on amphibian responses to chemicals in a small aquarium located within a controlled environmental chamber that contains only tadpoles, purified water, and the test chemical, many scientists are now using cattle tanks or other larger units located outside that may contain sediment, vegetation, and even other animals to more closely mimic real-life situations. Relyea et al. (2005) conducted a three-way factorial outdoor *mesocosm* experiment using 1200 L cattle tanks filled with 1000 L well water, an example of which can be seen in Fig. 10.3. All tanks contained tadpoles of gray treefrogs (*Hyla versicolor*), American toads (*Bufo [Anaxyrus] americanus*), and northern leopard frogs (*Rana [Lithobates] pipiens*), as well as zooplankton and algae. Some tanks also had predators, such as newts or larval diving beetles (*Dytiscus*

FIGURE 10.3   An example of an experimental setup using stock tanks as mesocosms. Mesocosm studies can be used to provide simplified ecosystems for cause and effect testing.

sp.), and some had the insecticide malathion or Roundup (the active ingredient was the herbicide glyphosate). With these microecosystems, the authors determined that Roundup had substantial direct negative effects on the tadpoles by reducing the total tadpole survival and biomass by 40%, compared with controls. They also found that Roundup had no indirect effects on the amphibian community, such as affecting predator survival or the abundance of algae. In contrast, malathion had few direct effects on the tadpoles and actually had a positive influence on tadpoles by killing the beetles (but not the newts). Simple laboratory chronic-exposure tests might have revealed some interactions, but not under the more natural presentations of the mesocosms. These results make it clear that pesticides can have both direct and indirect effects in natural communities and that these effects critically depend upon the composition of the community.

In another study, Boone and James (2003) tested the effects of the insecticide carbaryl and the herbicide atrazine on an amphibian community composed of southern leopard frogs (*R*. [*Lithobates*] *sphenocephala*), American toads, spotted salamanders (*Ambystoma maculatum*), and small-mouthed salamanders (*Ambystoma texanum*) under mesocosms conditions. Salamanders were virtually eliminated in carbaryl treatments, indicating that at realistic levels, this insecticide might cause population declines for salamanders in contaminated habitats. Carbaryl also negatively affected toad survival. Atrazine had negative effects on body size, development, and time to metamorphoses in the frog and toad species and these effects were associated with the chlorophyll measurements in water. Chlorophyll was produced by algae and its decrease suggested a decrease in food availability for these species. The authors also observed a significant atrazine-by-carbaryl interaction for spotted salamanders, resulting in smaller and less-developed spotted salamander larvae than in control ponds.

Field surveys can identify adversely affected populations and can provide information for assessing adverse ecological effects potentially caused by chemical stressors. However, field surveys alone cannot identify the specific causes of effects. Cause and effect can be more clearly established when field studies are accompanied by toxicity tests with appropriate chemicals. This is more easily done with mesocosms than in natural situations. The actual causes mediating adverse effects observed in the field may be chemical stressors, but these effects could also be caused or exacerbated by habitat alteration, offsite sources of toxic chemicals, and natural variability that spuriously suggests linkages between contaminant and ecological effects.

Yet it may be practically intractable to determine which chemical or chemicals are causing toxicity, and ecological endpoints (eg, reduced population abundance) may be signs that indicate cumulative effects associated with exposure to a complex chemical mixture. In part, outcomes of integrated field and laboratory investigations then gain interpretative context by our assessment of population-level effects, which is reliant on our observations and measures taken in the field.

Population Models in the Ecotoxicologist's Toolbox

As will be discussed in Chapter 12, we frequently rely on quantitative-modeling tools to characterize existing populations and develop population forecasts based on empirical data (eg, census data for individuals garnered from field studies). Published studies (eg, Snider and Brimlow, 2013) provide a variety of technical materials as background information for using modeling tools that can be used to estimate population-level effects potentially affected by chemical exposures. As such, models for evaluating population growth can become components of integrated field and laboratory approaches that contribute to our understanding of population-level effects.

Descriptions of population growth require consideration of mathematical models ranging from the simple to the complex. Yet, our interest here lies simply in introducing the concepts related to both population growth and the regulation of that growth through time. For example, a simple model may be posited to describe population growth as a simple algorithm Eq. (10.1):

$$\Delta N = \frac{N_{t+1}}{N_t}$$

(10.1)

wherein growth was considered at two times, N_t and N_{t+1}.

Then, from census data, we could simply describe how the population changed during the interval ΔN between these two points. We do not really know what events contributed to the difference between population size at these two periods—perhaps it was simply a matter of births and deaths, but emigration and immigration could also be factors. We also do not know how changes in survivorship occurred during the interval—exposures to disease agents, physical, or chemical hazards may have contributed to population declines if deaths exceeded reproduction or immigration during the interval between census data collections. If we simply count individuals without garnering additional data for attributes linked to those counts—for example, what sex and how old are the individuals we count at each census event—we also do not have data sufficient to say much about the change we observe in calculating ΔN; thus, population numbers are incompletely considered with respect to population growth.

Early recognition of this insufficiency or data gap encouraged development of models that captured population growth in age-structured populations (Leslie, 1945; Akçakaya et al., 1999; Caswell, 2001). Here, adults and offspring were assumed to coexist and age-specific contributions to population growth were shown with respect to recruitment and mortality. Thus, the projected population size was described as Eq. (10.2):

$$N_t = N_0 e^{rt} \tag{10.2}$$

wherein t is time, N_t is the population size at time t, N_0 is the initial size of the population, r is the relative and invariant rate of growth (expressed as a proportion of the population through time t), and e is the base of the natural logarithm (2.7182).

Although this equation is an improvement over the relatively simple difference equation that yielded ΔN, this model presents a different challenge with respect to forecasting population growth through time. Simply said, under these conditions, population growth (positive or negative) just never stops! Thus, our brief introduction to population modeling introduces the exponential growth model (Leslie, 1945; Vandermeer and Goldberg, 2013).

A population that experiences exponential growth increases as illustrated in Fig. 10.4. Yet, the possibility of such population growth over long time periods does not follow outcomes observed in the field or in the laboratory. Indeed, depending on the organism being considered, exponential growth may occur over small time intervals. For instance, when population size is very small and resources are abundant, such as when an invasive species enters into previously unoccupied habitats or in early growth phases in laboratory cultures of bacteria, populations may undergo rapid growth. Such growth in the absence of limited or no competition from other biota or physicochemical processes that affect the availability of otherwise abundant resources is characteristic of density-independent systems, but such system behavior cannot continue indefinitely. Instead, as time passes either the population consumes all of the available nutrients and crashes, or the systems fall under density-dependent mechanisms with respect to population growth and receive some regulation.

In general, factors that enhance or diminish population growth may be either density-dependent or density-independent, yet the literature suggests that hard and fast rules for assigning these descriptors vary from species to species. For example, competitive interactions with other species encountered in a shared habitat may strongly influence the extent to which populations might attain their maximum population levels. Other factors that tend to be density-dependent are predation and disease agents that change in strength with population density and can eventually limit population size (see Chapter 11). Other factors, including exposure to chemicals released to the environment and physical hazards, such as those linked to seasonal weather extremes and natural disasters, usually affect populations irrespective of their density and can limit population growth by severely reducing the number of individuals in the population.

In 1838, Pierre François Verhulst first posited that uninhibited exponential growth would eventually become limited with respect to resource availability and the growth of human populations. Mathematically, Verhulst (1838) described a logistic equation that limited exponential growth as the size of the population increased toward the carrying capacity, K, of its environment (Fig. 10.4) Eq. (10.3).

$$\frac{dN}{dt} = rN\left(\frac{K - N}{K}\right)$$

(10.3)

wherein dN is the change in numbers, dt is the change in time, dN/dt is the rate of growth, r is the intrinsic rate of natural increase, N is the population size at a given time, and K is a measure of environmental resistance known as *carrying capacity*.

As with exponential growth, logistic growth is commonly observed in laboratory studies. However, populations under more natural conditions often fluctuate in ways not predicted by simple logistic growth models. Such fluctuation is often ascribed to seasonal or other environmental cycles (eg, daily, lunar cycles). At other times, fluctuating population levels appear to be linked

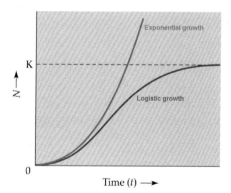

FIGURE 10.4   Idealized patterns for exponential growth compared to logistic growth. K on the ordinate axis is called the system's carrying capacity which serves as the upper bound on population size in systems that display logistic growth.

to density-dependent population growth factors or interactions among density-independent factors (eg, changing seasonal conditions linked to increased temperatures, decreased precipitation) and density-dependent factors (eg, exposure to disease agents that may be more abundant under certain weather conditions than in others). Populations may also display marked decreases in numbers if changing environmental conditions (including chemical releases that result in acutely lethal exposures) cause increased death rates that greatly exceed birth rates, which would place any population on a downward trend (see Beissinger and McCullough, 2002 for a discussion of evaluation of vulnerabilities and population viability analysis).

We're going out on a limb here, realizing that some exceptions will be found, but the effects of contaminants are almost always going to be density-independent. The severity of their effects will be largely determined by the average sensitivity of the population to the range of concentrations present in the exposure. The effects of the contaminant in and of itself should not be related to the density of the population being exposed. Given that, there may be mitigating factors such as "safe zones" where contaminant concentrations are lower and can hold only a certain number of animals, but those type of situations should be rare. We would expect that toxic levels of a pesticide or other chemical would claim a relatively constant percentage of organisms, regardless of population density. Since only density-dependent factors can actually regulate population numbers in animals (Lack, 1954) and plants, contaminants can increase population declines but cannot regulate them.

Given the strengths and weaknesses of exponential and logistic population models as tools applicable to understanding the effects of chemicals released in the environment, alternative models that leverage this modeling experience are available for the ecotoxicologist and amenable to exposure-effects analysis. Many of these alternative models are well-developed and used by population ecologists, but their application within the regulatory context is limited. Perhaps one of the most useful tools, and one widely applied to resource-management issues, is life-table analysis, which is implemented in its various forms.

Additive Versus Compensatory Mortality and Ecotoxicology

As we have stated, in population ecology it is important to identify how stressors affect a population's dynamics. Broadly, there are two ways a stressor can affect a population's growth (Nichols et al., 1984; Williams et al., 2002). The stressor can be *additive*, in which case it contributes to population loss in addition to all other factors and can have a real influence on the population. Alternatively, the stressor can be *compensatory* if other stressors would result in the same population loss even if the stressor of interest was not present. Take, for example, the case where a field investigator found a die-off of 200 mallards in a farmer's field and determines that the birds were killed by the pesticide malathion. If the malathion had an additive effect, those 200 mallards would have survived all other

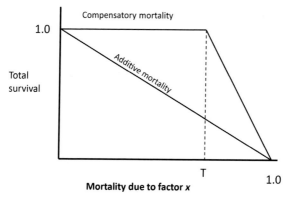

FIGURE 10.5   A conceptual model showing the difference between compensatory and additive mortality. With compensatory mortality, no real population effect occurs until some, probably very high, level. With additive mortality, actual population effects are seen even at the lowest effects level.

stressors including predation, hunting, food scarcity, disease, and the like. Thus, the loss of 200 birds was a true loss. If, however, those birds had not been poisoned and had died from any of the other stressors, then there is not a true loss to the population and the malathion could have been a compensatory stressor. Fig. 10.5 shows this graphically. In additive mortality, Factor X (eg, malathion) exerts mortality on the population regardless of how many individuals survive. In compensatory mortality, however, the factor has no influence on the population until some hypothetical point or threshold (T), when the concentration of malathion becomes sufficiently high to have an overwhelming effect on the population. In reality, that threshold may never be reached under field conditions.

It is extremely difficult to determine if a contaminant (or almost any other stressor for that matter) is actually having an effect on a population in and of itself. This simple model is made more complex because a factor can be compensatory at some point and additive at another unknown point, and multiple factors can interact to enhance mortality or ameliorate each other's effects. As discussed in Chapter 8, for example, the concentration of calcium or even zinc may alter the toxicity of other metals in an aquatic environment.

LIFE TABLE ANALYSIS

For the ecotoxicologist focused on characterizing exposure and effect relationships at the population level, life table analysis yields quantitative results that describe: (1) population age structure, including characterization of fecundity and reproductive endpoints; (2) population growth rate; and (3) patterns in population survivorship patterns. Life tables are built from empirical data garnered from field studies and provide a model of the age distribution of a population. Life tables consider the number of individuals of a given age or stage of development, let's call that n, at various times (t) during a sampling period, for example, during a long-term study or monitoring program. Among other population attributes, life tables allow us to characterize population growth, including an idea of how its structure changes with time. Whereas population ecology is keenly interested in understanding factors and processes that govern population growth, ecotoxicologists may extend the application of life table analysis to characterize chemical effects on populations based, in part, on observed results from laboratory toxicity tests. Perturbations, such as those linked to chemical exposure, become one of many environmental stressors at play in field settings, and serve as contributing factors influencing density-independent or density-dependent processes affecting individual survival and population growth to varying extents. Alternative methods, such as the Kaplan–Meier product-limit method (Skalski et al., 2005), are also widely reported in the literature as techniques for survival analysis. Life table models have long been available and may be deployed in integrated field and laboratory studies given empirical data likely gleaned from such investigations.

Completed life table models yield estimates of population abundance based on data assembled as a schedule of births and deaths for a population, or, more often than not, some portion of a population. Life table analysis frequently relies on models built on empirical data wherein year-class or stage-class cohorts—referred to as *cohort life tables* or *dynamic life tables*—are followed from birth to death. Such dynamic life tables are most often applied to short-lived organisms or long-lived biota amenable to longitudinal studies. As an alternative to cohort life table analysis, *static life table* or time-specific table analysis focuses on count data for all individuals entering a study during the same time interval and subsequently followed as a cross-section data. Static life table analysis is less frequently followed in field surveys than cohort tables.

As an illustration of cohort life table analysis, Table 10.2 summarizes the arithmetic derivation of typical life table outputs derived from empirical data used to develop such an analysis. Although life tables may vary with respect to empirical data used for the analysis, most follow a convention used for the following variables:

x = age (eg, 0, 1, 2, 3 years old) or stage (eg, egg, larvae, nymph, adult) class
n_x = number of individuals in each age or stage x
l_x = percent of the original cohort that survives to age or stage x ($= n_x/n_0$)
d_x = probability of dying during age or stage x ($= l_x - l_{x+1}$)
q_x = percent of dying between age or stage x and age or stage $x+1$ ($= d_x/l_x$)
b_x = number of offspring produced per individual in age/stage x

With data derived from literature or when integrated field and laboratory studies are available, summary information derived from cohort life table analysis includes estimates of several other important population variables:

R_0, or net reproductive rate (a measure of the change in population size), where $R_0 = \Sigma(l_x b_x)$

TABLE 10.2 Illustration of Data Assembled for Cohort Life Table Analysis

x (months)	n_x	l_x	d_x	q_x	b_x	$l_x b_x$
0	300	1.000	0.500	0.500	0	0.000
1	150	0.500	0.156	0.312	0	0.000
2	100	0.344	0.072	0.209	0	0.000
3	75	0.272	0.060	0.221	5	1.360
4	50	0.212	0.188	0.887	29	6.148
5	10	0.024	0.024	1.000	9	0.216
6	0	0.000	—	—	0	0.000
					Estimate, R_0	7.724

T, the generation time (the time between the birth of one cohort and the birth of their offspring) where: $T = (\Sigma x l_x b_x)/R_0$

r, the per capita rate of increase (as is R_0, r is a measure of the change in population size) and is calculated as: $r = \ln R_0/T$

Note that R_0 and r do not estimate the same population parameter; r tells us whether a population is experiencing more births than deaths, but R_0 tells us if the population is increasing or decreasing relative to some fixed point defined as 1. In other words:

If $R_0 > 1$, then population is increasing in size
If $R_0 < 1$, then population is decreasing in size
If $R_0 = 1$, then population size is constant

In contrast:

If $r > 0$, then population is increasing in size
If $r < 0$, then population is decreasing in size
If $r = 0$, then population size is constant

An example of using life tables to predict population effects of contaminants can be found in Sha et al. (2015), who sought to assess the relative toxicity effects of two polybrominated diphenyl ethers (PBDEs), BDE-47 and BDE-209 congeners, on the marine rotifer *Brachionus plicatilis*. They calculated rotifer population demographic parameters into life tables, including age-specific survivorship (l_x), age-specific fecundity (m_x), net reproductive rate (R_0), intrinsic rate of increase (r), finite rate of increase (λ), life expectancy (E_0), and generation time (T), and used them as measures of treatment effects. Though this was a laboratory study, rotifers are so small that they do not require large chambers or lakes to show population changes. Results from this study revealed increasingly intense negative effects on many of the rotifer demographic parameters with elevated PBDE concentrations. The population growth curves of control *B. plicatilis* showed almost instantaneous growth and reached peak abundances within 11 days, whereas *B. plicatilis* exposed to BDE-209 did not increase in numbers until after 5 days. PBDE exposure also reduced population abundances and peak population densities of the rotifer. The PBDEs suppressed carrying capacity (K) of the habitats in a dose-dependent fashion. The study also showed that the time for population growth to level off was shortened by PBDEs compared with controls. The authors concluded that life table demography and the population growth curve can be used to evaluate PBDE effects.

Another example might better demonstrate the value of life table parameters. Martinez-Jeronimo et al. (2013) exposed recently hatched *Daphnia schoedleri* to three sublethal concentrations of the pyrethroid α-cypermethrin (0.54, 5.4, and 54 ng/L) plus a control for 21 days. *Daphnia schoedleri* is another species that can be easily maintained in the laboratory for many generations. Effects were measured through a life table analysis for fecundity (m_x) and survivorship (l_x). In addition, the intrinsic rate of population growth (r), net reproductive rate (R_0), life expectancy at birth (e_x), generation time (G), and the average lifespan were also calculated for survivorship and fecundity values recorded during 21 days in each cohort. Survivorship curves showed substantial loss to the populations at 0.54 ng/L α-cypermethrin. Moreover, significant differences were found in all life table parameters. The α-cypermethrin negatively affected average life span, life expectancy at hatching (e_x), net reproductive rate (R_0), intrinsic rate of population growth (r), and, at the highest concentrations, generation time (G).

Regardless of our selection of tools and analytical models, as ecotoxicologists, the central question should be "How do exposure and effects findings derived from individual-based toxicity tests work with respect to our forecasting ecological effects from chemical releases in the field?" In this chapter and the next, we focus on extrapolations from toxicity tests commonly completed on individual organisms under controlled conditions to demographic endpoints characteristic of groups of individuals of the same species in a given place at a given time. In short, from a population ecologist's perspective, ecotoxicologists most often focus on acute exposures to reliably predict natality (specifically, birth rates) and mortality (specifically, death rates). Thus, they forecast population-level effects, perhaps gleaned from census data collected in the field with little, if any, empirical data collected that characterizes effects on other elements of population dynamics such as immigration and emigration and how those processes might influence interpretation of chemical effects on age distribution, genetic composition, or spatial distribution patterns of biota directly or indirectly exposed to chemical releases in field settings (see the section "Focus—Almost, But Still Not There: The Story of Contaminants and Population Status of California Amphibians," later in this chapter). As with other analytical methods, life table analysis can be a powerful tool, however these methods have their limits. One limitation is that there are few long-term studies available that follow cohorts through a lifetime. As such, our inability to conduct empirical "lifetime studies" generally precludes completion of a comprehensive cohort analysis. Our assumption of a stable-age distribution, though critical to the underlying analysis, is also difficult to fully characterize (Anderson et al., 1985; Skalski et al., 2005). However, as one of the tools available to ecotoxicologists focused on characterizing exposure and effects relationships in field settings, life table analysis serves to link empirical data derived from field or laboratory studies with modeling efforts intended to forecast alternative futures involving questions of resource management and releases of environmental chemicals.

STOCHASTICITY AND UNCERTAINTY

Given our brief overview of population growth and models built to project population growth through time, and density-independent and density-dependent factors potentially influencing population growth, we next briefly consider factors that promote caution in our interpretations of adverse effects from exposures to environmental chemicals in the field. In short, we focus on how stochastic (random) events occurring in the field might influence population growth and long-term viability (Beissinger and McCullough, 2002). Whether or not chemical exposures are likely stressors encountered in the field, stochastic events threaten the persistence of any population; the smaller the population, the greater the threat. Four general classes of threats may influence population growth and its long-term viability: demographic stochasticity, environmental stochasticity, natural catastrophes, and genetic stochasticity.

Demographic stochasticity accounts for variability in population growth rates linked to random differences in survival and reproduction among individuals within given time periods. As a process affecting any population, demographic stochasticity is usually more apt to influence relatively small populations—for example, with endangered species—and when jointly occurring with releases of chemicals to the environment, outcomes could be problematic, particularly when a catastrophic release of chemicals to habitats identified as critical to endangered species would represent an extreme event of significant consequence. In contrast to demographic stochasticity, *environmental stochasticity* includes abnormal events such as extremes in weather, such as prolonged winters characterized by unusually low temperatures or by cycles in drought or flooding that may affect populations of any size, even those of large populations.

Natural catastrophes might be considered an extreme case of environmental stochasticity. These catastrophes are usually unpredictable and may include local extirpation events such as major floods, volcanic eruptions, tsunamis, earthquakes, or flash fires that quickly consume acres of land. Perhaps the only defense that a population has against these events is a wide distribution, so that survivors can reinhabit an area, and the often-slim possibility of finding refugees from the event.

Genetic stochasticity refers to genetic drift; that is, changes in the genetic composition of a population unrelated to organism-level processes driving population growth, such as emigration, immigration, and other systematic events. Such stochasticity influences genetic structure of populations, reducing genetic diversity and population viability by adversely affecting the reproductive capacity of individuals. Genetic stochasticity becomes particularly problematic when initial population size is small, say, fewer than 100–300 individuals in the population. As with demographic stochasticity, genetic stochasticity is less likely to be a problem in populations large enough to buffer environmental stochasticity (eg, Lande, 1993; Dunham et al., 1999). Often, the species most vulnerable to these forms of stochasticity are those most often encountered on lists of sensitive species or species of concern to resource management organizations, such as threatened or endangered species. Among even these species, vertebrates with generally lower fecundity, may be more at risk than invertebrates or plants. Life-history attributes are critical to the evaluation of a population's growth and its long-term viability, with the most critical stages of an organism's life cycle likely yielding the greatest impact on its population dynamics.

As this chapter's preamble, courtesy of Bartholomew (1964), suggested, bridging the data and knowledge gaps inherent to extrapolation from individual-based models, such as those yielding toxicity data, to population, community, and ecosystem levels of biological organization continue to offer research challenges for ecotoxicologists and applied ecologists. In many respects, design and implementation of integrated field and laboratory studies addresses these scale-related issues given their focus on multiple stressors and the characterization of cause and effect relationships linked to the chemical exposures to biota under field conditions. In preparation for our focus on communities and ecosystems in the next chapter, and in recognition of the roles that humans play in releasing chemicals to the environment and the effects of these chemicals on human populations, integrated field and laboratory studies mirror the efforts of field epidemiologists whose primary focus lies with exposure and effects of multiple stressors, originally disease agents, on human health. Indeed, the process is identical; only the names of the stressors may have changed. Indeed, exposure is the common thread that ties together all receptors, particularly within exposure networks encountered in field settings.

Segue

Here, we begin our overview of integrative processes, building from individual-based studies, such as toxicity testing and assessment introduced in Chapter 2 and setting the stage for our focus on populations in this chapter. In future chapters, we will subsequently extend our overview to "populations of populations," more commonly called communities and ecosystems (see Chapter 11).

In essence, we will become more aware of the inter-relationships among individuals, populations, and communities, all part of larger ecosystems. As such, human health becomes entangled in our overview of ecotoxicology as an issue in which we discover dependencies throughout the larger "system of systems" through the discipline of field epidemiology. Although beyond the scope of our overview, field epidemiology captures disciplines in social sciences, such as medical sociology or psychology and other biological sciences (microbiology, pathology, or nutrition). Although we leave those disciplines to their own devices, we have so far presented brief overviews focused on essentials of ecotoxicology and also provided a brief overview of selected chemicals released to the environment. In addition, the principal focus of this chapter has been on introducing spatiotemporal scale as an underlying ecological principle that we will consider next in our snapshot of communities and ecosystems, and the effects of chemical exposures in highly networked populations whose individuals serve as receptors and transfer agents.

Focus—Almost, But Still Not There: The Story of Contaminants and Population Status of California Amphibians

Populations of several species of amphibians in the Sierra Mountains of California have been declining at alarming rates. For example, among 16 sites in Lassen National Park that had recent records of the Cascades frog (*Rana cascadae*), only one site had frogs, and even there only two individuals were found (Fellers and Drost, 1993). Farther south at Yosemite National Park, five of seven species of frogs and toads had suffered serious declines (Drost and Fellers, 1996). One of the most common species had shrunk to a few remnant populations. Even attempts to repopulate areas that had depleted populations of amphibians failed (Fellers et al., 2007). Although these studies occurred 20 years ago, the status of most California anurans has not improved in recent years (Brown et al., 2014). At the start of concern for amphibian species, several factors including drought, habitat loss, and introduced fish were thought to be the causes for the declines. Eventually, introduced trout were shown to be important in some lakes, but not in much of the areas of decline (Knapp and Matthews, 2000). What was causing these widespread declines in California anurans? Salamander populations did not seem to be at risk, so what was particular to frogs and toads?

Eventually, two possible causes came to the forefront of attention: contaminants and disease. We'll review some of the data supporting contaminants first, then return to disease. First, a little bit of geography: California has four principal national parks in the Sierra Nevada Mountains, from north to south, they are Lassen Volcanoes National Park, Yosemite National Park, and Sequoia/ Kings Canyons National Park. West of the Sierra Nevada is the Central Valley, a huge agricultural region that uses thousands of tons of active-ingredient pesticides each year, although the amount of pesticide used has declined over the past decade or more (California Department of Pesticide Regulation, 2013). Prevailing winds from the coast carry volatilized pesticides into the foothills and mountains with concentrations of contaminants in general decreasing with distance from the coast and elevation—the two measures being highly related (Bradford et al., 2010). These winds also tend to blow in a southerly direction and circulate the lower end of the Valley or the San Joaquin Valley around Sequoia/ Kings Canyon National Parks, carrying heavier loads of pesticides. In contrast, coastal and more northern populations are declining less precipitously. Using the still abundant Pacific treefrog (*Pseudacris regilla*) as a sentinel species, Sparling et al. (2001) found that cholinesterase (ChE) activity in tadpoles was depressed in mountainous areas east of the Central Valley compared with sites along the coast or north of the Valley. Cholinesterase was also lower in areas where ranid population status was poor or moderate compared with areas that had good ranid status. Up to 50% of the sampled population in areas with reduced ChE had detectable organophosphorus pesticide residues, mostly diazinon, malathion, and chlorpyrifos with concentrations as high as 190 ppb wet weight. In addition, nearly 90% of some populations had measurable endosulfan concentrations and 40% had detectable p,p'-DDT and o,p'-DDT residues. Angermann et al. (2002) found polychlorinated biphenyls (PCBs), PBDEs, and other persistant organic pollutants (POPs) in the tissues of *P. regilla* in a pattern consistent with wind-blown contaminants.

Laboratory experiments showed that some California frogs could be very sensitive to pesticides. *H. regilla* was less sensitive to chlorpyrifos (LC50 365 μg/L) than the more threatened mountain yellow leg frog (*Rana boylii*, LC50 65.5 μg/L) (Sparling and Fellers, 2009). Treefrogs were also less sensitive to endosulfan (LC50 15.6 μg/L) than *R. boylii* (LC50 0.55 μg/L). However, by the time contaminant studies had begun on Californian frogs, field residues of these pesticides were in the ng/L range. Recall from Chapter 5 that many organophosphorus pesticides are chemically altered to their oxon or sulfon forms which have half-lives of a few days in the environment. Oxon forms of diazinon and malathion were one to two orders of magnitude more toxic than parent forms and chloroxon (degradate of chlorpyrifos) was at least three orders of magnitude more toxic (Sparling and Fellers, 2007). Thus, at the very least, endosulfan and chloroxon were determined to be lethal in the laboratory at concentrations found in the environment.

An extensive 2-year study was led by Dr Deborah Cowman in 2001 and 2002 that raised *P. regilla* in Teflon-net structures in situ within Lassen Volcanoes, Yosemite, and Sequoia National Parks (Sparling et al., 2015). Twenty contaminants were identified in tadpoles with an average of 1.3–5.9 and a maximum of 10 different contaminants per animal. In line with the prevailing wind patterns, Sequoia tadpoles had the greatest variety and concentrations of contaminants in 2001. They also experienced the greatest mortality and the slowest developmental rates in both years and the lowest cholinesterase activities of all three parks in 2001. In addition, Yosemite and Sequoia tadpoles had greater genotoxicity than those in Lassen during 2001. In 2001, tadpoles at Yosemite showed a high rate of hind limb malformations. Although concentrations of most contaminants were below known lethal concentrations, simultaneous exposure to multiple chemicals and other stressors could have resulted in lethal and sublethal effects. Unfortunately, we do not know much about the interactive effects of chemicals.

Despite all of these and other studies and data on chemicals and amphibians' sensitivity to them, we still cannot claim that we have shown a population effect. Recall the difference between additive and compensatory mortality. Whereas we have seen mortality in the field that would appear to be related to contaminants, we have not studied other stressors such as predation or disease simultaneously with contaminants. We also cannot distinguish if the mortality was additive or compensatory. Furthermore, there is doubt because the field concentrations tended to be much lower than what were found to be lethal in the laboratory. Perhaps the combination of chemicals are more toxic than any one alone. It is likely that because the use of organochlorine pesticides ceased in the Central Valley 40 years ago, concentrations were much higher in the past and these and other POPs may have historically devastated populations of frogs. However, we do not have data from back then. Perhaps (there is that word again) the current-use pesticides do not last sufficiently long in the environment and field collections failed to catch their peak concentrations. The bottom line is that the hypothesis that contaminants have caused the long-term population declines seen in several species of California frogs still has not been scientifically proven.

Another cause for some population declines has been found. The fungus *Batrachochytrium dendrobatidis* infects anurans and feeds on keratin in the skin, causing the disease chytridiomycosis (chytrid). This disease infiltrates the dermis of the frogs and toads, reducing dermal respiration and water transport, resulting in die-offs. The fungus is global in distribution and is reportedly widespread throughout much of California (Fellers et al., 2011). No doubt chytrid has been responsible for some die-offs. Unfortunately, there is a difference between the way chytrid and contaminants affect amphibians that plays a major role in attracting interest. Chytrid kills relatively quickly, resulting in many dead bodies of frogs strewn along shores or shallow waters of a wetland and making a very conspicuous display. Contaminants, on the other hand, seldom produce acute mortality. Rather they work quietly by reducing reproductive rates, altering endocrine systems, producing slow-acting cancers, and possibly even facilitating death by conspicuous diseases resulting from impaired immune systems.

STUDY QUESTIONS

1. Did you notice any differences between the study of population ecotoxicology and population ecology while reading this chapter?
2. Discuss in your own words what is meant by spatiotemporal scales.
3. Are contaminants more likely to cause an extirpation or an extinction? Why?

4. How can controlled laboratory experiments on cause and effect with a single chemical and a single species help in understanding possible population effects under more natural conditions? Can they?

5. In this chapter, we mentioned that aversion to contaminants may affect population size or density, but that we could not find any specific examples in a scan of the scientific literature. Can you imagine a way in which chemical aversion could have population effects? Provide details.

6. What problems might occur if you conducted field surveys without having run laboratory tests or access to studies that did run tests with the chemicals and species of concern?

7. What are your thoughts on mesocosm experiments? Can mesocosms completely mimic a natural environment? What are some of the advantages and disadvantages of using mesocosms for testing chemicals in an aquatic environment?

8. Do you imagine that a contaminant could be a density-dependent regulating factor for a population? If so, what conditions would have to exist for that to happen?

9. In cooperation with your class, design a study to determine if a contaminant has an additive or compensatory effect on a population. For simplicity's sake, you can use a mesocosms design.

10. Have you used or built life tables in any other course you have taken, such as an ecology course? What do you think of the idea of using life table techniques to determine if a contaminant is exerting a population effect?

11. We will discuss stochasticity in greater detail when we get to risk assessment and we'll see that randomness can be a major factor in making good decisions. How can stochasticity affect the interpretations of a scientific study?

12. What did you learn from the story in the "Focus" section? What more would have to be done to show that contaminants are having a true population effect?

REFERENCES

Akçakaya, H.R., Burgman, M.A., Ginzburg, L.R., 1999. Applied Population Ecology: Principles and Computer Exercises Using RAMAS EcoLab 2.0, second ed. Sinauer Associates, Sunderland, MA, p. 285.

Albers, P.H., Heinz, G.H., Ohlendorf, H.M. (Eds.), 2000. Environmental contaminants and terrestrial vertebrates: effects on populations, communities and ecosystems. Society of Environmental Toxicology and Chemistry, Pensacola, FL, p. 351.

Anderson, D.R., Burnham, K.P., White, G.C., 1985. Problems in estimating age-specific survival rates from recovery data of birds ringed as young. J. Anim. Ecol. 54 (1), 89–98.

Angermann, J.E., Fellers, G.M., Matsumura, F., 2002. Polychlorinated biphenyls and toxaphene in Pacific tree frog tadpoles (*Hyla regilla*) from the California Sierra Nevada, USA. Environ. Toxicol. Chem. 21, 2209–2215.

Bartholomew, G.A., 1964. The roles of physiology and behaviour in the maintenance of homeostasis in the desert environment. In: Hughes, G.M. (Eds.), Homeostasis and Feedback Mechanisms. Symposium, Soc. Exper. Biol. 18, 7–29.

Beissinger, S.R., McCullough, D.R., 2002. Population Viability Analysis. University of Chicago Press, Chicago, p. 593.

Bonham, C.D., 2013. Measurements for Terrestrial Vegetation, second ed. John Wiley & Sons, Inc., Hoboken, NJ, p. 264.

Boone, M.D., James, S.M., 2003. Interactions of an insecticide, herbicide, and natural stressors in amphibian community mesocosms. Eco. Appl. 13, 829–841.

Bradford, D.F., Stanley, K., McConnell, L.L., Tallent-Halsell, N.G., Nash, M.S., Simonich, S.M., 2010. Spatial patterns of atmospherically deposited organic contaminants at high elevation in the southern Sierra Nevada mountains, California, USA. Environ. Toxicol. Chem. 29, 1056–1066.

Brown, C., Wilkinson, L.R., Kiehl, K.B., 2014. Comparing the status of two sympatric amphibians in the Sierra Nevada, California: insights on ecological risk and monitoring common species. J. Herpetol. 49, 74–83.

Cain, N.L., Bowman, W.D., Hacker, S.D., 2014. Ecology, third ed. Sinauer Associates, Inc, Sunderland, MA, p. 596.

California Department of Pesticide Regulation, 2013. Databases. <http://www.cdpr.ca.gov/dprdatabase.htm> (accessed 25.12.15).

Caswell, H., 2001. Matrix Population Models, second ed. Sinauer Associates, Inc, Sunderland, MA, p. 722.

Drost, C.A., Fellers, G.M., 1996. Collapse of a regional frog fauna in the Yosemite area of the California Sierra Nevada, USA. Conserv. Biol. 10, 414–425.

Dunham, J., Peacock, M., Tracy, C.R., Nielsen, J., Vinyard, G., 1999. Assessing extinction risk: integrating genetic information. Conserv. Ecol. 3 unpaginated.

Elzinga, C.L., Salzer, D.W., Willoughby, J., 1998. Measuring and monitoring plant populations. BLM Technical Reference 1730-1. U.S. Dept. of the Interior, Bureau of Land Management, Denver, CO, p. 477.

Fellers, G.M., Drost, C.A., 1993. Disappearance of the cascades frog *Rana cascadae* at the southern end of its range, California, USA. Biol. Conserv. 65, 177–181.

Fellers, G.M., Bradford, D.F., Pratt, D., Wood, L.L., 2007. Demise of repatriated populations of mountain yellow-legged frogs (*Rana muscosa*) in the Sierra Nevada of California. Herpetol. Conserv. Biol. 2, 5–21.

Fellers, G.M., Cole, R.A., Reinitz, D.M., Kleeman, P.M., 2011. Amphibian chytrid fungus (*Batrachochytrium dendrobatidis*) in coastal and montane California, USA anurans. Herpetol. Conserv. Biol. 6, 383–394.

Flotemersch, J.E., Stribling, J.B., Paul, M.J., 2006. Concepts and Approaches for the Bioassessment of Non-wadeable Streams and Rivers. EPA 600-R-06-127. US Environmental Protection Agency, Cincinnati, OH, p. 245.

Gimme, H.W., Hengeveld, R., 2014. Autecology: Organisms, Interactions and Environmental Dynamics. CRC Press, Boca Raton, FL, p. 484.

Gregory, R.D., Gibbons, D.W., Donald, P.F., 2004. Bird census and survey techniques. In: Sutherland, W.J., Newton, I., Green, R.E. (Eds.), Bird Ecology and Conservation: A Handbook of Techniques. Oxford University Press, Oxford, pp. 17–56.

Hodkinson, I.D., Jackson, J.K., 2005. Terrestrial and aquatic invertebrates as bioindicators for environmental monitoring, with particular reference to mountain ecosystems. Environ. Manage. 35, 649–666.

Kristensen, H.H., Burgess, L.R., Demmers, T.G.H., Wathes, C.M., 2000. The preferences of laying hens for different concentrations of atmospheric ammonia. Appl. Anim. Behav. Sci. 68, 307–318.

Knapp, R.A., Matthews, K.R., 2000. Non-native fish introductions and the decline of the mountain yellow-legged frog from within protected areas. Conserv. Biol. 14, 428–438.

Lack, D., 1954. The Natural Regulation of Animal Numbers. Clarendon Press, Oxford, p. 353.

Lande, R., 1993. Risk of population extinction from demographic and environmental stochasticity and random catastrophes. Am. Natur. 142, 911–927.

Leslie, P.H., 1945. On the use of matrices in certain population mathematics. Biometrika 33, 183–212.

Linder, G., Krest, S.K., Sparling, D.W. (Eds.), 2003. Amphibian decline: an integrated analysis of multiple stressor effects. Society of Environmental Toxicology and Chemistry, Pensacola, FL, p. 345.

Martinez-Jeronimo, F., Arzate-Cardenas, M., Ortiz-Butron, R., 2013. Linking sub-individual and population level toxicity effects in *Daphnia schoedleri* (Cladocera: Anomopoda) exposed to sublethal concentrations of the pesticide alpha-cypermethrin. Ecotoxicology 22, 985–995.

Newman, M.C., 2001. Population Ecotoxicology. John Wiley & Sons, Inc., New York, p. 242.

Newman, M.C., Clements, W.H., 2007. Ecotoxicology: A Comprehensive Treatment. CRC Press, Boca Raton, FL, p. 880.

Nichols, J.D., Conroy, M.J., Anderson, D.R., Burnham, K.P., 1984. Compensatory mortality in waterfowl populations a review of the evidence and implications for research and management. Trans. N. Am. Wildl. Nat. Resour. Conf. 49, 535–554.

Peck, D.V., Averill, D.K., Herlihy, A.T., Hughes, R.M., Kaufmann, P.R., Klemm, D.J., et al., 2007. Environmental Monitoring and Assessment Program—Surface Waters Western Pilot Study: Field Operations Manual for Non-wadeable Rivers and Streams. Draft. U.S. Environmental Protection Agency, Office of Research and Development, Washington, DC.

Relyea, R.A., Schoeppner, N.M., Hoverman, J.T., 2005. Pesticides and amphibians: the importance of community context. Ecol. Appl. 15, 1125–1134.

Sha, J., Wang, Y., Lu, J., Wang, H., Chen, H., Qi, L., et al., 2015. Effects of two polybrominated diphenyl ethers (BDE-47, BDE-209) on the swimming behavior, population growth and reproduction of the rotifer *Brachionus plicatilis*. J. Environ. Sci. 28, 54–63.

Shugart L.R., Theodorakis C.W., Bickham A.M., Bickham, J.W., 2003. Genetic effects of contaminant exposure and potential impacts on animal populations. In: Hoffman, D.J., Rattner, B.A., Burton Jr. G.A., Cairns Jr., J. (Eds.), Handbook of Ecotoxicology, second ed. Society of Environmental Toxicology and Chemistry, Pensacola, FL, pp. 1129–1148. 1290 p.

Silvy, N.J. (Ed.), 2012. The Wildlife Techniques Manual, vol. 1 Johns Hopkins University Press, Baltimore, Maryland, pp. 136.

Skalski, J.R., Ryding, K.E., Millspaugh, J., 2005. Wildlife Demography: Analysis of Sex, Age, and Count Data. Academic Press, New York, p. 656.

Snider, S.B., Brimlow, J.N., 2013. An introduction to population growth. Nature Education Knowledge 4, 3. Available at: <http://www.nature.com/scitable/knowledge/library/an-introduction-to-population-growth-84225544> (accessed 23.12.15).

Sparling, D.W., Fellers, G.M., McConnell, L.L., 2001. Pesticides and amphibian population declines in California, USA. Environ. Toxicol. Chem. 20, 1591–1595.

Sparling, D.W., Fellers, G., 2007. Comparative toxicity of chlorpyrifos, diazinon, malathion and their oxon derivatives to larval *Rana boylii*. Environ. Poll. 147, 535–539.

Sparling, D.W., Fellers, G.M., 2009. Toxicity of two insecticides to California, USA, anurans and its relevance to declining amphibian populations. Environ. Toxicol. Chem. 28, 1696–1703.

Sparling, D.W., Bickham, J., Cowman, D., Fellers, G.M., Lacher, T., Matson, C.W., et al., 2015. In situ effects of pesticides on amphibians in the Sierra Nevada. Ecotoxicology 24, 262–278.

Sutherland, W.J., 2006. Ecological Census Techniques: A Handbook. Cambridge University Press, Cambridge, p. 448.

USDA, USFS, 2005. Threatened, Endangered and Sensitive Plants Survey Field Guide. United States Department of Agriculture, Forest Service, Rangeland Management Staff, Washington, DC, p. 32.

US EPA, 2002. Methods for Evaluating Wetland Condition: Using Amphibians in Bioassessments of Wetlands. EPA-822-R-02-022. Office of Water, U.S. Environmental Protection Agency, Washington, DC, p. 41.

US EPA, 2007. National Rivers and Streams Assessment: Field Operations Manual. EPA-841-B-07-009. U.S. Environmental Protection Agency, Washington, DC, p. 354.

US EPA, 2011. National Wetland Condition Assessment: Field Operations Manual. EPA-843-R-10-001. U.S. Environmental Protection Agency, Washington, DC, p. 485.

Vandermeer, J.H., Goldberg, D.E., 2013. Population Ecology: First Principles, Second edition Princeton University Press, Princeton, NJ, p. 288.

Verhulst, P.F., 1838. Notice sur la loi que la population suit dans son accroissement. Correspondance mathématique et physique 10, 113–121.

Williams, B.K., Nichols, J.D., Conroy, M.J., 2002. Analysis and Management of Animal Populations: Modeling, Estimation, and Decision Making. Academic Press, New York, p. 817.

Chapter 11

Community-Level and Ecosystem-Level Effects of Environmental Chemicals

Terms to Know

Load
Airshed
Watershed
VOC
SVOC
Biological Assessments
Biological Indicators (from a community perspective)
Community
Ecosystem
Groundwater
Index of Biological Integrity

INTRODUCTION

As we introduced in Chapter 10, hierarchical models—those capturing scale differences across time and space, or those representing different levels of biological organization—can reflect actual circumstances and be implemented practically to evaluate chemical effects on biological components of many ecosystems. This chapter extends our previous consideration of how chemicals affect populations by examining the populations of many species that occupy shared habitats within a given landscape; that is, we briefly consider communities, and at larger scales, ecosystems. Indeed, scale differences are encountered in many contemporary issues and are related to chemical releases into the environment that directly result in soil, water, and air pollution to lesser or greater extents. Indirectly, these chemical releases may also serve as contributing factors in exposures potentially linked to other adverse outcomes, such as biodiversity loss and climate change (Fig. 11.1).

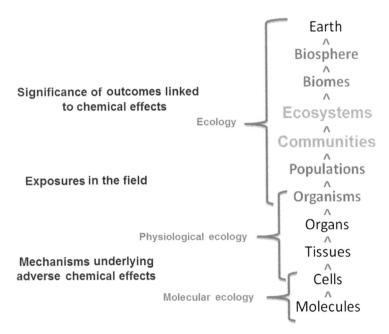

FIGURE 11.1   Recall our conceptual models of levels of biological organization from Chapter 10, which focused on populations. Here we focus on the interrelatedness of levels of biological organization beyond organisms and populations that extend our linear array of "receptors" that potentially interact with chemicals and other environmental stressors during exposures in the field.

COMMUNITIES, ECOSYSTEMS, AND SPATIOTEMPORAL SCALES

Multiple stressors—in many cases strongly dominated by chemical stressors—impact ecological systems from local to global scales. Thus, understanding these environmental issues requires perspectives that consider interactions among physicochemical and biological processes at multiple spatial and temporal scales. In addition, as suggested in the previous chapters of this book, we encourage the following of an analytical path that employs integrated field and laboratory studies with field surveys or remote sensing to determine what occurs on the ground and laboratory toxicity studies to test which hypotheses developed in the field are valid. If there is a regulatory concern, results of these findings can then be applied to informed resource management practices, such as remediation.

Our focus in this chapter resides on large-scale interactions that are intractably linked to higher levels of biological organization. These higher-order levels of ecosystem processes, such as those linked to carbon and nutrient cycling, are affected by releases of chemicals (Cain et al., 2014; Gimme and Hengeveld, 2014). For example, the release of greenhouse gases and other atmospheric chemicals, such as particulate matter, ozone, NO_x, and SO_4, will inevitably be complicated because of air, land, and water-use policies and practices that potentially affect habitats as an outcome of exposures to multiple environmental stressors (eg, air pollution, climate change, land mass fragmentation) on ecosystem function. Even with integrated field and laboratory studies, there will be inherent complexity and uncertainty in understanding the relationships between chemical exposure and effects at the community or ecosystem levels (Luo and Magee, 2011). Therefore, investigators may have to opt for less-than-absolute certainty when it comes to characterizing cause and effect linkages between environmental chemicals and biological effects. Regardless of this complexity and uncertainty, ecotoxicologists and applied ecologists have tools and methods available to evaluate these higher levels of biological organization and develop an understanding, at some level, of the possible chemical effects encountered in field settings within the context of these larger spatial scales. Such tools and methods are frequently deployed under the auspices of biological assessment.

SCIENTIFIC SUPPORT FOR COMMUNITY EFFECTS FROM CONTAMINANTS

There has been an extensive list of scientific studies showing that contaminants can affect entire communities. The greatest number of studies have occurred with aquatic micro- and macroinvertebrates because, compared with vertebrates, these organisms are easily maintained in the laboratory and can be tested in multiple-species combinations to derive some understanding of how a chemical may impact an entire community. Communities of fish and soil organisms have also been studied, but few if any studies have examined entire communities of birds or mammals. In this regard, scientists often find that some species are quite sensitive to the chemical of concern whereas others are less so and can tolerate higher concentrations before they show adverse effects, such as death.

To illustrate a scientific, nonregulatory investigation of how contaminants may affect communities, we turn to a study conducted by Schmidt et al. (2013) along streams in the Colorado Rockies. Their objective was to determine the relationship between metal contamination from mining activities and either insect larval abundance or the frequency of adults emerging from these streams. They identified linear gradients in total metal concentrations through the course of the streams and then sampled both larval abundance and adult emergence. The authors developed a complicated index based on the relative toxicity of metals and their concentration in the insects (see the study for more details) to assess a community-wide toxicity quotient. They found that larval abundance remained relatively constant and high until a threshold concentration of metal bioavailability was reached, and then it fell precipitously. In contrast, adult emergence fell drastically, even at very low metal concentrations and the authors concluded that adult emergence was a more sensitive marker of metal contamination than larval abundance. From an ecological perspective, a low frequency of adult emergence could result in fewer eggs laid for subsequent generations or a lower availability of food for terrestrial insectivores because these adults can be an important source of nutrition to animals living near or in the streams.

BIOLOGICAL ASSESSMENTS

Biological assessments are evaluations of the condition of an area of interest using surveys of the structure and function of a community's resident biota (eg, fish, benthic macroinvertebrates, periphyton, amphibians for aquatic habitats; see Karr and Chu, 1997; Barbour et al., 1999). As a long-used method, biological assessment has been routinely applied during the process of evaluating water quality. For example, water quality of streams or rivers flowing through a landscape may be affected by external conditions such as a leaky landfill or erosion because of connectivity among the different components (land, water, air) of a human-dominated landscape. Biological assessments provide resource managers with information on the condition of the biological community and help document the response of this community to environmental stressors. For surface waters, such as rivers, streams, wetlands, ponds, or lakes, biological assessments evaluate the condition of a waterbody by sampling species that spend all or part of their lives in that waterbody (Karr and Chu, 1997; Flotemersch et al., 2006; US EPA, 2004, 2007, 2011a). As such, field studies collect a snapshot biological survey of an area, wherein they collect a representative sample of the biological community found in a waterbody. A biological survey is a systematic method for collecting a consistent, reproducible, and reliable sample of the aquatic biological community in a waterbody. The same may be said of terrestrial and wetland habitats (Cooperrider et al., 1986; Pilz et al., 2006; US EPA, 2002, 2011b).

As an example of bioassessments, Martínez-Lladó et al. (2007) studied sediment-dwelling macroinvertebrates in Barcelona Harbor in the Port of Barcelona, Spain. Barcelona Harbor is one of the busiest harbors in the Mediterranean and the investigators wanted to see if either polycyclic aromatic hydrocarbons from oil or tributyltin, an antifouling biocide applied to ships' hulls to ward off barnacles and the like, affected the macroinvertebrate communities. They collected chemical and biological samples throughout the harbor and found a significant trend in community characteristics from the inner harbor outward to reference areas outside of the harbor. They used species richness, total organism density, the Shannon-Weaver (H') diversity index, and a biotic index that was a weighted composite of major invertebrate groups. Species richness was not always measurable, but species diversity increased from the inner harbor to the reference sites and the biotic index decreased in the same direction. This particular index gave higher values to taxa that were less sensitive to contamination and more likely to inhabit polluted sites. The species diversity and biotic indices correlated with tributyltin concentrations, but not polycyclic aromatic hydrocarbons, which led the authors to conclude that tributyltin exerted a toxic effect on invertebrate communities in the harbor sediments.

Recall the discussion of bioindicators in Chapter 3. These bioindicators are useful to determine if individual organisms are being affected by exposure to contaminants. There are also *biological indicators* to assess the heath of communities and ecosystems. When an ecotoxicologist is working with a new or poorly known species, he or she may compare the plasma characteristics of individuals thought to be exposed to those from reference sites and see if there is a difference, even when the appropriate values are unknown. Ecotoxicologists working at a community or ecosystem level will also compare community structure or species diversity between a reference site and one that is suspected or determined to be contaminated. Data collected at reference sites provide a basis or benchmark on the biological condition of a healthy site. Ideally, reference sites have not been perturbed, particularly within the context of anthropogenic stress. Unfortunately, human activities on aquatic systems have become increasingly widespread. Thus, reference sites likely display some degree of impact due to human activities, but they still represent a best approximation of natural conditions for comparison with other sites under investigation.

Although biological indicators were initially developed for assessments of surface water quality, other biological indicators have been developed to assess the condition of any environmental setting—freshwater, estuarine, or marine water bodies; wetlands of various types; or the wide range of terrestrial ecosystems. For example, if we look toward biological assessment of surface waters, an historic glimpse of biological indicators used in biological surveys includes:

- Fish (eg, trout, sunfish, perch, salmon)
- Benthic macroinvertebrates (eg, insects, snails, crayfish, worms)
- Periphyton (eg, algae)
- Amphibians (eg, frogs, salamanders)
- Macrophytes (eg, aquatic plants)
- Birds (eg, residents or seasonal migrants)

Biological assessments provide an empirical basis to evaluate the impacts of chemicals released to the environment, particularly those with no emission quality standards. In addition, the results of biological assessments address adverse effects potentially linked to exposures of chemical mixtures, conditions typically encountered in field settings. Biological assessments provide data to better characterize physical stressors, such as flow and siltation, or weather and indicate the extent to which biological stressors (eg, invasive species) may be contributing to reduced conditions of the community. In part, these biological indicators are evaluated in parallel with assessments of habitat condition; for example, for lotic habitats instream, riparian data, soil characteristics, or overall habitat quality are collected simultaneously with those gathered on biota in the system. Jointly, then, biological data and habitat data reflect the overall ecological integrity of an area under investigation, and provide a direct measure of both present and past effects of stressors on the biological integrity of an ecosystem.

As such, the biological assessment process integrates cumulative effects of different stressors from multiple sources, providing an ecosystem-level measure of their aggregate effect, and contributes empirical data regarding biological responses that support development of stressor-response models (see Chapters 12 and 13).

As another example of bioassessments, this time a semiaquatic, semiterrestrial bioassessment, Cesar et al. (2014) evaluated a site that was used to dump dredged sediments from the Santos Bay in Brazil. In this case, the authors did not use a reference site, but employed an integrated field and laboratory approach. The sediments came from the marine environment and were now transposed to land so the assessment could be said to affect both land and water. The authors evaluated sediment quality by quantifying both the structure (eg, sand, silt, clay) and chemical composition, including polycyclic aromatic hydrocarbons, chlorinated hydrocarbons such as polychlorinated biphenyls, pesticides, phthalates, metals, and nutrients. By now you should have a pretty good idea of why these chemicals were chosen. The investigators also conducted acute and chronic laboratory toxicity tests using amphipods (*Tiburonella viscana*) and sea urchins (*Lythechinus variegatus*), both of which were found in the bay. They found evidence of toxicity through all the methods they used, which suggested that the area for disposal of the dredged material was significantly altered with respect to sediment quality and could probably generate harmful effects on the local biota. After further analysis of the contaminant data, the researchers determined that most contaminant concentrations were below the limits established by Brazilian regulation. Now several questions could be asked, such as:

(1) Were the Brazilian standards set too high?

(2) Was there something about the dredging and other handling of the sediments that increased their toxicity above what would occur with undisturbed sediments?

(3) Is it germane to use marine aquatic organisms to test the toxicity of what are now soil samples?

Overall, biological assessments contribute to technical findings regarding an area's condition. If the area under investigation displays conditions reduced relative to those of the reference site, then the area is considered impaired. Yet, as critical as biological assessments are for detecting impacts on a system, such technical findings alone do not likely identify the cause or causes of impairment. Indeed, given the multiple stressors at play in these larger spatial scales and through time periods much greater than those achieved in traditional laboratory toxicity tests, the number of factors involved with exposure preclude such likelihoods. Take, for example, the study cited previously (Cesar et al., 2014). The authors did not really do an assessment of the macroinvertebrates on impacted site; that is, they did not sample the macroinvertebrate community in the sediment or soil to see if it showed signs of perturbation. Rather, they used organisms from the source of the sediment or soil. Did these organisms really reflect the multiple stressors in the impacted site?

Another study, conducted by Aazami et al. (2015), took a different approach in an aquatic environment. The authors wanted to assess the biological status of the Iranian Tajan River using a measurement (sometimes referred to as metric in these types of studies) on macroinvertebrates and a series of 28 physiochemical parameters and 10 habitat factors for 17 sites. The Shahid-Rajaee dam divides the Tajan River into upstream and downstream parts, with different land uses. A total of 18 metrics were used to represent four components of ecosystem quality: tolerance, diversity, abundance (total number of taxa, individuals, and members of the orders Ephemeroptera, Plecoptera, and Trichoptera), and composition of assemblages (percent of Ephemeroptera, Plecoptera, and Trichoptera). Results showed that the macroinvertebrate index decreased from upstream (with very good water quality) to downstream (with poorer water quality due to human activities). There were more industrial activities such as pulp mills and sand mining in the downstream part of the river. The authors admitted that, even with all of the variables measured, there was uncertainty in the relationships of habitat quality, pollution, and the physiochemical properties of highland versus lowland rivers.

There might be some differences in the concepts of exposures or effects compared with organisms or populations because levels of biological organization also capture differences in spatial and temporal scale. However, we can extend concepts of exposure and effects with only limited angst. Indeed, we need to only modify our expressions of exposure, less so effects on time and space, to begin integrating across levels of organization.

COMMUNITIES AND ECOSYSTEMS: EXPOSURES AND EFFECTS

Although an oversimplification, conceptually, a community refers to the collection of species interacting in a particular area. Indeed, some community ecologists debate whether a community is nothing more than an aggregation of species co-occurring in one area, or whether a community exists as an integrated unit (Newman and Clements, 2007; Mittelbach, 2012). Yet from an analytical perspective, when considered within the context of communities or ecosystems, exposure and effects are characterized by endpoints that are principally derivatives of those we considered at the organism level of biological organization, and that we subsequently linked to endpoints commonly characterized at population level of organization (Newman, 2001; Newman and Clements, 2007). For example, toxicity tests measure outcomes such as survival and fecundity, which easily translate to inputs called life table analysis—deployed by ecotoxicologists interested in characterizing effects on populations exposed to chemicals released to the environment (Rowland, 2003; Skalski et al., 2005; Barnthouse et al., 2007; also see Chapter 10). Similarly, much of our focus related to endpoints from exposure and effects simply acknowledge the spatial and temporal attributes captured by communities or ecosystems (Gurney and Nisbet, 1998). These community or ecosystem endpoints, however, are likely to be confounded by factors other than exposure to environmental chemicals. Thus, outcomes of toxicity tests, based on their relatively simple experimental designs, yet prominent in characterizing exposure and effects relationships at the organism level, may reveal only a portion of the effects of contaminants on communities and ecosystems.

Exposure Within a Community and Ecosystem Context

When we discuss exposure at levels of biological organization that display larger spatial coverage than that experienced by biota exposed to chemicals in laboratory beakers and when we consider exposures that span lifetimes and multiple generations—well beyond most time periods examined during laboratory tests—we begin to see the complexity and difficulties in interpretations of how to discern the effects of contaminants on communities and ecosystems separately from all other stressors that might affect these levels of organization. However, such outcomes are critical to our understanding of effects that may very well be inconsistent with those effects observed in laboratory toxicity tests. Only the interpretative context has changed.

At community and ecosystem levels, exposure is commonly considered relative to *load*, a term that goes beyond the equally common terms of exposure that usually occur as a measure of concentration; that is, a mass of chemical in a given volume of exposure medium (eg, mg/L in water, mg/kg in soil). In contrast, load is a quantitative estimate of exposure to a chemical released to the environment whose volume is much greater than that encountered in laboratory toxicity tests and generally absent from the experimental controls in place. A critical load simply extends the term's definition as a quantitative estimate of exposure to one or more chemicals below which significant harmful effects do not occur to sensitive biota in the environment. Thus, you can see how terms of concentration and load are linked and how distinctions between estimates of exposure follow from conditions specified by units of mass and volume applied in their derivation.

Although we will not cover the arithmetic for derivation of environmental loads, we need to briefly characterize environmental compartments frequently applied to the derivation of environmental loads—airshed and watershed—that identify frames of reference when considering exposure at the community or ecosystem levels of biological organization.

Airshed. Airshed finds much of its formal definition within regulatory contexts; however, that definition is consistent with underlying experimental and observational data applicable to the exposure considered by ecotoxicologists working with air quality. Simply stated, an airshed is the volume over an area of land in which airborne chemicals travel to reach a particular river, lake, bay, or other body of water given the area of the land surface. Not surprisingly, atmospheric loads of chemicals are usually expressed volumetrically. Airshed is roughly analogous to the concept of the watershed and also considers air basins as geographical areas that share the same air because of topography, meteorology, and climate. Ideally, an airshed is a portion of the atmosphere that behaves coherently with respect to dispersion of chemicals released to emissions. Most often, an airshed is considered to be a volume of air receiving emissions that affects a specific watershed or catchment. Thus, chemicals released into the air may be linked to their potential deposition to terrestrial ecosystems involving organisms dependent on soils and waterbodies for habitats (NRC/NAS, 2004; Sportisse, 2010). Although conceptually related, the definitions and characteristics of a watershed differ, sometimes subtly, other times markedly from those of the atmosphere.

Watershed. Everyone knows what a watershed is, right? Local definitions of watershed, however, may or may not agree with a commonly applied definition. A watershed is an area from which all precipitation flows to a single stream or set of streams (Brooks et al., 2012). Depending on local terminology, individual users might use terms often considered as synonyms of watershed: *catchment, catchment area,* or *drainage basin.* However, *watershed* will be the term of choice in this discussion. Watersheds are bound by drainage divides or boundaries between drainage basins wherein all the precipitation on opposite sides of a drainage divide will flow into different drainage basins. For example, the total land area drained by the Mississippi River constitutes its drainage basin, whereas that part of the Mississippi River drained by the Ohio River is the Ohio River watershed, and that part drained by the Missouri River is the Missouri River watershed (Fig. 11.2). A watershed consists of surface water—lakes, streams, reservoirs, and wetlands—and all underlying groundwater. As in our example of the Mississippi River drainage basin, larger watersheds may consist of large subunits, which in turn contain many smaller watersheds. It all depends on the frame of reference, principally the point of outflow; all land that drains water to the outflow point is the watershed for that outflow location. There are subtle differences between the terms *catchment* and *watershed;* the latter is usually used in hydrologic applications and has a specific characterization that reflects physical attributes such as elevations, whereas the former simply refers to an amount of water caught in a given area of a watershed. Catchments occur as land features or built structures (eg, a dam) that catch and hold water. Some catchments are open, connected to adjacent geographic features that assure linkages and flows of surface water. Other catchments may be closed; that is, limited with respect to surface runoff as you would observe with depressional wetlands or with groundwater seepage lakes that have no inlets or outlets. As such, watersheds simply reflect topographically-limited surface areas that link physical attributes or processes to their hydrology.

In the field, watersheds are significant geographic features; streamflow and water quality of waterbodies are affected by various factors, some human-induced, some not. Climatic variables, such as those related to precipitation, and variables describing water and sediment discharge, water storage, and evapotranspiration are commonly measured. Various hydrologic models may be used

Major US Watersheds

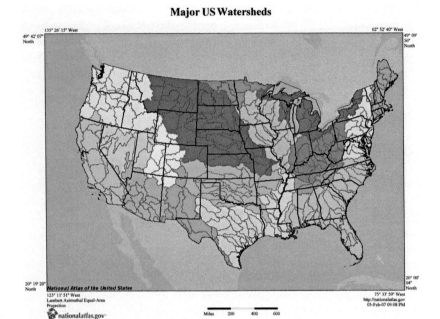

FIGURE 11.2   Major watersheds in the United States. Each of these major blocks of land contain many smaller watersheds and catchments with rivers or streams that eventually flow together. *Courtesy U.S. Geological Survey.*

to characterize a watershed's role within a larger landscape. Simply said, however, what happens on lands above a river-outflow point makes a difference in what we collect from discharge points at watershed boundaries. Rather like Las Vegas, don't you think?

Loads. Recall our definition of load. Indeed, from an ecotoxicological perspective, critical loads might be considered a landscape-based value analogous to those measures derived from toxicity tests performed in laboratory exposures. For example, atmospheric critical loads as characterized through various models, all based on soil health, represent a quantitative estimate of exposure to environmental chemicals below which significant harmful effects on specified sensitive elements of the environment do not occur. Critical loads for particulates and chemicals—metals, *volatile organic chemicals (VOCs),* and *semivolatile organic chemicals (SVOCs;* eg, phenols and phthalates), and *persistent organic pollutants (POPs)*—released to the atmosphere have been developed by various regulatory agencies throughout the world. Although the concept of critical loads is becoming increasingly apparent in the environmental literature, much of the regulatory focus, particularly historic efforts to establish air quality criteria in the United States, have been expressed in terms of concentration per unit volume, not in loads. Regardless of the atmospheric chemicals, critical loads for chemical constituents of ambient air are usually derived from one of three general model systems—empirical, simple mass balance, or dynamic. Each is based to varying extents on observational studies or the compilation and assembly of existing data rather than on integrated field and laboratory studies. However, data from these studies have been used in the technical development of these model systems. Critical loads may be derived from observational data for a component of the landscape; for example, a surface water system such as a lake and its adjacent habitats. For such a system, physical habitat attributes such as surface water pH and soil physiochemical characteristics—say, nitrate and acid-neutralizing capacity—are measured along with vegetation biomass, cover, and species composition, and then observed vegetation responses are linked to habitat attributes. In this example, then, we might characterize a critical load for nitrate based on observed vegetative biomass. Regardless of methods deployed to estimate critical loads, each depends on available observational data. Empirical critical loads require observational data of known quality for both exposure (eg, deposition values for chemicals of concern) and effects (eg, spatially linked response data collected for vegetation during the period of deposition).

Similarly, chemical loads to surface waters are often managed through setting limits for an entire watershed. For example, within the United States, *total maximum daily loads (TMDLs)* are the daily load of a given contaminant that a given environment can be subjected to each day without showing any degradation (NRC/NAS, 2001). Groundwater is also regarded as an integral component of a watershed because watersheds are defined by their natural hydrology, which reflects the potential interconnectedness of groundwater and surface water, perhaps best displayed by riparian wetlands and groundwater recharge to rivers and streams.

There is a very complex relationship among the various components of water (Fig. 11.3). Water can evaporate from surface waters and land and enter the atmosphere. Alternatively, water from the atmosphere comes down as precipitation, initially striking land or immediately becoming part of the surface water found in rivers, streams, lakes, ponds, wetlands, etc. Much of this surface water enters larger bodies of water and eventually makes its way to oceans or seas. However, a good portion also percolates through the soil or sediments into deeper layers known as groundwater. *Recharge areas* are sufficiently elevated so that water tends to enter soil and recharge the aquifers. Some of this groundwater is close enough to the surface in unconfined aquifers that it can return to surface waters during drought or through tapping into the aquifers with wells. Discharge areas tend to be lower and promote the movement of water in unconfined aquifers back to the surface as seeps or upwellings. Deeper groundwater may be beyond the reach of all but the deepest wells and be stored for hundreds of thousands of years (USGS, 2015). All forms of water except perhaps those in very deep, confined aquifers are potentially open to contamination. Even this water might be contaminated through deep in-ground activities, such as mining.

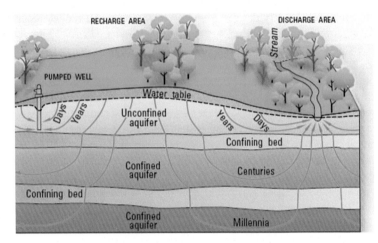

FIGURE 11.3   A graphic model of the relationships among atmospheric, surfacewaters, and groundwaters. See text for explanation. *Courtesy U.S. Geological Survey.*

Effects at Community and Ecosystem Levels

Results of any biological assessment provide descriptions—qualitative or quantitative—of relationships between environmental chemicals and biological or ecological endpoints. For our purposes, we are interested in endpoints as quantitative measures of an observed or measured effect; it is a measurable environmental characteristic that may be related to a characteristic interpretable within the context of resource management. As we have mentioned, work focused on communities and ecosystem endpoints may be measured in field, laboratory, or integrated studies. To ease the potentially long list of activities that could be pursued in field studies, ecotoxicologists have identified a menu of generic endpoints—ecological entities and attributes that have been shown to indicate stress or exposure at higher levels of ecological organization.

Community-Level Effects

Examples of community-based endpoints that are routinely incorporated in biological assessments are listed in Table 11.1 as general classes that can be refined, depending on the community being surveyed. Among the most commonly used community characteristics are the number of species, species evenness, and species diversity. Sometimes key indicator species may receive particular attention. Indicator species are those known to live in a region and are particularly sensitive to perturbations. For instance, in streams and small rivers, insects, either larvae or adults, within the orders Ephemeroptera (mayflies), Trichoptera (caddisflies), and Plecoptera (stoneflies) are frequently used as bioindicators of good water quality (Fig. 11.4; see Smith, 2001 for life history backgrounds

TABLE 11.1 Endpoints Commonly Evaluated in Field Surveys Completed as Part of Biological Assessments

Level of Biological Organization	Endpoints Potentially Serving to Measure Effects
Community	• Number of species • Species evenness/dominance • Species diversity • Pollution indices • Community quality indices • Community type
Ecosystem	• Biomass • Productivity • Nutrient dynamics

(A)   (B)   (C)

FIGURE 11.4 Some sensitive stream macroinvertebrates often used as indicator species in contaminant studies: (A) Ephemeroptera (mayflies), (B) Trichoptera (caddisflies), and (C) Plecoptera (stoneflies). In each set, an adult is shown on the top and a larva or immature individual on the bottom. Caddisfly larvae build small tubes from pebbles that they use to hide themselves.

for these orders and their representative species). These and other parameters are convenient for summarizing data collected from biotic surveys, and are easily measured, appropriate for wider geographic scales typically linked to areas of interest, and serve to integrate outcomes from acute and chronic exposures to environmental chemicals. For most biota considered in designed field sampling efforts—vertebrates, invertebrates, or vegetation—sample statistics, particularly variance, benefit interpretation of effects on environmental health and possible contaminant exposure. The season may be very important in affecting abundance or life-stage dynamics; it would not be meaningful for investigators to survey woodland birds in one forest or macroinvertebrates in a wetland during May and another site in November in temperate areas, for example.

While these metrics are widely used for a variety of habitats, they may not be very diagnostic of specific problems nor well-standardized with respect to sampling methods. For instance, stoneflies, mayflies, and caddisflies are sensitive to many water-quality factors including pH, oxygen concentration, and turbidity in addition to contaminant exposures. Differences between two sample sites in indicator species may identify that something is happening, but not what that something may be. Thus, attempts to parallel reference and contaminated sites for as many factors as possible is often essential for meaningful comparisons. However, indicator species have been chosen for their ease of sampling as well as for their sensitivity to perturbations, so intensive sampling and data summarization will likely not be necessary to describe community changes.

Another type of community-level endpoint, generally termed an index of community quality, may be indicative of chemical effects or of habitat quality. For instance, observational studies may suggest that the replacement of one set of species with another was linked to conventional organic pollution (ie, sewage and similar effluents (Hynes, 1960)). The concept of indicator species is similar to community indices in that they are both intended to describe the state of communities relative to anthropogenic effects. The presence or abundance of a species considered to be either pollution-sensitive or pollution-tolerant has been used to indicate the status of a community. Other indices of generic community quality, such as the *index of biological integrity (IBI)* may serve as indicators of the state of communities because they are sensitive to physical habitat quality as well as to chemicals released to surface waters (see Karr, 1987; Barbour et al., 1999). All of these community-quality indices reduce to one number in the information obtained from a biotic survey. While this may sound simplistic, if an index is well characterized for a region, such as IBI, then it can be used to indicate how chemical effects compare with effects from other disturbances in similar communities. IBI is frequently implemented under the auspices of biological assessment and will be briefly considered as an example of a community-level evaluation procedure.

Index of Biological Integrity

In its simplest characterization, IBI is a scoring system that is often applied to biological assessments for rivers, streams, and wetlands. For example, in wetlands, depending on geographic location and wetland type, both plant and macroinvertebrate communities will be studied, wherein a selected number of metrics will be used to frame an empirical basis of wetland quality. There have been some efforts to develop IBIs for terrestrial environments (Diffendorfer et al., 2007). As such, each metric is evaluated based on specimens collected in the field, and then they are identified and their abundance is estimated. Subsequently, a score will be determined, based on the results of these metrics.

Scores derived from multiple metrics are then combined to yield a total score—the IBI. The higher the score, the greater the likelihood an area is in excellent condition. Some would say, "the wetland or upland is healthy." The level of analysis required in such an assessment depends on the extent to which a habitat was sampled as part of the biological assessment process; for example, an IBI score developed for macroinvertebrates and vegetation would require additional analysis than one based solely on macroinvertebrates, given the relationships between the IBI for macroinvertebrates and that IBI based on plants might also be considered jointly. Depending on the level of analysis—birds and amphibians may also be considered in completing a biological assessment for wetlands—ecotoxicologists and applied ecologists may characterize habitats as being in "excellent," "moderate," or "poor" condition, and the system's health status is recorded so subsequent investigations could better gauge status and trends with respect to long-term outcomes potentially linked to environmental perturbations, such as a release of chemicals in the adjacent lands within the watershed. With our increased attention to an ecosystem level of biological organization, we begin to appreciate the complexity of the systems with which we work, as our next section suggests.

Glennon and Porter (2005) developed an IBI based on forest birds in Adirondack Park, New York, along a gradient of human impact. They created the IBI by placing birds into 12 different guild categories and scoring sample areas according to relative representation of specialist versus generalist guild types. The divisions were made along four categories: food, origin (native or exotic), nest placement, and primary habitat, with subdivisions within most categories. The authors found significant differences in total, functional, compositional, and structural integrity on five land-use types ranging from hamlet to wilderness. In all cases, integrity was lowest in towns and increased in wilderness areas. They also found that biotic integrity showed strong groupings amid five land-use classes. Biotic integrity was strongly related to roadlessness and level of human developments. Specific contaminants were not examined in this study, but it would not have been difficult to include a contaminant metric, which probably would have agreed with the human development versus wilderness impact.

However, before moving to our next section, we should briefly consider the analysis tools used to complete any biological assessment. Besides the metrics of IBI, many other tools of data analysis are available and may be applied to characterize communities in general, including those evaluated with respect to effects of occurring chemical exposures. For example, in any community, measures of species abundance, species richness, and species diversity may be observed. These are community-level endpoints commonly derived by ecologists in their studies of systems, be those perturbed or largely unaffected by anthropogenic effects. If an ecotoxicologist is interested in evaluating effects on species diversity that are potentially associated with chemical exposure, a commonly used diversity index may be applied in the community analysis, such as the Shannon-Weaver Index (H′), which provides an understanding of community structure. In general, a diversity index provides information about the number and distribution of species in a community. Various measures of community characteristics have been developed and can be applied to ecotoxicological investigations (Pielou, 1977).

In general, community ecologists interested in evaluating chemical effects focus on long-established endpoints, just as population ecologists do if chemical exposures are of interest for their sustainability of populations evaluations. Thus, population ecologists and community ecologists focus their ecological study on individual populations and interactions of species within natural communities, respectively. Ecotoxicologists focus on ecological patterns and dynamics at population, community, and ecosystem levels of organization, which enables them to link responses in populations, underwritten by organism-level observations for individual species, to communities and ecosystem responses linked to ecological patterns and dynamics of community assemblages that co-occur in the same geographic area.

Effects at the Ecosystem Level

Whereas biological communities are composed of species that occur together, ecosystems include biological communities and abiotic properties that are more generally related to the exchange of energy and nutrients among functionally defined groups of organisms and their environments. The most commonly measured ecosystem properties are the biomass of the system or its components (eg, trophic levels), productivity of the system or its components (eg, primary and secondary production), and nutrient dynamics (eg, nitrogen-mineralization rates).

In an extensive review of the effects of contaminants on ecosystem functioning in marine and estuarine habitats, Johnston et al. (2015) identified 264 relevant papers published across a wide range of contaminants. They found that toxic contaminants generally altered marine ecosystem functioning by reducing productivity and respiration. Effects varied, however, according to the type of contaminant and the component(s) of the system studied (eg, particular trophic levels, functional groups, or taxonomic groups). These studies were highly biased toward planktonic communities rather than overall biodiversity. The studies rarely included a measure of biodiversity and rarely interpreted their findings within an ecosystem function context. Many studies that included multiple communities within an ecosystem were more likely to find no effect of contamination, and the authors suggested that this might be due to ecological interactions among members of these communities. A strong majority of studies identified negative impacts of contaminants on primary production, which led the reviewers to believe that contaminants may greatly affect the ecosystem services and benefits provided by these systems. They recommended that future studies should focus on the relationships between biological diversity and ecosystem functioning. The understanding of chemical contaminant effects will remain incomplete until direct measures of both variables are undertaken within multiple ecosystems, and Johnston et al. (2015) suggested that productivity and respiration may serve as key endpoints.

We have spent a lot of this book discussing vertebrate and macroinvertebrate ecotoxicology. However, we have not spent much effort on the ecotoxicology of bacteria or microscopic algae. Yet both are incredibly important components of any ecosystem and lend themselves well to studies on respiration and productivity. They can also be strongly affected by contaminants such as pharmaceutical and personal care products (PPCP) and pesticides designed as microbiocides. In aquatic systems, a whole bacteria-driven ecosystem exists in a micrometer-thin film and is often surrounded by a self-produced matrix that keeps it together. Rosi-Marshall et al. (2013) looked at the effects of PPCP on the organisms living in these biofilms. They measured in situ responses of stream biofilms to six common pharmaceutical compounds: caffeine; cimetidine (a histamine treatment for gastrointestinal disorders such as ulcers); ciprofloxacin (a broad spectrum antibiotic); diphenhydramine (an antihistamine used to treat seasickness and some symptoms of Parkinson's disease); metformin (used in treating diabetes); ranitidine (used for treating acid reflux and ulcers); and a mixture of each by deploying pharmaceutical-diffusing substrates in streams in Indiana, Maryland, and New York. They found fairly consistent results across seasons and regions. On average, algal biomass was suppressed from 4% to 22% relative to controls by caffeine, ciprofloxacin, diphenhydramine, and the mixed treatment. In addition, respiration of the biofilm was significantly suppressed by caffeine (53%), cimetidine (51%), ciprofloxacin (91%), diphenhydramine (63%), and the mixed treatment (40%). In autumn in New York, photosynthesis was also significantly suppressed by diphenhydramine (99%) and the mixed treatment (88%). Compared with controls, diphenhydramine exposure significantly altered bacterial community composition and resulted in significant increases in *Pseudomonas* sp. and decreases in *Flavobacterium* sp. in all three streams; both of these genera comprise hundreds of species and no attempt was made to identify them. From a purist attitude, the effects on respiration, biomass, and photosynthesis would be considered ecosystem effects, but the changes in bacterial species composition would be more appropriately classified as community effects and you can see how these measurements are related. The authors concluded that PPCP, alone or in combination, influenced stream biofilms, which could have consequences for higher trophic levels and important ecosystem processes.

Ecosystem properties can be difficult to measure within the time frame most often available for a biological assessment, and unless field-study design criteria allow sufficient time for implementation, empirical data will likely be highly variable and not particularly diagnostic. Interpretation will likely require a better grasp of uncertainties captured in a snapshot of a complex system at a single point in time, but may be broadly applicable and sufficient for informing resource management decisions.

Regardless of technical issues, largely dependent on time or, better said, the lack of time, evaluation of complex systems such as those observed at community and ecosystem levels of biological organization, resource managers in the regulatory arena frequently incorporate biological assessments in environmental quality programs. For example, as we noted previously in this chapter, water quality has a relatively well-established set of tools such as IBI for conducting biological assessments. Although the level of complexity likely precludes routine deployments at the ecosystem level of biological organization, such methods can be used to assess aquatic life and characterize surface waters as either impaired or not, provided sample designs are sufficient. Regardless of the scale, biological assessments provide empirical measures of stress and exposure by directly assessing the response of communities in the field. Applying tools such as IBI at large spatial scales, such as those typically encountered with ecosystems, likely require the development of study designs that are focused on community types across a series of ecosystems, wherein measurable changes in the biotic community—for example, the return of native species, decreases in anomalies and lesions in fish and amphibians, or decreases in pollution-tolerant species paired with an increase in pollution-sensitive species—can readily be communicated to resource managers and an interested public.

In addition, as response-stressor relationships are documented, biological assessments in combination with chemical stressor data can be used to help predict and track environmental outcomes of management actions. As such, a cause and effect analysis may be initiated, initially through a process targeted to identify a stressor or combination of stressors that cause biological impairment. The ability to identify stressors and evidence supporting those findings is a critical step in developing strategies that will improve the quality of resources vulnerable to those stressors, such as chemicals released to the environment. To identify stressors principally linked to adverse effects observed in biota across the range of exposures at play, the identification necessarily draws upon a wide variety of disciplines in a number of environmental areas: ecology, biology, geology, geomorphology, chemistry, statistics, and ecotoxicology. In completing such a stressor identification study, new avenues open that benefit our understanding of complex adaptive systems.

Communities and Ecosystems: Network Analysis and Modeling

Recent developments in the quantitative analysis of complex, nonbiological networks, based largely on the mathematics of graph theory, have been rapidly translated to studies of ecological networks. Many of the networks stem from conceptual food-web models and can be applied to exposure analysis. Studies of these exposures are often developed to evaluate the transfer of chemicals from foods to animals that consume those foods to deposition into the environment as waste products. Other conceptual models that cover atmospheric deposition or other networks are common. Ecological networks captured through conceptual models such as these consider the more complex structural and functional systems of communities and ecosystems as they demonstrate small-world topology, highly connected hubs, and modularity. Network analysis of biological systems, from subcellular to ecosystem levels of biological organization will only increase in their application to ecotoxicological studies.

Ecological networks are representations of the interactions that occur between species within a community or between communities within an ecosystem (Fig. 11.5). Traditionally, interactions encountered in such networks include symbiotic relationships such as competition, mutualism, and predation. Network properties of particular interest are generally focused on stability and structure. Yet, ecotoxicologists originally deployed such conceptual models to better understand how chemicals moved through the environment from abiotic compartments such as soils, sediments, and waters to biological compartments—terrestrial and aquatic plants and animals (eg, Cantwell and Forman, 1993). Such conceptual models as in Fig. 11.5 clearly suggest that tracking flows of chemicals between exposure matrices from abiotic compartments to biological components of the system will require forecasting models that might predict how chemicals would be "seen" by any lifeform during exposure. Such a goal has proven elusive, yet building from our increased understanding of ecotoxicology in general, and tracking chemicals as they move through the environment has benefited from developments in other fields. For an increasing number of ecotoxicologists, graph theory and network analysis have become integral tools in the analysis of a flow network, and because such analysis has benefited from empirical data derived from integrated field and laboratory studies and

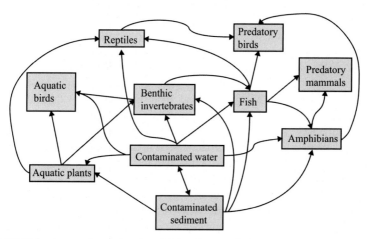

FIGURE 11.5   An aquatic conceptual model that illustrates linkages among biota, contaminated sediments, and surface waters.

biological assessments in the past, present, and future. Ecotoxicology can also benefit from these recent advancements in ecological network research.

As such, ecological network analysis considers complexity, especially that observed in ecosystems and communities, and how that affects ecosystem functioning. Depending on the analyst's perspective, ecological networks may be categorized and subdivided into various types. For ecotoxicologists, much of their work builds on traditional food webs, as illustrated in Fig. 11.5. Among a long list of items to pursue, future research in application of network analysis to ecotoxicological questions should focus on increasing our understanding of similarities and differences in chemical fate and effects among various communities that occur across ecosystems. Not surprisingly, based on what we already have discovered, these questions would inevitably require matching efforts in comparative ecotoxicology, so we can better address issues entangled with outcomes of organism- or population-centered studies. As such, ecotoxicologists would gain from developing an increased understanding of mechanisms underlying observations in the field. For example, analysis of compartments within network structures would allow ecotoxicologists to gain an increased understanding of food webs and mechanisms underlying chemical transfers between compartments within a network, or we could enhance our characterization of differences in chemical transfers linked to the underlying differences in abiotic factors that influence such transfers. Such refinements will enable us to improve our capability to forecast chemical transfers between compartments and reduce or mitigate impacts and consequences of increasing environmental perturbations, such as release of chemicals to the environment, and, of equal significance, improve our capability to discriminate among multiple stressors.

Biological assessment, particularly when completed within the context of integrated field and laboratory study, is essential to our understanding of chemical effects on biological systems at any level of organization. Of equal importance, biological assessment or monitoring provides critical empirical data that benefits management and conservation of ecosystems. However, current biological assessment and monitoring approaches are incompletely developed and tools, such as those of ecological network analysis, need to be developed and placed into service, particularly in fields of study focused on stressors—chemical, physical, or biological—and their interactions with biota when exposed in complex natural systems.

The analysis of ecological networks potentially offers insights into community and ecosystem responses to chemicals released to the environment; such network approaches likely yield capabilities for our interpretation of current ecological conditions. More importantly, if we augment biological assessment and monitoring activities by implementing those studies within a framework of ecological network analysis that relies on empirical data derived from direct observation and underwritten by designed laboratory studies targeted to data gaps identified in preliminary studies, this would potentially allow us to characterize ecological networks across environmental gradients encountered within various ecosystems. Future enhancements to the analysis of ecological networks may require our adopting emerging technologies and analytical approaches (eg, remote sensing capabilities to biosensors to nanosensors), enabling biological assessment, and monitoring to move beyond today's implementation practices and address the many ecological responses that can only be understood from a network-based perspective.

STUDY QUESTIONS

1. What types of questions would ecotoxicologists interested in communities ask that would be different from those who are more interested in individual organisms?
2. True of False. In some ways, community ecotoxicology is an extension of ecotoxicology at the population level.
3. In general, what happens to the role of multiple stressors as we venture from population to community to ecosystem scales?
4. True or False. Lab studies no longer have a function in studying toxicity for scales larger than the population because we cannot manufacture anything as complex as an ecosystem in the laboratory.
5. If ecologists and ecotoxicologists are interested in such things as emigration, immigration, births, deaths, and age structures, what are comparable things that interest community-based scientists?
6. Pick a group of organisms (eg, macroinvertebrates, plants, fish, amphibians, frogs) and develop a design for a biological assessment of this group in an ecotoxicological context.
7. What is the difference between a bioindicator used for studying the responses of individual organisms to contaminant exposure and a biological indicator used for the same purpose but in a community context?
8. From a geographic perspective, is there a 100% overlap between airsheds and watersheds? Why or why not?
9. True or False. Some sources of groundwater may spend millions of years underground without ever surfacing.
10. Identify some of the environmental measures that would be of interest in the study of how contaminants affect ecosystems.

REFERENCES

Aazami, J., Sari, A.E., Abdoli, A., Sohrabi, H., Van den Brink, P., 2015. Assessment of ecological quality of the Tajan River in Iran using a multimetric macroinvertebrate index and species traits. Environ. Manage. 56, 260–269.

Barbour, M.T., Gerristsen, J., Snyder, B.D., Stribling, J.B., 1999. Rapid Bioassessment Protocol for Use in Streams and Wadeable Rivers: Periphyton, Benthic Macroinvertebrates, and Fish, second ed. U.S. Environmental Protection Agency, Office of Water, Washington, DC, p. 344, EPA 841-B-99-002.

Barnthouse, L.W., Munns Jr., W.R., Sorensen, M.T., 2007. Population-Level Ecological Risk Assessment. CRC Press, Boca Raton, Florida, p. 376.

Brooks, K.N., Ffolliott, P.F., Magner, J.A., 2012. Hydrology and the Management of Watersheds, fourth ed. Wiley-Blackwell, New York, p. 552.

Cain, N.L., Bowman, W.D., Hacker, S.D., 2014. Ecology, third ed. Sinauer Associates, Inc., Sunderland, Massachusetts, p. 596.

Cantwell, M.D., Forman, R.T.T., 1993. Landscape graphs: ecologication modeling with graph theory to detect configurations common to diverse landscapes. Lands. Ecol. 8, 239–251.

Cesar, A., Lia, L.R.B., Pereira, C.D.S., Santos, A.R., Cortez, F.S., Choueri, R.B., et al., 2014. Environmental assessment of dredged sediment in the major Latin American seaport (Santos, Sao Paulo—Brazil): an integrated approach. Sci. Total Environ. 497, 679–687.

Cooperrider, A.Y., Boyd, R.J., Stuart, H.R. (Eds.), 1986. Inventory and Monitoring of Wildlife Habitat. U.S. Department of the Interior, Bureau of Land Management, Denver, Colorado, p. 858.

Diffendorfer, J.E., Fleming, G.M., Duggan, J.M., Chapman, R.E., Rahn, M.E., Mitrovich, M.L., et al., 2007. Developing terrestrial, multi-taxon indices of biological integrity: an example from coastal sage scrub. Biol. Conserv. 140, 130–141.

Flotemersch, J.E., Stribling, J.B., Paul, M.J., 2006. Concepts and Approaches for the Bioassessment of Non-Wadeable Streams and Rivers. US Environmental Protection Agency, Cincinnati, Ohio, p. 245, EPA 600-R-06-127.

Gimme, H.W., Hengeveld, R., 2014. Autecology: Organisms, Interactions and Environmental Dynamics. CRC Press, Boca Raton, Florida, p. 484.

Glennon, M.J., Porter, W.F., 2005. Effects of land use management on biotic integrity: an investigation of bird communities. Biol. Conserv. 126, 499–511.

Gurney, W.S.C., Nisbet, R.M., 1998. Ecological Dynamics. Oxford University Press, Oxford, p. 352.

Hynes, H.B.N., 1960. The Biology of Polluted Waters. Liverpool University Press, Liverpool, England, p. 202.

Johnston, E.L., Mayer-Pinto, M., Crowe, T.P., 2015. Chemical contaminant effects on marine ecosystem functioning. J. Appl. Ecol. 52, 140–149.

Karr, J.R., 1987. Biological monitoring and environmental assessment: a conceptual framework. Environ. Manage. 11, 249–256.

Karr, J.R., Chu, E.W., 1997. Biological monitoring: essential foundation for ecological risk assessment. Hum. Ecol. Risk Assess. An Intl. J. 3, 993–1004.

Luo, J., Magee, C.L. 2011. Detecting evolving patterns of self-organizing networks by flow hierarchy measurement. Complexity. http://dx.doi.org/10.1002/cplx.20368. Published online in Wiley Online Library.

Martínez-Lladó, X., Gibert, O., Marti, V., Diez, S., Romo, J., Bayon, M., et al., 2007. Distribution of polycyclic aromatic hydrocarbons (PAHs) and tributyltin (TBT) in Barcelona harbour sediments and their impact on benthic communities. Environ. Pollut. 149, 104–113.

Mittelbach, G.G., 2012. Community Ecology. Sinauer Associates, Inc., Sunderland, Massachusetts, p. 400.

NSA/NRC National Research Council/National Academy of Science, 2001. Assessing the TMDL Approach to Water Quality Management. Committee to Assess the Scientific Basis of the Total Maximum Daily Load Approach to Water Pollution Reduction, Water Science and Technology Board, National Research Council; Board on Environmental Studies and Toxicology; Board on Atmospheric Sciences and Climate; Division on Earth and Life Studies. National Academy Press, Washington, DC, p. 122.

NRC/NAS National Research Council/National Academy of Science, 2004. Air Quality Management in the United States. Committee on Air Quality Management in the United States;

Board on Environmental Studies and Toxicology; Board on Atmospheric Sciences and Climate; Division on Earth and Life Studies. National Academy Press, Washington, DC, p. 426.

Newman, M.C., 2001. Population Ecotoxicology. John Wiley & Sons, Inc., New York, p. 242.

Newman, M.C., Clements, W.H., 2007. Ecotoxicology: A Comprehensive Treatment. CRC Press, Boca Raton, Florida, p. 880.

Pielou, E.C., 1977. Mathematical Ecology. John Wiley & Sons, Inc., London, p. 385.

Pilz, D., Ballard, H.L., Jones, E.T. 2006. Broadening Participation in Biological Monitoring: Handbook for Scientists and Managers. Gen. Tech. Rep. PNW-GTR-680. U.S. Department of Agriculture, Forest Service, Pacific Northwest Research Station, Portland, OR, p. 131.

Rosi-Marshall, E.J., Kincaid, D.W., Bechtold, H.A., Royer, T.V., Rojas, M., Kelly, J.J., 2013. Pharmaceuticals suppress algal growth and microbial respiration and alter bacterial communities in stream biofilms. Ecol. Appl. 23, 583–593.

Rowland, D.T., 2003. Demographic Methods and Concepts. Oxford University Press, Oxford, England, p. 546.

Schmidt, T.S., Kraus, J.M., Walters, D.M., Wanty, R.B., 2013. Emergence flux declines disproportionately to larval density along a stream metals gradient. Environ. Sci. Technol. 47, 8784–8792.

Skalski, J.R., Ryding, K.E., Millspaugh, J., 2005. Wildlife Demography: Analysis of Sex, Age, and Count Data. Academic Press, New York, p. 656.

Smith, D.G., 2001. Pennak's Freshwater Invertebrates of the United States: Porifera to Crustacea, fourth ed. John Wiley & Sons, Inc., New York, p. 664.

Sportisse, B., 2010. Fundamentals in Air Pollution: From Processes to Modelling. Springer, New York, p. 299.

US EPA US Environmental Protection Agency, 2002. Methods for Evaluating Wetland Condition: Using Amphibians in Bioassessments of Wetlands. EPA-822-R-02-022. Office of Water, U.S. Environmental Protection Agency, Washington, DC, p. 41.

US EPA U.S. Environmental Protection Agency, 2004. Wadeable Streams Assessment: Benthic Laboratory Methods. EPA 841-B-04-007. U.S. Environmental Protection Agency, Office of Water and Office of Environmental Information, Washington, DC.

US EPA U.S. Environmental Protection Agency, 2007. National Rivers and Streams Assessment: Field Operations Manual. EPA-841-B-07-009. U.S. Environmental Protection Agency, Washington, DC, p. 354.

US EPA U.S. Environmental Protection Agency, 2011a. 2012 National Lakes Assessment. Field Operations Manual. EPA 841-B-11-003. U.S. Environmental Protection Agency, Washington, DC, p. 234.

US EPA U.S. Environmental Protection Agency, 2011b. National Wetland Condition Assessment: Field Operations Manual. EPA-843-R-10-001. U.S. Environmental Protection Agency, Washington, DC, p. 485.

USGS U.S. Geological Survey, 2015. Groundwater discharge – the water cycle. <http://water.usgs.gov/edu/watercyclegwdischarge.html> (accessed 29.12.15).

Chapter 12

Modeling in Ecotoxicology

Terms to Know

Model
Modeling
Conceptual Model
Mathematical Model
Empirical Model
Deterministic Model
Stochastic Model
Simulation
Statistical Model
Quantitative Structure–Activity Relationship (QSAR)
Physiologically Based Toxicokinetics (PBTK)
Trimmed Spearman–Karber
Probit Analysis
Logistic Regression

INTRODUCTION

Models and modeling are two words that often strike fear in the hearts of many scientists, including biologists and even ecotoxicologists. As a discipline, ecotoxicology is highly dependent on a more than passing grasp of biological systems at any level of organization, and modeling these systems is critical to our understanding and better living within the bounds of that system. Regardless of this common ground in science, models and the process of building or simply working with models is something we all do to varying degrees. In part, our environmental futures lie in developing predictive and forecasting tools—models—that yield outcomes which benefit resource management. For some of us, the use of models is subtle, more similar to using tools as one would in crafting sentences with a word processor or developing spreadsheets for compiling or assembling data collected in laboratory or observational studies. Others working within the discipline jump eagerly at research opportunities focused on developing models in their various guises, and inevitably discovering the range of outcomes that such modeling efforts yield.

Ecotoxicology Essentials.
© 2016 Elsevier Inc. All rights reserved.

In this chapter, we focus on models and modeling within the context of problem solving in general, and we provide an overview of how models and modeling are utilized as tools in ecotoxicology. We will also present an example of how models and modeling may be applied to evaluating environmental contaminants. In a case study, we will consider polychlorinated biphenyls (PCBs) in freshwater habitats and entanglements among fish and humans exposed to PCBs as a result of recreational fishing.

What is modeling—or more precisely for our focus on ecotoxicology, what is scientific modeling? An Internet search (ie, the Encyclopedia Britannica, http://www.britannica.com/science/scientific-modeling) yields the definition that *scientific modeling* is characterized as a process yielding conceptual, physical, or mathematical representations of real phenomena that are difficult to observe directly. Modeling is a central component of modern science. Simply stated, scientific models are at their best approximations of the objects and systems that they represent—they are not exact replicas. Rather, models are our "best guesses" and we work toward improving and refining models so their performance matches our goals, which are often resource management goals. Models cover the spectrum of real-world representations. In so doing, models take on various guises, often as visualizations of an object or system, that are built from experimental or observational data. Models attain varying degrees of success as abstractions or hypothetical constructs of natural phenomena or processes in an effort to forecast future states of that process or system. In other words, some models project real-world outcomes quite accurately, while others do not. Weather forecasts are perhaps the most familiar outcomes of scientific modeling efforts, and represent just one type of model built upon a foundation of data and knowledge of phenomena; for weather forecasting, we rely on recent and historic weather data and on mathematical analyses of this data or information sources to forecast future, hypothetical outcomes of these phenomena. In other words, we make guesses for tomorrow's weather by looking back at historic weather data and using those data to forecast upcoming weather using models that look ahead and project future temperatures and precipitation during a forecast period.

Models used by ecotoxicologists are not as widely appreciated by many outside the discipline, nor do they typically display a developmental history that matches those models deployed by meteorologists, but nonetheless, models of various types remain key components of ecotoxicological "toolboxes." In the next section, we consider a categorization of models in general, with emphasis on those of interest to ecotoxicologists. We then present an overview of the models deployed to characterize biological effects linked to exposures for chemicals released to the environment.

DIFFERENT MODELS, DIFFERENT REALIZATIONS

To grasp the range of model types, categorization tools are frequently used to characterize the different kinds of models, or technical activities involved in developing these models and using them as tools for data analysis or data integration. Though the categories are few, the terminology may be nuanced to elaborate on them, reflecting in part the discipline that is initiating model development. An ecotoxicologist, a computer scientist, a statistician, or a socioecologist may each see a model as something different, even if they are working on the same environmental problem. For example, we can consider a set of recreational fishing activities enjoyed by the public that was then directly linked to consumption of fish exposed to PCBs in the sediments and surface waters of freshwater lakes. We will visit that example in the FOCUS section later in this chapter, but first we look at those categories of models that are crosscutting with respect to models deployed in a range of scientific disciplines.

Most frequently, terms such as conceptual, physical, mathematical, statistical, and computer characterize categories of models; thus, for our immediate application, these terms provide a common currency shared among environmental scientists of various disciplines, particularly when working on studies of interest to resource managers. These broad categories reflect differences in levels of abstraction versus detail in a model construct. Cognitive differences among individuals working on developing and applying models to address environmental issues in general, and ecotoxicological questions in particular require each group of investigators to find common ground on how data and knowledge are shared among different scientific disciplines. Although there is no authoritative agreement that defines these terms, for ecotoxicologists, particularly those challenged to analyze and interpret data collected to characterize toxicity and how effects are linked to ecological structure and function, these terms provide the approximate scope of each category and are generally understood.

Conceptual Models

Conceptual models are largely qualitative and external representations built as cognitive structures and subsequently translated to wider audiences as schematic diagrams or graphic pictorials, the latter ranging from cartoon-like storyboards to highly artistic representations that characterize the state of a biological system, often having many biotic and abiotic components. Conceptual models are also frequently referred to as "mental models" in cognitive science, educational psychology, machine-learning literature, or computer science. Examples of conceptual models in this book include Fig. 4.7 in Chapter 4, which illustrates the principal of biomagnification, Fig. 8.10 in Chapter 8, which depicts the mercury cycle, and Figs. 12.4 and 12.7 in this chapter.

As conjured by scientists, conceptual models are representations that are shared by a given discipline and gain their coherence with the scientific knowledge of that discipline. Conceptual models are a common ground for all models that follow, be those external representations following mathematical or statistical formulations, constructed of material objects as in a physical model, or developed as analogies, schematic, or graphic objects with quantitative links to existing or empirical data gathered as part of ongoing investigations. Conceptual models may simply be considered idealized representations of real objects, phenomena, or situations that often serve as starting points for more detailed models that are subsequently developed in higher resolution for the system under study.

Conceptual models are often dynamic and adaptive constructs, particularly when a system is likely to respond to outside perturbations. Ecotoxicologists are usually interested in understanding relationships between biota exposed to environmental chemicals and their responses potentially linked to these exposures. Similarly, if a community-level response is considered in a conceptual model, the possible effects displayed by biota would necessarily require our understanding of community structure, within community relationships, and linkages among communities to more completely characterize chemical and effects relationships. In an ecosystem-level context, conceptual models might attempt to depict a representation of a system's biotic and abiotic components, as well as processes critical to a system's performance and sustainability that incorporates our assumptions about how components and processes are related. As such, a conceptual model, or potentially numerous, interrelated conceptual models when we are working with complex systems, may benefit from our gaining a more complete understanding of chemical effects by helping us identify gaps in our knowledge and refine working hypotheses regarding a system's form and function. Depending on the level of biological organization of primary concern—individuals within a species, populations, or communities and ecosystems—conceptual models built as storyboards may capture hierarchical structures representing chemical exposure and effects among individuals of a species, among populations of potentially exposed species, or at higher levels of biological organizations involving communities and ecosystems.

Physical Models

A *physical model* represents a physical construct whose characteristics resemble the physical characteristics of the modeled system. In the broad interests of ecotoxicology, an example of a physical model might be a three-dimensional representation of a proposed sewage treatment plant. Physical models often support and guide the work of environmental engineers, hydrologists, foresters, or agricultural scientists, whose focus lies in land use and water use or in developing artificial environments related to such purposes as using natural resources for energy development or agricultural use, whereas a biologist or an ecologist considers the same system as something to manage for the benefit of fish and wildlife under the auspices of habitat. Ecotoxicologists and applied ecologists work in the "in-between," most often addressing conflicts or outcomes linked to competing uses of a common resource.

Mathematical Models

Biologists, applied ecologists, or ecotoxicologists often rely on mathematical models to approach their work on investigating outcomes linked to competing uses of resources. Distinguishing between mathematical and physical models is awkward at times, particularly when a mathematical model underscores the design and operation of a physical model. In short, a mathematical model is simply a symbolic model whose properties are expressed in mathematical symbols and relationships that are sometimes critical in the construction of a physical model, particularly when we consider relationships between science and engineering within the context of resource development and environmental applications over a wide range of technologies. An example of a mathematical model familiar to anyone who has taken an ecology course is the equation that describes unrestrained or exponential growth of a population (12.1):

$$rN = dN/dt \tag{12.1}$$

where dN is the change in number of individuals, dt is the change in time, and r is the intrinsic rate of natural increase.

Mathematical models are commonly used to quantify results, solve problems, and predict behavior—a practice most often encountered by ecotoxicologists when statistical tools are applied to data analysis and interpretation. Thus, mathematical models and statistical models share a common language—mathematics—but may be quite distinct in their value as tools commonly used by ecotoxicologists. Mathematical models provide a description or summarization of important features of a real-world system or phenomenon in terms of symbols, equations, and numbers. Nonetheless, mathematical models are approximations and do not always yield what is actually measured or observed in the field or laboratory. Mathematical techniques are useful to model situations and understand real-world problems, often through a process that initially simplifies and translates a problem into mathematical language and then later translates a mathematical solution back into the real-world situation. Setting up the problem then forecasting and validating outcomes of the model are important steps toward solving a problem in the modeling process. Simply stated, *mathematical modeling* is the process of applying mathematics to a real-world problem with the goal of greater understanding the latter. The modeling process may or may not result in a complete solution of a real-world problem, but, if well implemented, outcomes of the modeling process will shed more light on the problem under investigation.

Mathematical modeling may be categorized into four broad approaches: empirical models, deterministic models, stochastic models, and simulation models. *Empirical models* are built by characterizing mathematical relationships among variables in the problem based on available data, focusing largely on crafting a model or perhaps multiple models that are best suited for describing real-world settings. For example, if you plotted data on population growth over time and determined a mathematical equation to fit that growth, you would have an empirical model. More often than not, such models may be developed as *deterministic models* wherein model outcomes are fixed by use of an equation or set of equations to forecast outcomes of an event or the value of a variable of interest; the equation for exponential growth is a deterministic model.

In contrast, we may apply tools of *stochastic modeling*, which simply takes deterministic modeling another step closer to real-world situations. In stochastic models, outcomes are not fixed, but rather are subject to randomness and probabilities of events occurring. For example, an exposure and effects setting can be incorporated into the equations within the model construct. If you modified an equation for exponential growth to include a factor of predation that could take a range of values based on a random number generator, you would have created a stochastic model. Practically speaking, stochastic models reflect the realities of real-world settings that occur with some probability rather than with certainty.

Simulation models most often occur as iterative implementations of an existing mathematical model or set of mathematical models—commonly stochastic constructs—that generate outcomes for "what-if" scenarios that are intended to portray alternative environmental settings. For example, ecotoxicologists considering exposure-effects between PCBs in sediment and accumulation of PCBs in fish would query a mathematical model to better understand such relationships. Simulation modeling relies on translating mathematical models into computer models, wherein a selected computer language, such as R or another comparable programming tool, enables us to consider nearly any possibility for a system under study. Computer models attempt to simulate a system under investigation, and may use a mathematical model to find analytical solutions to problems. These solutions will hopefully allow the investigator to predict the behavior of a complex system from a set of parameters and initial conditions. As such, computer models allow us to develop numerical models of real-world complexities using a modeling system or simulation language. The discipline of risk assessment often uses simulation models that contain stochastic factors of weather, population status, or other elements that can use randomized but restrained values and the investigator can tweak these factors to see which are most important to exposure.

Considering mathematical models under the auspices of simulation studies, we can shift from mathematical models that predominantly yield solutions at a point of time to those yielding numerical solutions through time. The latter models often stem from equations that determine how a system changes from one state to the next or how one variable depends on the value or state of other variables in the model.

As a discipline increasingly inhabited by specialists, computer modeling and simulation modeling have become invaluable in evaluating complex systems. As with any tool, however, we can get "lost in the weeds" of detail, if that detail does not address the issues and problems we are tasked with. However, when applied with restraint, mathematical models and computer simulations can yield greater understanding of a process and identify problem areas or obstacles in processes, contribute to evaluations of effects of perturbations on systems, or process changes linked to perturbations. They can also identify potential remedies to mitigation effects associated with system perturbations and evaluate impacts or changes associated with perturbations. These benefits of mathematical modeling and computer simulation can be accomplished via various types of models, each of which may be used depending on the type of available data applied ecologists or ecotoxicologists have. As such, discrete models reflect computer implementation focused on snapshots of a system at specific times (eg, a given age-sex class structure of a population), and continuous models would represent systems that change continuously over time (eg, modeling a population that is growing or diminishing with time).

Statistical Models

In contrast to the relatively unbounded universe of mathematical models, statistical models are a specific group of models focused on the characterization of actual data. Mathematical models may be statistical in their implementation, but not necessarily. For example, the idealized mathematical model of population growth is deterministic but not statistical because it is not derived from actual data. If we instead collected data on an expanding population over several years and used regression to describe the trend in growth, that would be a statistical model. However, ecotoxicologists, along with biologists and ecologists in general, are most often primarily concerned with statistical characterization of data gathered as part of their work. For example, an ecotoxicologist might be interested in characterizing estimates of growth of plants occurring in areas with increased concentrations of metals such as zinc and lead in soils, or their focus may be on characterizing the levels of PCBs in fish tissues collected from a lake bearing these chemical contaminants in sediments. There are a wide range of statistical tools available to ecotoxicologists that allow them to quantitatively characterize their field observations or technical findings from laboratory studies; these tools are adequately described in the many textbooks available on statistics (eg, D'Agostino and Sullivan, 2005).

BRIEF EXPLANATIONS OF SOME COMMON MODELS IN ECOTOXICOLOGY

Following this description of modeling, it might be useful to discuss a few of the major efforts to employ models in the discipline of ecotoxicology. As mentioned previously, many tools or methods are available to generate statistical models and these undoubtedly represent the most common form of modeling in ecology and ecotoxicology. However, there are some modeling efforts that are particularly useful in an ecotoxicological context.

Quantitative Structure–Activity Relationship Models

Hopefully when you were studying the earlier chapters that discussed specific families of contaminants you realized that there are commonalities in structure within these families. PCBs have the same backbone of two phenyl groups even though the number of chlorines vary and the behavior of the chemicals can be predicted based on the location and number of these chlorines.

Recall also, the increased toxicity that comes from some PCBs with a bay region. Even the structurally more diverse organochlorine and organophosphorus compounds share reactive groups that define their behavior and toxicity, but there are also other reactive elements such as hydroxyls that alter the specific behavior of the molecule. In other words, we can predict the toxicity of a contaminant based on the fine details of its structure to some extent. Only 300–400 or so of the more than 70,000 chemicals have had any toxicity testing at all. *Quantitative structure–activity relationship (QSAR)* models could be a great tool to predict, to some extent, the toxicity of the other 98–99% of the chemicals.

The main objective of QSAR is to observe and measure the biological responses of a set of molecules and relate the measured activity to some molecular structure on their surface. The product of QSAR will then produce useful regression equations, images, or classification models in either two-dimensional or three-dimensional form that relate their biological responses or physical properties to their molecular structure. One simplified way of thinking about QSARs is a regression approach following the relationship:

$$\text{Biological activity} = c1 + c2 + c3 + \ldots + d$$

where the c factors are physical predictors of the chemicals and d is the uncertainty in the equation.

In QSAR modeling, the predictors consist of physicochemical properties or theoretical molecular descriptors of chemicals. The QSAR response-variables in our context are measures of toxicity of the chemicals. QSAR models first summarize supposed or known relationships between chemical structures and biological activity in a group of chemicals and then predict the activities of new chemicals. QSAR is possible due to extensive databases that have cataloged the relationships between structural elements of organic chemicals and the responses they induce. These databases can be used by sophisticated software programs (eg, Escher and Hermens, 2002) to derive the predicted outcomes. In a very crude way, an environmental chemist can input data on the types and concentrations of chemicals found in a contaminated area and the software will produce an estimate of the general toxicity of the mixture. In some cases, the toxicity estimates can be refined if target organisms are well characterized as to their responses to contaminants.

A bit more on theory behind QSARS—biological activity can be expressed quantitatively as the concentration of a substance required to give a certain biological response, such as death—the well-known lethal concentration (LC50 or EC50). In addition, when physicochemical properties or structures are expressed by numbers (eg, mg/kg), one can find a mathematical relationship or quantitative structure–activity relationship, between the two. The mathematical expression, if carefully crosschecked, can then be used to predict the modeled response of other chemical structures to a generalized organism. Models can be refined by adding species-specific differences in sensitivity which sometimes may be obvious (eg, photosynthesis inhibitors will affect plants differently than animals) or they can be subtler, such as accounting for the greater dermal absorption of contaminants by amphibians or the ability of female vertebrates to pass chemicals to their offspring through maternal transfer. QSAR is often used in risk management to predict the possible effects of a mixture of chemicals on specific target groups (ie, fish, invertebrates, birds, etc.) This modeling approach is complex and, as much as we hate to say this, a thorough review of these methods is beyond the scope of this book. We refer interested students to a few good reviews of this model (Schultz et al., 2003; Christen et al., 2013).

Physiologically Based Toxicokinetics Modeling

Another group of models employed in medicine and toxicology is *physiologically based toxicokinetics modeling (PBTK)*. Toxicokinetics is the study of the uptake, distribution through the body, metabolism, and loss or depuration of chemicals, often on an organ by organ (or compartment) basis. PBTK modeling is a mathematical modeling technique that utilizes information on absorption, distribution, metabolism and excretion of synthetic or natural chemical substances in humans and other animal species.

PBTK can help ecotoxicologists and risk managers reach more accurate assessments of toxicity and risk due to exposures to contaminants. We most often base assessments of toxicity on external concentrations of chemicals in environmental matrices. However, it is the internal concentration that actually gives rise to the biological effective dose and, if we better understand how organisms handle these internal concentrations, we may be able to more fully understand toxicity. Thus, an understanding of the relationship between the external and internal concentrations of chemicals and how the internal concentrations exert responses can greatly aid risk assessments. From a practical perspective, PBTK may be used to explain differences in sensitivity due to species, gender, or age (Schultz and Hayton, 1999; Herr et al., 2010).

The premise behind PBTK is that the toxicity of chemicals may be related to how quickly they move through the body, how rapidly they are broken down, what their metabolites are, and what tissues they tend to invade. In medicine, it is useful to know how long a drug persists in the body so that optimal timing for taking medication can be determined, for example. PBTK models use differential equations to describe the behavior of a given chemical in the body. These equations use the principles of mass transport, fluid dynamics, and biochemistry to simulate the fate of a chemical in the body and then relate these dynamics to toxicity. They can also address questions concerning: (1) the relation between effects at high doses often given under controlled experimentation to the much lower field exposures, (2) expected effects due to differences in the duration of exposure as, for example, from continuous to discontinuous, (3) what differences may occur due to routes of administration, (4) interspecific differences in sensitivity or responses, and (5) differences within a species due to age, sex, or reproductive status. See Nichols et al. (1990) and Campbell et al. (2012) for more information and an application of PBTK.

Statistical Models

Perhaps the greatest occurrence of model use in ecotoxicology are the almost limitless statistical models generated by a plethora of scientific studies. Some statistical methods, such as analysis of variance or a student's t-test are used to determine if differences occur between two or more groups (eg, one group of plants exposed to lead in their soil versus a control group). These are not model builders per se. However, correlation and regression and their family of associated methods can be used to develop models with respect to the relationships among independent and dependent variables. Here we might ask the question, "what is the relation between the concentration of chemical x and the rate of juvenile growth in laboratory mice?" The higher the regression coefficient or r^2, the greater the amount of variance in the dependent variable (growth) accounted for by the independent variable (concentration). Other methods, such as multiple linear regression, Akaike's information criterion, or principal component analysis can aid in model development by providing information on which independent variables are predictive of changes in response. Any search of the scientific literature will reveal scores upon scores of studies performing these types of model building.

MATHEMATICAL TOOLS AND ECOTOXICOLOGY

In general, demands for the application of mathematical and statistical tools in data analysis have long been used in many disciplines within the biological and environmental sciences and are now considered essential elements of any serious investigation. Our focus here lies with those tools primary to the demands placed on ecotoxicologists. Today's biologists, ecologists, and environmental scientists need to have command of linear algebra, and an increasing awareness of calculus, differential equations, and discrete mathematics with particular reference to graph theory and its role in network analysis. Ecotoxicologists should have more than a casual awareness of population ecology, given our interest in characterizing exposure and effects relationships at the population or higher level. Of course, a simple exponential model only works for a limited time period, for example, when there are unlimited resources for the population. However, once resources become limited, other models prove more informative, including the logistic model, given by Eq. (12.2):

$$\frac{dN}{dt} = rN\left(1 - \frac{N}{K}\right) \tag{12.2}$$

which models the situation where population size (N) at any specific time (t) is a function of intrinsic rate of population increase (r) and the system's maximum value for carrying capacity (K). When population growth is density-dependent, a more complicated version of the model is given by Eq. (12.3):

$$\frac{dN}{dt} = rN\left[1 - \left(\frac{N}{K}\right)^{\theta}\right] \tag{12.3}$$

which models a population that tends to extinction if its population size becomes too low; that is, theta (θ) enters the model as a parameter that characterizes the relationship between r and N relative to K (Fig. 12.1). Even if ecotoxicologists primarily focus on characterizing chemical toxicity for individuals within a species, being familiar with population-level consequences of toxicity-dependent events can improve interpretations of effects associated with acute and chronic outcomes from traditional toxicity tests. Indeed, ecotoxicologists primarily concerned with characterizing toxicity ply their own mathematical and statistical models that should be appreciated by colleagues whose primary interests focus on estimating population-level consequences linked to exposures of environmental chemicals.

Modeling Dose–Response Relations

A variety of graphical and computational methods can be used to derive a median LC50 from concentration and mortality data produced by an acute mortality test. Three statistical methods have been widely used to calculate LC50 or EC50 values—probit analysis, logistic regression and trimmed Spearman–Karber. In the selection of any method of data analysis, practical and statistical considerations should be noted.

Probit analysis has a long history in statistical applications for evaluating binary data (Finney, 1978). Probit analysis is a parametric procedure (Morgan et al., 1982) that relies on linear regression following transformation of toxicity data. As such, probit is well-suited for characterizing binomial response variables (eg, live or dead, diseased or healthy) such as those derived from dose–response experiments. Data derived from toxicity tests are generally proportion or percent-response data (eg, percent survival, portion affected based on number of respondents relative to number exposed) at corresponding exposure

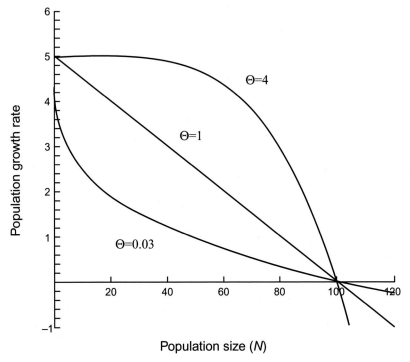

FIGURE 12.1 A simple graphic presentation of relationships between population size and population growth rate under different levels of density-dependent regulation.

concentrations (or doses, depending on the type of toxicity test). Ideally, concentration-response data express themselves as a sigmoidal relationship (Fig. 12.2), but through probit analysis, these data are transformed into a linear form upon which regression analysis is conducted Output from probit analysis yields median effect estimates, such as LC50s with accompanying 95% confidence intervals about this value. A parametric approach such as probit analysis is efficient provided the distribution of responses follows the assumed model of a normal (or bell-shaped curve) distribution. We refer readers to the previous sources for an explanation on how to conduct a probit analysis and to statistical packages such as SAS or R that have routines for conducting these tests.

Logistic regression was also designed to handle binary data and has many similarities to probit analysis. The method was initially formulated by David Cox (1958). Logistic regression is based on the cumulative logistic distribution which distinguishes it from probit analysis. While this may be a substantial difference for statisticians, it is of less relevance to the general practitioner of dose/response curves and either method is treated through canned statistical routines. We include this method primarily to round out the methods used to calculate dose/response models.

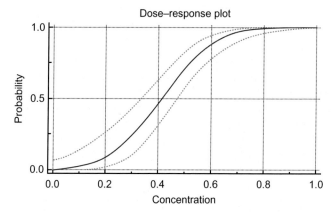

FIGURE 12.2   Idealized plot of a sigmoidal concentration-response curve where probability of survival is plotted relative to exposure concentration.

One other method, the trimmed Spearman–Karber (Hamilton et al., 1977) is often applied when the data do not fit a normal distribution. Trimmed Spearman–Karber analysis is a nonparametric procedure, wherein the only data requirements are: (1) at least one mortality or effects proportion (eg, number dead/number exposed) must be less than or equal to 50%, and (2) at least one must be greater than or equal to 50%. It is not possible to calculate the LC50 by the trimmed Spearman–Karber method if the mortality proportions are all less than 50% or are all greater than 50%.

As suggested by our overview of models and their application to problems in ecotoxicology, practitioners need to be conversant with the tools of mathematics, particularly with those concepts related to system's structure and function encouraged throughout career development. Practitioners should also be able to implement problem-solving methods in order to analyze exposure-effects using analytical methods and appropriate computational tools. However, as relatively recent public debates suggest (AMS, 2013), no one answer is forthcoming regarding how much mathematics a biological scientist (or an ecotoxicologist for that matter) really needs. This is because the answer depends on you. Depending on your area of interest in ecotoxicology, you may simply need to understand and use the mathematics underlying evaluations of concentration-response or dose–response, yet understanding them requires a short jump down (what you might think) a rabbit hole where you encounter a range of mathematical experiences that yield an understanding for how and why a particular data analysis works. Similarly, applied mathematicians whose work lies in ecotoxicological or toxicological studies would benefit from gaining more than a casual understanding of the principles of toxicity testing and the data analysis following from that field or laboratory investigation.

Regardless of your path into ecotoxicology, an appropriate course of study would include experience with probability and statistics, discrete models, linear algebra, calculus and differential equations, modeling, and programming. For example, becoming familiar with statistics and data analysis would benefit improved understanding of ecotoxicological principles, which likely would encourage additional development of a "math toolbox" personalized for areas of your specialization. Ideally, modeling biological phenomena in ecotoxicological studies would place these tools of mathematics and statistics into perspective. For example, investigators should generate hypotheses and then test those hypotheses using the mathematical and statistical tools that have become part of their tradecraft.

BEYOND MATHEMATICAL AND STATISTICAL MODELS: UNCERTAINTY AND ITS ROLE IN MODELING

When all is said and done, models are simply representations of real-world events and processes that are inevitably entangled with uncertainty. There are two general types of uncertainty—*aleatory* uncertainty (also referred to as random uncertainty or stochastic uncertainty) and *epistemic* uncertainty—affect interpretation of model outcomes within environmental settings. Aleatory uncertainty deals with the randomness (or predictability) of an event, while epistemic uncertainty reflects our "state-of-knowledge." The concept of uncertainty when applied in a scientific context contains a complexity that is often inadequately appreciated. In a perfect world, exposure and effects models perform best when based on the results of combined observational and experimental studies. Yet, a perfect world is often not the case when we develop exposure-response models, particularly those based solely on empirical data collected in observational field studies wherein ambient exposures are determined through field measurements of chemical concentrations and effects are largely characterized through observations of biota exposed to these toxicants in the field. In contrast to either experimental or observational approaches or the ideal of data derived from integrated field and laboratory studies, forecasts of exposure and effects are frequently crafted based on expert judgement. Thus, a model can incorporate subjective measures into its construction and forecasting that may or may not fully capture the empirical basis of either experimental or observational approaches. However, pursuit of an expert opinion–based approach to derivation of forecasting models should include measures to better characterize the reliability of the estimates in the face of contributing sources of uncertainty (see Pearl, 2009; Oden et al., 2010).

Uncertainty in forecasting models, regardless of their application and implementation, is an unavoidable property of empirical or data-dependent, science-based analysis of exposure and effects. Indeed, computational science has generated many forecasting models intended to ease the decision-making process regardless of their applications in ecotoxicology or other science disciplines. Oden et al. (2010) identified issues common to a wide range of computational models that are currently implemented and whose outcomes contribute to presumably informed management decisions. The vast majority of computer models in this arena are intended to accurately describe reality, and for many of these computer models, we simply seek accurate depictions of events in the near or distant future, often conditioned on past events. Such is the case for computational models common to ecological sciences. Increasingly, computational science is focused on "predictive simulation," wherein models are systemically evaluated and data uncertainties and their propagation through a computational model are characterized to produce forecasts for events of interest with quantified uncertainty. Thus, if models and data are sufficient, scientifically based forecasts of reality will contribute to informed decisions.

We close this chapter with a real-world encounter with a relatively widely distributed environmental chemical—PCBs—and consider how exposures are manifested in freshwater habitats, and potentially contribute to chemical exposures across a wide range of fish and wildlife, including the potential for effects on public health.

FOCUS—PCBs AND CRAB ORCHARD LAKE

Background

Crab Orchard Lake (COL) in southern Illinois was created in 1940 to supply water and provide opportunities for recreational activities to local communities in the region (Fig. 12.3). During World War II, the Department of Defense gained temporary ownership of the land and constructed the Illinois Ordnance Plant to produce explosives. After the war, the property was transferred to the U.S. Department of the Interior and the U.S. Fish and Wildlife Service (FWS) became a steward of the land as a resource-management area part of their refuge systems. Production of explosives continued to be the main industry and more than 200 tenants have operated a variety of manufacturing plants (ie, plated metal parts, ink, electrical components, machined parts, painted products, and boats) under lease from the refuge. As a result of these mixed land uses, soils, aquatic sediments, and water in various areas of the refuge were contaminated, and its PCB contamination was identified in the 1970s. A fish-consumption advisory has been in effect since 1988 for COL on common carp (*Cyprinus carpio*), channel catfish (*Ictalurus punctatus*), and largemouth bass (*Micropterus salmoides*, Illinois Dept. Public Health, Undated). Remediation of lake sediments has occurred, and restoration projects are also underway, yet the fish-consumption advisory remains in place. Among the many resource questions asked by the U.S. FWS and the public is: how long will the fish-consumption advisory remain in place? Although there is no simple answer at this time, models do offer insights to the biological and physicochemical processes at play and aid in our understanding of the issues involved in both the original posting

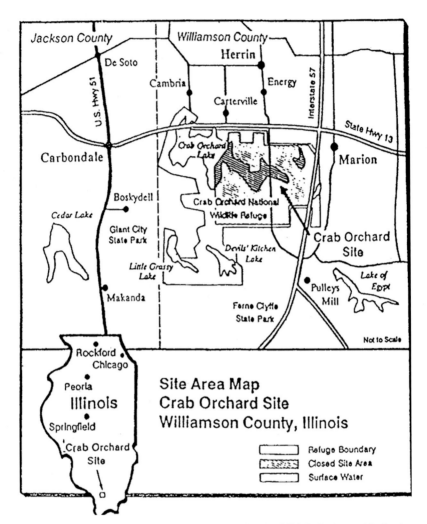

FIGURE 12.3   Map of Crab Orchard Lake and the National Wildlife Refuge around it, Southern Illinois.

of the advisory and suggest methods for developing resource-management strategies that respond to a complex exposure-effects issue.

Recall from Chapter 6 that production of PCBs was banned in late 1970s, except for use in totally enclosed systems that do not release the chemicals into the environment. Despite this regulatory measure, PCBs are still widespread in soil, sediments, water, and air, primarily because these chlorinated organics degrade slowly once released into the environment. Historic releases of PCBs

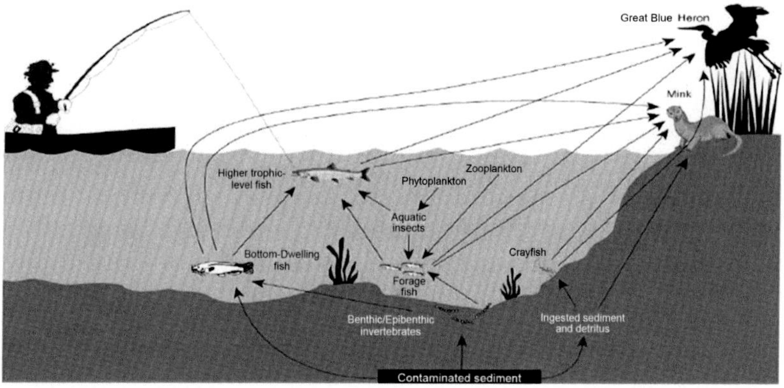

FIGURE 12.4   Conceptual model linking bioaccumulation process from the vantage of numerous receptors.

at COL were associated with industrial processes, and PCB residues in the lake remain, despite previous remediation.

PCB contamination presents short-term and long-term toxicity risks to ecological systems, as well as human health. In many aquatic systems including COL, PCB-contaminated sediments are expected to be the primary source of PCB exposure (Gewurtz et al., 2009). Exposure to PCB-contaminated sediments may result in adverse effects on the structure and function of sediment-dwelling benthic communities, both in the short term (acute toxicity) and long term (chronic toxicity). Such adverse effects may subsequently drive reduction or loss of populations of fish and wildlife dependent on benthic organisms as a primary food source. Ecosystem services, such as the decomposition of organic matter and water filtration, may also be adversely affected by PCBs, contributing to a loss of environmental integrity and sustainability. For human health concerns, adverse effects linked to PCB exposures include impairments of immune, neurological, and reproductive systems, including development of young. As with other chlorinated organics (eg, legacy pesticides, dioxins, and furans), exposure to PCBs may also be linked to an increased risk of cancer in exposed populations. Therefore, the presence of PCB-contaminated sediments in COL represents a risk to sediment-dwelling biota, fish health, and ultimately human health.

Fishes, Recreational Fishery, and the Fish-Consumption Advisory

Consider the conceptual model in Fig. 12.4. Here, we graphically represent the essence of exposure in sediments and surface waters to PCBs, with recreational fishing representing one component in the environmental exposure picture. Within the context of multiple stressors, physical habitats at COL have been impacted by sedimentation and upland erosion, and water quality has been affected by excessive nutrient loading from nonpoint source inputs and municipal wastewater discharges. Remedial actions, however, have been implemented in the past; most notably for our current narrative, sediments contaminated by PCBs were dredged from COL's areas of concern in 1996.

Despite these remedial actions, PCB exposures continue, in part, because residual PCBs remain in sediments. Thus, PCBs continue to bioaccumulate consequent to food-chain-linked transfers. Biomagnification of these chlorinated organic chemicals in fatty tissues of foraging COL fishes and any organisms feeding on those fishes also occurs. Even sediments marginally contaminated with PCBs may yield high concentrations of PCBs in aquatic or terrestrial organisms—including humans.

Biodegradation, Environmental Fate, and Bioavailability of PCBs

Recall from Chapter 6 that physical and microbial activity may degrade PCBs either aerobically or anaerobically in soils or in aquatic environments, particularly in sediments. In conjunction with microbial transformation, volatilization of PCBs is also a critical physicochemical process that potentially reduces PCB concentrations in environmental matrices such as freshwater sediments. Photolysis is one more way that PCBs are broken down in the environment, but this process may be seasonal and limited to those PCBs that are near or at the surface of the water or in the atmosphere.

In addition to decomposition, we should also consider mechanisms that affect the bioavailability of PCBs. In general, environmental fates of hydrophobic organic compounds such as PCBs in water are driven by partitioning among dissolved, suspended, or adsorbed phases of the chemical. For example, when adsorbed to dissolved organic matter (DOM), PCBs become less bioavailable. In contrast, adsorption to particulate organic matter (POM) may increase the chemical's availability to detrital food webs or bottom-feeding fishes. In detrital compartments within aquatic habitats, the fate of PCBs is dependent on the structure and origin of PCB congeners (Baker et al., 1991).

In general, a combination of binding processes (sorption) and mass transport processes (diffusion) determine PCB partitioning between aqueous and solid phases. Partitioning influences exposure of fishes to PCB residuals in sediments, subsequent incorporation of PCBs into fish tissues, and ultimately the derivation of fish-consumption advisories.

Developing Models of Bioconcentration and Bioaccumulation for COL

When developing fish-consumption advisories, regulatory agencies derive input values for PCB tissue residues in selected fish conditioned on risk scenarios or models of fish consumption by target human populations (eg, individuals involved with recreational angling or subsistence lifestyles, and females of child-bearing age (Scherer et al., 2008; US EPA, 2015). Ideally, model inputs underlying fish-consumption advisories are derived from field-based characterizations of fish-tissue residues. Through food chain interactions, invertebrates, fish, and eventually humans can bioaccumulate and even biomagnify PCBs. Yet to start our overview of such a process, we need to take a step back and characterize bioconcentration.

To model bioconcentration, we develop a bioconcentration factor (BCF), which is defined as the ratio of the chemical concentration in biota (C_b) to a measure of chemical concentration in water (C_w). Both C_b and C_w may be specified by form, such as tissue residue concentration or as water concentrations measured as total or dissolved, respectively. As a proportionality constant of the chemical concentration, BCF ideally represents steady-state conditions for C_b and C_w (Neely et al., 1974; Veith et al., 1979) Eq. (12.4):

$$\text{BCF} = C_b / C_w \qquad (12.4)$$

Conceptually, BCF is equally applicable to any aquatic species, yet our focus here resides with fish, principally because of their importance as food for humans. In the context of fish-consumption advisories, however, BCF may show limited utility because the calculation ignores chemical exposures mediated through the diet. Isolating water-only exposures from other exposure routes represents an oversimplification most easily achieved under laboratory conditions, as early work has demonstrated (see Neely et al., 1974; Veith et al., 1979). Given that most fish species rely on multiple pathways for gaining food and necessary nutrients, modeling food-chain exposures in humans based on water-only exposures in fish likely underestimates fish-tissue residues and may undermine the derivation of protective fish-consumption advisories. Therefore, bioaccumulation provides a conceptual context for more completely characterizing tissue residues for PCBs in fish species of interest in COL.

In contrast to bioconcentration, bioaccumulation is linked to the uptake of environmental chemicals by all exposure routes, including diet (Fig. 12.5; Arnot and Gobas, 2006). An evaluation of uptake, then, considers chemical uptake and transport across respiratory and other cutaneous surfaces (eg, lining of buccal cavity), as well as uptake of chemical via ingestion of food or other particulate materials wherein absorption through the gut lining is critical to mediating the chemical accumulation process. Chemical exchanges across gills of fish and from their food are generally considered to occur by passive diffusion of chemicals between a fish's internal aqueous phase and its external aqueous environment, whether it be the surrounding ambient water or the aqueous phases of the fish's intestinal contents. Bioaccumulation factors (BAFs) represent a ratio estimator of concentration of a chemical in biota. As a first approximation, bioaccumulation considers a fish as a system of solvents where a fish's body burden of chemical is estimated from its live-weight considered as fractions of whole fish. These fractional estimates are comprised of compartments or phases: water,

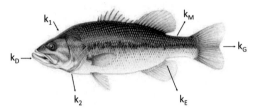

FIGURE 12.5 A conceptual diagram representing the major routes of chemical uptake and elimination in an aquatic organism. k_D, dietary uptake rate constant; k_1, gill uptake rate constant; k_2, gill elimination rate constant; k_M, metabolic transformation rate constant; k_E, fecal egestion rate constant; k_G, growth dilution rate constant. *Modified from Arnot and Gobas (2006).*

lipid, and nonlipid organic material. These phases are subsequently linked to corresponding chemical concentrations of, in our case, PCBs. Provided uptake of PCBs follows simple partitioning processes assumed for each phase in such a compartment model (Arnot and Gobas, 2006), and depuration rates of chemicals from different fish tissues do not differ significantly (Norheim and Roald, 1985; Kleeman et al., 1986), then equilibration among water, lipid, and nonlipid organic phases of a fish occurs rapidly in comparison with external exchanges. Thus, for organic chemicals such as PCBs, bioaccumulation is simplified as a sum of chemical residues in each compartment or phase of individual fish, which can then serve as input to model (eg, the chemical bioaccumulation and the growth of populations within an age-structured fish community). Such individual-based models may then be scaled up by incorporating population-level attributes, preferably supported by empirical data sufficient to reduce uncertainties associated with such scaling (see Bioaccumulation and Aquatic System Simulator (BASS); Barber, 2008). Although BASS and other bioaccumulation models represent advances over a simple bioconcentration approach, these models do not adequately capture the complexity of natural systems.

On an individual basis, Fig. 12.5 provides a graphic representation of biological events, notably various types of rate constants that are critical to characterizing bioconcentration and bioaccumulation in fishes that lead to subsequent characterizations of processes potentially occurring in trophic transfers of PCBs. Yet, uncertainties linked to our estimating bioaccumulation of PCBs in fishes is entangled with scale-dependent factors, principally the interrelationships among environmental fate and transport processes between the environment and the food chain, progressing up the food chain (Fig. 12.6). Such a conceptual model represents field settings, yet we ask that you focus on that—the interconnectedness of the system—rather than specific compartments depicted in the illustration. The abiotic and biotic components at COL are interdependent and uncertain with respect to estimates of rates in uptake through ingestion or diffusion across epithelial tissue, and loss of chemical or its metabolites through, for example, fecal or excretory routes of elimination. Moreover, the amount of existing information from COL and from previously published studies may be relatively limited. If

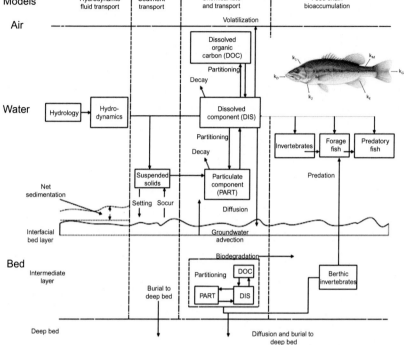

FIGURE 12.6   Compartment model indicating relationships among abiotic and biotic components of sediment-water systems such as that of COL, Illinois, with fishes as a prominent member among biotic compartments considered in this conceptual model. *Modified from NRC/NAS (2001).*

available, data gleaned from other locations with similar environmental settings and common frames of reference with respect to exposure can be incorporated into the modeling process. For example, investigators may develop estimates of uptake and depuration rate constants based on incompletely to poorly supported assumptions and with relatively limited site-specific data. As such, Best Professional Judgment (BPJ), which is a close conceptual relative to a Scientific Wild-Ass Guess (SWAG), derives an estimate from past experience, general impressions, available data from existing literature, and heuristic or approximate calculations until or if actual data can be gathered and interpreted with regards to PCBs, including its fate within the abiotic and biotic system of COL captured by the underlying conceptual or scientific models (Fig. 12.6).

Fortunately, we do not have to operate completely in the dark. For many chemicals, much work has been published and because PCBs have been well studied for many years, much of the data is derived from a wide range of field and laboratory studies. For COL, these existing data can help us understand the current problem of how sediment-bound PCBs enter the food chain.

Bioaccumulation in Sediment-Water Systems

As we have already stated, bioaccumulation results from organisms absorbing environmental chemicals by all exposure routes including transport across respiratory and cutaneous surfaces and through the diet. While BCF and BAF are mathematically similar, a third factor may be useful in characterizing relationships between tissue residues of PCBs and their sediment concentrations (Burkhard, 2009); hence, the *biota-sediment accumulation factor* (BSAF) Eq. (12.5):

$$\text{BSAF} = \frac{(C_b / f_{lipid})}{(C_{sed} / f_{soc})} \qquad (12.5)$$

where C_b is the contaminant concentration in biota, f_{lipid} is the lipid fraction of the biota, C_{sed} is the contaminant concentration in sediment, and f_{soc} is the sediment organic carbon fraction. Although they are not as fully developed as models focused on BAF as an estimator of uptake from water and dietary sources, BSAF approaches have been used with a variety of PCB-exposed freshwater organisms (Babut et al., 2012; Burkhard et al., 2013). If empirical data are sufficient, BSAF may be applicable to the evaluation of the relationships among residual PCBs in COL sediments, fishes in COL, and derivation of fish-consumption advisories.

Modeling Biomagnification

Biomagnification is an outcome of transfer up a food chain. As such, biomagnification results from a sequence of biologically-driven processes—bioconcentration or bioaccumulation occurs in producers (such as phytoplankton or other primary producers), that are subsequently consumed by primary consumers. Secondary consumers subsequently accumulate chemical residues by way of their foraging and preying activity. If the chemical of concern—here, PCBs—is highly lipophilic and poorly metabolized, then tertiary consumers accumulate even greater concentrations of the chemical of concern when foraging or preying on producers or consumers for their diet. Biomagnification nearly always results in higher concentrations of tissue residues at trophic level increases (eg, in fishes) than would be expected if water was the only exposure mechanism. Following a path similar to that for BCF, BAF, or BSAF, a *biomagnification factor* (BMF) is a simple ratio estimator of concentration of a chemical in the consumer organism (C_{bc}) to that in their prey (C_{bp}), and can be expressed as Eq. (12.6):

$$\text{BMF} = C_{bc} / C_{bp} \qquad (12.6)$$

Depending upon the conditions linked to its characterization, BMF is frequently expressed as an observed, lipid-normalized ratio of numerator and denominator. A lipid-normalized BMF value equal to or less than one has been interpreted as characteristic of a chemical having not been biomagnified; values greater than 1.0 are considered indicative of a chemical that has been biomagnified. When deriving fish-consumption advisories, the potential utility of biomagnification factors is incompletely understood. Although some authors have demonstrated the importance of fish trophic level in determining tissue residues of PCBs (Rasmussen et al., 1990), others have argued that feeding location and individual foraging strategies may be more important than trophic levels for some fish species (Lopes et al., 2011). This uncertainty is compounded by the many sources of variability that influence the fate of chemicals ingested by organisms. For example, a chemical's water solubility, lipid solubility (as estimated by k_{ow}), molecular weight, and volume influence its disposition once consumed in the diet. Similarly, diffusion rates and transfers among gut, blood, and lipid pools, as well as rates of assimilation, metabolism, egestion, and organism growth influence chemical concentrations in tissues (recall Fig. 12.5).

As secondary or tertiary consumers within food chains or food webs, humans occupy a relatively high trophic level. This position leaves them particularly vulnerable to chemicals that biomagnify and to attendant potentially adverse health effects (Bierman, 1990). However, confounding occurs when any organism has several food choices or sources of preferred food. Not surprisingly, chemical concentrations may differ among foods and among sources of these foods. All chemicals in a human diet that potentially bioaccumulate do not necessarily biomagnify. Across the range of biota exposed to PCBs in aquatic and sediment habitats, there are a range of outcomes related to the bioaccumulation and biomagnification of PCBs, which should be considered in evaluations of fish-consumption advisories based on total PCBs. In field settings, there are potentially multiple abiotic and biotic sources of PCBs, and the properties of individual PCB congeners substantially affect accumulation or degradation pathways. Depending on the scale of analysis—biochemical and molecular, organismal, population and community, or system level of organization—compartment models as described next can provide conceptual frameworks to support derivation of fish-consumption advisories.

Compartment Models Focused on Accumulation of PCBs by Fish

Compartment models are often used to describe transport of material, such as chemicals, in biological systems at various levels of organization. If the compartment is in an individual, this could be part of a previously described PBTK model or the compartment may be a species within a community or an abiotic part of the environment, such as sediment. Regardless of scale, a compartment model consists of discrete components (referred to as compartments) interconnected with other components within a system. Materials within each compartment are assumed to be well mixed, and compartments are potentially involved in exchanges of material with each other, following certain rules. Regardless of their origins, bioaccumulation models simply lay out a process wherein a

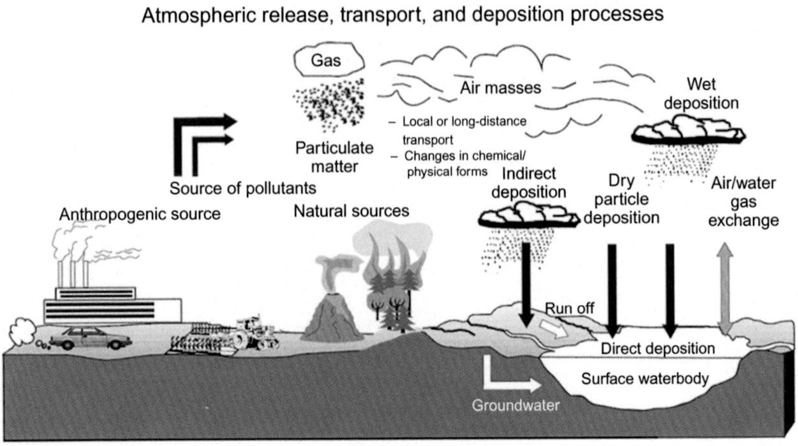

FIGURE 12.7   Some hydrophobic organic chemicals such as PCBs and polybrominated diphenyl ethers (PBDEs) are increasingly identified as atmospheric chemicals subject to relatively long-distance transport; thus, a complete accounting for this nonpoint source contribution to load should be included during evaluations of natural recovery as a remediation method targeted on PCBs in areas such as COL, Illinois. *From US EPA (2000).*

system of compartments is considered with the context of material (and potentially energy) flows. Our primary concern here is with PCB tissue residues and their bioaccumulation by biota inhabiting aquatic habitats that have contaminated sediments.

However, variability in PCB tissue residues in fishes are potentially confounded by alternative pathways for PCB inputs to the system. For example, the Great Lakes region has a well-documented history of atmospheric inputs of PCBs (US EPA, 2007) in addition to point sources, such as those at COL (Fig. 12.7). Indeed, net fluxes of PCBs suggest increasing volatilization from surface waters to the surrounding atmosphere, which makes waterbodies such as the Great Lakes sources of PCBs to the atmosphere when historically such areas were largely characterized as sinks. Regionally, COL will likely be downwind from source areas wherein volatilization and combustion processes will continue contributing to PCB loads in the lake, in addition to local contributions linked to historic releases. For example, both Chicago and especially St Louis serve as sources of aerial PCB deposition that may confound interpretation of transfers of PCBs from sediments to biota—especially vertebrates, including humans.

Models Underlying Fish-Consumption Advisory Implementation

Derivation of fish-consumption advisories generally follows a relatively simple, but yet-to-be-validated model focused on human consumption of fish, conditioned on fish-tissue residue concentrations. For application to COL, the model relied on inputs for tissue residue values for total PCBs in species of interest—common carp, channel catfish, and largemouth bass. In regulatory settings, model implementation would focus on deriving estimates of chronic daily intake (CDI), which follows an expansion of US EPA (2000) guidance wherein Eq. (12.7):

$$CDI \ (mg/kg - d) = \frac{C \times IR \times FI \times ED \times EF}{BW \times AT} \qquad (12.7)$$

where

d = dermal absorption,
C = concentration of PCB in tissue (mg/kg),
IR = ingestion rate (kg/day or kg/meal),
FI = fraction ingested from contaminated source,
ED = exposure duration (year),
EF = exposure frequency (day/year or meals/year),
BW = body weight (kg), and
AT = averaging time (day)

Under regulatory implementation, default values may be applied for selected human receptors such as females of child-bearing age, infants, and preadolescent children. Binnington et al. (2014) considered the effectiveness of fish-consumption advisories regarding persistent organic chemicals such as PCBs, and observed that exposure to most persistent organic pollutants occurred mainly through the ingestion of contaminated food. Turyk et al. (2012) identified that fish are a major means of exposure to bioaccumulating contaminants such as PCBs. For example, fish-consumption advisories based on PCBs in fish tissue have been developed for the Great Lakes sport-fishing community, and for some European Union countries (eg, Finland and Sweden; see European Commission, 2006; Kiljunen et al., 2007). Human exposure to PCBs has been linked to neurocognitive effects (Walkowiak et al., 2001; Stewart et al., 2008). Reproductive effects have also been characterized, such as abnormal development and decreased fecundity consequent to exposures to nondioxin-like PCB exposures (Toft et al., 2004). Dioxin-like congeners of PCBs have also been classified as probable human carcinogens (Group 2A) by the International Agency for Research on Cancer (IARC, 2009); PCB-126 (3,3′,4,4′,5-pentachlorobiphenyl) has been further identified as carcinogenic to humans (IARC, 2012). Thus, PCB-based fish-consumption advisories encourage reduced consumption of lipid-rich fish such as carp and catfish by sensitive populations and members of sport-fishing communities. Binnington et al. (2014) concluded that fish consumption guidelines are helpful for protecting human health, particularly for chemicals with human elimination half-lives of less than 1 year, but that such guidelines should be continually evaluated and improved.

PCBs have long been recognized as important environmental contaminants, given their environmental persistence, their relatively poor metabolization rate by biota ranging from microbes to invertebrates to vertebrates, including humans, and their tendency to accumulate in lipids of exposed biota. PCBs are bioaccumulated by aquatic and sediment-dwelling invertebrates and vertebrates (eg, fishes and amphibians), which enable these chemicals to enter food chains and food webs across a wide range of habitats. In fishes and in humans that consume contaminated organisms, PCBs will accumulate in their tissues. Such accumulation is troubling, because it may lead to body burdens of PCBs that could have adverse health effects in humans and wildlife.

Summary: Models and Modeling

Although most often quoted out of context, Box and Draper (1987, p. 424) characterized a polynomial model they considered as:

> *an approximation [that] does not necessarily detract from its usefulness because all models are approximations. Essentially, all models are wrong, but some are useful. However, the approximate nature of the model must always be borne in mind.*

In general, their observations apply to any model or the process of model development. Modeling is an iterative process of building and rebuilding, particularly in the early stages of a model's development for biological or ecological applications. This iterative process involves revisions, periodic updates, and, more often than not, destructive steps linked to a varying number of "lessons learned" about the system being modeled. As Sjödin et al. (2005, p. 3862) noted:

> *[S]implifying assumptions are necessary in order to turn complex biological systems into caricatures that are, on the one hand, simple enough to analyze, and on the other hand, realistic enough to capture key features of the process under investigation.*

In order to address questions such as those considered in the preceding case study or in any such studies involving issues of risk and uncertainty, we developed multiple models that captured resource managers concerns that helped focus the analysis. Data analysts, whether detailed to scientific or engineering problems, all concede to Puccia and Levins (1985, p. 2):

> *Every model distorts the system under study in order to simplify it. ... There are two dangers in model building: one is that the model does not tell us about the world; the other is that is a faithful representation, and therefore we are overwhelmed. Simplification is both legitimate and necessary as long as we are cautious, are willing to change the original underlying assumptions as necessary and build new models, and carefully interpret the model's predictions.*

STUDY QUESTIONS

1. What are the values of models? When might you use scientific models?
2. Develop a conceptual model that would outline what you have learned in this course through drawing, computer image generation, or another process.
3. Explain in your own words the difference between mathematical models. Are all statistical models mathematical models?
4. What are some practical uses of QSARs? Of PBTK models? Why would they be used in risk analysis?
5. Have you had an opportunity to analyze data that you or someone else collected? Did you run any modeling routines in that analysis? Share with your class some personal examples of using statistical modeling.
6. A good exercise would be for your instructor to provide data or actually run probit analysis, logistic regression, and trimmed Spearman–Karber analyses and provide output for the class to examine.
7. Take another look at Fig. 12.5. Explain in your own words what the arrows refer to in a context involving any contaminant.
8. What factors, other than toxic residues in a sample of fish, lead to declaring a food advisory in edible fish? Why is it that seafood (including freshwater fish) are the substance of these food advisories and not waterfowl or consumable wildlife such as deer?

REFERENCES

AMS, 2013. Opinion: two views: how much math do scientists need? Not. AMS 60, 837–838.

Arnot, J.A., Gobas, F.A.P.C., 2006. A review of bioconcentration factor (BCF) and bioaccumulation factor (BAF) assessments for organic chemicals in aquatic organisms. Environ. Rev. 14, 257–297.

Babut, M., Lopes, C., Pradelle, S., Persat, H., Badot, P.-M., 2012. BSAFs for freshwater fish and derivation of a sediment quality guideline for PCBs in the Rhone basin, France. J. Soils Sed. 12, 241–251.

Baker, J.E., Eisenreich, S.J., Eadie, B.J., 1991. Sediment trap fluxes and benthic recycling of organic-carbon, polycyclic aromatic-hydrocarbons, and polychlorobiphenyl congeners in Lake Superior. Environ. Sci. Technol. 25, 500–509.

Barber, M.C., 2008. Bioaccumulation and Aquatic System Simulator (BASS) User's Manual Version 2.2. U.S. Environmental Protection Agency, National Exposure Research Laboratory, Ecosystems Research Division, Athens, GA. EPA/600/R-01/035 update 2.2.

Bierman Jr., V.J., 1990. Equilibrium partitioning and biomagnification of organic chemicals in benthic animals. Environ. Sci. Technol. 24, 1407–1412.

Binnington, M.J., Quinn, C.L., McLachlan, M.S., Wania, F., 2014. Evaluating the effectiveness of fish consumption advisories: modeling prenatal, postnatal, and childhood exposures to persistent organic pollutants. Environ. Health Perspec. 122, 178–186.

Box, G.E.P., Draper, N.R., 1987. Empirical Model Building and Response Surfaces. John Wiley & Sons Inc., New York, NY, p. 688.

Burkhard, L., 2009. Estimation of Biota Sediment Accumulation Factor (BSAF) from Paired Observations of Chemical Concentrations in Biota and Sediment. U.S. Environmental Protection Agency, Ecological Risk Assessment Support Center, Cincinnati, OH. EPA/600/R-06/047.

Burkhard, L.P., Mount, D.R., Highland, T.L., Hockett, J.R., Norberg-King, T., Billa, N., et al., 2013. Evaluation of PCB bioaccumulation by *Lumbriculus variegatus* in field-collected sediments. Environ. Toxicol. Chem. 32, 1495–1503.

Campbell Jr., J.L., Clewell, R.A., Gentry, P.R., Andersen, M.E., Clewell, H.J., 2012. Physiologically based pharmacokinetic/toxicokinetic modeling. Mol. Biol. 929, 439–499.

Christen, V., Hickmann, S., Rechenberg, B., Fent, K., 2013. Highly active human pharmaceuticals in aquatic systems: a concept for their identification based on their mode of action. Aquat. Toxicol. 96, 167–181.

Cox, D.R., 1958. The regression analysis of binary sequences (with discussion). J. Roy. Stat. Soc. B. 20, 215–242.

D'Agostino, S.R., Sullivan, L., 2005. Introductory Applied Biostatistics. Brookes Cole Publ., Boston, MA, p. 672.

Escher, B.I., Hermens, J.L.M., 2002. Modes of action in ecotoxicology: their role in body burdens, species sensitivity, QSARs, and mixture effects. Environ. Sci. Technol. 36, 4201–4217.

European Commission, 2006. Commission Regulation (EC) No 199/2006 of 3 February 2006 Amending Regulation (EC) No 466/2001 Setting Maximum Levels for Certain Contaminants in Foodstuffs as Regards Dioxins and Dioxin-like PCBs.

Finney, D.J., 1978. Statistical method in biological assay Mathematics in Medicine Series, third ed. Macmillan Publishing Company, Inc., New York, p. 505. (original publisher Charles Griffin and Company, Ltd, London).

Gewurtz, S.B., Gandhi, N., Christensen, G.N., Evenset, A., Gregor, D., Diamond, M.L., 2009. Use of a food web model to evaluate the factors responsible for high PCB fish concentrations in Lake Ellasjoen, a high Arctic Lake. Environ. Sci. Poll. Res. 16, 176–190.

Grimm, V., Railsback, S.F., 2005. Individual-Based Modeling and Ecology. Princeton University Press, Princeton, NJ, p. 448. (Chapter 2, A Primer to Modeling).

Hamilton, M.A., Russo, R.C., Thurston, R.V., 1977. Trimmed Spearman–Karber method for estimating median lethal concentrations. Environ. Sci. Technol. 11, 714–719.

Herr, D.W., Mwanza, J.-C., Lyke, D.F., Graff, J.E., Moser, V.C., Padilla, S., 2010. Relationship between brain and plasma carbaryl levels and cholinesterase inhibition. Toxicology 276, 172–183.

IARC International Agency for Research on Cancer, 2009. Identification of Research Needs to Resolve the Carcinogenicity of High-Priority IARC Carcinogens. IARC Technical Publication No. 42, International Agency for Research on Cancer France.

IARC International Agency for Research on Cancer, 2012. Chemical Agents and Related Occupations. International Agency for Research on Cancer, France.

Illinois Department of Public Health. Undated. Fish advisory, Crab Orchard Lake, Williamson County. <http://www.idph.state.il.us/envhealth/fishadvisory/craborchardlake.htm> (accessed 16.11.15.).

Kiljunen, M., Vanhatalo, M., Mantyniemi, S., Peltonen, H., Kuikka, S., Kiviranta, H., et al., 2007. Human dietary intake of organochlorines from Baltic herring: implications of individual fish variability and fisheries management. Ambio 36, 257–264.

Kleeman, J.M., Olson, J.R., Chen, S.M., Peterson, R.E., 1986. 2,3,7,8-tetrachlorodibenzo-para-dioxin metabolism and disposition in yellow perch. Toxicol. Appl. Pharmacol. 83, 402–411.

Lopes, C., Perga, M.E., Peretti, A., Roger, M.C., Persat, H., Babut, M., 2011. Is PCB concentration variability between and within freshwater fish species explained by their contamination pathways? Chemosphere 85, 502–508.

Morgan, R.M., Kundomal, Y.R., Hupp, E.W., 1982. SAS probit analysis for cadmium mortality. Environ. Res. 29, 233–237.

National Research Council/National Academy of Science (NRC/NAS), 2001. A Risk-Management Strategy for PCB-Contaminated Sediments. Committee on Remediation of PCB-Contaminated Sediments, Board on Environmental Studies and Toxicology, Division on Life and Earth Studies, National Research Council. National Academies Press, Washington, DC, p. 452.

Neely, W.B., Branson, D.R., Blau, G.E., 1974. Partition coefficient to measure bioconcentration potential of organic chemicals in fish. Environ. Sci. Technol. 8, 1113–1115.

Nichols, J.W., McKim, J.M., Andersen, M.E., Gargas, M.L., Clewell III, H.J., Erickson, R.J., 1990. A physiologically based toxicokinetic model for the uptake and disposition of waterborne organic chemicals in fish. Toxicol. Appl. Pharmacol. 106, 433–437.

Norheim, G., Roald, S.O., 1985. Distribution and elimination of hexachlorobenzene, octachlorostyrene and decachlorobiphenyl in rainbow trout, *Salmo gairdneri*. Aquat. Toxicol. 6, 13–24.

Oden, T., Moser, R., Ghattas, O., 2010. Computer predictions with quantified uncertainty, Part I. SIAM News 43, 1–3.

Pearl, J., 2009. Causality: Models, Reasoning, and Inference, second ed. Cambridge University Press, Cambridge, p. 464.

Puccia, C.J., Levins, R., 1985. Qualitative Modeling of Complex Systems: An Introduction to Loop Analysis and Time Averaging. Harvard University Press, Cambridge, Massachusetts, p. 259.

Rasmussen, J., Rowan, D., Lean, D., Carey, J., 1990. Food chain structure in Ontario lakes determines PCB levels in lake trout (*Salvelinus namaycush*) and other pelagic fish. Can. J. Fish. Aquat. Sci. 47, 2030–2038.

Scherer, A.C., Tsuchiya, A., Younglove, L.R., Burbacher, T.M., Faustman, E.M., 2008. Comparative analysis of state fish consumption advisories targeting sensitive populations. Environ. Health Perspec. 116, 1598–1606.

Schultz, I.R., Hayton, W.L., 1999. Interspecies scaling of the bioaccumulation of lipophilic xenobiotics in fish: an example using trifluralin. Environ. Toxicol. Chem. 18, 1440–1449.

Schultz, T.W., Cronin, M.T.D., Walker, J.D., Aptula, A.O., 2003. Quantitative structure–activity relationships (QSARs) in toxicology: a historical perspective. J. Mol. Struct. Theochem. 622, 1–22.

Sjödin, M., Styring, S., Wolpher, H., Xu, Y.H., Sun, L.C., Hammarstrom, L., 2005. Switching the redox mechanism: models for proton-coupled electron transfer from tyrosine and tryptophan. J. Am. Chem. Soc. 127, 3855–3863.

Stewart, P.W., Lonky, E., Reihman, J., Pagano, J., Gump, B.B., Darvill, T., 2008. The relationship between prenatal PCB exposure and intelligence (IQ) in 9-year-old children. Environ. Health Perspec. 116, 1416–1422.

Toft, G., Hagmar, L., Giwercman, G., Bonde, J.P., 2004. Epidemiological evidence on reproductive effects of persistent organochlorines in humans. Repro. Toxicol. 19, 5–26.

Turyk, M.E., Bhavsar, S.P., Bowerman, W., Boysen, E., Clark, M., Diamond, M., et al., 2012. Risks and benefits of consumption of Great Lakes fish. Environ. Health Perspec. 120, 11–18.

US EPA U.S. Environmental Protection Agency, 2000. Guidance for Assessing Chemical Contaminant Data for Use in Fish Advisories, Volume 2, Risk Assessment and Fish Consumption Limits, third ed. EPA 823-B-00-008, United States Environmental Protection Agency, Office of Water (4305), Washington, DC, p. 383.

US EPA U.S. Environmental Protection Agency, 2007. Survey of New Findings in Scientific Literature Related to Atmospheric Deposition to the Great Waters: Polychlorinated Biphenyls. U.S. Environmental Protection Agency, Office of Air Quality Planning and Standards, Health and Environmental Impacts Division. Climate. International and Multimedia Group, Research Triangle Park, NC, p. 33.

US EPA U.S. Environmental Protection Agency, 2015. Advisories and technical resources for fish and shellfish consumption. <http://www2.epa.gov/fish-tech> (accessed 16.11.15.).

Veith, G.D., DeFoe, D.L., Bergstedt, B.V., 1979. Measuring and estimating the bioconcentration factor of chemicals in fish. J. Fish. Res. Board Can. 36, 1040–1048.

Walkowiak, J., Wiener, J.A., Fastabend, A., Heinzow, B., Kramer, U., Schmidt, E., et al., 2001. Environmental exposure to polychlorinated biphenyls and quality of the home environment: effects on psychodevelopment in early childhood. Lancet 358, 1602–1607.

Chapter 13

Chemical Stressors
and Ecological Risk

Terms to Know

(Ecological) Risk Assessment
Risk Analysis
Risk Management
Hazard
Stressor
Receptor
Toxicity Reference Value
Hazard Quotient
Competing Risk
Aggregate Risk
Vulnerability Analysis
Sufficient
Necessary
Precipitating Factors
Predisposing Factors
Reinforcing Factors
Enabling Factors

INTRODUCTION

What do we mean by *ecological risk assessment*? Whenever contaminants are present in the environment, they pose some degree of potential harm to that environment, its plants and animals, and the people that live there. This potential harm is called *risk* or *ecological risk* (also see the following discussion). The risk from any occurrence of contaminants in the environment may be negligible, indeed below the level that humans can measure, or it may be substantial. The instances of environmental catastrophes described in this book are examples of the latter.

A branch of ecotoxicology, called *risk assessment*, evaluates the harm that may come from using or storing contaminants in the environment. Risk analysts

Ecotoxicology Essentials.

391

use mathematical, biological, and chemical models to predict what may happen under various scenarios with the goal of choosing an optimal program of use or storage. Sometimes, they have to make a scientific wild-ass guess (SWAG) when data are not available. For instance, risk from a landfill may vary depending on what type of underlayment or capping is used to seal the contents of that landfill from discharging into the environment. The risk analyst may be involved in making choices of these materials and in predicting what may happen to nearby rivers, streams, or air should the landfill fail in preventing a discharge. This is just one of many different functions of risk assessment.

Many journal reviews and reference books that focus on practices and methods for evaluating risks associated with chemicals in the environment have been published recently (see Suter, 2006; Suter et al., 2000). Not surprisingly, risk analysts have not achieved complete agreement on how technical findings of ecotoxicology should be applied to questions related to risks associated with environmental chemicals. Indeed, beyond this introduction, your future study will discover that ecotoxicology affords a range of opportunities to apply its technical findings and methods to the risk evaluation process. How those technical findings and methods are applied to questions focused on chemicals in the environment vary widely, and each implementation of a risk-based analysis will have its champions and critics. However, here we focus on conceptual common ground and do not enter the fray of expert opinion regarding the technical analysis of risks associated with chemicals in the environment. At least that is our goal.

Our introductory overview of the evaluation of environmental risks begins with a brief review of terms and some historic context regarding evaluation of risks within environmental settings, and conceptual frameworks for such evaluations focused on environmental chemicals. We then consider a snapshot of sorts focused on the technical processes deployed for an evaluation of risks of naturally occurring chemicals or those released to the environment directly or indirectly through human.

TERMS OF THE TRADECRAFT

For this introduction we will opt for terms that follow consensus-based definitions, often having regulatory or legal precedence. *Risk* is characterized as the probability of occurrence of an adverse effect on human health or the environment as a result of exposure to a hazard. In general, a hazard is defined as a condition, process, or state that adversely affects the environment. We focus here on environmental hazards that manifest themselves as physical or chemical pollution in air, water, and soils and can cause widespread harm to a range of life forms, including humans. Environmental hazards may take many forms, ranging from occupational and work-related hazards to those typically considered as environmental hazards involving fish and wildlife, as a category of events focused on ecological outcomes. Hazards for our present purposes are regarded as physical events or processes (eg, landslides), or chemical releases to air, water, and soils. In general then, environmental hazards are frequently linked as potential causes of harm to biotic components of the environment, including adverse effects on humans. As such, environmental hazards range from those associated with events that are linked to intentional or accidental release materials to the environment (eg, routine applications of agrichemicals or mishaps resulting in chemicals released to soil, water, or air), natural hazards (eg, earthquakes, floods, or extreme weather events such as hurricanes or tornados), and events linked to use of a hazardous technology (eg, nuclear power for generation of electricity). Throughout this book, we have focused on environmental chemicals and the role that ecotoxicology plays in describing the relationships among those chemicals and the abiotic and biotic components of the environment. Our primary considerations for the balance of this chapter are characterizing how hazards and risks may be linked to chemicals released to soil, water, and air.

HISTORICAL PERSPECTIVES ON CHEMICAL STRESSORS AND ECOLOGICAL RISKS

Risk assessment can deal with any type of *stressor*. A stressor is any physical, chemical, or biological "agent" potentially capable of inducing a negative response from exposure. Stressors may adversely affect specific natural resources or entire ecosystems, including plants and animals, as well as the environment with which they interact. The history related to the evaluation of hazards and risks of environmental stressors—chemical, physical, and biological—is relatively short, yet the concepts of hazard evaluation and risk assessment have long been in practice, including evaluations of multiple stressors (eg, Menzie et al., 2007; Piggott et al., 2015). Although hazards and risks of chemicals in the environment initially brought the practice of risk assessment to a wide range of users, much of the practice advanced today will be considered in a relatively general context, providing guidance rather than prescriptions for implementation (Suter, 2006; US EPA, 2015a). These generic approaches are process driven, and their implementation necessarily involves specific application of a wide range of analytical tools determined, in part, by the nature of the risks being considered. Many of those tools and underlying technical findings that frame risk-based questions are directly linked to outcomes of ecotoxicology. Most practitioners focused on questions related to chemicals in the environment consider the risk-assessment paradigm initially advanced by the National Academy of Science (1983). This was a starting point that motivated numerous investigations, including those focused initially on ecological risks associated with chemical stressors (Warren-Hicks et al., 1991; Suter et al., 2000) and those applications concerned with multiple stressors occurring across a range of spatiotemporal contexts (eg, Graham et al., 1991).

With these terms in place for hazards and risks, we will characterize *risk analysis* as a process that strives to qualitatively or quantitatively estimate environmental risk, whereas *risk assessment* more often than not incorporates elements of administrative and management policy that reflect, in part, an individual's or organization's level of risk tolerance or risk acceptability when interpreting outcomes of an analysis of risks. Risk analysis incorporates qualitative or quantitative estimates of selected attributes of environmental hazards—for example, adverse effects and events or conditions that lead to or modify such adverse effects—in order to characterize how environments or exposed populations of humans or fish and wildlife might respond by manifesting adverse effects linked to exposure. Risk analysis becomes entangled with risk assessment, particularly within the settings of *risk management*, wherein: (1) decisions are developed regarding hazards of environmental chemicals, biota exposed to those hazards, and adverse effects associated or potentially associated with those exposures; (2) mitigation options are crafted as part of the implementation of those decisions; and (3) outcomes of those decisions are evaluated through monitoring and surveillance activities (Fig. 13.1). We will return to the process represented in Fig. 13.1 later in this chapter and illustrate how topics in previous chapters may be viewed within the context of this process.

For environmental chemicals, outcomes of a comprehensive evaluation of risks potentially serve to better inform resource managers and others regarding the nature and magnitude of adverse effects to humans (eg, residents, workers, and recreational visitors), and fish or wildlife exposed to chemicals occurring in soil, sediment, water, or air. Indeed, increased awareness and interest in managing environmental risks associated with chemicals and other environmental

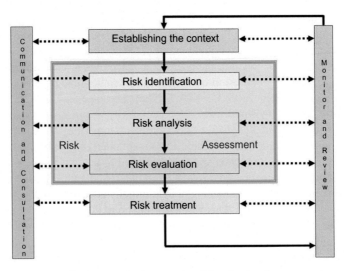

FIGURE 13.1 Generalized process for incorporating risk evaluations into resource management practices with particular applications to releases of environmental chemicals. '*Courtesy of US EPA*'.

stressors has gained visibility across a range of governmental and nongovernmental organizations, industrial, and other interests, as well as the general public. Information regarding risks gained by risk managers benefits their decision-making processes with respect to protecting humans and the environment from chemicals and a wide range of other environmental stressors. Indeed, we are all "risk managers" regardless of our roles as regulators (eg, federal or state officials serving to protect the environment), industrial or business authorities whose practices potentially impact the environment, or private citizens making decisions regarding chemical use, directly or indirectly contributing to the release of chemicals to the environment, and through their participation in the environmental regulatory process, contributing to development of risk and environmental policies governing their use.

GENERALIZED PROCESS FOR EVALUATING ENVIRONMENTAL RISKS

An evaluation of environmental risks relies heavily on scientific format. However, an increasing awareness of the sociopolitical elements of the environmental risk assessment and risk management processes has become well-established as part of the overall process (eg, Gowdy and Erickson, 2005; Brown and Timmerman, 2015).

The focus at this point will be on ecotoxicological ties that shape a generalized process for evaluating ecological risks. Chapter 12 will demonstrate the linkages between ecological and human-health risks, particularly in environmental settings that illustrate potential for sharing common frames of exposure to chemicals in soil, sediment, water, or air. Regardless of the context, whether it is conditioned on ecotoxicological frames of reference or on toxicological frames of reference specific to human-health applications, risks linked to environmental chemicals are strongly conditioned by: (1) chemical quantities occurring in exposure matrices—that is, what concentration of chemical(s) occur in a given matrix; (2) the toxicity of the chemical(s); (3) pathways linking a receptor and chemical(s) in an exposure matrix or, more commonly, multiple environmental matrices with multiple chemical constituents; and (4) duration of exposure between receptor and matrix contaminated with environmental chemical(s). A *receptor* in this context is any biological agent, usually plants or animals, considered representative of organisms that are likely to be encountered at the location of interest and likely exposed to environmental chemicals of concern. As we have seen from previous chapters, each of these conditioning factors will shape chemical risks that are linked to concentration and to duration of exposure. For example, risk analysis is dependent on ecotoxicological data that characterize a chemical's acute and chronic toxicity through the application of methods previously considered in Chapter 2.

Steps in Risk Assessment Processes

Based on the four conditioning qualities identified previously, we will revisit our generalized framework for implementing an evaluation of risks summarized in Fig. 13.1. Regardless of the language and terms applied to characterizing the assessment, evaluations of ecological and human health risks may be considered as a relatively simple four-component process that consists of establishing the context, identification, analysis, and evaluation. Establishing the context involves characterizing the site of concern—is it aquatic, terrestrial, atmospheric, etc.; are we concerned with soil, sediment, water, or air? Risk identification categorizes the species of concern and the chemicals that may be involved. Risk analysis attempts to determine to what degree the environment has or is likely to be impaired and whether biota are being affected. Risk evaluation brings in other factors such as economics, cost versus benefit scenarios, and determines whether and to what extent remediation should occur. The last category on the diagram, risk treatment may or may not occur depending on the results of the evaluation. As such, the analytical process assures technical outcomes that inform risk-based, resource management decisions with a level of detail required by the resource manager. Depending on the resource management questions and technical issues driving the evaluation of risks, assessment and monitoring needs may be satisfied through analysis of existing data and information, or designed field and laboratory studies may be undertaken as part of the assessment and monitoring activities (Suter, 1991; Karr and Chu, 1997).

In its simplest summary, the analysis, assessment, and management of risks are captured by a stepwise, iterative process wherein: (1) questions are formulated; (2) observations or experiments are conducted to provide answers that address those questions; and (3) decisions are made given the answers to the questions that initiated the process. Either decisions that result from the initial assessment may yield sufficient information so that management can evaluate potential outcomes of a particular management action, or the analysis process might be augmented to address critical data gaps identified as outcomes of the initial process. In risk-assessment language, data developed during the first round of investigation were not sufficient to support management decisions within the level of uncertainty reflected by the risk-tolerance of the decision-makers; thus, a second round of investigative studies would be indicated. The assessment process might also uncover a need to conduct a follow-up study to determine the effectiveness of remediation or perhaps a monitoring strategy for a defined period of time would be seen as being necessary.

In general, and regardless of the specific risk evaluation process being initiated, be that regulatory (Suter, 2006; US EPA, 2015b) or nongovernmental in kind (Suter, 2006), a planning and scoping stage serves to define the purpose and scope of any evaluation of risk. For risk evaluations focused on environmental chemicals, the risk analysis and assessment process often begins by collecting existing data or initiating reconnaissance investigations intended to characterize the nature and extent of chemical release to the environment, as well as to gather existing information and data regarding the fate and effects of the chemicals of concern. In pursuing existing data and information, we gain a better understanding of how these chemicals might behave in the future. Based on these preliminary assessments, a risk assessor then considers the frequency and magnitude of exposures, and identifies pathways potentially linking sources of environmental chemicals to humans and ecological receptors. This phase of the generalized process is most frequently termed "analysis of exposure," which is then considered relative to an "analysis of effects."

Analysis of effects is given different names by regulatory agencies or in peer-reviewed literature (Suter, 2006; National Academy of Science, 2009; Kapustka and Landis, 2010). Regardless of the name, however, each implementation of an analysis of effects is linked to the chemicals present and their toxicities, particularly because they serve as estimates of the nature and magnitude of possible adverse effects. Ideally, evaluations of risks would be based on reliable and complete data and existing information regarding the nature and extent of contamination, fate and transport processes, the magnitude and frequency of exposures to biological receptors in the field, and a chemical's toxicity. Unfortunately, having all of the desired information is the exception. In general, existing data and information on actual exposures and effects are limited. Thus, all risk estimates are uncertain to varying extents. When risk estimates are so uncertain that they exceed the tolerance of managers, those conducting the assessment may have to delve deeper into the literature or obtain additional relevant data in an iterative process, as indicated in the generalized process summarized in Fig. 13.1.

Once an analysis of exposures and effects is completed and risk has been assessed, risk management (risk management can combine risk evaluation and treatment) takes center stage in any project's lifecycle. Risk management is a relatively straightforward process that focuses on how to protect human health and ecological receptors. For example, identifying risk-based wastewater discharge limits that are acceptable for mining activities or identifying clean-up objectives for hazardous materials previously released to the environment can be elements of risk management.

Our overview of an evaluation of risk has been generalized. Yet, much of the practice implemented today follows regulatory guidance initially developed in the late 1970s and early 1980s with a focus on human health (Suter, 2006), which was followed shortly after by a similar process for evaluating risks associated with environmental chemicals when ecological receptors, principally fish and other aquatic biota, wild birds, mammals, and species of concern were the principal receptors of interest to federal and state regulatory agencies in United States, Canada, or Europe (Suter, 2006). For example, the US EPA provides guidance for conducting risk assessments with some processes focusing on human health, others on ecological concerns. Although the guidance for each of these regulatory-driven practices differs due to the receptors of primary concern, and there are relatively subtle distinctions between stages in the processes called for evaluating risks to human health and ecological concerns, our generalized process summarized in Fig. 13.1 easily accommodates such nuanced regulatory differences (Table 13.1).

Within regulatory contexts, ecological risk assessments are more widely implemented and support a variety of regulatory and resource management programs, including the regulation of hazardous waste sites, industrial chemicals, and pesticides, as well as the management of watersheds or other ecosystems affected by multiple chemical, physical, or biological stressors. As such, ecological risk assessment has been applied to projects that forecast likelihoods of future effects (prospective) or evaluate the likelihood that effects are caused by past exposure to stressors (retrospective). Information from ecological risk assessments may then be used by risk managers for follow-up, such as communicating to interested parties and the general public, limiting activities related to the ecological stressor, limiting use of a given chemical, or developing a monitoring plan to determine if risks have been reduced or whether an ecosystem is recovering. Thus, the current practice of ecological risk assessment can be applied to many "what-if" analyses related to forecasts of changes associated with various mitigation options considered for development under the auspices of risk management.

Assigning Risk

However, trying to simplify risk management to its core requires our brief return to the underlying characteristics of risk. Mathematically, risk is expressed in terms of the probability of hazard (as a numeric value bound by 0 and 1, where an event with "probability 0" never happens and an event with "probability 1.0" always happens, or as categorical estimates wherein ordinal assignments of risks are characterized based on experience or survey outcomes). For example, we might judge a given action as having risk rated as high, moderate, or low, and we subsequently gauge our risk management decision accordingly.

All risks are accompanied by consequences, most often characterized by frequency of occurrence (such as a numeric or categorical estimate of time) and severity (such as a numeric or categorical estimate of adverse impact, US EPA, 2009). A risk-based decision hinges on the definition of acceptable risk associated with a management action intended to offset risks linked to particular hazards. For example, management alternatives may range from no action to control chemicals released to the environment to costly management actions undertaken to attain a level of acceptable risk associated with release of chemicals, say, to a surface-water body such as a stream or river. Risk management, then, is the process of identifying and controlling hazards, where controls are intended to eliminate hazards or, more often than not, reduce their risk and attendant consequences to an acceptable level.

TABLE 13.1 Comparison of Generalized Framework With Regulatory Frameworks Guiding Evaluations of Risks for Human-health or Ecological Concerns

Generalized Process for Evaluating Risks (Fig. 13.1)	Regulatory Process Evaluating Risks Related to Human Health[a] (National Academy of Science, 1983)	Regulatory Process for Evaluating Risks Related to Ecological Concerns[b] (US EPA 1992, 1998)
Problem identification and specification: Information collected to help determine what biological systems are at risk, including special status species (eg, Endangered Species Act)	*Hazard identification:* Examines whether a stressor has the potential to cause harm to humans and/or ecological systems, and if so, under what circumstances	*Problem formulation:* Information is gathered to help determine what, in terms of plants and animals, is at risk and what needs to be protected
Analysis: Determines chemical concentrations in exposure matrices (abiotic and biotic), target receptors, and biological effects in receptors; determines exposure, particularly frequency, timing, and levels of contact with a stressor	*Dose-response assessment:* Examines the numerical relationship between exposure and effects *Exposure assessment:* Examines what is known about the frequency, timing, and levels of contact with a stressor	*Analysis:* Determines what plants and animals are exposed and to what degree they are exposed, and if that level of exposure is likely or not likely to cause harmful ecological effects
Characterization, synthesis, and interpretation: Determines numerical relationships between exposure and effects; development weight and strength of evidence about nature and extent of the risk from exposure to environmental stressors; complete sensitivity and uncertainty analysis	*Risk characterization:* Examines how well the data support conclusions about the nature and extent of the risk from exposure to environmental stressors	*Risk characterization:* Involves risk estimation, risk description, wherein risk estimation combines exposure profiles and exposure-effects and risk description provides important information for interpreting the risk results and identifies a level for harmful effects on the plants and animals of concern

[a]*Human health risk assessment includes four basic steps and is generally conducted following various EPA guidance documents, with early phases of the process involving preliminary assessments and a range of planning and scoping activities.*
[b]*Ecological risk assessments focus on evaluating how likely it is that the environment may be impacted as a result of exposure to one or more environmental stressors such as chemicals, land change, disease, invasive species, and climate change. Ecological risk assessment includes three phases, and is generally conducted following the Guidelines for Ecological Risk Assessment (US EPA, 1998).*

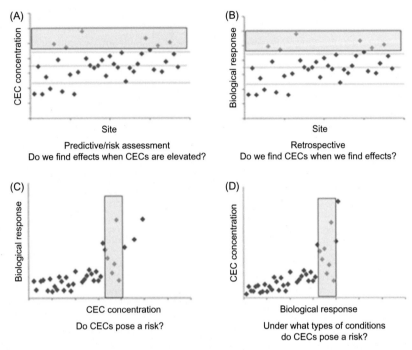

FIGURE 13.2 Some ways of looking at relationships between discharges from WWTP and managerial decisions. (A) Concentrations of contaminant discharges vary among WWTP and managers may decide to take action on those in the yellow bar. (B) Rather than contaminant concentrations, managers identify sites where effects occur and may decide to focus on those which display effects. (C) Instead of collecting data on effects, managers rely on published literature and make decisions based on previously identified threshold data. (D) Managers look at published data and identify the range of concentrations where a particular effect is seen and may then select sites to improve (Diamond et al., 2015).

A clear example of the interactive role between assessment and management of risk was presented by Diamond et al. (2015). Their focus was on chemicals of emerging concern (CECs), such as pharmaceuticals and personal care products that can be discharged from wastewater treatment plants (WWTP) because there is no currently available technology to remove these CECs from discharge water. However, their concepts can apply to other contaminants as well. Consider collecting data on CECs from many different WWTPs, some of which discharge high concentrations of CECs and others that discharge lesser amounts (Fig. 13.2A). Risk managers may decide to take steps to remediate the most polluted sites, as indicated by the yellow bar.

Alternatively, data may be gathered on the probability that an area is subject to risk (Fig. 13.2B). Signs to look for might include physiological responses, or signs at the population or community level. For example, one common community risk would be reduced species richness with the species known to be most sensitive absent from collections. The risk-management team might choose to remediate only those sites where risk is greatest. In the diagram, the most polluted sites (Fig. 13.2A) coincide with sites that show effects (Fig. 13.2B), but this need not be the case in all examples. Perhaps the most polluted sites still have concentrations that are below effects levels. Finding sites with no ecological impairment and relatively high concentrations of certain CECs, or those that are ecologically impaired with low CEC concentrations, could provide evidence that specific CECs are unlikely to pose a risk to aquatic life in similar situations. When concentrations of CECs are low but ecological damage is occurring, the managers may send out an assessment team to explore the possibility that other, non-CEC stressors are having an impact. When the concentration and effects data coincide (ie, there is evidence of harm at the sites with the highest concentrations), risk managers are asked if that harm is acceptable or not.

A third approach that has been used in some assessments is to develop information on potential thresholds based on measured CEC levels (or class) associated with documented ecological effects (Fig. 13.2C). In this case, water concentrations of selected CECs are measured and then compared with data from laboratory or previous field studies to determine if there is an overlap between known effect concentrations and measured concentrations. This process could lead to a threshold-setting exercise in which a regulatory agency, such as the US EPA, established acceptable standards of contamination. We caution that it may be too early to establish acceptable standards for most CECs due to a lack of information at this time.

Finally, a fourth approach that might be used to screen sites is to compare the range of a specific physiological effect, such as the production of vitellogenin in male fish or the average fecundity of fish for screening for endocrine-disrupting effects in relation to CEC concentration (Fig. 13.2D). This fourth approach could evaluate under what conditions certain physiological indicators of exposure, such as endocrine responses, could be translated into higher-level impacts or could be used to identify characteristics of water bodies (in addition to the CECs of interest) that correlate with relevant biological measures. In this approach, managers attempt to identify sites where CEC exposures would present a risk or identify stressors, in addition to the CECs of interest, that predict sites with potential biological impacts.

Hazards and Risks

Depending on the contingencies that drive the environmental risk-assessment process—here, our focus has been environmental chemicals—the evaluation of risks may be categorized as hazard-based or as risk-based to inform resource management decisions. The distinction between these ends of the technical-analysis spectrum are largely dependent on empirical data and information, with a hazard-based process generally reflecting limited empirical data, while a risk-based analysis will be more data rich. Risk-based evaluations tend to be more probabilistic in their derivation, whereas hazard-based evaluations tend to be more experience-dependent expressions of risk, based largely on expert judgements and a variety of analytical tools. In practice, a mix of hazard-based and risk-based tools are frequently used for evaluating environmental chemicals, particularly when the analysis focuses on chemicals that are not well known with respect to their toxicity or exposure potentials, or when multiple stressors are considered and chemicals are only a subset of the possible hazards. Given this brief overview of hazard-based and risk-based approaches likely underlying risk management decisions, we close our brief consideration of hazards and risks with a short consideration of a frequently deployed method for applying technical findings derived from ecotoxicological research to assessment and management of risks linked to exposures of environmental chemicals; namely, the analysis of exposure through food-chain analysis and the subsequent estimation of risk by comparing this estimate of exposure to a toxicity reference value.

EXPOSURE MODELS AND FOOD-CHAIN ANALYSIS FOR BIRDS AND MAMMALS

Exposure models for terrestrial and semiaquatic vertebrates, principally birds and mammals, usually focus on materials transfer and remain a favorite regulatory application for evaluating exposure comparable to the process implemented in human-health risk evaluations (Linder and Joermann, 2001; Linder et al., 2004). As indicated in Chapter 12, models are part and parcel of the skillset commonly used in risk evaluation and have become more commonly used as the disciplines of ecotoxicology and risk analysis have developed in recent years (Pastorok et al., 2001). Here, we simply ask you, the reader, to entertain some mathematical reasoning without fear and loathing because conceptually, Eq. (13.1) simply provides a mathematical explanation of how a wild bird or mammal can be exposed to sources of environmental chemicals:

$$E = \frac{1}{T}\sum C_{ijk}t_k \tag{13.1}$$

where
  $E =$ Exposure concentration or exposed dose
  $T =$ Total time and space over which the concentrations in various microenvironments or habitats are to be averaged
  $C_{ijk} =$ Concentration in microenvironment k that is linked to environmental matrix i by pathway j
  $t_k =$ Time and space that accounts for a receptor's contact with specific microenvironment or habitat k

This equation simply translates to "you are what you eat and drink"; that is—and please recall, risk is an iterative process—as a first estimate of exposure for some chemicals, drinking water and ingesting food likely serve as dominant sources in pathways of exposure. Most often, Eq. (13.1) is simplified, and becomes a unitless narrative (Eq. (13.2)):

$$ED = 3Der_{ed} + Inh_{ed} + f(Inh_{ed}) + Ing_{ed}$$
$$+ f(Ing_{ed}) + DW_{ed} + f(DW_{ed})$$
(13.2)

where

$ED =$ Exposed dose
$Der_{ed} =$ Dermal- or cutaneous-exposed dose
$Inh_{ed} =$ Inhalation-exposed dose
$f(Inh_{ed}) =$ Exposed dose coincidental to inhalation
$Ing_{ed} =$ Ingestion-exposed dose
$f(Ing_{ed}) =$ Exposed dose coincidental to ingestion
$DW_{ed} =$ Drinking water consumption exposed dose
$f(DW_{ed}) =$ Exposed dose coincidental to drinking water consumption

Ultimately, given practical matters and all too frequently, the relatively sparse to nonexistent data available, risk analysts further simplify, which yields a food-chain dominated exposure model (Eq. (13.3)):

$$ED = 3Ing_{ed} + f(Ing_{ed}) + DW_{ed}$$
(13.3)

wherein chemicals contributed by dermal and inhalation routes of exposure are considered negligible to the receptor. In the first iteration, the fraction of exposed dose that is absorbed is frequently assumed to be 100%, or otherwise estimated using available literature values to account for efficiencies in the uptake of material across surfaces such as the walls of the gut or other barriers between external and internal environments or across cell membranes. Although physiologically oversimplified, food-chain models lump together "whole-animal" functions of water consumption, foraging rate, food intake, and food processing through parameters that have been predefined by the regulatory agency. These functions provide an estimate of exposure referred to as exposed dose or site-exposure level.

For our first iteration estimate of risk, we need an estimate of toxicity—here, referred to as a *toxicity reference value (TRV)*—derived in a similar manner to exposed dose or site-exposure level, which regulatory agencies such as US EPA have developed and published in support of their regulatory programs (eg, US EPA, 2003). These TRVs can provide a better understanding of how exposure to selected chemicals in the environment translates into risks.

ESTIMATING RISKS USING SIMPLE RATIO ESTIMATORS

A commonly applied approach for evaluating ecological risks from environmental chemicals relies on a simple ratio estimator (US EPA, 2015c) that estimates potential for adverse effects to an ecological receptor, such as wild bird or mammal, by comparing the receptor's estimated levels of exposure—recall exposed dose or site-exposure level from above—with appropriate TRVs. Most TRVs for chemicals commonly encountered in environmental settings of regulatory interest have been derived from published toxicity studies, wherein toxicity values such as *no-observed-adverse-effect level (NOAEL)* or *lowest-observed-adverse-effect-level (LOAEL)* (see Chapter 1) were identified and compiled. Then a representative value was selected as a benchmark TRV for regulatory applications. TRVs are both chemical-specific and species-specific, and the comparison to estimate risk takes the form of a ratio, referred to as the hazard quotient (HQ) (13.4):

$$HQ = \frac{\text{Site Exposure Level}}{\text{TRV}} \qquad (13.4)$$

If the value of HQ is less than or equal to 1.0, risk is considered negligible to low, and no unacceptable effects will occur in the exposed population of receptors. In contrast, if the value of the HQ exceeds 1.0, then an unacceptable impact may occur, and risk will be moderate to high. Within regulatory settings, food-chain modeling has become a primary tool for evaluating exposure in birds and mammals. However, depending on the animal-at-risk and its life history (eg, early-life stages tend to be more sensitive to exposure than adults), the tool may be relatively ineffective for evaluating potential adverse effects and risks under field conditions. Animals with life history attributes that markedly differ from wild birds and mammals may not fit well with the HQ concept. For example, if dermal exposures are assumed to be unimportant, this may be very misleading for amphibians whose semipermeable skin plays a critical role in respiration and may be an important route of exposure for some environmental contaminants (Linder et al., 2010).

CAUSE-AND-EFFECT RELATIONSHIPS, MULTIPLE STRESSORS, UNCERTAINTY, AND RISKS

As became apparent in our brief overviews of effects of chemicals on population (see Chapter 10) and community and ecosystem (see Chapter 11) levels of organization, the more pieces of the puzzle, the more likely we are to encounter questions that defy simple answers. Indeed, complexities in biological systems across levels of organization or spatial scales through time will make the use of models that rely on technical findings (see Chapter 12) more difficult. Regardless of the innovation that continues to develop in the ever-expanding discipline of ecotoxicology, and the publication of new methods and high-quality data that fill data gaps critical to evaluating risks, environmental and resource managers focused on evaluating effects of environmental chemicals on biological systems, especially for large spatial scales or for systems considered of "special status," uncertainty must be considered in lockstep with characterization of risks. Thus, to close our chapter focused on risk and chemical stressors we briefly identify issues related to risks and uncertainty, particularly because those are entangled with cause-and-effect analysis.

With multiple stressor exposures, there are increased potentials for confounding analyses of cause-and-effect relationships between environmental chemicals and biological responses—especially at population and community levels of organization (see Chapter 11). Confusion increases as the number of chemicals at low concentrations increases. This can happen, for example, with the discharge of permitted wastewater releases following treatment at a municipal water treatment facility. Consequently, the simplicity of laboratory-derived characterizations of cause-and-effect relationship will likely be inadequate to explain field situations.

The number of examples of such entanglements involving multiple stressors, exposures in the field, and confounded cause-and-effect characterizations for environmental effects are many. However, we have conceptually identified categories of risks that, in part, help address these circumstances and enable our efforts to characterize uncertainties to follow suit. For example, risks may be viewed as being of various types: competing risks, aggregate risks, cumulative adverse effects, and cumulative risks, residual risks, and landscape or regional risks. We can also speak of vulnerability analysis, most often within the context of proactive risk management.

Types of Multiple Risks

Competing Risks

Competing risks—regardless of the level of biological organization considered—are very common in natural resources. These exist in land or water resources that are generally challenged by a range of chemical, physical, and biological stressors in the field. Each of these stressors may potentially be characterized as a hazard capable of resulting in a loss of system integrity and may trump chemical hazards such as, for example, when physical aspects of wildfire overwhelm the hazards from chemicals (eg, dioxins, furans) released by the fire. Simply said, regardless of a system's complexity or its level of biological organization, if system sustainability is the desired outcome of a resource-management plan, managers must consider multiple, interacting strategies such as chemical-management practices, land-use and water-use practices, or management of biological agents, such as invasive species or disease organisms. This is because a system's health may be at risk from numerous hazards that act independently or in combination to adversely impact that system.

Cumulative Adverse Effects and Cumulative Risks

When evaluating chemical and physical stressors as part of an environmental risk assessment, cumulative risk is narrowly defined as risk associated with exposure to all chemical and physical stressors with a common mechanism of toxicity. For chemical hazards, cumulative risk assessments combine risks from multiple exposure pathways and routes for all substances that have a common mechanism of toxicity. For example, within a multiple-stressors context, such risks would be identified by cumulative adverse effects associated with a loss of system integrity yielding similar, if not identical, end states.

Aggregate Risks

Aggregate risks share attributes of competing and cumulative risks and may be regarded as the sum total of all risks from multiple sources. Simply defined, aggregate risks are those that are due to single stressors that come from many different sources and collectively add to the total stress on a system. For example, chemicals may weaken the immune systems of organisms and make them more susceptible to viral or bacterial infections. Weakened animals may be easier for predators to catch and kill. The bacteria, viruses, and predators may have always been in the environment, but it was not until after the additive stress from chemicals that they could harm the biota.

For evaluating hazard-specific competing risks or aggregate risks, all possibilities may be seen as racing to see which occurs first. For biological and ecological systems, as well as complex systems in general, networks of stressors may occur in series (one after another), parallel (occurring together), or in a variety of other patterns with each other (Westbury et al., 2007; Olff et al., 2009). In addition, many systems may be characterized by marked interdependence and interconnectedness, which means biological or ecological systems will necessarily reflect operational structures that make management difficult or even impossible to analyze (Luo and Magee, 2011).

Residual Risks

Residual risk is the risk remaining after control measures have been identified and selected for managing hazards. To manage residual risks, a system's multiple stressors and their cumulative risks must be well-characterized. Similarly, the uncertainties associated with residual risks must be understood because management of these risks generally responds to uncertainties and those risks considered acceptable and manageable. Assessment of residual risks is an important step in the iterative process of risk assessment.

To address these residual risks, resource management often focuses on the system at risk through a top-down, holistic approach and seeks to resolve problems by identifying proximal causes related to maintaining system integrity. In contrast, much of the analysis focused on risks associated with multiple stressors and much of the regulatory focus-guiding risk management activities related to

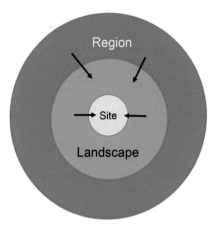

FIGURE 13.3   A simplified conceptual model of how higher levels of spatial hierarchies can affect lower levels. The regional level, which might be a watershed, will affect landscape areas such as agricultural fields in a county and these may affect the site of interest, which could be a stream.

natural resources relies on bottom-up approaches. A bottom-up approach is generally characterized by efforts to identify whether suspected causes are key to resolving resource management problems, then moves to eliminate those risks without determining the extent to which that solution will resolve the management problem. For multiple risks potentially requiring conflicting solutions, a bottom-up approach contributes to management conflicts when priorities are set among risks or to evaluate whether a risk management action is likely to succeed.

Landscape and Regional Risk

Landscape and regional risk assessments have increasingly been incorporated into the environmental decision-making process (eg, Westbury et al., 2007; Olff et al., 2009; Kapustka and Landis, 2010). These risks occur at spatial scales larger than the area of interest as, for example, an entire watershed at the regional level and surrounding agricultural fields at the landscape level (Fig. 13.3). Each higher level has inputs into the lower levels and understanding the risks at higher levels will facilitate understanding of all the risks at the site level. While the methods and tools applied in such analyses vary depending on the scope and spatial scale of the problem, decision making may benefit from the melding of risk disciplines focused on evaluating multiple stressors and their effects on biological and ecological structure and function (eg, Chen et al., 2013). Much of the impetus for developing these larger spatial-scale processes resulted from increasing applications of the environmental risk-assessment process, which provides a systematic approach to evaluating undesired effects linked to exposures and environmental stressors, including chemicals, land-use changes, altered hydrology, and climate change. Larger-scale approaches to evaluating multiple stressors can benefit from an ecological risk assessment framework, since the analysis can reflect (1) explicit consideration of scale and spatial organization during problem formation; (2) spatial heterogeneity in exposure characterization; and (3) extrapolation from small-scale studies to broad-scale effects, representing adverse effects from widely dispersed chemicals, such as those released to the atmosphere. From a risk-management perspective, opting into a nested, spatial hierarchy for evaluating risks could also provide maps or other visualization techniques for risk communication, while informing environmental decision makers.

Given this integrated spatiotemporal setting, a vulnerability analysis is more easily developed, again at various spatiotemporal scales.

Vulnerability Analysis

Within the context of environmental risk assessment, vulnerability is defined as a measure of a system's likelihood to experience adverse effects due to exposures from environmental perturbations or stressors, such as chemicals (Ippolito et al., 2010). Not surprisingly, these environmental perturbations are referred to as *threats*. Landscape or regional vulnerability analysis is comparable to chemical stressor evaluations focused on individual species or their populations. For instance, benthic invertebrates may be more vulnerable than open-water biota to poorly water-soluble materials released in aquatic environments. Sensitivities to environmental stressors, such as chemical contaminants may predispose some organisms as receptors or target species that are regarded as representative species for guilds of similarly exposed biota. Parallel to risk analysis, vulnerability analysis focuses on the quality of a resource at risk, including the frequency of occurrence and intensity of the hazards.

Regardless of the spatial scale, for many environmental stressors a fundamental limitation to analysis of risk is lack of data. Yet, in the likely event that data are not sufficient for a fully implemented quantitative analysis, reliance on categorical tools to evaluate hazards can be used in risk-management practices to develop a better understanding of the uncertainties associated with sparsely available empirical data; for example, as part of a hazard-based evaluation or risks.

Uncertainty

As Brewer and Gross (2003) observed, the desire to forecast risks and consequences linked to systemwide change encounters many challenges, prominent among them is accounting for uncertainly in models of ecological and physical processes. Such challenges confront risk analysts and natural resource managers who work within a network of multiple stressors and multiple hazards. Two general types of uncertainty—aleatory uncertainty and epistemic uncertainty—affect the characterization of risks, especially within their roles in influencing risk management. *Epistemic uncertainty* is also referred to as *subjective uncertainty, parameter uncertainty,* or *model-specific uncertainty,* and more simply stated, reflects uncertainty linked to our state of knowledge. On the other hand, *aleatory uncertainty*, frequently referred to as *stochastic uncertainty*, relates to the inability to fully characterize a model of a system that represents higher levels of development than those detailed by basic events in a process. Often, such sources of uncertainty are confused with statistical uncertainty, which is largely a complex of measurement error, observational error, and model-dependent assumptions underlying any particular method applied to data analysis. In any process, basic events more often than not consist of lower-level events, each potentially characterized by processes subject to failure or loss of integrity.

When applied to science-based discussions regarding risks, uncertainty often reflects a level of complexity that is inadequately appreciated. For example, in analysis of complex adaptive systems, such as ecosystems, the evaluation of model, parameter, or aleatory uncertainty is often based on expert opinion out of necessity (Krueger et al., 2012). Some types of uncertainty are more easily quantified than others, although a complete quantitative treatment of all types of uncertainty is often not achievable. For example, uncertainty arising through error, bias, and imprecise measurement, and uncertainty arising through inherent variation in natural parameters can be addressed through sampling in the field or in other forms of acquiring data. Uncertainty can be estimated by the quality and quantity of data obtained and can be regulated to some degree by creating study designs wherein the amount and type of data that will be accepted are defined prior to the study. For example, do we need 10 samples or 100? Do we measure to 1 g/L or 1 μg/L? In contrast, epistemic uncertainty arises through a lack of knowledge or scientific ignorance, which reflects uncertainty related to the state of knowledge (or rather lack of knowledge); thus, epistemic uncertainty is frequently termed *irreducible uncertainty.* Using the previous examples, we might not have any idea of how many samples will be needed or what level of accuracy in measurements are required. Each of these types of uncertainty undoubtedly exists in every analysis or forecast that forms part of risk characterization. Regardless of resource management needs, and in many respects the driver behind much of the research focused on uncertainty analysis as part of risk characterization, we close with a brief overview of cause-and-effect analysis. Although a longtime research interest in many disciplines, cause-and-effect analysis remains a critical element linked to the evaluation of environmental risks.

CAUSE-AND-EFFECT ANALYSIS

Robert Koch (1843–1910) was a prominent physician and pioneering microbiologist/epidemiologist who developed four postulates, all of which must be true for an infectious organism to be responsible for a disease: (1) the organism must always be present, in every case of the disease; (2) it must be able to be isolated from a host containing the disease and grown in pure culture; (3) samples of the organism taken from pure culture must cause the same disease when inoculated into a healthy, susceptible animal in the laboratory; and (4) the organism must be isolated from the inoculated animal and be identified as the same original organism first isolated from the originally diseased animal. To some degree, these postulates pertain to ecotoxicology with a few modifications, mostly dealing with the terminology of replacing organism with chemical or contaminant, and replacing pure culture with something like exposed population.

With exposures to multiple stressors, cause-and-effect analysis frequently calls for the application of Koch's postulates (Rothman, 2012) as part of the analytical process, but the evaluation of cause and effect for complex systems goes well beyond that process which was critical to identifying numerous causative agents of infectious disease (eg, cholera, tuberculosis, typhoid, diphtheria, and others) in the late 19th century. Indeed, exposures to multiple stressors complicate the process outlined by Koch's postulates. However, early foundations of cause-and-effect analysis that preceded those efforts focused on communicable diseases that demonstrated the value of a systematic, problem-solving approach. When applied to exposures involving multiple stressors, evaluations of cause-and-effect relationships may still be accomplished but they require a degree of complexity similar to that of the system being addressed.

Integrated field and laboratory investigations and manipulative studies have contributed to the process of cause-and-effect analysis for evaluating multiple-stressor exposures. With varying levels of effect, these studies moved from exploratory studies to those more focused on establishing relationships among specific stressors in environmental exposures, as, for example between cancer and the presence of polycyclic aromatic hydrocarbons (PAHs). Through experience and many studies, scientists have developed postulates similar to those of Koch and formalized these into a discipline called "ecoepidemiology" (Susser and Susser, 1996a,b; Susser, 2004).

Cause-and-effect analyses lead, in part, to an emphasis on risk management practices that ultimately stem from both estimates of exposures to chemicals in the field and from toxicity benchmarks derived in the laboratory as inputs critical to risk characterization. Cause-and-effect analyses assumes various guises, depending upon their specific applications and implementations For example, they may lead to identification of a disease agent or identification of a root cause of failure when your computer operating system malfunctions (Pearl, 2009). However, a common foundation binds these various applications because all are characterized by *response* and *explanatory variables*, and *risk factors* that influence the expression of response (or nonresponse, as it may be). Explanatory and response variables may be *direct* or *indirect* in their association (Fig. 13.4), with *multifactorial systems* being complicated by interactions among component factors that influence the expression of response. The degree of complication in multifactorial systems yields various categories of cause. For example, a frequent categorization of cause used in epidemiological studies identifies factors as necessary or sufficient. These terms are most easily illustrated in simple systems, wherein a

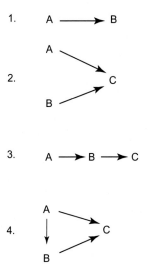

FIGURE 13.4 Simplified illustrations of direct and indirect cause-and-effect relationships. We consider simple two-factor or three-factor systems and network flows connecting these factors when we consider identifying cause-and-effect relationships. For example, in (1) A is the direct cause of B, while in (2) A and B are independent, but both are direct causes of C. In (3) and (4), relatively simple multifactorial cases are illustrated. In (3), A and C are indirectly linked by B, which is the direct cause of C. A and B are direct causes of C in (4), although A may also occur as an indirect cause if B serves as a contributing or intervening factor.

cause is sufficient if it inevitably yields an effect. In multifactorial processes such as disease or an otherwise compromised state of a complex system, sufficient cause nearly always occurs as a set of interacting component causes, where one component is commonly described as *the* cause. A *necessary cause* must always be present to produce a specific effect, outcomes classically observed in epidemiological studies focused on human health or investigations of fish or wildlife disease outbreaks (eg, *Myxobolus cerebralis* must always be present for a diagnosis of whirling disease in salmonid fishes). However, *M. cerebralis* may be present in very low titers and no disease is expressed, so the mere presence of *M. cerebralis* is not *sufficient* to cause the disease; it must be at some higher level to induce the signs of the disease. Yet many infectious and noninfectious diseases—the latter frequently of concern to chemical exposures—may be produced by different sufficient causes that may or may not have component causes in common. Uncomplicated infectious disease is frequently characterized by a disease agent that serves as a necessary cause, and in some instances, a factor may be *"necessary and sufficient,"* depending on the specific process being considered. In ecotoxicology. a particular contaminant, say p,p′-DDE, may be necessary in all species to induce a specific kind of eggshell thinning, but the concentration of DDE sufficient to produce thinning is species-specific. Infectious diseases are often host-specific.

In multifactorial processes such as those characteristic of environmental exposures to chemicals, factors may be necessary, sufficient, neither, or both. Component causes in a multifactorial system are generally characterized as *predisposing factors, precipitating factors, reinforcing factors*, and *enabling factors*. In characterizing failures in biological systems, predisposing factors are those that increase susceptibility (eg, that of a host to a disease agent or process). For example, the manifestation of disease in a host is frequently influenced by its immune status. Similarly, within an ecotoxicological context, a predisposing factor might be age, as that factor may be critical to the development of metabolic detoxification systems. Precipitating factors are those that are associated with the definitive onset of response, but are not sufficient in the absence of a necessary cause (eg, whirling disease may be precipitated by infection with *M. cerebralis* in susceptible individuals). Reinforcing factors are those that aggravate the expression of response, which in the case of disease agents might be repeated exposures in a genetically predisposed, susceptible individual or in an ecotoxicological context, to a behavioral factor that exacerbates exposure to chemically contaminated prey (eg, failure of avoidance behaviors). Enabling factors tend to be less clearly characterized than other categories of component factors because they are those components of exposure that facilitate the expression of response (eg, dry years may enable disease outbreaks to occur in grassland habitats that usually have a low incidence of disease, such as those reliant on arthropod vectors, or reduced prey-base may enable disease outbreaks predicated on malnutrition of the host). An example in an ecotoxicological context would be high uptake of heavy metals in waters with low calcium concentrations.

Epidemiological cause-and-effect models approach ecological complexity when disease processes are considered within a field setting where multiple stressors are common features of exposure. In such settings, simple linear models of cause and effect may be of limited usefulness, since multifactorial systems are characterized by having factors with varying intensity that interact at various levels in the system. These multifactorial systems are adaptive and highly dynamic, yielding a "web of causation" (Krieger, 1994), in which both epidemiology and ecotoxicology wrestle with exposures to multiple stressors of various types—biological, chemical, or physical. Indeed, beyond simple cause-and-effect analysis focused on identification of a single disease-causing agent or characterization of effects linked to a single toxicant, exposures in the field must necessarily acknowledge the ever-present role of confounding factors that inevitably produce spurious associations among variables and potentially mask real cause-and-effect relationships.

STUDY QUESTIONS

1. What is meant by risk in ecotoxicology? What poses the risk, and what is at risk?
2. Identify at least five types of stressors on populations of organisms. Can we say that any one stressor is more important than others for organisms in the same group (eg, terrestrial mammals)?

3. What is the difference between risk analysis, risk assessment, and risk management?
4. Outline the steps of a generalized risk analysis process.
5. Under what conditions might a manager accept some degree of risk? (Ie, if the manager decides that she can tolerate some risk.)
6. What are the four ways of making management decisions based on risk assessment as described by Diamond et al. (2015)?
7. What is a hazard-based assessment? How does it differ from a risk-based assessment?
8. Discuss with your class the concept of multiple stressors. What are some of these stressors? Which model, single or multiple stressors, better reflects real-world situations? How do multiple stressors interact to make an assessment more complex?
9. Describe how competitive, cumulative, aggregate, and residual risks are similar and different from each other.
10. What contributes to uncertainty in a risk assessment? How does this uncertainty affect the development of management decisions?

REFERENCES

Brewer, C., Gross, L., 2003. Training ecologists to think with uncertainty in mind. Ecology 84, 1412–1414.

Brown, P.G., Timmerman, P. (Eds.), 2015. Ecological Economics for the Anthropocene: An Emerging Paradigm. Columbia University Press, New York.

Chen, S., Chen, B., Fath, B.D., 2013. Ecological risk assessment on the system scale: a review of state-of-the-art models and future perspectives. Ecol. Model. 250, 25–33.

Diamond, J., Munkittrick, K., Kapo, K.E., Flippin, J., 2015. A framework for screening sites at risk from contaminants of emerging concern. Environ. Toxicol. Chem. 34, 2671–2681.

Gowdy, J., Erickson, J.D., 2005. The approach of ecological economics. Camb. J. Econ. 29, 207–222.

Graham, R.L., Hunsaker, C.T., O'Neill, R.V., Jackson, B.L., 1991. Ecological risk assessment at the regional scale. Ecol. Appl. 1, 196–206.

Ippolito, A., Sala, S., Faber, J.H., Vighi, M., 2010. Ecological vulnerability analysis: a river basin case study. Sci. Total Environ. 408, 3880–3890.

Kapustka, L.A., Landis, W.G. (Eds.), 2010. Environmental Risk Assessment and Management from a Landscape Perspective. John Wiley & Sons, Inc., New York.

Karr, J.R., Chu, E.W., 1997. Biological monitoring: essential foundation for ecological risk assessment. Hum. Ecol. Risk Assess 3, 993–1004.

Krieger, N., 1994. Epidemiology and the web of causation: has anyone seen the spider? Soc. Sci. Med. 39, 887–903.

Krueger, T., Page, T., Kubacek, K., Smith, L., Hiscock, K., 2012. The role of expert opinion in environmental modelling. Environ. Model. Softw. 36, 4–18.

Linder, G., Joermann, G., 2001. Assessing hazard and risk of chemical exposures to wild mammals: food-chain analysis and its role in ecological risk assessment. In: Shore, R., Rattner, B. (Eds.), Ecotoxicology of Wild Mammals. John Wiley & Sons, Inc., New York, pp. 635–670.

Linder, G., Harrahy, E., Johnson, L., Gamble, L., Johnson, K., Gober, J., et al., 2004. Sunflower depredation and avicide use: a case study focused on drc-1339 and risks to non-target birds in

North Dakota and South Dakota. In: Kapustka, L.A., Galbraith, H., Luxon, M., Biddinger, G.R. (Eds.), Landscape Ecology and Wildlife Habitat Evaluation: Critical Information for Ecological Risk Assessment, Land-use Management Activities, and Biodiversity Enhancement Practices, ASTM STP 1458. ASTM Intl, West Conshohocken, PA, pp. 202–220.

Linder, G., Palmer, B.D., Little, E.E., Rowe, C.L., Henry, P.F.P., 2010. Physiological ecology of amphibians and reptiles: life history attributes framing chemical exposure in the field. In: Sparling, D., Linder, G., Bishop, C.A., Krest, S.K. (Eds.), Ecotoxicology of Amphibians and Reptiles, second ed. CRC Press, Boca Raton, FL, pp. 105–166.

Luo, J., Magee, C.L., 2011. Detecting Evolving Patterns of Self-organizing Networks by Flow Hierarchy Measurement. Complexity http://dx.doi.org/10.1002/cplx.20368. Published online in Wiley Online Library.

Menzie, C.A., MacDonell, M.M., Mumtaz, M., 2007. A phased approach for assessing combined effects from multiple stressors. Environ. Health Perspect. 115, 807–816.

National Academy of Science, 2009. Science and Decisions: Advancing Risk Assessment. Committee on Improving Risk Analysis Approaches Used by the U.S. EPA, Board on Environmental Studies and Toxicology. Division on Earth and Life Studies, National Research Council, Washington, DC, p. 424.

National Academy of Science, 1983. Risk Assessment in the Federal Government: Managing the Process. Committee on the Institutional Means for Assessment of Risks to Public Health, National Research Council, Washington, DC, p. 191.

Olff, H., Alonso, D., Berg, M.P., Eriksson, B.K., Loreau, M., Piersma, T., et al., 2009. Parallel ecological networks in ecosystems. Phil. Trans. Royal Soc. B 364, 755–1779.

Pastorok, R.A., Bartell, S.M., Ferson, S., Ginzburg, L.R., 2001. Ecological Modeling in Risk Assessment: Chemical Effects on Populations, Ecosystems, and Landscapes. CRC Press, Boca Raton, FL, p. 328.

Pearl, J., 2009. Causal inference in statistics: an overview. Stat. Surv. 3, 96–146.

Piggott, J.J., Townsend, C.R., Matthaei, C.D., 2015. Reconceptualizing synergism and antagonism among multiple stressors. Ecol. Evol. 5, 1538–1547.

Rothman, K.J., 2012. Modern Epidemiology, second ed. Oxford University Press, New York, p. 280.

Susser, E., 2004. Eco-epidemiology: thinking outside the black box. Epidemiology 15, 519–520.

Susser, M., Susser, E., 1996a. Choosing a future for epidemiology: I. Eras and paradigms. Am. J. Public Health 86, 668–673.

Susser, M., Susser, E., 1996b. Choosing a future for epidemiology: II. From black box to Chinese boxes and eco-epidemiology. Am. J. Public Health 86, 674–677.

Suter II, G.W., 1991. Applicability of indicator monitoring to ecological risk assessment. Ecol. Ind. 1, 101–112.

Suter II, G.W., 2006. Ecological Risk Assessment, second ed. CRC Press, Boca Raton, FL, p. 680.

Suter II, G.W., Efroymson, R.A., Sample, B.E., Jones, D.S., 2000. Ecological Risk Assessment for Contaminated Sites. CRC Press, Boca Raton, FL, p. 460.

US EPA, 1986. Hazard Evaluation Division. Standard Evaluation Procedure, Ecological Risk Assessment. EPA 540/9-85-001. Prepared by Urban DJ, Cook NJ. Office of Pesticide Programs. US Environmental Protection Agency. Washington, DC, p. 103.

US EPA, 1992. Framework for Ecological Risk Assessment. EPA/630/R-92/001. Risk Assessment Forum. U.S. Environmental Protection Agency, Washington, DC, p. 57.

US EPA, 1998. Guidelines for Ecological Risk Assessment. EPA/630/R-95/002F. Risk Assessment Forum. U.S. Environmental Protection Agency. Washington, DC, p. 188. (Published on May 14, 1998, Federal Register 63(93): 26846–26924).

US EPA, 2003. Generic Ecological Assessment Endpoints (GEAEs) for Ecological Risk Assessment EPA/630/P-02/004F. Risk Assessment Forum. U.S. Environmental Protection Agency. Washington, DC, p. 67.

US EPA, 2009. Risk Management Program Guidance for Offsite Consequence Analysis. EPA 550B99009. Office of Solid Waste and Emergency Response. U.S. Environmental Protection Agency. Washington, DC. p. 134.

US EPA, 2015a. Conducting an Environmental Risk Assessment. <http://www.epa.gov/risk/conducting-ecological-risk-assessment> (accessed 12.12.15).

US EPA, 2015b. Risk Assessment. <http://www.epa.gov/risk> (accessed 12.12.15).

US EPA, 2015c. Eco Risk Characterization. <http://www.epa.gov/region8/eco-risk-characterization#hq> (accessed 12.12.15).

Warren-Hicks, W., Parkhurst, B., Baker Jr., S.S., 1991. Ecological Assessment of Hazardous Waste Sites: A Field and Laboratory Reference. EPA/540/R-92/003. United States Environmental Protection Agency. Office of Research and Development. Washington, DC, p. 299.

Westbury, A.-M., Tiller, D., Metzeling, L., 2007. Environmental flows and ecological health of the lower Wimmera River. In: Wilson, A.L., Dehaan, R.L., Watts, R.J., Page, K.J., Bowmer, K.H., Curtis, A. (Eds.), Proceedings of the 5th Australian Stream Management Conference. Australian Rivers: Making a Difference. Charles Sturt University, Thurgoona, New South Wales, pp. 449–454.

Chapter 14

Contaminant Considerations in Humans

Terms to Know

Occupational Safety and Health Administration (OSHA)
National Institute for Occupational Safety and Health (NIOSH)
American Conference of Governmental Industrial Hygienists (ACGIH)
Chloracne
Hypospadias
Cryptorchism
International Agency for Research on Cancer (IARC)
Mesothelioma
Acrodynia
Itai-itai Disease

INTRODUCTION

When it comes to contaminant exposure, humans are pretty much like any other animal. We can generally suffer the same maladies when exposed to certain contaminants—cancer, genetic mutations, liver and other organ damage, suppressed immune systems, and many of the other problems that we have discussed in previous chapters. There is one major difference between us and other animals, however. Whereas animals tend to be the innocent victims of pollution and contaminant dispersal, we are the primary cause of these problems. There are certainly some sources of naturally occurring contaminants such as forest fires and volcanoes that expel polycyclic aromatic hydrocarbons and sporadic instances of toxic concentrations of metals. However, the vast majority of contaminants originate due to human activities such as mining, inadequate control of industrial wastes, or overuse of pesticides. Some 35 years ago, the cartoonist Walt Kelly developed a classical cartoon with his foremost character, Pogo the possum, saying "We have met the enemy and he is us" (Fig. 14.1). He was referring to our role in creating pollution, and the saying is just as true today as it was then.

As we have discussed in previous chapters, some nations have made major advances in reducing contaminants in many parts of the world. Unfortunately,

FIGURE 14.1   The cartoon *Pogo* by Walt Kelly, featuring the possum Pogo laying it on the line. *Courtesy Wikimedia Commons.*

other industrial and developing countries are just starting to become concerned about the amounts and types of pollutants that they are expelling. Globally, we still have a long way to go before we can truly say that the global environment is clean.

In this chapter, we discuss some relationships between humans and contaminant issues. We will revisit some topics we have previously mentioned in passing, but will primarily focus on human health and risk.

HOW ARE HUMANS EXPOSED TO CONTAMINANTS?

Contaminant exposure to people comes in many forms. We may ingest them through food or water. Those who spend considerable time outdoors, especially farmers, may be exposed to pesticides through dust and absorption through the lungs or skin. City dwellers inhale toxins from smog and other forms of air pollution. An area of particular concern is maternal transfer of contaminants through nursing by very young children. For those in several areas of industry, inhalation of manufacturing byproducts can be a serious problem. In the United States, the US Environmental Protection Agency (EPA) and the Occupational Safety and Health Administration (OSHA) within the Department of Labor are the principal agencies concerned with regulating exposure to toxic chemicals from the environment and in the workplace; similar agencies serve this purpose in all developed countries. The U.S. Food and Drug Administration (FDA, 2014) is concerned about contaminants in our food supply.

Contaminants in Food and Water

Contaminants are an added cause for concern along with allergens and pathogens that may be transferred in our water and food supplies. Currently, the FDA is particularly concerned about a short list of contaminants in addition to monitoring hundreds of other chemicals that pose risks to humans. Among the contaminants currently under study are (FDA, 2014):

1. *Acrylamide*—A crystalline compound with the chemical formula of C_3H_5NO that is formed in starchy foods such as grains, potatoes, and coffee under high cooking temperatures from frying, roasting, and baking, for example. At very high doses of acrylamide, some laboratory animals developed cancers and the FDA is concerned about the possibility of the same effect in people.
2. *Benzene*—We have already stated that benzene forms the backbone for several organic pollutants and is a recognized human carcinogen. The Maximum Concentration Level (MCL) for benzene is 5 ppb and the FDA found concentrations higher than that in soft drinks preserved with benzoic acid. Benzoic acid can apparently convert to benzene. The soda industry is taking steps to reduce or eliminate benzoic acid.
3. *Dioxins, furans, and PCBs*—The FDA is concerned because these chemicals are extremely persistent and dioxins and furans may be produced through natural processes. These chemicals may be found in fatty meats and fish because they are lipophilic. Dioxins and furans can also be produced through high temperatures used in cooking.
4. *Ethyl carbamate (Urethane)*—Carbamates can be pesticides but ethyl carbamate is a naturally occurring compound formed during the fermentation of wines. At high doses, it can cause cancer in laboratory animals. The wine industry is interested in reducing the chemical's presence by altering how wine is produced.

5. *Melamine*—Recall that melamine is a triazine and therefore related to atrazine and simazine herbicides. It and its analogs have been added to milk produced in China that has been used to manufacture infant formulas and other dairy-based products that were shipped to other countries. In 2008, several companies were implicated in a scandal involving milk and infant formula which had been adulterated with melamine. The FDA (2009) provided advisories for more than 25 human and pet foods in addition to infant formulas that contained melamine. Most of the detected concentrations were very low, less than the level thought to be of concern, but 300,000 people were affected. Other sources of melamine are old, chipped dishware made of Melmac and some ceramics made in China.

6. *Perchlorate*—Perchlorate (ClO_4^-) is a potent thyroid inhibitor that can occur naturally and is also produced as an additive to explosives and rocket fuels. It has been identified in some foods and drinking water in western states. It is of particular concern to young children.

In addition to these compounds, the FDA monitors the presence or absence of 484 pesticides in domestically produced and imported food products. In 2012, for example, the FDA identified 195 pesticides in food products (FDA, 2012). A total of 43% of domestic products had pesticides, but only 2.8% had concentrations in excess of those recognized to be a risk. Fruits had the greatest occurrence of pesticide residues with up to 92% of peaches and 85% of apples, grapes, and raspberries having some pesticides detected. In addition, 4% of pear samples and 20% of a small number of eggplant samples had concentrations that were higher than the maximum allowable limits. Among imported products, only 33.6% had pesticides, but 11.1% of all products sampled had concentrations that were considered to be in violation of accepted concentrations. Similar to domestic findings, fruits and fruit products had the highest percentages of detected contaminants among many imported commodities. Many commodities including blackberries, limes, lemons, papaya, snap peas, basil, paprika, and coffee had pesticides in at least 20% of their samples with concentrations that exceeded allowable levels. Up to 50% of tea samples were in violation of FDA standards. Do you like Spanish food? A total of 69% of cilantro samples exceeded acceptable concentrations of pesticides.

The US EPA has set drinking water criteria for more than 75 different primary chemicals and radionuclides and 15 additional contaminants of secondary interest (US EPA, 2014). Nitrates, sulfates, lead, and fluoride are among the most commonly occurring contaminants in drinking water, but several pesticides, heavy metals, and other compounds have been identified in wells that are connected to aquifers and receive no treatment prior to consumption.

Exposure to Agricultural Chemicals

Farm workers are primarily subjected to pesticides and other agricultural chemicals through contacting sprays and drinking tainted water. Approximately 10,000 to 20,000 cases of pesticide poisoning in the United States are reported to physicians annually by migrant workers, pesticide applicators, other farm workers, pet groomers, and groundskeepers (CDC, 2015a), The United Nation's World Health Organization estimates that 186,000 deaths occurred globally from pesticides in 2002; deaths that could have been prevented (WHO, 2015).

Urban Air Pollution

Breathing air pollution such as ground level ozone, particulates, carbon monoxide, nitrogen oxides, sulfur dioxide, and lead can have numerous effects on human health, including respiratory problems, hospitalization for heart or lung disease, and even premature death. Children are most sensitive to ozone and may experience long-term health effects because of their still-developing lungs. In addition to smog, urbanites tend to be more exposed than rural inhabitants to other types of air pollution including more than 175 hazardous chemicals such as benzene, methylene chloride, mercury, and dioxins (EPA, 2015a). Some of these chemicals are known or suspected carcinogens. Others may cause respiratory effects, birth defects, and reproductive and other serious health effects. Some can even cause death or serious injury if accidentally released in large amounts. Fortunately, air pollution has been decreasing in the United States but other countries suffer very high concentrations of air contamination. China, especially its major cities such as Beijing, are infamous for their poor air quality (Fig. 14.2). During the 2008 Olympics, the only thing that permitted the games to occur in Beijing was a massive but short-lived effort to close down factories and limit vehicle traffic.

FIGURE 14.2　A smoggy day in Beijing, China. *Courtesy China.org.cn.*

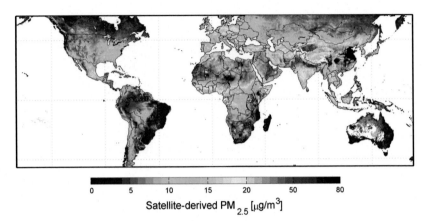

$$\text{Satellite-derived PM}_{2.5}\,[\mu g/m^3]$$

FIGURE 14.3   Satellite-derived imagery showing the global distribution of fine particulate matter. *Courtesy NASA (2010). http://www.nasa.gov/topics/earth/features/health-sapping.html.*

One particular contaminant of global concern are called *ultrafine particles*; air pollutant particles that are smaller than 2.5 µm in diameter and abbreviated $PM_{2.5}$ (see Chapter 9). These ultrafine particles can penetrate deep into the lungs and cause several different kinds of respiratory ailments. Epidemiologists suspect that millions of premature deaths occur each year due to these diseases (NASA, 2010). Investigators (van Donkelaar et al., 2010) developed a satellite-based map illustrating the distribution of $PM_{2.5}$ on a global basis (Fig. 14.3). $PM_{2.5}$ can occur naturally due to loose soils. Northern Africa and the Middle East have very high concentrations of these particulates due to their sandy soils. The particulates can also be produced through combustion of fossil fuels, vehicle exhausts, and other sources that are especially common in urban areas. There are high concentrations of $PM_{2.5}$ in the urban areas of China, India, and a few industrial regions in the United States.

Maternal Transfer

If you recall, we have previously discussed maternal transfer of chemicals which involves, in mammals, chemicals in milk being transmitted to offspring. Maternal transfer can also occur when the infant is in the womb and in egg-laying animals. Whereas lipophilic chemicals such as many persistent organic pollutants may be transmitted through the fatty portion of milk, hydrophilic components can be transmitted along with the liquid fraction. In developing embryos and fetuses, the placenta provides a barrier to some contaminants. However, the placenta is at least partially permeable to heavy metals such as lead, mercury, and cadmium, particularly when they are combined with organic molecules. These metals can also be found in amniotic fluid and in umbilical cords (Caserta et al., 2013). Fetuses are particularly sensitive to contaminant exposure because their organ systems are under rapid development. Similarly, neonates are also sensitive to chemical exposures. This is why there are so many concerns about pregnant and nursing women becoming exposed to contaminants, even medicines, through ingestion.

Speaking of the placenta as a partial barrier, Needham et al. (2011) investigated maternal and cord blood, cord tissue, placenta, and milk in connection with human births in the Faroe Islands, where exposure to marine contaminants can be high. In 15 sample sets, they measured 87 environmental chemicals, almost all of which were detected both in maternal and fetal tissues. Lipid-based concentrations of organohalogen compounds (eg, PCBs, dioxins, PBBs) in maternal serum averaged 1.7 times those of cord serum, 2.8 times those of cord tissue and placenta, and 0.7 times those of milk. For organohalogen compounds detectable in all samples, a high degree of correlation between concentrations in maternal serum and the other tissues investigated was generally observed ($r^2 > 0.5$). Transfer from the mother to the infant was reduced as the degree of halogenation increased. Concentrations of some organochlorines, and several PCB congeners with low chlorination were higher in fetal samples than in the mother and showed poor correlation with maternal levels. Perfluorinated compounds (eg, some flame retardants) occurred in lower concentrations in cord serum than in maternal serum. Cadmium, lead, mercury, and selenium were all detected in fetal samples, but only mercury showed close correlations among concentrations in the mothers and offspring. The authors concluded that although the environmental chemicals examined passed through the placenta and were excreted into milk, there was variation in the passage of chemicals from mother to fetus. This variation would add uncertainty to any risk assessment conducted for infants.

There can be strong associations between contaminant concentrations in mothers and their infants. For example, in Amazonia, people who live along the rivers consume more fish than city dwellers, or rural inhabitants (Marques et al., 2013). Mercury concentration is also higher in the hair (a noninvasive tissue for sampling) of riverine people than in the other two groups, and mercury in the hair of newborns correlates very highly with that of the mother, both among and within groups of people (Table 14.1, within-group ranged from 0.6744 to 0.8952; $p = 0.0001$ in each case).

Workplace Exposures

Workers can be exposed to harmful chemical compounds in the form of solids, liquids, gases, mists, dusts, fumes, and vapors. These chemicals can exert toxic effects through inhalation, absorption through the skin, or ingestion. OSHA sets enforceable permissible exposure limits (PELs) to protect workers against the health effects of exposure to hazardous substances, including limits on the

TABLE 14.1 Relationships Among Median (Min-Max) Fish Consumption, Mercury Content in Maternal Hair, Neonate Body Weight, and Mercury Content in Neonate Hair Among Groups of Natives Living in the Madeira River Basin, Brazil

Characteristic	Riverine People	Urban	Rural	P[a]
Maternal hair Mercury (μg/g)	12.12 (1.02–130.72)	5.36 (0.73–24.14)	7.82 (2.56–41.1)	0.00001
Fish meals/month	5 (0–7)	2 (0–7)	3 (2–7)	0.00001
Newborn body weight (kg)	3.15 (2.01–5.25)	3.21 (2.2–5.9)	3.01 (2.0–4.3)	0.048
Neonate hair mercury (μg/g)	3.01 (0.09–18.5)	1.5 (0.11–4.81)	1.98 (0.29–8.77)	0.00001

Source: From Marques et al. (2013).
[a]*Based on the median test.*

airborne concentrations of hazardous chemicals in the air. Most OSHA PELs are 8-h time-weighted averages (TWA), although there are also ceiling and peak limits, and many chemicals include a skin designation to warn against skin contact. Approximately 500 chemicals have had PEL assignments. Most of these PELs are for air contaminants because it is easier to protect workers from direct contact with hazardous materials than it is to protect them from fumes and airborne particulates that may be inhaled. Unfortunately, OSHA admits that many of these PELs are outdated because they were established in the 1970s and have not been updated. However, they still provide some protection for workers because companies can be fined for exceeding these standards. Separate lists are provided for general industry and marine environments. Other agencies that have established PELs or their equivalents include the National Institute for Occupational Safety and Health (NIOSH) under the Centers for Disease Control (CDC) and the American Conference of Governmental Industrial Hygienists (ACGIH), an association sponsored by industries and professional hygienists. California has established its own set of PELs which, like other state regulations, can be stricter but not more lenient than federal regulations. In many cases, these agencies have similar standards (Table 14.2).

SOME EFFECTS OF ENVIRONMENTAL CONTAMINANTS ON HUMANS

When discussing the effects on contaminants on wildlife we frequently included humans. We do not want to rehash all of that material. Instead, we will cover some of the more salient information in this section.

TABLE 14.2 Permissible Limits of Selected Contaminants in Industrial
Settings as Established by Regulatory Agencies

Chemical	8-h Federal OSHA	8-h California OSHA	10-h NIOSH	8-h ACGIH
Acetone	1000 ppm	500	250	250
Aldrin	0.25 µg/m³	0.25	0.25	0.05
Aluminum dust	5 µg/m³	5	5	1
Carbaryl	5 µg/m³	5	5	0.5
Carbon dioxide	5000 ppm	5000	5000	5000
Chlorine	1 ppm	0.5	0.5	0.5
DDT	1 µg/m³	1	0.5	1
Dieldrin	0.25 µg/m³	0.25	0.25	0.1
Ethyl alcohol	1000 ppm	1000	1000	1000
Grain dust	10 µg/m³	10	4	4
Hydrogen cyanide	10 ppm	4.7	4.7	4.7
Lindane	0.5 µg/m³	0.5	0.5	0.5
Ozone	0.1 µg/m³	0.1	0.1	0.05–0.2
Pyrethrum	5 µg/m³	5	5	5
Sulfuric acid	1 µg/m³	0.1	1	0.2
Zinc chloride	1 µg/m³	1	1	1

Source: U.S. Department of Labor, OSHA, 2015.
Note: 1 µg/m³ = 0.001 ppm.

Dioxins, Furans, and Polychlorinated Biphenyls

These chlorinated hydrocarbons are found everywhere in the environment due to their long-term persistence. The compounds that fall within the middle of the degree of chlorination are most toxic due to their lipophilic nature and mobility. Short-term exposure of humans to high levels of dioxins may result in skin lesions, such as chloracne (Fig. 14.4), patchy darkening of the skin, and altered liver function. Long-term exposure is linked to impairment of the immune system, the developing nervous system, the endocrine system, and reproductive functions. According to the United Nations' World Health Organization, dioxins are a known human carcinogen.

Cardiovascular effects have been reported in some groups of factory workers exposed to dioxins. These effects included atherosclerosis and deterioration of

FIGURE 14.4 Viktor Yushchenko, the former president of the Ukraine, was reportedly poisoned with dioxins in 2004 and subsequently developed a serious case of chloracne. The picture on the left shows him prior to the poisoning, the middle several months after the incident, and on the right several years afterward. *Courtesy Wikimedia Commons.*

blood vessels in a small group of workers more than 35 years after occupational exposure to herbicides in the Czech Republic (ATSDR, 2012). This group had one of the highest body burdens of the highly carcinogenic 2,3,7,8-tetrachlorod-ibenzo-p-dioxin (TCDD) ranging from 3300 to 74,000 pg TCDD/g lipids. Liver damage including increased levels of β-lipoproteins, cholesterol, and triglycerides were associated with these increased body burdens of TCCD (Pelclova et al., 2001).

In Italy, a group of workers exposed to an industrial accident involving dioxins experienced an increase in chronic obstructive pulmonary disease which resulted in some mortality among affected individuals. In another occupationally exposed group, increased chronic liver disease was associated with high exposures; exposed individuals also had abnormal concentrations of several hepatic enzymes compared with controls (Neuberger et al., 1999).

Endocrine effects have been reported with industrial exposures to dioxins. For example, the incidence of Type II diabetes and increased secretion of thyroid stimulating hormone were found in Vietnam veterans who were exposed to Agent Orange (ATSDR, 2012).

Some studies have shown an increased immunological response including elevated levels of T-cells and decreased concentrations of Immunoglobulin G IgG, a form of antibody that is key to fighting off infections. The effects of dioxins on human immune systems have not been definitively established because these symptoms are contradictory.

With regards to neurotoxicological effects, in the period between 1965 and 1968, approximately 350 workers were accidentally exposed to TCDD during the production of trichlorophenoxyacetic acid-based herbicides in Czechoslovakia (ATSDR, 2012). Nerve and brain damage were reported in some workers in a

10-year follow-up. The estimated mean plasma concentration of TCDD at the time of exposure was approximately 5000 pg/g of plasma lipids, many times greater than control populations. In 1996, about 30 years later, the mean TCDD plasma concentration was still elevated at 256 pg/g. Abnormal electromyography (EMG) and electrocardiography (ECG) (ie, muscle and heart scans) were observed in a fraction of exposed workers (Pelclova et al., 2001), but the sample size available after 30 years was very small (13 individuals).

Thus, dioxins can cause many different types of problems in humans. The related compounds of furans, PCBs, and brominated hydrocarbons produce similar types of problems because they are so similar chemically. Production of PCBs ceased in the late 1970s, but some of the other chemicals are still being produced, either naturally or through industry, and are likely to be present for many years to come. Fortunately, people who are not in the risk classes involving industries or exposure to Agent Orange, although undoubtedly having some of these persistent compounds in their bodies, are not likely to encounter a sufficient amount of chemical to cause problems.

Current-Use Pesticides

Sanborn et al. (2012) presented an extensive review of current-use pesticides, predominately organophosphorus pesticides, on human health. They enumerated an impressive number of ailments due to these chemicals in infants, children, and adults. Organophosphorus pesticides can cross the placental barrier and can be measured in both cord blood and amniotic fluid so fetuses are exposed to these chemicals. Frequent problems that have been reported with neonates whose mothers were exposed to pesticides include low birth weights, small head circumferences, and lower-than-average body lengths. Conditions called *hypospadias*, in which the urethra exits the penis in babies below its normal location at the tip, and *cryptorchism*, where one or both testes fails to descend, have been commonly associated with maternal pesticide exposures, although other factors can result in the same conditions.

Neurological problems that have been associated with pesticide exposures in young children include behavioral problems, cognitive issues such as poor reading ability and poor memories, impaired coordination, and lower scores on standardized developmental tests. Those who have been exposed to organophosphorus pesticides in the womb may experience reduced IQs and other issues. Adolescents who worked as pesticide applicators in agricultural fields had higher incidence of depression, numbness, lower acetylcholinesterase activity, and poorer ability to concentrate than those of similar age that did not work as applicators (Callahan et al., 2014).

There have been many studies that have looked at a relationship between pesticide exposure in children and either attention deficit disorder (ADD) or attention deficit hyperactive disorder (ADHD), and there appears to be a clear cause-and-effect relationship between the two. In one study (Bouchard et al., 2010), for each tenfold increase in urinary concentrations of organophosphorus metabolites, children experienced a 2.3-times-higher risk of having ADHD. Chlorpyrifos and malathion are two of the pesticides that have been associated with ADD or ADHD.

In addition to neurological problems, organophosphorus pesticides have been related to asthma in both children and adults and other respiratory problems. Coumaphos, 2,4-D, glyphosate, parathion, and metaloxyl are specific pesticides that have been linked to lung issues (Hernandez et al., 2011).

Emerging Contaminants

In this section we discuss contaminants that are more fully examined in Chapter 9, such as plasticizers, pharmaceuticals, and the like.

Pharmaceuticals and Personal Care Products (PPCP). In addition to medicines, cleansers, and sanitizers used by humans, pharmaceuticals also include veterinary medicines. Pharmaceuticals generally enter the environment through sewage systems. The most likely way that humans come into contact with these contaminants is through drinking water because water treatment plants and sewage treatment plants are usually not equipped to remove these chemicals.

One group of chemicals that is of increasing concern and controversy are the antimicrobials given to livestock. Producers use a variety of antimicrobials to treat sick animals or to prevent disease but the greatest use is to promote growth and increase food efficiency. From 17.8 to 24.6 million pounds (8090 to 11,188 tons) of antimicrobials were used just a few years ago for nontherapeutic purposes in chickens, cattle, and swine, compared with just 3 million pounds (1364 tons) used in human medicine. These antibiotics include arsenicals, polypeptides, glycolipids, tetracycline, elfamycins, macrolides, lincosamides, polyethers, beta-lactams, quinoxalines, streptogramins, and sulfonamides (Landers et al., 2012). The greatest concern then, is that antibiotic-resistant strains of bacteria can evolve. Since humans are subject to many of the bacteria found in livestock, our health may be at risk from these resistant bacteria (Landers et al., 2012; Holman and Chenier, 2015). Antibiotic-resistant bacteria can become incorporated into meat and can also infect farm workers and meat handlers through contact. Antibiotic-resistant bacteria of animal origin have been found around livestock farming operations, on meat products available for purchase in retail food stores, and have been identified as the cause of clinical infections and subclinical colonization in humans (Landers et al., 2012). Recently, many meat producers are voluntarily reducing the use of antimicrobials as growth enhancers and are restricting their use to treat sick animals or to reduce the risk of disease largely in response to consumer demands (eg, Jones et al., 2015).

As we noted in Chapter 9, human and veterinary medicines may pose another set of risks to human health. Fortunately, the concentrations of these chemicals in water are typically several orders less than those used in therapeutic treatments, but we really do not understand the implications of long-term exposure to low concentrations of endocrine disruptors, just as one example. More than 50 pharmaceuticals were identified in drinking water in 2004 (Boxall, 2004) and by now others have certainly been identified. The truth is, we really do not know what long-term risks these chemicals pose to humans (Kümmerer, 2010).

Plasticizers. Plastics, more correctly chemicals that are used in making plastics, are known to produce endocrine disrupting effects. The most potent of these is bisphenol-A (BPA), which has been related to diabetes, obesity, cardiovascular disease, and hypertension (Ranciere et al., 2015). BPA is a known endocrine disruptor in several different ways. It binds with estrogen receptors on cells and has estrogenic effects equal to 17-beta estradiol, a naturally occurring estrogen. BPA can also act as an antiestrogen, blocking the estrogenic response by competing with endogenous 17-beta estradiol. The chemical also binds to androgen receptors and may be an antiandrogen, blocking natural androgen activity in males. It also binds with thyroid receptors and can block the activity of that hormone. Numerous studies have shown direct connections between BPA and all of these effects (eg, Rochester, 2013).

Phthalates are found in many PPCP in addition to plastics, especially polyvinyl chloride (PVC) plastics. They have been associated with several endocrine effects in humans. The key word here and in most studies on humans is "associated." For ethical reasons, controlled studies on humans and contaminants are seldom done; as a result, these studies cannot completely rule out other factors that might affect human physiology. If the number of humans in these studies is sufficiently large, the better studies can statistically control for other variables such as age, sex, ethnicity, and the like. Phthalates have been significantly associated with shorter gestational periods in fetuses, antiandrogen effects in infant males from prenatal exposures, early onset of breast development in prepubertal girls, lowered sperm motility and concentration in men, and reduced follicle stimulating hormone in women (Hauser and Calafat, 2005).

Nanoparticles. Just as nanoparticles are incredibly and increasingly diverse, so too are their great benefits and potential harm to human health. In Chapter 9, we briefly mentioned how nanoparticles can be constructed to carry specific medicine or radioisotopes into cancer cells or to areas of the body that may be otherwise difficult to treat. They will also potentially be of great benefit to medical imaging, disease diagnoses, drug delivery, and gene therapy. Possible harmful effects are essentially only guesswork at this time. These very small particles and other ultrafine particles could become part of nuisance dust, carrying their products around the world and into our bodies. For example, ultrafine particles, which are in the same size range of nanoparticles, have been linked to the increase in respiratory ailments around the world. The science is simply too new to determine what could happen. This is an area that requires great care and ethical wisdom as it develops (Gwinn and Vallyathan, 2006).

Polycyclic Aromatic Hydrocarbons (PAHs)

For most people, exposure to polycyclic aromatic hydrocarbons (PAHs) comes from breathing contaminated air. Several PAHs are found in cigarette smoke and contribute to the risk of smoking. Just living in an urban environment can place a person at risk to airborne PAHs in smog. Exposure to fumes from generators, motors, and the like including automobile exhausts can increase one's risk to these contaminants (see the FOCUS section later in this chapter). PAHs may also be ingested with certain foods and absorbed dermally.

The acute effects of PAHs on human health will depend mainly on the extent of exposure, which is determined by the concentration of PAHs during exposure, the toxicity of the PAHs, and the route of exposure (eg, via inhalation, ingestion, or skin contact; Kim et al., 2013). Symptoms can include eye and skin irritation, nausea, vomiting, diarrhea, and reduced lung functioning in asthmatics. Chronic exposure can be more serious, although most of what we know comes from animal studies and can only be surmised for humans (Kim et al., 2013). Some PAHs, such as benzo[a]pyrene, benz[a]anthracene, benzo[b]fluoranthene, benzo[k]fluoranthene, chrysene, dibenz[ah]anthracene, and indeno[1,2,3-cd]pyrene, are recognized as probable human carcinogens by the US EPA (2015b). Organs usually susceptible to PAH-induced cancers include lungs, bladder, skin, and gastrointestinal tracts. Most evidence for cancer comes from workers exposed to relatively high concentrations of PAHs in their workplaces. Other studies have associated PAH exposure in the fetus (such as smoking by the mother during pregnancy) with low birth weights, premature births, delayed child development, low IQ at age three, and increased incidence of behavioral problems at ages six to eight (reviewed by Kim et al., 2013).

Metals

Lead and mercury are widely recognized as the main metals of human health concern because they are widely distributed and demonstrate a high level of toxicity. Which of the next set of metals poses the greatest risk to humans is arguable. Asbestos, a metalloid, has caused cancer in thousands of people in the mining and construction industries. Arsenic, another metalloid, is very toxic under industrial situations. Cadmium, chromium, zinc, and other so-called heavy metals can also be toxic to humans, again usually in certain industries like smelters and metal refining operations. In this section we will discuss only a few metals that have demonstrated ill effects on humans.

Cadmium. Cadmium exposure may cause kidney damage, mostly in the renal tubules. This results in the excretion of low molecular weight proteins. According to Järup (2003), tubule degeneration caused by Cd is irreversible. The initial tubular damage may progress to more severe kidney damage such as affecting the filtering apparatus or glomeruli of the kidney and even renal failure. An excess risk of kidney stones, possibly related to an increased excretion of calcium in urine following the tubular damage, has been demonstrated in several studies (Järup, 2003). Long-term cadmium exposure may result in bone disease, as witnessed in the 1950s in Japan due to Cd-contaminated water used for the irrigation of rice paddies and people working directly in flooded paddies (see the FOCUS section later in this chapter).

Cadmium has been designated a human carcinogen by the International Agency for Research on Cancer (IARC, 2012), but is considered a probable human lung carcinogen by the US EPA (2015c) which says that some of the studies were equivocal. The principal mode of exposure that can lead to cancer is inhalation, either from industrial sources or through smoking cigarettes. Cadmium has also been association with kidney and prostate cancers, but Järup (2003) believed that the case for cadmium causing cancers other than lung cancer is weak.

Based primarily on animal studies, Cd is also considered to have estrogenic properties. It has been demonstrated that it can bind with estrogen receptors and stimulate the production of estrogen. Similarly, Cd may stimulate progesterone receptors in the body and is thus a double agent for some estrogen-dependent diseases such as breast and endometrial (lining of the uterus) cancer, endometriosis, and spontaneous abortions (Rzymski et al., 2015).

Lead. The symptoms of acute lead poisoning are headache, irritability, abdominal pain, and various symptoms related to the nervous system (Järup, 2003). More serious conditions are usually neurological and include lead encephalopathy (brain disease) characterized by sleeplessness and restlessness, lack of attention leading to psychosis, confusion, and unconsciousness. Children may be affected by behavioral changes, learning, and concentration difficulties. Chronic exposures can lead to memory deterioration, prolonged reaction time, and reduced ability to comprehend. Other neurological changes affect the peripheral nerve through reduced nerve conduction velocity and reduced dermal sensibility. Severe cases may be permanent. The classical picture includes a dark blue lead sulfide line at the gum line in the mouth (Fig. 14.5). As in

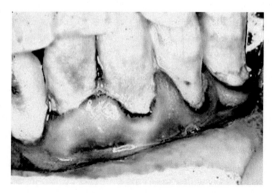

FIGURE 14.5   Lead poisoning can be identified by a characteristic blue line at the tooth/gum intersection.

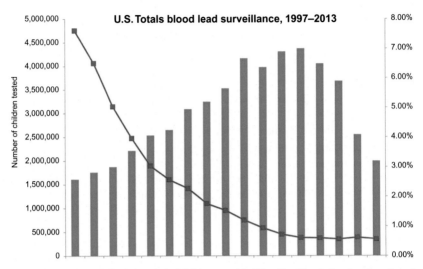

FIGURE 14.6   Trends in the number of children tested for blood lead levels (bars) and number of confirmed concentrations greater than 10 µg/dL, concentrations above which can lead to toxicosis, as determined by the CDC. *Courtesy Centers for Disease Control.*

animals, lead poisoning also leads to reduced hemoglobin titers, anemia, and severely reduced ALAD levels (see Chapter 8).

Children are more sensitive to lead toxicity than adults because lead cannot penetrate the blood–brain barrier in adults, but does in children. One of the most severe routes of lead toxicity is eating old paint. Prior to 1978, lead was used in paint to speed up drying, increase durability, maintain a fresh appearance, and resist moisture that causes corrosion. In that year, the United States placed a ban on using lead to reduce poisoning of children eating paint chips from walls and other surfaces. In older homes and apartments that have not been refurbished, these leaded paint chips can still be found. In 1988, the Lead Contamination Control Act authorized the CDC to initiate program efforts to eliminate childhood lead poisoning in the United States. Since the late 1990s, enforcement and public awareness of the dangers of lead poisoning to children have led to a marked decline in the number of children with blood levels greater than 10 µg/dL, the level above the CDC's lowest threshold for possible toxicosis (Fig. 14.6; CDC, 2015b). Despite this, lead poisoning remains a problem in urban slums where little attention is paid to refurbishing old, worn-out lead paint with more modern latex or acrylics.

Asbestos. Asbestos is a group of naturally occurring silicate minerals characterized by long, think fibrous crystals each composed of millions of microscopic fibrils. Whereas asbestos mining has occurred for thousands of years, large-scale mining did not start until the late 1800s, when it was discovered that asbestos can be used in a wide variety of products including building insulation, sound absorption materials, heat insulation around pipes and boilers, and

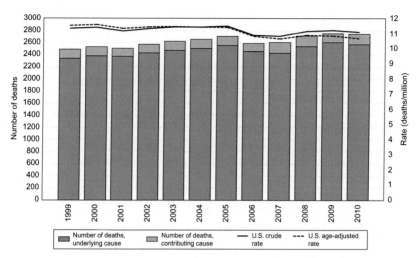

FIGURE 14.7   Incidence of mesothelioma from 1999 to 2010 in the United States. *Courtesy of Centers for Disease Control.*

fire proofing. If asbestos is left undisturbed it causes little problem. The risk is sharply elevated when the fibrous blankets of asbestos are broken, frayed, or otherwise disturbed. Then the microscopic fibrils can become airborne and inhaled. Once inside the lungs, the fibrils penetrate tissue and cause irritation, which results in various forms of respiratory ailments, ultimately leading to lung cancers, one of which is mesothelioma (Ismail-Kahn et al., 2006). The fibrils can also be carried through the bloodstream to other tissues, causing cancers in the lining or pleura of other organs. Compared with other forms of cancer, mesothelioma is rare but its incidence has remained stable over the past decade or longer (Fig. 14.7; CDC, 2014). It may take decades for malignant mesothelioma to express itself so the deaths observed at any given time may have been due to exposure 40 or 50 years ago; the use of asbestos was greatly abated by the late 1980s (ATSDR, 2001). Malignant mesothelioma is closely associated to industrial exposures such as in mining, certain manufacturing operations, and in building trades. Thus, the cancer has become a legal issue leading to lawsuits and a fund for victims, which is why we frequently hear about the disease.

Mercury. We discussed mercury and methylmercury in some detail, including Minamata disease, in Chapter 8 but because it is arguably the most toxic of metals, especially in its organic form, now we'll mention some of the highlights reported by the ATSDR review (1999). Everyone is exposed in sublethal background concentrations to the elemental form of the metal because mercury occurs naturally. The use of mercury in dental amalgams is another exposure route. Estimates of the amount of mercury released from dental amalgams range from 3 to 17 μg/day. This may contribute up to 75% of the total daily mercury exposure for humans, depending on the number of amalgam fillings one has. Dental amalgams alone are not considered to be a serious risk to health (ATSDR, 1999). Fish consumption can be of concern, particularly fish taken from polluted water, as a risk to fetuses and young children. Sensitive populations to eating fish taken from contaminated waters include pregnant women, children under the age of six, people with impaired kidney function, and people with hypersensitive immune responses to metals.

The nervous system is particularly sensitive to mercury toxicosis. Problems include personality changes such as irritability, shyness, nervousness, or tremors; changes in vision such as narrowing of the visual field; deafness; loss of muscle coordination; loss of sensation; and memory loss. These problems might be reversible after exposure is eliminated. The brain of adults is not usually affected by toxicosis from metallic mercury because the blood–brain barrier prevents the metal from passing. However, the blood–brain barrier of fetuses and infants is not completely formed and thus they may be subject to significantly more harm than adults (Järup, 2003). Mercury vapors and organomercury can move past the blood–brain barrier and cause permanent damage to the central nervous system even in adults (ATSDR, 1999).

Other organs that may be affected by mercury poisoning include the kidneys, mouth linings, and lungs in the case of mercury vapors, and cardiovascular system. Mercury may be present in a mother's milk, especially if the woman consumes considerable amounts of contaminated fish or works in one of the industries where mercury vapors may be found. Therefore, milk can be a significant route of exposure to nursing young children (ATSDR, 1999). Children who are exposed to elemental or organomercury for extended periods of time may develop a condition called *acrodynia*, or pink disease, which can result in severe leg cramps, irritability, and redness of the skin, leading to peeling of extremities. Other symptoms of acrodynia include itching, swelling, fever, accelerated heart rate and elevated blood pressure, rashes, sleeplessness, and weakness.

FOCUS—EXAMPLES OF MAJOR CONTAMINANT—HUMAN DISASTERS

In previous chapters, we discussed the disasters associated with the leakage of methyl isocyanate, a chemical used in the manufacture of carbamate pesticides in Bhopal, India in 1984. We also discussed Minamata disease that was caused by the discharge of methylmercury-contaminated waters in Japan over a period of 20 years before it was discovered in the 1950s. These are only two of several tragedies that have occurred due to contaminants. Next, we will discuss several other examples. The examples only include those disasters that involved a significant number of people. They do not include other human-caused environmental disasters such as many oil spills or radioactive meltdowns at Three Mile Island, Pennsylvania, in 1979, Chernobyl, Russia, in 1986, or Fukushima,

Japan, in 2011. Nor do they include the great number of mining incidents that did not involve humans to any great extent.

1. *Cadmium toxicity resulting in Itai-itai (ouch-ouch) disease, Jinzu River Basin Japan, 1912–1959.* In traditional Japanese practice, rice is planted in flooded impoundments or paddies. Water levels are carefully maintained until harvest, and most of the work is manually conducted by farmers in sandals or bare feet. This method promotes considerable contact between the water in the paddy and the skin of the farmers. In addition, the main staple of these farmers' diets is rice, which can accumulate many of the metals in the soil and fish from these paddies.

   In the Jinzu River Basin, mining of metals had occurred for hundreds of years but the first report of itai-itai occurred in 1912 and the disease was officially recognized in 1968 as the first disease induced by environmental pollution in Japan after legal proceedings (Kanazawa n.d.). An upstream Cd mine discharged its tailings in the Jinzu River for many years and the Cd concentrations in the sediments of the rice paddies gradually increased to toxic proportions.

   The principal and most obvious symptom of itai-itai disease is a chronic and often severe pain in the spine and joints. The primary targets of this Cd toxicosis are the kidneys and bone. Thousands of people were affected by this disease with symptoms ranging from mild pain and some kidney damage to death. Women were proportionately more affected because they generally tend to retain Cd longer than men.

   Maruzini et al. (2014) conducted a follow-up survey of afflicted people in 2005, some 26 years after the peak of the itai-itai problem. They surveyed 7529 inhabitants of the Cd-polluted Jinzu River basin and 2149 controls from nonpolluted areas who participated in an earlier study conducted from 1979 to 1984. The authors found that over time the mortality risk for residents of the area who had protein (proteinuria) or protein and sugars (glucoproteinuria) in their urine significantly increased compared with controls that had no urinary issues and even in controls that had similar renal problems. Respiratory, renal, cardiovascular diseases, and diabetes in men and women were also higher in the exposed population and the mortality rates in women over a 26-year period were higher than in those from outside of the basin. The most severely impacted and exposed women but not men had higher mortality risks for cancer of the colon, rectum, uterus, kidneys, and urinary tract compared with controls. Women without obvious kidney damage also had higher rates of heart disease. Since the development of mechanized cultivation of rice, the problem of itai-itai disease has diminished substantially.

2. *Smog, Donora Valley, Pennsylvania, 1948.* One of the first documented cases of smog-related deaths occurred October 30–31, 1948. Unfortunately, this occurred many years before air pollution was a watchword, and well

before the Clean Air Act, so records were not well kept and detailed air samples were not taken, even if they could have been analyzed for the chemicals that we now know exist in smog. What we know is that atmospheric conditions in the vicinity of Donora, Pennsylvania, contributed to the deaths of 19 people within a 24-h period. Of the fatalities, two had active pulmonary tuberculosis and the others had chronic heart disease or asthma; all were between 52 and 85 years old. While mortality due to poor air quality is not unknown for those with cardiac or lung problems, an additional 500 people out of a population of 14,000 became ill with symptoms of respiratory problems. Most likely, many others also suffered but did not seek medical help (Pennsylvania, 2015). At that time, Donora was surrounded by heavy industry, largely steel smelters. In 1902, the Carnegie Steel Company completed a facility that consisted of several smelting furnaces and a second wire producing company was erected. In 1916, another mill opened and in 1917 the Donora Zinc Works began production. These industries attracted rail service and shipping, all of which contributed to air pollution.

For several years prior to the great smog, residents of Donora complained of industrial pollutants that "eats the paint off your houses" and prevents fish from living in the river (Pennsylvania, 2015). The State Bureau of Industrial Hygiene measured an extraordinarily high level of sulfur dioxide, soluble sulfates, and fluorides in the air on the days of the smog. These sulfur compounds were attributed to the zinc smelting plant, steel mills' open hearth furnaces, a sulfuric acid plant, and to slag dumps, coal-burning steam locomotives, and river boats. Donora sits in a bowl surrounded by hills which undoubtedly contributed to retaining the smog around the city and kept the pollutants within inhalation range of the residents.

Largely as a result of this disaster, Pennsylvania established the Division of Air Pollution Control in 1949 to study the matter. The situation was also one motion of many that eventually led to the state to pass laws mandating clean air and water.

3. *The Great Smog, London, 1952.* London is well known for its fog and smog, with some events even leading to increased mortality among residents, but the events of December 5–9, 1952, were unprecedented and have not been repeated since (MetOffice, 2015). The weather in November and early December 1952 was very cold and snowy, so Londoners burned copious amounts of coal to keep warm. Normally the smoke from the chimneys would rise into the sky and disperse, but at this time, an anticyclone weather event hung over the region. An anticyclone is a high-pressure system invoking a circular air pattern that can push air downward, warming it as it descends, causing a thermal inversion and inducing fog. Thus, the chimney exhausts were trapped close to the ground. The inversion of 1952 also trapped particles and gases emitted from factory chimneys in the London area and brought additional plumes from industrial areas in Europe.

Early on December 5, the weather conditions in the London area became excellent for producing fog (MetOffice, 2015). During the fog, the atmosphere at the human level was inundated with tremendous amounts of PAHs, gases, metal vapors, and other contaminants. The daily pollution consisted of 1000 tons of smoke particles, 2000 tons of carbon dioxide, 140 tons of hydrochloric acid, 14 tons of fluorine compounds, and tons of other unanalyzed chemicals (MetOffice, 2015). In addition to the hydrochloric acid, it is estimated that 370 tons of sulfur dioxide were converted into 800 tons of sulfuric acid.

The Great Smog lasted until December 9. During that time, approximately 4000 people died directly, with an estimated 8000 dying later due to complications such as severe asthma and respiratory or circulatory ailments. Thousands of others experienced health problems but survived. Even cattle at a stockyard were killed by the smog and travel was disrupted for days. This event led to several new laws being passed, including the Clean Air Acts of 1956 and 1968 in the United Kingdom. These acts banned emissions of black smoke and mandated that residents of urban areas and operators of factories had to gradually convert to smokeless fuels. The laws have helped, but Londoners continue to be plagued by smog. For example, 750 people died from a smog event in 1962. In more modern times, the conditions leading to the Great Smog have nearly vanished due to a widespread application of central heating and diminished use of coal as a fuel source.

4. *Seveso, Italy—Dioxins, 1976.* On July 10, 1976, the Icmesa Chemical plant, a relatively small factory that manufactured various chemicals, created the largest human exposure to TCDD, the most toxic of dioxins, ever recorded. The details that resulted in the explosion are well known, but suffice to say that due to inadequate controls and safe checks, a reactor used in the making of 2,4,5 trichlorophenol from 1,2,4,5 tetrachlorobenzene overheated above a critical temperature, initiating an exothermic reaction that resulted in the reactor exploding. As a result, 6 tons of chemicals were quickly released into the atmosphere and settled over $18 \, km^2$ ($6.9 \, mi^2$) of the surrounding area. Among the substances released was 1 kg of TCDD. While 1 kg does not seem like a lot of chemicals, TCDD normally is produced in trace amounts less than 1 ppm of the solution in the reactor, but the elevated temperatures of the reactor concentrations exceeded more than 100 ppm (Hay, 1979).

More than 37,000 residents were potentially affected although the vast majority were in an area with low levels of contamination (Homberger et al., 1979). Within days, 3300 animals, mostly poultry and domestic rabbits, were found dead. Eventually, over 80,000 livestock animals were slaughtered to prevent TCDD from getting into the human food supply, and 15 children were quickly hospitalized with skin inflammation. By the end of August, the most contaminated zone had been completely evacuated and fenced, 1600 people of all ages had been examined and 447 were found to suffer from skin lesions or chloracne. An advice center was set up for pregnant women and 26 opted for an abortion, which was legal in special cases at the time. Another 460 women gave birth without problems and their infants did not show any sign of teratogenesis or pathology. Two government commissions were established to develop a plan for quarantining and decontaminating the area, budgeting approximately40 billion lire, or $47.6 million.

In retrospective studies, Bertazzi (1991) surveyed thousands of potential victims and found that choracne was the most evident result. But some instances of peripheral neuropathy and liver enzyme inductions had occurred. Other early effects noted were peripheral neuropathy and liver enzyme induction. The determination of other, possibly severe results of dioxin exposure (eg, birth defects) was hampered by inadequate information; however, generally, no increased risks were evident. There was some increased mortality due to cardiovascular and respiratory diseases and an increase in the incidence of diabetes among exposed people. Bertazzi et al. (2001) later confirmed that those who had been exposed had elevated rates of cancer, cardiovascular effects, and endocrine issues.

5. *Love Canal, New York—A Soup of Contaminants, 1978.* "Quite simply, Love Canal is one of the most appalling environmental tragedies in American history" (Beck, 1979). It started in the early 1900s when a developer, William Love, wanted to build his version of utopia on the shores of Lake Niagara Falls. He dug a canal between the upper and lower falls to install a hydro-powered electrical generator. Unfortunately, a downturn in economics and the development of alternating current to drive electricity caused an abrupt end to Love's dreams. The partial canal stood unused from 1910 to 1920, when it was used as a waste disposal landfill. More than 22,000 lbs (9.979 kg) and 80 chemicals were dumped into the landfill, 11 of which were suspected carcinogens.

The canal was filled and covered with dirt in 1953 and about six years later houses and a school were built on the site. Once the canal was filled, other developers created a subdivision. By 1978 a series of wet winters and above-average rainfall raised the water table in the region and seepage of the landfill began. According to Beck (1979), who was the director of the EPA at the time, trees and gardens began turning black and dying, a swimming pool popped up out of its foundation and was floating in "a sea of chemicals… Everywhere the air had a faint, choking smell. Children returned from play with burns on their hands and faces." Ultimately the federal government declared Love Canal a disaster area and relocated 239 families, leaving about 700 families in "less-contaminated areas." In retrospect, five children were born with birth defects that included clubbed feet, cleft palate, and supernumerary digits. Chromosomal damage in some adult residents was suspected but never confirmed. Retrospective studies (Gensburg et al., 2009a,b) demonstrated that in exresidents of the Love Canal, the incidence of bladder and kidney cancer were about 1.5 times more than the surrounding population, but the difference was not significant. Similarly, the mortality rate of Love Canal residents did not differ from the surrounding population. In terms of humans seriously impacted by contaminants, the Love Canal story is relatively mild. However, those who lived there at the time do not considered it trivial and, given the number of Superfund sites in the country, such events could happen again.

In the big picture of anthropogenic and natural disasters, contaminant-caused catastrophes arguably have not played a major role. Many more people have become ill or died due to more subtle events such as long-term exposure to sublethal concentrations of toxins. How many more people have developed cancer and died, not from a major event such as the Great Smog in London or the Love Canal, but from simply living and breathing low levels of contaminants is very difficult to assess.

STUDY QUESTIONS

1. What are some of the major ways that humans can be exposed to contaminants?
2. What is acrylamide and how is it formed in foods?
3. What chemical in wine is currently under investigation by the FDA?
4. According to the FDA, which class of foods has the greatest concentration and frequency of pesticide concentration?
5. How do ultrafine particles and potentially nanoparticles physically impair human health?
6. True or False. The human placenta has no filtering capability for contaminants which is why there are food consumption advisories for pregnant women.
7. True or False. OSHA is the only organization that sets exposure guidelines for work place environments.
8. Describe some of the effects of dioxins and related compounds on human health.
9. According to this chapter, which pesticides have been associated with a common childhood behavioral problem?
10. Why is BPA of concern to human health?
11. What is the name of the human cancer caused by asbestos? Why has it received considerable legal attention?
12. If you suspected that someone had lead toxicosis why would you look in that person's mouth?
13. What dental practice puts mercury directly in your mouth?
14. True or False. Over the span of many years disasters caused by contaminants have resulted in far more human deaths than common long term exposure to pollution.
15. What weather conditions facilitated the Great Smog of London?

REFERENCES

ATSDR [Agency for Toxic Substances and Disease Registry], 1999. Toxicological Profile for Mercury. ATSDR, Public Health Service, Atlanta, GA.

ATSDR [Agency for Toxic Substances and Disease Registry], 2001. Toxicological Profile for Asbestos. ATSDR, Public Health Service, Atlanta, GA, <http://www.atsdr.cdc.gov/toxprofiles/tp61.pdf>.

ATSDR [Agency for Toxic Substances and Disease Registry], 2012. Chlorinated Dibenzo-p-dioxins (CDDs). <http://www.atsdr.cdc.gov/substances/toxsubstance.asp?toxid=63>.

Beck E.C., 1979. The Love Canal Tragedy. <http://www2.epa.gov/aboutepa/love-canal-tragedy> (accessed 27.11.15.).

Bertazzi, P.A., 1991. Long-term effects of chemical disasters. Lessons and results from Seveso. Sci. Total Environ. 106, 5–20.

Bertazzi, P.A., Consonni, D., Bachetti, S., Rubagotti, M., Baccarelli, A., Zocchetti, C., et al., 2001. Health effects of dioxin exposure: a 20-year mortality study. Am. J. Epidemiol. 153, 1031–1044.

Bouchard, M.F., Bellinger, D.C., Wright, R.O., Weisskopf, M.G., 2010. Attention-deficit/hyperactivity disorder and urinary metabolites of organophosphate pesticides. Pediatrics 125, E1270–E1277.

Boxall ABA, 2004. The environmental side effects of medication. Embo. 5, 1110–1116.

Callahan, C.L., Al-Batanony, M., Ismail, A.A., Abdel-Rasoul, G., Hendy, O., Olson, J.R., et al., 2014. Chlorpyrifos exposure and respiratory health among adolescent agricultural workers. Int. J. Environ. Res. Public Health 11, 13117–13129.

Caserta, D., Graziano, A., Lo Monte, G., Bordi, G., Moscarini, M., 2013. Heavy metals and placental fetal-maternal barrier: a mini-review on the major concerns. Euro. Rev. Med. Pharmacol. Sci. 17, 2198–2206.

CDC [Centers for Disease Control and Prevention], 2014. Malignant mesothelioma (all sites): Number of deaths, crude and age-adjusted death rates, U.S. residents age 15 and over, 1999–2010.

CDC [Centers for Disease Control and Prevention], 2015a. Pesticide illness & injury surveillance. <http://www.cdc.gov/niosh/topics/pesticides/> (accessed 27.11.15.).

CDC [Centers for Disease Control and Prevention], 2015b. Lead. <http://www.cdc.gov/nceh/lead/> (accessed 27.11.15.).

van Donkelaar, A., Martin, R.V., Brauer, M., Kahn, R., Levy, R., Verduzco, C., et al., 2010. Global estimates of ambient fine particulate matter concentrations from satellite-based aerosol optical depth: development and application. Environ. Health Perspect. 118, 847–855.

FDA U.S. [Food and Drug Administration], 2009. Melamine contamination in China. <http://www.fda.gov/newsevents/publichealthfocus/ucm179005.htm> (accessed 27.11.15.).

FDA U.S. [Food and Drug Administration], 2012. Pesticide monitoring program fiscal year 2012 pesticide report. <http://www.fda.gov/downloads/Food/FoodborneIllnessContaminants/Pesticides/UCM432758.pdf> (accessed 27.11.15.).

FDA U.S. [Food and Drug Administration], 2014. Chemical contaminants. <http://www.fda.gov/food/foodborneillnesscontaminants/chemicalcontaminants/default.htm> (accessed 27.11.15.).

Gensburg, L.J., Pantea, C., Fitzgerald, E., Stark, A., Hwang, S.A., Kim, N., 2009a. Mortality among former love canal residents. Environ. Health Perspect. 117, 209–216.

Gensburg, L.J., Pantea, C., Kielb, C., Fitzgerald, E., Stark, A., Kim, N., 2009b. Cancer incidence among former love canal residents. Environ. Health Perspect. 117, 1265–1271.

Gwinn, M.R., Vallyathan, V., 2006. Nanoparticles: health effects—pros and cons. Environ. Health Perspect. 114, 1818–1825.

Hauser, R., Calafat, A.M., 2005. Phthalates and human health. Occup. Environ. Med. 62, 806–818.

Hay, A., 1979. Séveso: the crucial question of reactor safety. Nature 281, 521.

Hernandez, A.F., Parron, T., Alarcon, R., 2011. Pesticides and asthma. Curr. Opin. Allergy Clin. Immunol. 11, 90–96.

Holman, D.B., Chenier, M.R., 2015. Antimicrobial use in swine production and its effect on the swine gut microbiota and antimicrobial resistance. Can. J. Microbiol. 61, 785–798.

Homberger, E., Reggiani, G., Sambeth, J., Wipf, H.K., 1979. The Seveso accident: its nature, extent and consequences. Ann. Occup. Hygiene 22, 327–370.

IARC [International Agency for Research on Cancer], 2012. Cadmium and cadmium compounds. Monograph 100C. <http://monographs.iarc.fr/ENG/Monographs/vol100C/mono100C-8.pdf> (accessed 23.11.15.).

Ismail-Khan, R., Robinson, L.A., Williams Jr, C.C., Garrett, C.R., 2006. Malignant pleural mesothelioma: a comprehensive review. Cancer Cont. 13, 255–263.

Järup, L., 2003. Hazards of heavy metal contamination. Brit. Med. Bull. 68, 167–182.

Jones, P.J., Marier, E.A., Tranter, R.B., Wu, G., Watson, E., Teale, C.J., 2015. Factors affecting dairy farmers' attitudes towards antimicrobial medicine usage in cattle in England and Wales. Prevent. Vet. Med. 121, 30–40.

Kanazawa [Kanazawa Medical University], n.d. Itai-itai disease. <http://www.kanazawa-med. ac.jp/~pubhealt/cadmium2/itaiitai-e/itai01.html> (accessed 23.11.15.).

Kim, K., Jahan, S.A., Kabir, E., Browh, R.J.C., 2013. A review of airborne polycyclic aromatic hydrocarbons (PAHs) and their human health effects. Environ. Intl. 60, 71–80.

Kümmerer, K., 2010. Pharmaceuticals in the environment. Ann. Rev. Environ. Resour. 35, 57–75.

Landers, T.F., Cohen, B., Wittum, T.E., Larson, E.L., 2012. A review of antibiotic use in food animals: perspective, policy, and potential. Public Health Rep. 127, 4–22.

Marques, R.C., Bernardi, J.V.E., Dorea, J.G., Brandao, K.G., Bueno, L., Leao, R.S., et al., 2013. Fish consumption during pregnancy, mercury transfer, and birth weight along the Madeira River Basin in Amazonia. Int. J. Environ. Res. Public Health 10, 2150–2163.

Maruzini, S., Nishijo, M., Nakamura, K., Morikawa, Y., Sakurai, M., Nakashima, M., et al., 2014. Mortality and causes of deaths of inhabitants with renal dysfunction induced by cadmium exposure of the polluted Jinzu River basin, Toyama, Japan; a 26-year follow-up. Environ. Health 13 <www.ehjournal.net>.

MetOffice, 2015. The Great Smog of 1952. <http://www.metoffice.gov.uk/learning/learn-about-the-weather/weather-phenomena/case-studies/great-smog>. (The MetOffice is the website for the Meteorological Service of the United Kingdom) (accessed 24.11.15.).

NASA [National Aeronautics and Space Administration], 2010. New map offers a global view of health-sapping air pollution. <http://www.nasa.gov/topics/earth/features/health-sapping.html> (accessed 27.11.15.).

Needham, L.L., Grandjean, P., Heinzow, B., Jorgensen, P.J., Nielsen, F., Patterson, D.G., et al., 2011. Partition of environmental chemicals between maternal and fetal blood and tissues. Environ. Sci. Technol. 45, 1121–1126.

Neuberger, M., Rappe, C., Bergek, S., Cai, H., Hansson, M., Jager, R., et al., 1999. Persistent health effects of dioxin contamination in herbicide production. Environ. Res. 81, 206–214.

Pelclova, D., Fenclova, Z., Dlaskova, Z., Urban, P., Lukas, E., Prochazka, B., et al., 2001. Biochemical, neuropsychological, and neurological abnormalities following 2,3,7,8-tetrachlorodibenzo-p-dioxin (TCDD) exposure. Arch. Environ. Health 56, 493–500.

Pennsylvania [Pennsylvania Historical and Museum Commission], 2015. The Donora Smog Disaster, October 30–31, 1948. <http://www.phmc.state.pa.us/portal/communities/documents/1946-1979/donora-smog-disaster.html> (accessed 25.11.15.).

Ranciere, F., Lyons, J.G., Loh, V.H.Y., Botton, J., Galloway, T., Wang, T.G., et al., 2015. Bisphenol A and the risk of cardiometabolic disorders: a systematic review with meta-analysis of the epidemiological evidence. Environ. Health 14, 46.

Rochester, J.R., 2013. Bisphenol A and human health: a review of the literature. Rep. Toxicol. 42, 132–155.

Rzymski, P., Tomczyk, K., Rzymski, P., Poniedziałek, B., Opala, T., Wilczak, M., 2015. Impact of heavy metals on the female reproductive system. Ann. Agric. Environ. Med. 22, 259–264.

Sanborn, M., Bassil, K., Vakil, C., Ker, K., Ragan, K., 2012. Systematic Review of Pesticide Health Effects. Ontario Coll Fam Phys, Toronto.

US EPA [U.S. Environmental Protection Agency], 2014. National primary drinking water standards. <http://water.epa.gov/drink/contaminants/> (accessed 27.11.15.).

US EPA [U.S. Environmental Protection Agency], 2015a. Air toxics - urban air toxics monitoring program. <http://www3.epa.gov/ttnamti1/uatm.html> (accessed 27.11.15.).

US EPA [U.S. Environmental Protection Agency] 2015b. TRI-Listed Chemical. <http://www.epa. gov/toxics-release-inventory-tri-program/tri-listed-chemicals> (accessed 30.12.15.).

US EPA [U.S. Environmental Protection Agency], 2015c. Cadmium compounds. <http://www3. epa.gov/airtoxics/hlthef/cadmium.html> (accessed 27.11.15.).

WHO [World Health Organization], 2015. Pesticides. <http://www.who.int/topics/pesticides/en/> (accessed 27.11.15.).

Chapter 15

Regulation of Environmental Chemicals and Damage Assessment

Terms to Know
United Nations Environment Programme (UNEP)
The United Nations Framework Convention on Climate Change
European Union Directorate for the Environment
Natural Resource Damage Assessment
MARPOL 73/78
Total Maximum Daily Load
National Pollutant Discharge Elimination System
National Priorities List
Superfund Sites

INTRODUCTION

Given the worldwide problems caused by environmental contaminants and the billions of dollars spent each year to clean them, many agencies have been established from the international to the municipal level to regulate the use of these chemicals and try to make sure that they are used in an ecologically sound fashion or remediated if necessary. We have discussed the US Environmental Protection Agency (US EPA) throughout this book as the chief watchdog agency in the United States. The US EPA has its counterparts in virtually every state in the country and its national counterparts in most countries. In addition, several agencies and United Nations accords operate at international levels. In this chapter, we will discuss the regulation of environmental contaminants at all of these levels.

INTERNATIONAL AUTHORITIES

United Nations

The United Nations (UN), headquartered in New York City (Fig. 15.1), is an international organization founded in 1945. It is currently made up of 193 member states. In 2000, the organization established its Millennium Development

FIGURE 15.1 The United Nations, through its Environment Programme, develops protocols and holds conventions to help reduce global pollution.

Goals (UN, 2015) which laid out eight primary goals to be accomplished during the next 15 years, to 2015. At least three of these goals: eradicating extreme poverty and hunger; fighting diseases (including malaria); and ensuring environmental sustainability are deeply involved with environmental contaminants. Eradicating hunger cannot be feasibly done without pesticides to support crop production. Combating disease, especially malaria, involves continued use of DDT to ward off mosquito vectors. Ensuring environmental sustainability focuses on climate change, much of which has been attributed to anthropogenic sources of greenhouse gases, but also includes issues dealing with clean water, clean air, and the survival of species.

A principal office within the UN for contaminant issues is the United Nations Environment Programme (UNEP). The UNEP is a leading global environmental authority that helps set the global environmental agenda, promotes implementation of the environmental dimension of sustainable development for the world, and serves as an authoritative advocate for the global environment. Its mission is: "To provide leadership and encourage partnership in caring for the environment by inspiring, informing, and enabling nations and peoples to improve their quality of life without compromising that of future generations" (UN, undated). UNEP cannot unilaterally impose laws or regulations, instead it calls on its member nations to develop protocols (eg, the Kyoto Protocol) whose agreements can sometimes be enforced through international cooperation. In conventions (eg, the Stockholm Convention), the nations make recommendations that are binding to the point of international cooperation, but may not have the power of law.

Disputes between countries concerning these protocols and conventions can be brought before the International Court of Justice, otherwise known as the World Court, which is the United Nation's primary judicial body. A criticism of the International Court is that its judgments are also not necessarily binding.

A nation, even a member nation, may refuse to acknowledge the court's authority. In addition, any of the permanent members of the Security Council (China, France, Russia, the United Kingdom, and the United States) may veto any enforcement of the cases and thus avoid responsibility.

Contaminants are a running theme through most of UNEPs major programs. A few major contaminant-related events that were led by UNEP include:

The United Nations Framework Convention on Climate Change (UNFCCC), an international environmental treaty established at a conference held in Rio de Janeiro in 1992. The objective of the treaty was to "stabilize greenhouse gas concentrations in the atmosphere at a level that would prevent dangerous anthropogenic interference with the climate system." The UNFCCC specified an objective of stabilizing greenhouse gas emissions in developed countries at their 1990 levels by the year 2000. This objective was not even approached. The treaty is not legally binding and contains no enforcement mechanisms. Instead, the treaty provides a framework for negotiating specific international treaties (called "protocols") that may set binding limits on greenhouse gases. An early protocol of some fame was the *Kyoto Protocol* in 1997, which established legally binding obligations for developed countries to reduce their greenhouse gas emissions. The United States never ratified the protocol (had it voted on and approved by the legislature) and withdrew from the agreement in 2001, seeking instead to follow its own timeline of greenhouse gas reductions. Canada withdrew its participation in 2012. The Protocol went into effect in 2005 and was extended until 2020. Several other protocols have adjusted the deadlines and target emissions for the countries that remain members (UN, 2014). In mid December 2015 188 nations signed the Paris Climate Accord. This accord applies to all nations, not just the developed ones as did the Kyoto Accord. The goals of the accord are to reduce the emission of green house gases so that the global air temperature does not exceed 2°C or 3.6°F by 2025. At present there is considerable hope that the objectives will be met.

We have already discussed the *Stockholm Convention* in previous chapters, but some details bear repeating. The Stockholm Convention is a global treaty to reduce persistent organic pollutants (POPs) in the environment. As we mentioned, POPs are chemicals that remain intact in the environment for long periods, become widely distributed geographically, accumulate in the fatty tissue of living organisms, and are toxic to humans and wildlife. POPs circulate globally and can cause damage wherever they travel. More than 150 countries signed the Convention in 2001 and it entered into force, on May 17, 2004. In implementing the Convention, governments agreed to eliminate or reduce the release of POPs into the environment. An initial list of the "dirty dozen" was drawn up for elimination or significant reduction. These included the organochlorine pesticides of aldrin, dieldrin, endrin, heptachlor, toxaphene, DDT, hexachlorobenzene, and mirex; polychlorinated biphenyls (PCBs); dioxins; and furans. In subsequent years, 11 other contaminants made the list with the most recent addition being hexabromocyclododecane, a brominated flame retardant in 2013. Other chemicals are under consideration as the list grows longer. The United Nations has had several other programs, conventions, and protocols which we cannot go into detail about because we could fill a book on what the organization has done. Suffice to say that the ones we have discussed are only a few of the highlights.

Organisation for Economic Co-operation and Development (OECD)

The stated mission of the Organisation for Economic Co-operation and Development (OECD) is: "to promote policies that will improve the economic and social well-being of people around the world" (OECD, 2015a). The OECD is an international organization consisting of 34 member nations and is head-quartered in Paris, France (Fig. 15.2). It provides a forum in which governments can work together to share experiences and seek solutions to common problems. The OECD has a strong focus on economic issues, however, because economics covers a huge span of topics they also have substantial input on global contaminant issues. In their own words: "We set international standards on a wide range of things, from agriculture and tax to the safety of chemicals."(OECD, 2015b). In this regard, OECD has a branch that covers chemical safety and biosafety. They are not an enforcement agency, but they do conduct studies on contaminants and provide guidance to member nations on related issues. The work of OECD includes establishing guidelines for contaminant testing, good laboratory practices, quantitative structure activity relationships (QSARs); issuing legal opinions; and providing environmental, health, and safety briefs (OECD, 2015b). Since its founding in 1961, OECD has issued scores of opinion papers, study results, and decisions on a broad range of contaminant issues.

FIGURE 15.2 The Organisation for Economic Co-operation and Development, headquartered in Paris, provides testing and guidelines for its member nations around the world.

European Union

The European Union (EU) is a political and economic union of 28 member states that are located primarily in Europe (the union includes Turkey which is partially in Europe and partially in Asia). It was founded in 1993 and operates through a system of institutions and intergovernmental-negotiated decisions by the member states. Among these institutions is the European Commission and within that is the Directorate for the Environment. Unlike the other international organizations discussed thus far, the EU has regulatory and disciplinary authority and hence does more than simply advise.

The objectives of the Directorate-General for Environment are to protect, preserve, and improve the environment; to propose and implement policies that ensure a high level of environmental protection; preserve the quality of life of EU citizens; enforce environmental regulations among member states; and represent the European Union in environmental matters at international meetings. The Directorate for the Environment has a complex program covering all of the major contaminants, their sources and effects. Specific groups of chemicals, such as biocides, pesticides, pharmaceuticals, or cosmetics, are covered by their own legislation.

NATIONAL REGULATION OF CONTAMINANTS

Without doing a nation by nation check, it is safe to say that most if not all developed countries have their own environmental and contaminant programs including offices or directorates dedicated exclusively to chemical pollution. Some contaminant issues such as greenhouse gases and climate change affect virtually all nations. Other issues such as oil spills mostly affect oil-producing nations, but all countries have some involvement as they import or export oil products. Underdeveloped nations such as the Sudan, Somalia, and others are also concerned about pollution although they may not have the resources to do very much about it and have to rely on international assistance, especially through UNEP. A few countries such as China, North Korea, and Myanmar are very secretive and the extent of their concern for the environment is not very apparent. We cannot do an exhaustive study of individual nations so we will only focus on the United States.

In the United States, contaminant-related issues crosscut many government departments and offices. We cannot neglect the role of Congress in passing laws, establishing agencies, and setting their missions, but the real, on the ground work occurs within the Executive Branch and its departments. Of course, the US EPA is the chief enforcer and policy maker within the Executive Branch and we will cover that agency. However, oil spills, federal lands, clean oceans, and mine-related contaminants are of concern in many other departments. What follows is a concise narrative of agencies directly involved with contaminant issues.

State Department

The State Department is the nation's primary contact with other nations. Part of this contact is developing and entering treaties, protocols, and conventions to promote global concern for the environment. In this regard, the State Department has programs to develop and implement energy, agricultural, oceans, environmental science, and technology policies to address global challenges including environmental stewardship. The department is particularly concerned about the global causes and effects of climate change and helps other nations by providing scientific and economic support to reduce their emissions. The Bureau of Oceans and International Environmental and Scientific Affairs takes the lead in this department to promote these interests.

Department of Defense (DOD)

As mentioned in Chapter 9, the Department of Defense (DOD) is by far the largest land owner in the United States with millions of acres under its jurisdiction. Each branch (Army, Navy, Marines, Air Force) of the department has its own environmental office. A major issue in the DOD are the 141 Superfund sites (see next), more than 6000 hazardous waste dumps, and a total of 39,000 contaminated areas on its properties. Many of these sites are particularly problematic because they reside on lands that the DOD wants to be rid of due to downsizing the number of installations under its purview. Highly contaminated areas such as Superfund sites prevent the department from unloading these areas until the contamination is remediated. In most cases, the department hires third parties such as consulting firms to carry out the remediation but they still require site managers who are DOD employees to oversee the activities. The EPA works closely with DOD to make sure the sites are properly cleaned.

Department of the Interior

The Department of the Interior has considerable involvement with contaminant issues. Historically, scientists within the Fish and Wildlife Service conducted groundbreaking research on the effects of contaminants on wildlife from the 1970s through today. In particular, those stationed at the Patuxent Wildlife Research Center in Maryland and the Columbia Environmental Research Center in Missouri, working primarily with birds and fish, respectively, were early leaders on the effects of organic contaminants to wild animals. Today, the centers are under the jurisdiction of the US Geological Survey. Their studies involve laboratory and field research.

The Fish and Wildlife Service—This agency has charge of and works closely with the EPA and the National Marine Fisheries Service (NMFS) on Natural Resource Damage Assessment (NRDA) issues. NRDA steps in after large-scale environmental crises such as the BP oil spill, a blow out of a mining retention pond, or some other releases of hazardous substances, to calculate the monetary cost of restoring natural resources including wildlife that were damaged by these releases. The objective of NRDA is to try and recover some of the costs from potentially responsible parties to restore these populations or their habitats. Damages to natural resources are evaluated by identifying the functions or "services" provided by the resources, assessing the immediate value of these services, and quantifying the future reduction in service levels as a result of the contamination. For example, sometimes populations of commercially harvested fish are quickly decimated by contaminant exposure, but other times there is loss even when populations seem to be healthy due to persistent contaminants in their tissues and subsequent risk to humans. NMFS becomes involved when marine oil spills occur.

One other important function of the US Fish and Wildlife Service with regards to contaminants exists within the Ecological Services Branch of the agency. Within each of the 50 states, Ecological Services offices staff specialists in wetlands, water resources, endangered species, conservation planning assistance, natural resource damage assessment, and contaminants. The offices provide scientific assistance to wildlife refuges, other federal and state agencies, tribes, local governments, the business community, and private citizens.

Contaminant specialists in these offices review project plans including environmental assessments prior to the start of major projects and review license applications, proposed laws, and regulations to avoid or minimize harmful effects on wildlife and habitats. In cases of significant releases of hazardous waste, they often lead the efforts for NRDA.

Other agencies within the Department of the Interior and thumbnail sketches of their involvement with contaminants are as follows:

Bureau of Land Management—Second only to DOD in the amount of land owned, the BLM has many open pit and underground mines and oil derricks on its properties which it oversees to make sure that environmental laws are being followed. Just as some examples, the Bureau has jurisdiction over 258 million acres (60 million ha) of surface lands and 700 million acres (157 million ha) of subsurface minerals primarily in the 12 western states. On these lands, there are more than 46,000 oil leases and 308 coal leases totaling 475,000 acres (106 million ha). Most of the subsurface lands can be mined. In addition, hundreds of abandoned mines also pose risks for water pollution. The Bureau must follow federal guidelines on these activities to make sure that mining operations, landfills, surface waters, and atmospheric releases are in compliance.

Bureau of Safety and Environmental Enforcement—This agency works to promote safety, protect the environment, and conserve resources offshore through vigorous regulatory oversight and enforcement. Its range of operations is the outer continental shelf and it is mainly concerned with oil exploration and drilling. While this agency provides guidance on safety and environmental stewardship, it is also a regulatory agency with the ability to impose fines and other incentives to industry. An important objective for the Bureau is to decrease the number and magnitude of oil spills in deep oceanic waters.

Office of Surface Mining and Reclamation and Enforcement—Historically, surface mining for coal has earned a bad reputation. Prior to federal and state laws, particularly the federal Surface Mining Control and Reclamation Act in 1977, mining companies had little regulation and owners often strip-mined an area, turning the soil horizons upside down and when the seam wore out, simply left everything behind. Acid mine drainage permanently destroyed habitat, and alien-appearing landscapes were the rule. After regulations were established to reclaim stripped mine land in the United States, much of the environmental damage has ceased. In the late 1970s and 1980s, the Office of Surface Mining did most of the monitoring and enforcement of strip-mine operations. Today, most coal-producing states have their own programs for those purposes and the Office focuses on overseeing and assisting state programs and developing new tools to help the states and groups get the job done. The burning of coal and long-term storage of coal residues or fly ash continues to present problems, but the Office of Surface Mining along with cooperative mining companies have mitigated the negative impacts of the initial extraction of coal. In addition to overseeing current coal mining operations, the Office also helps funds the restoration of lands that were strip mined prior to 1977 (Fig. 15.3).

US Geological Survey (USGS)—The mission statement of the USGS is: "The USGS serves the Nation by providing reliable scientific information to describe and understand the Earth; minimize loss of life and property from natural disasters; manage water, biological, energy, and mineral resources;

FIGURE 15.3 What reclamation can do. Left: Church Creek in West Virginia was mined before the 1977 reclamation laws and was abandoned. It remained as an ecological blight for many years until it was reclaimed. Right: What the site looks like today after West Virginia, with the aid of the federal Office of Surface Mining and Enforcement, restored the land. *Photo courtesy of West Virginia Department of Environmental Protection. http://www.dep.wv.gov/aml/Pages/default.aspx.*

and enhance and protect our quality of life." Its emphasis is on providing scientific information, not necessarily on conducting research. In its role of providing information, the USGS is very good at monitoring and surveying natural resources. In the area of contaminants, the USGS has extensive water quality monitoring programs such as National Water Quality Assessment Program (NAWQA) that began in the 1980s, Source Water Quality Assessment that monitors the sources of drinking water, National Stream Quality Accounting Network, the National Atmospheric Deposition Program, and several others. These programs monitor basic water quality parameters such as flow rates of rivers, nutrients, chlorophyll, and pH and they also sample for a broad range of contaminants including pharmaceuticals, pesticides, heavy metals, volatile organics (light molecular PAHs, alkanes, and alkenes). Their atmospheric program monitors sulfur dioxide, nitrous oxides, and other contaminants depending on region of the country and need. As with many federal agencies, the USGS is also active in monitoring greenhouse gases and global climate change. The USGS does not have regulatory authority, but it regularly reports its findings to agencies at the local, state, and national levels that do.

A previous chapter briefly mentioned Patuxent Wildlife Research Center and the Columbia Environmental Research Centers as pioneers in conducting research on the effects of contaminants to wildlife and fish. Scientists at these and other USGS research centers continue to conduct research on the effects of a wide range of contaminants on individuals and naturally occurring populations. Suffice to say that many of the research studies we cited throughout this book were conducted by scientists at these centers. Also in line with research studies, the USGS oversees the Cooperative Research Units that are at land grant universities scattered across the country and sponsors and advises graduate student research. Some of their research focuses on the effects of contaminants on fish and wildlife and on the fate and transport of contaminants.

Department of Commerce

National Oceanic and Atmospheric Administration (NOAA)—NOAA is the chief federal agency with regards to oceans and the atmosphere. Whereas the Department of Interior has the Bureau of Safety and Environmental Enforcement that oversees some activities, especially oil drilling on the continental shelf, various offices within NOAA cover from the shoreline up to the exclusive economic zone, 200 nautical miles from shore. Contaminant-related concerns include tracking and predicting the direction of oil spills, monitoring the fate of toxic chemical spills in the ocean, helping to reduce microplastics and other debris, and monitoring the quality of tidal water. The agency also tracks the effects of water-borne contaminants on marine sea life such as corals, benthic

FIGURE 15.4   Oceanic gliders for underwater (left) and surface (right) monitoring of water quality and other parameters are new innovations used by NOAA and other agencies.

organisms, and shellfish. The agency also hosts mussel watches along coastlines and in the Great Lakes in which mussels are monitored to determine status and trends of aquatic contaminants. Some of this monitoring is conducted with high tech, unmanned oceanic gliders that can dive to ocean bottoms or sample surface waters (Fig. 15.4).

NOAA also monitors air quality in the United States and globally. Through their Earth Systems Research Laboratory, NOAA regularly monitors levels of air quality elements such as tropospheric ozone, carbon monoxide, and aerosol particles across the globe. The ability to determine the status and development of global climate change is of major concern in these endeavors.

US Coast Guard—One of the many functions of the paramilitary US Coast Guard is to make sure that there are no illegal discharges of pollutants into US waters. Under the National Contingency Plan, the Coast Guard is the predesignated federal coordinator for oil and hazardous substance incidents in all coastal and some inland areas.

As such, the US Coast Guard is the principal enforcer for the international agreement, MARPOL 73/78—the International Convention for the Prevention of Pollution from Ships, (1973, 1978)—and for US federal regulations regarding the discharge of wastes and other contaminants by ocean-going ships. MARPOL 73/78 is an important international marine environmental convention which was developed to minimize pollution of the oceans and seas, including dumping, oil, and air pollution. Membership in the current convention consists of 152 nations, representing 99%of the world's shipping tonnage. The Coast Guard enforces regulations on marine sanitation devices and therefore on discharges. Years ago, ships emptied their bilge tanks untreated directly into ocean waters under the misguided concept that any harmful effects would be diluted by the vast waters of the seas. Today, due to extensive maritime shipping we realize that dilution is not the solution to this type of pollution and it is now illegal to discharge untreated wastes and sewage into US waters.

US Environmental Protection Agency (EPA)

As the chief watchdog for contaminants, the US EPA has offices across the country to oversee remediation efforts on hazardous landfills and enforce federal regulations concerning emissions, effluents, water treatment plants, and virtually all things of a contaminant nature. Congress created the US EPA in 1972 and it has passed several acts which provide marching orders to the EPA in its efforts to reduce pollution and to protect human health. We cover a few of these acts next.

Federal Food, Drug, and Cosmetic Act (FFDCA, 1938)—This is among the oldest of the acts dedicated to regulating the purity of food, cosmetics, homeopathic medicines, and bottled water. It stems from the older Pure Food and Drug Act (1906) which illustrates a long history of Congressional concern for untainted food and drugs. The US Department of Agriculture through the Food and Drug Administration has jurisdiction over most of the Act and that makes sure of such things as requiring that the coloring and additives in food are identified and standardized, that medical devices generally do what they are intended for, and that nothing toxic enters bottled water or homeopathic medicines. The FFDCA also authorizes the US EPA to set tolerances, or maximum residue limits, for pesticide residues in foods. Once a tolerance is established, any food containing more than that concentration is subject to seizure. As new information on any of the products or pesticides is acquired, the tolerance levels are changed.

Clean Air Act (1970)—This is one of the first Acts that fell under the jurisdiction of the EPA although it was passed prior to the existence of the agency. It was written to control air pollution on a national level and was initially managed by the Public Health Department. It is one of the United States' first and most influential modern environmental laws. It follows and very significantly strengthens the Air Pollution Control Act (1955) and an older, 1963 version of the Clean Air Act. While those older acts defined air pollution and provided funding for research, the Clean Air Act set out to actually control air pollution through directives to increase monitoring of point and nonpoint sources and providing enforcement through fines and even jail sentences for offenders. Major amendments to the current Clean Air Act were passed in 1977 and 1990. These amendments included acid rain, ozone depletion, standards for stationary or point sources (eg, incinerators, electric power plants), and toxic air pollutants; and increased enforcement authority. The Clean Air Act was the first major environmental law in the United States to include a provision for citizens to press legal action against violators of the law and against the government should it be lax in enforcing the law (Fig. 15.5).

Numerous state and local governments have enacted similar legislation, either implementing federal programs or filling in locally important gaps in federal programs. The amendments also established new auto gasoline reformulation requirements, set standards to control evaporative emissions from gasoline, and mandated new gasoline formulations sold from May to September in many

FIGURE 15.5 Los Angeles in heavy smog, indicating that the Clean Air Act still needs to be enforced. *Wikimedia Commons.*

states. There are some 20 different formulations of gasoline depending on federal and state regulations. A big difference is between the summer and winter blends. Winter-blended gasoline is required in northern states because it has a higher vapor pressure than summer blends to make sure that the gasoline vaporizes in the cylinders. The switch from summer to winter blends and back again can be a cause for refineries to be temporarily shut down and for corresponding hikes in the price of gasoline.

The Act focused and set standards for the six major air pollutants or *criteria pollutants* of ozone, particulate matter, carbon monoxide, nitrogen oxides, sulfur dioxide, and lead. The EPA has the authority to monitor these contaminants and invoke fines to companies that violate the emission standards and the standards can be updated as necessary. *De facto* additions to this list include greenhouse gases such as carbon dioxide, methane, and water vapor that are involved in global climate change although standards have not been established for some of these pollutants. We still have a way to go before we reach the air quality standards set by the Clean Air Act.

Clean Water Act (1972)—Following more or less on the heels of the Clean Air Act, Congress passed the Clean Water Act two years later. This act is actually a major amendment to the Federal Pollution Control Act (1948) which had a primary focus on safe drinking water. The current law, which was again amended in 1984, monitors and regulates water pollution. Its objectives are to restore and maintain the chemical, physical, and biological integrity of the nation's waters by preventing point and nonpoint pollution sources; providing assistance to publicly owned treatment works for the improvement of wastewater treatment; and maintaining the integrity of wetlands. It is the last of these objectives that has been the most controversial. Over the years, wetlands and streams under the jurisdiction of the law have been variously defined and currently have a narrow definition in that they must be connected to navigable rivers. In 2006, the Supreme Court decision in *Rapanos v. United States* defined protected waters as "only those relatively permanent, standing or continuously flowing bodies of water 'forming geographic features' that are described in ordinary parlance as 'streams… oceans, rivers, [and] lakes." In 2014, Congress voted to expand the protection provided by the Act to many more streams and wetlands, but at the time of this writing the amendment has been stalled by the courts.

At present, the Clean Water Act consists of six programs or titles: research, grants for water treatment plants; standards and enforcement; permits and licenses; general provisions (eg, whistleblower protection, citizens right to press legal charges against the government); and other funding to states. Standards have been set for essentially all forms of water pollution including silt and other particulates, contaminants, biologicals, biosolids from treatment plants, thermal pollution, and other organic and inorganic pollutants. The benchmark for each of these standards is called the *Total Maximum Daily Load (TMDL)* and fines can be given to point-discharge violators. It is often difficult to identify responsible parties for nonpoint sources of contamination, but the EPA also has programs to try and mitigate those sources.

The EPA recognizes that not all sources of water pollution can be controlled or prevented. Municipalities, for example, may need to discharge rainwater that contains street debris and few, if any, water treatment plants are 100% effective. Therefore, the EPA issues National Pollutant Discharge Elimination System (NPDES) permits that allow some discharges into public water sources.

Resource Conservation and Recovery Act (RCRA, 1976)—This is the principal federal law covering solid wastes and disposable wastes. Solid waste includes all garbage or refuse; sludge from wastewater or water supply treatment plants; wastes from air pollution control facilities; and other discarded material including solid, liquid, semisolid, or contained gaseous material resulting from industrial, commercial, mining, agricultural, and community activities (EPA, 2015a). These wastes include both hazardous and nonhazardous materials. RCRA is the result of major amendments to the Solid Waste Disposal Act of 1965. The current law focuses on controlling the ongoing generation and management of solid waste streams in what has been called a cradle to grave oversight. This means that the EPA requires detailed information on solid wastes from their point of generation, through their transportation and treatment, to storage and their ultimate disposal or destruction. Special provisions exist for agricultural wastes, underground storage tanks, and municipal waste streams. RCRA deals with ongoing processes involving waste disposal in contrast with CERCLA, which deals with waste sites that have been abandoned or neglected for long periods of time. Many states assist the EPA by implementing RCRA provisions within their own borders.

Toxic Substances Control Act (TSCA, 1976)—This law authorizes EPA to require reporting, record-keeping and testing requirements, and restrictions relating to chemical substances and/or mixtures. Substances such as food, drugs, cosmetics, and pesticides are generally covered by other laws but TSCA includes substances such as radon, polychlorinated biphenyls, asbestos, lead-based paint, and an ever-growing list of chemicals (EPA, 2015b). Provisions of TSCA require:

- That chemical companies notify the EPA prior to any manufacturing of new chemical substances;
- That companies implement toxicity testing on new chemicals or when new uses of existing chemicals are being proposed;
- Maintenance of the TSCA Inventory which contains more than 83,000 chemicals. As new chemicals are commercially manufactured or imported, they are placed on the list.

Contrary to its title, TSCA includes chemicals that are highly toxic and some that have very little toxicity in its inventory and the same regulations pertain to both. While TSCA requires that manufacturers conduct toxicity tests, especially to determine if a chemical is carcinogenic, some 62,000 chemicals were grandfathered in (meaning that they existed before TSCA was enacted and are not subject to the same regulations as "new" chemicals); by law these chemicals are not considered to pose "unreasonable risk" although we do not know if they are safe. There are many other exceptions to the law including substances used in small quantities for research purposes, foods, many food additives, drugs, cosmetics, radioactive materials, tobacco, and pesticides, but these are covered by other laws or federal agencies.

While this law sounds like it should be a powerful deterrent to unregulated production of hazardous chemicals, it has been frequently criticized for weaknesses in operation. Of all the chemicals in its inventory, only 250 have had toxicity testing by the EPA as of 2015. Of these, 140 were tested by regulatory order, 60 were tested after voluntary consent by the manufacturer, and the others were voluntarily discontinued by their manufacturers (Wilson and Schwarzman, 2009). In addition, there are 3000 high production volume chemicals produced or imported in quantities exceeding one million pounds per year that have not been tested. These high volume chemicals raise concern about the lack of basic hazard information. Altogether, manufacturers in the United States import or make 12.2 trillion tonnes of chemicals each year (Wilson and Schwarzman, 2009).

Comprehensive Environmental Response, Compensation, and Liability Act (CERCLA, 1980)—The CERCLA was passed to remediate the huge number of abandoned or unmanaged hazardous waste sites scattered throughout the county. If, after a hazard review of a site demonstrates that it is a significant problem, the site is registered on EPA's National Priorities List (NPL) as a Superfund Site. Priority for remediation is based on the score the site receives during its hazard review. There are tens of thousands of dump sites across the country (see Fig. 15.6 for an example), but as of late 2015 there were 1320 active Superfund sites, 51 proposed sites, and 389 deleted sites where no further remediation was warranted (US EPA, 2015c). When possible, the EPA identifies a Potentially Responsible Party or culprit to help or take over the costs of clean-up which can amount to millions of dollars. When such a party (usually a corporation) cannot be found, the US EPA pays for clean up through a special Superfund trust. Several of these Superfund sites exist on government lands such as military bases so finding a PRP might be easy, but many others have been abandoned for decades and the companies that were originally responsible for the mess have disappeared long ago. Remediation also entails a five-year observation period after the clean up to make sure that nothing pops up. The regulations and protocols for evaluating, listing, and remediating sites maintain a tight control of the entire process. Identification and information on the Superfund sites are maintained for public viewing through the US EPAs Superfund Enterprise Management System (SEMS).

Federal Insecticide, Fungicide, and Rodenticide Act (FIFRA, 1996)—Pesticides are one of the group of chemicals not covered by TSCA. This act fills that gap. FIFRA provides for federal regulation of pesticide distribution, sale, and use. All pesticides distributed or sold in the United States must be registered by the US EPA. FIFRA replaced the Federal Insecticide Act (1910).

FIGURE 15.6   Valley of the Drums in Bullitt County, KY. This is a 23 acre (9.3 ha) Superfund site containing more than 100,000 drums of industrial wastes and chemicals. *Wikimedia Commons.*

The original act was written primarily to assure that farmers were receiving high quality pesticides for application in their fields, secondarily to protect the public, with no concern paid to the potential damage to the environment. The current Act places emphasis on public and environmental safety. Before the US EPA may register a pesticide under FIFRA, the applicant must show, among other things, that using the pesticide according to specifications "will not generally cause unreasonable adverse effects on the environment."(US EPA, 2015d). This requires that the manufacturer conduct studies on the effectiveness of the proposed pesticide, its chemical fate and transport, and its toxicity. If during toxicity trials the chemical seems to pose some risk of causing cancer, a more extensive battery of tests must be conducted. Unreasonable risk in this definition is further explained as: "(1) any unreasonable risk to man or the environment, taking into account the economic, social, and environmental costs and benefits of the use of any pesticide, or (2) a human dietary risk from residues that result from a use of a pesticide in or on any food inconsistent with the standard under section 408 of the Federal Food, Drug, and Cosmetic Act."

After testing has been conducted, the manufacturer may submit the information and request that the chemical be registered for use. If registration is approved, the company must then prepare a label for approval that describes how the pesticide is to be used and other information such as the active ingredient and signs of toxicity. Under commercial application, failure to follow the label instructions may result in fines and loss of commercial applicator license if the infraction is sufficiently severe. There are separate regulatory processes for conventional pesticides, antimicrobials, and biopesticides such as *Bacillus thuringensis*. The US EPA maintains a large list of restricted use pesticides that are not available to the general public because they pose too great of a risk. These restricted use pesticides are only available to certified applicators.

It costs millions of dollars to register a pesticide so manufacturers are highly selective about the chemicals that they intend to register and many smaller volume or specialty products are not submitted for registration. A frequent criticism is that the US EPA allows chemical companies to conduct their own testing rather than having them submit the chemical to independent laboratories and this practice has been likened to having the fox guard the chicken house.

The 1988 amendments to FIFRA authorized the US EPA to conduct a complete review of the human health and environmental effects of pesticides first registered before November 1, 1984, to make decisions about these pesticides' future use. The goal of the reregistration program is to mitigate risks associated with the use of older pesticides while preserving their benefits. Reregistration may require the same type of testing that registration of new chemicals must undergo. Initially 1150 pesticides fell under this mandate, but 229 of these were terminated before being reregistered. The agency also reassessed existing tolerances (pesticide residue limits in food) to ensure that they met the safety standard established by the Food Quality Protection Act (FQPA) of 1996. FQPA amended the Federal Food, Drug, and Cosmetic Act (FFDCA), requiring the US EPA to reassess within 10 years the existing 9721 tolerances and tolerance exemptions to ensure that they met the new "reasonable certainty of no harm" safety standard.

In addition to reregistering specific pesticides, the agency is undergoing reregistering processes for specific products. A given pesticide may occur in many formulations or products based on concentration of pesticide and the other chemicals that are used as carriers, surfactants, and activity enhancers. After the US EPA has completed a pesticide reregistration process and determined that the pesticide could be reregistered, the agency can call for data on the individual products. The agency then reviews any testing data and updated labeling. The US EPA may reregister products that meet current standards. As of the end of fiscal year 2014, 24,584 products were subject to product reregistration (US EPA, 2015e). Of these, more than 10,000 products were canceled either by the EPA or voluntarily by the manufacturer and more than 5000 products were reregistered. This product-based reregistration continues.

Under FIFRA, the US EPA can also call for special reviews when it has reason to believe that the use of a registered pesticide may result in unreasonable adverse effects on people or the environment due to new data or interpretations of existing data. Recent special reviews were called for aldicarb, atrazine, propazine, simazine, and ethylene oxide. The review for aldicarb came due to concern for risks to young children and infants and prompted the voluntary withdraw of the pesticide by the manufacturers. The special review of the triazines (atrazine, propazine, and simazine) was called because new evidence linked these chemicals to breast cancer. Among other purposes, ethylene oxide is used as to fumigate medical equipment, bird seed, and bee hives. Relatively recent information and literature reviews suggested that the chemical produces risk of cancer, genotoxicity, and other effects in medical personnel. The EPA decided after conducting the special review that outdoor use and bird seed treatment with ethylene oxide did not pose serious risk and that medical chambers used for sterilization need to be properly vented (US EPA, 2008).

Regulation at the State and Municipal Levels

Every state in the nation has an agency that regulates environmental quality. A few states have their agencies embedded in their conservation department, but most have separate agencies specifically for monitoring and enforcing the state's laws concerning contaminants. These agencies go by various names: Department of Environmental Quality, state Environmental Protection Agencies, Department of Environmental Management, and others, but the bottom line is that they all do similar things—set standards, monitor, educate, enforce, and remediate. Even US territories such as Puerto Rico and the Marshall Islands have their own regulatory agencies.

States are able to establish their own pollution standards. However, while they can be more stringent that federal guidelines, they cannot be more lax. California has among the highest standards with regards to air pollution. For example, California state standards are stricter for atmospheric particulate matter and ozone than standards set by the Environmental Protection Agency. For larger particles, those greater than 2.5 mm in diameter, the US EPA standard is $150 \, \mu g/m^3$, while that of California is $50 \, \mu g/m^3$.

Major cities such as New York City and Los Angeles also have municipal offices that issue permits, monitor air and water quality, and issue fines when necessary or are members of regional consortiums that do the same. Most cities follow state and federal standards.

STUDY QUESTIONS

1. In what way does the United Nations have the ability to enforce international laws on contaminants? Is this a very strong form of enforcement?
2. The agreements that the United Nations oversees and the specific methods that the EPA develops to conduct analyses are both called protocols. Why, do you think, the term applies to both of these rather dissimilar conditions?
3. Review the various agencies and their functions from the international to the national levels. What provisions are being made by these agencies to study or restrict greenhouse gases and combat global climate change?
4. On what issue(s) are the US DOD and the US EPA most likely to interact?
5. Part of the Natural Resource Damage Assessment is establishing the economic value of natural resources. Discuss with your class how the US Fish and Wildlife Service might go about determining the value of a duck or an uncontaminated wetland during an NRDA of a major oil spill.
6. The Bureau of Land Management include extensive areas of coal mining, mineral extraction, and siliviculture. Should the federal government be in the business of selling off natural resources? Defend your opinion with your class.
7. We mentioned that NOAA hosts mussel watches in both marine environments and the Great Lakes where they monitor the contaminants in mussels. Based on information in other chapters of this book, why do you suppose NOAA selected mussels as their subjects to monitor?
8. Choose one of the following Acts and develop a short paper covering its origins, purposes, and successes: Clean Air Act, Clean Water Act, FIFRA, TSCA, RCRA. Be sure to go into greater detail than what is presented in this chapter.
9. What is the name of the agency that has jurisdiction over your state's contaminant issues? What is its mission statement?
10. What does your town or city do to control pollution?

REFERENCES

European Union, 2015. European Commission, Environment. <http://ec.europa.eu/dgs/environment/index_en.htm>.

OECD Organisation for Economic Cooperation and Development, 2015a. OECD. <http://www.oeced.org/>.

OECD Organisation for Economic Cooperation and Development, 2015b. Chemical safety and biosafety. <http://www.oecd.org/chemicalsafety/>.

United Nations, undated. United Nations Environment Programme. <http://www.unep.org/about/>.

United Nations, 2014. Framework convention on climate change. <http://unfccc.int/2860.php>.

United Nations, 2015. The Millennial Development Goals Report. <http://www.un.org/millenniumgoals/2015_MDG_Report/pdf/MDG%202015%20rev%20(July%201).pdf>.

US EPA U.S. Environmental Protection Agency, 2008. Reregistration Eligibility Decision for Ethylene Oxide. EPA 738-R-08-003 Office of Pesticide Programs. U.S. Environmental Protection Agency, Washington, DC.

US EPA U.S. Environmental Protection Agency, 2015a. Resource conservation and recovery act (RCRA). <http://www.epa.gov/agriculture/lrca.html>.

US EPA U.S. Environmental Protection Agency, 2015b. Summary of the Toxic Substances Control Act. <http://www2.epa.gov/laws-regulations/summary-toxic-substances-control-act>.

US EPA U.S. Environmental Protection Agency, 2015c. CERCLA overview. <http://www.epa.gov/superfund/policy/cercla.htm>.

US EPA U.S. Environmental Protection Agency, 2015d. Summary of the Federal Insecticide, Fungicide and Rodenticide Act. <http://www2.epa.gov/laws-regulations/summary-federal-insecticide-fungicide-and-rodenticide-act>.

US EPA U.S. Environmental Protection Agency, 2015e. Reregistration and other review programs predating pesticide registration review. <http://www2.epa.gov/pesticide-reevaluation/reregis-tration-and-other-review-programs-predating-pesticide-registration#Product-reregistration>.

Wilson, M.P., Schwarzman, M.R., 2009. Toward a new US chemicals policy: rebuilding the foundation to advance new science, green chemistry, and environmental health. Environ. Health Perspect. 117, 1202–1209.

Chapter 16

Wrap Up

INTRODUCTION

This is one of those chapters that virtually every book, especially textbooks, are expected to have. It's almost a requirement that there be a chapter that sort of summarizes everything that has been presented in some thoughtful, reflective way. Well, here it is for this book. We will recap where the discipline of ecotoxicology has been, describe its current status, more or less, and provide the author's reflections on what more is needed, at least in the next several years. The good thing that you, the student, should know is that there is no "Terms to Know" section and no study questions to consider. Sit back, enjoy the chapter, and know that some of you will be the ones who address the ecotoxicological problems of the future.

From Whence It Came

We can do a bit of historical snooping into contaminant ecology by consulting search engines to see what pops up as scientific articles. A widely used search engine for scientific articles, *Web of Science,* reports that the first papers on insecticides were published around 1875. One author wrote in 1889 (Anonymous, 1889) on the use of insecticides in those days. He (women were not included as experts in agriculture in those days) highly praised arsenic-based compounds, such as Paris Green and London Purple. Other chemicals included carbolic acid (phenol) soaps, bisulfide of carbon, coal tar, lime, and plaster. He also mentioned extracts of hellborne, tobacco, and pyrethrum. That's it. Surprisingly, given the rate of inflation ($1 in 1889 = $25 today), he quoted that the annual loss in crops to insects was between $200 and $250 million. That would translate to around $5–$6.2 billion today. Keep in mind that the annual production of crops in those days was significantly smaller, so that $200 million would have been a very substantial percent of the nation's agricultural output.

For the next several years, a publication appeared every few years up to the late 1930s, when two or three papers began to appear on insecticides per year. In 1943, an article in *Science* cited 5000 different compounds used as insecticides or fungicides (Frear, 1943). No real mention was made of DDT until 1944. Some farmers may have been using it by then, but science's interest was not stirred. By 1945, more than 200 papers were published on some topic related to pesticides. Among these papers, 62 (29%) were related to DDT, at that time its technical name was 2,2-bis (parachlorophenyl-1, 1, 1-trichlorethane). Of these papers, 59% were totally in praise of the insecticide, 11% were mostly in favor of using the chemical but issued a bit of caution, and only 6% gave negative opinions or warnings. The other 24% were neutral, often describing some technique for analyzing for DDT in one environmental matrix or another but not providing any judgment of the chemical. Other insect-control agents included a few naturally derived compounds, pyrethrins, oils, rotenone, and a couple of other organochlorines (OCs). Papers advocating the use of natural parasites accounted for 17% of the publications. Hardly the 5000 touted by Frear (1943). No doubt farmers were using control agents that science did not consider. Farmers still do so today in small farming operations and some organic farms.

By 1965, organophosphorus (OP) and carbamates were on the scene and garnered some attention from researchers. That year, 725 papers were published on pesticides and, of a subsample of 175, the two groups of pesticides accounted for 27% and 13% of the publications, respectively. This was a year of comparisons where studies on several pesticides, most of which were looking for the greatest efficacy, accounted for 12% of the subsample. Organochlorine pesticides, other than DDT were still in the running and accounted for another 20% of the research publications that year. In contrast, DDT was on the way out due primarily to the buildup of resistance to this pesticide and only 6% of the papers were exclusively concerned with DDT.

After 1965, research interest in pesticides exploded (Fig. 16.1). The number of research papers on pesticides continues to grow exponentially. Other contaminants that were known at that time such as polycyclic aromatic hydrocarbons (PAHs) show a similar trend although not as domineering as pesticides. Polychlorinated biphenyls (PCBs) and their relatives, plastics, and personal care products did not make it onto the scientific radar until a few to many years later. What caused this tremendous interest in pesticides? Well, back in Chapter 1, we mentioned in passing Rachel Carson's *Silent Spring* (1962). The publication of this book is marked with a red *X* on Fig. 16.1. This book had a dramatic effect of awakening the American public to the threats posed by indiscriminate use of pesticides. Once the public was roused, they gave a clear signal to Congress that they wanted something done. I would not say that this book was the sole cause for the interest in pesticides, but it certainly helped. (By the way, I highly recommend this book as required reading for anyone in the least interested in ecotoxicology.)

I recall those years leading up to 1970 (the first Earth Day). Universities across the nation began offering courses in environmental science and the media (nowhere near what we have today) were filled with articles on green living, recycling, and tending to Mother Nature. This was also the era of "hippies," and many, but a minority, of young adults were moving into communes where they could "tune in" to nature and "drop out" of conventional society. In the 1960s and early 1970s, several key pieces of federal legislature were passed along

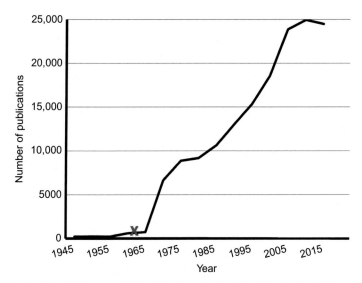

FIGURE 16.1   The growth in the number of scientific publications dealing with pesticides has exploded since 1965. Note that the earliest record of a paper dealing with insecticides was in the late 1800s. The red *X* represents when Rachel Carson's *Silent Spring* (1962) was written.

with a significant federal budge for research on the environment. Scientists at universities, government agencies, and private corporations across the country began very important base-line studies on water, air, and terrestrial environmental science. The foundations for ecotoxicology were laid in the 1950s and 1960s and the infrastructure arose quickly after that. Judging by the growth of published scientific papers since then, the interest has not waned. Beyond a doubt, the United States was the principal driver of this research, followed by India (surprisingly?), Canada, and the United Kingdom; the United States, India, Canada, and the United Kingdom were leaders from the start in the 1950s.

During this time, the basic concepts of median lethal dose, LC_{50}, LD_{50}, and even EC_{50} and ED_{50}, which were unknown to toxicologists before the advance of ecotoxicology became common topics for studies. These concepts are based on controlled experimentation, something that is just not easily accomplished with humans, particularly when one is interested in lethal doses. Somehow society does not appreciate using humans as test animals. At first, those studying the effects of contaminants were primarily interested in how much of a chemical was necessary to kill organisms. Later, as information on environmentally realistic concentrations of chemicals and sublethal effects was obtained, attention turned to effects below outright death such as growth, behavior, immunotoxicity, genotoxicity. Only recently in the big picture of things have ecotoxicologists really turned toward multiple stressors.

WHERE ARE WE NOW?

In 2014, the most recent year with a full accounting of publications, 29,084 articles were published on pesticides, 7488 on PAHs, and a whopping 84,405 on heavy metals (Fig. 16.2). These papers reflect all areas of ecotoxicology—scientific publications, articles on risk assessment or management, methods development in analytical chemistry, federal and state regulations, and, to some extent, public attitudes about the presence of contaminants in the environment. Just a side note—we cannot say that the numbers provided all represent unique, independent sources of information. Often investigations will include more than one type of contaminant such as PCBs, dioxins, and furans with maybe polybrominated diphenyl ethers (PBDEs) thrown in as well because they all can be analyzed on the same instruments in the same analytical lab; this would lead to one paper being recorded multiple times. However, I was not going to comb through more than 30,000 titles to see how many included more than one chemical, sorry. Looking back on the history of ecotoxicology based on research and publication, it is easy to say "We've come a long way, baby" in a relatively short period of time.

The discipline of ecotoxicology rests on the shoulders of toxicology, ecology, chemistry, and civil law, all of which are much older fields of inquiry.

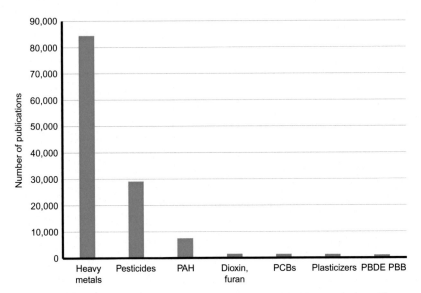

FIGURE 16.2 Each year, thousands of scientific papers are published in ecotoxicology. The year 2014 was the most recent one that we could be sure had a complete record at the time that this book was written. At that time, more than 84,000 papers were published with a focus on heavy metals and 29,000 papers were written on pesticides. The other common contaminants were represented less frequently but still had many papers written on them. These numbers do not necessarily represent unique contributions. In many cases, studies would involve more than one contaminant, such as PCBs and dioxins, and could be counted more than once across the range of contaminants.

I would like to briefly discuss how ecotoxicology has given back to each of these areas.

Toxicology

Today we continue to unravel the responses of organisms to environmental contaminants, come to a deeper appreciation of the intricacies of multiple exposures and multiple stressors working at all scales, and continue to incorporate these concerns into our research perspectives. There are both applied and basic benefits from this research. From an applied perspective, we can now evaluate risk to organisms and environments more reliably than 20 years ago, and much of our acquired knowledge has contributed to advances for the medical treatment of contaminant-related diseases in both humans and wildlife. These endpoints and the studies that develop them have contributed immeasurably to our understanding of the effects of chemicals on both ecological and human subjects. Through the use of animal and plant models and hundreds of contaminants, we have developed groundbreaking advances in understanding cancer, genotoxicity, cellular mechanisms, immunosuppression, endocrinology and endocrine disruption, teratology, and physicokinetics.

In basic science, we have unraveled much about contaminant mechanisms in many groups of organisms, and have developed a clearer understanding of the genetic principals behind contaminant toxicity. We have advanced modeling the exposure and response axis. We have also developed new tools such as genomics, proteomics, and advanced modeling techniques, such as quantitative structure-activity relationship (QSAR) models.

Ecology

We have come to a clearer understanding that contaminants can be a serious threat to some populations of animals and perhaps plants. We also know that, for the most part, contaminants are most likely to exert negative effects in the environment in the realm of multiple stressors. In many situations animal and plant populations are affected by a few to many contaminants acting together in ways that are as yet unknown completely. In other situations, contaminants coexist with other stressors such as disease, parasites, intra- and interspecific competition that impinge on a population's or even a community's probability of surviving. Organismal and population effects have been studied far more often than effects on communities or ecosystems, but that does not mean that effects at higher scales are not important. By knowing that chemicals most often act in combination with other stressors, ecologists have been able to come to a deeper understanding on how multiple stressors operate in general.

Funding for studies dealing with contaminants has also helped ecologists and ecotoxicologists develop a better understanding of some of these affected populations. For instance, Cothran et al. (2013) substantiated what others have found—that amphibians (and by extension, other organisms) which have been living in chemically exposed areas develop increased genetic tolerance over time to those chemicals; such genetic modification may also apply to other stressors, such as disease. Speaking of genetics, Monsanto and other chemical companies are using genetic modification to make crops tolerant to herbicides. Genetic modification has received a lot of attention by the public and the scientific community because, while it may appear beneficial in the short term, its long-term risks are not well known. Can incorporated genes prove harmful to humans and organisms that consume genetically modified foods? Can the genes become incorporated in other plants, making them "superweeds" (Union of Concerned Scientists, 2013)? Can the genes used in genetic modification become the next class of contaminants? These are questions to which we have no answers at this time.

Chemistry

The field of chemistry has received many benefits from ecotoxicology. A plethora of new analytical methods have been developed to extract and quantify environmental chemicals at the parts per billion or even trillion concentrations (US EPA, 2015). Along with the need to identify chemicals at low concentrations in the environment, analytical instrumentation has been improved. Much of what is now accepted as good laboratory practices, quality assurance, and quality control were developed in response to the need for secure and extremely reliable practices for the regulatory community, but have also found recognition in many research laboratories.

Regulation and Civil Law

In Chapter 15, we discussed some of the agencies and laws that have been established at the international, national, state, and local levels that pertain to ecotoxicology. Such laws as the Clean Air Act and Clean Water Act have been amended, usually to make them stronger, but Congress is perpetually debating how strong our environmental regulations should be. Should the federal government protect all wetlands or only those connected to navigable waters? What are the maximal tolerable concentrations of each of the many contaminants found in drinking water, flowing water, and standing water? We often hear debates about whether the limitations set by the US EPA are sufficient or overly protective. Today the United States is a global leader in environmental law and enforcement and we should be proud that our legal system protects our environment.

WHAT REMAINS TO BE DONE?

We have come a long way over a little more than 50 years, but we have not achieved everything we need to do. Smith et al. (2010) convened a conference of ecotoxicologists from around the world to discuss and collate a list of ecotoxicological needs on a regional and global basis. The results were a set of scientifically based opinions and, as such, can be disputed by others but nevertheless represent best-professional judgments from around the world. What follows here is an encapsulation of those opinions. Taken together, this discussion highlights where research in ecotoxicology might go in the short and medium term (1–10 years).

Africa

Africa has a rich biodiversity of megafauna, particularly reptiles, birds, and mammals. It is also one of the least known continents in terms of contaminant issues. There is very little research coming out of the nations that make up the continent. We do know, however, that there is a gross overuse of pesticides, including OC and OP compounds. DDT is still used to ward off mosquitos that might be carrying malaria and comparatively high OC concentrations can be detected in humans, wildlife, soil, and water. Other legacy compounds are also applied to crops and are widely spread throughout many of the countries. Another concern is that several African nations are developing countries where new problems, such as pharmaceutical and personal care products (PPCP) may increase in waterways with further development. Africa poses a lot of questions about ecotoxicology, but not many answers. Research is sorely needed on many aspects of contaminants: what pesticides are being used in what concentrations and where? Are there serious exposure risks? Are new chemicals coming onto the scene, especially in urban areas? Is there sufficient monitoring of these problems?

Middle Eastern Countries

The Middle East is a region in the southwestern portion of Asia characterized by low rainfall, abundant petroleum, and much civil strife. Water is a huge environmental issue because this region is served only by three major rivers and much of the land is desert. Clean, potable water is scarce. The people are poorly educated in environmental sciences or even in agricultural practices—leading to the indiscriminate use of pesticides and fertilizers, and putting the limited water resources at great risk. Other than water, the chief environmental issue is, of course, petroleum. Due to frequent wars, oil wells have become frequent targets of one group or another, which causes abundant oil spills and air pollution from PAHs, dioxins, and furans. There is essentially no research being conducted in this area and no environmental regulation to speak of, thus little information about the environment is available.

Asia

Asia is the largest continent in terms of area and people. It is dominated by China, the most populous nation in the world, and eastern Russia, but also consists of Indochina, Japan, and other island nations. Only recently has the government of China become interested in the environmental health of the nation. Smog is a very significant problem in most of the larger cities in the country, especially Beijing. China's economy is rapidly growing, relying in part on an expanding manufacturing and exporting infrastructure. Environmental regulations are not keeping pace with this growth, which leads to greater pollution. Among the chemicals of greatest concern are brominated flame retardants (BFRs), including polybrominated diphenyl ethers (PBDEs). In many ways, BFRs behave like PCBs—they are long-lasting and can easily disperse through the atmosphere, especially those BFRs that are not highly brominated (see Chapter 6). They are known endocrine disruptors and may affect both the hormones associated with reproduction and the thyroid. Therefore, additional research on BFRs, including their distribution and effects on biota as well as quantitative risk assessments, is sorely needed.

Australia

Australia, an island continent, covers a very large area but the majority of its people live along the eastern coast. A large proportion of central Australia is desert and dry grasslands. Australia has been isolated from other continents for millennia and has a unique mammalian fauna. The country is an enigma because it is a modern, developed country with large cities and extensive agriculture, and yet, there has been very little research conducted on contaminants. This is particularly surprising because the native fauna are unique and may respond to chemicals very differently from the species on other continents. Most regulations on permissible concentrations of contaminants are based on species in North America or Europe, which may be very misleading. In the western part of the continent, locusts can be a problem and land owners use either the OP fenitrothion or the phenylpyrazole fipronil to control these grasshopper-like pests. While the toxicity of fenitrothion to nontarget organisms has been adequately characterized, that of fipronil has not been, so its effects on native fauna remains largely an unknown.

India

This subcontinent of Asia is the second most populous nation on Earth and has a very high population density. India is beginning to come into the modern world in terms of economics, although just trying to provide the basics to its millions of people impedes its progress. India is the second largest producer of pesticides in Asia and the third largest user in the world. DDT is still used to control malaria. Other OCs that are used extensively include endosulfan and lindane. India provides an exception to the general rule that pesticides seldom cause large die-offs of wildlife. Due to high concentrations of persistent organic pollutants (POPs) and unregulated use of other pesticides, there have been many die-offs of birds, both native and migratory. Research is desperately needed on the effects and concentrations of OC pesticides on birds and there is an urgent need for regulations on these chemicals and on enforcement of these regulations. Smog also remains a problem in many cities of India; Delhi is said to have air quality that is at least as poor as that in Beijing, China.

Europe

Geographically, Europe can be divided into four main parts based on the cardinal directions of the compass. Politically, we could arguably separate Europe into those 27 countries that are part of the European Union (EU) and the 23 that are not. Many of the countries that are not part of the EU are in Eastern Europe and are generally not as well-off as those that are, but there are some countries struggling economically in the Union. The EU has enacted a program of integrated pollution prevention and control which appears to be proactive and effective. Collectively, Europe is plagued by habitat loss and deterioration, encroachment of human beings on or abutting natural areas, climate change, invasive species, legacy and new contaminants, PAHs, and trace elements. Endocrine disruption due to chemical exposure has been identified in field studies of mammals, birds, reptiles, fish, and molluscs. Among emerging chemicals for which limits or regulations have not yet been set, Europeans are concerned about PPCPs, nanoparticles, and BFRs. Research is needed on all three of these high priority groups, especially BFRs, which are so widespread.

North America

The authors limited their discussion of North America to the United States and Canada. Both nations have well-established environmental laws and regulatory agencies (US EPA and Environment Canada, respectively). Like Europe, environmentalists in North America are most concerned with emerging contaminants. Nanoparticles, PPCPs, BFRs, and e-waste (waste associated with the production of cyber equipment) are among the leading concerns. As in Europe, BFRs head the list of concerns and a need for more research on these chemicals is called for. One particular issue associated with PBDEs is their presence in human breast milk. The average concentration of PBDEs in milk was 10 times greater than the mean in other parts of the world (Lorber, 2008); to date, we really don't know what this means in terms of health effects in neonates and young children.

Latin America

This includes Mexico, Central America, and all of South America. There is some research being conducted on contaminants in Latin American environments, but laws and their enforcement are highly variable among countries. Legacy compounds are still being used, such as DDT for malaria, endosulfan, and lindane (note that lindane is still of limited use in the United States for treatment of head lice and scabies in humans). Brazil is the world's leader in producing surfactants and detergents for agriculture and cleaning products. The release of these chemicals into the environment poses some concern. Other chemicals include legacy compounds, such as PCBs and OCs that are still commonly found in the environment, although production has mostly ceased.

Arctic and Antarctica

Although these areas are at opposite ends of the planet, they share many of the same contaminant issues. No agriculture or industry occurs in these lands so any contaminants that are present come from external sources. Polar bears and arctic birds have had PCBs and OCs in their tissues for decades. Now the group of concern is again the BFRs, such as PBDEs. The climate of these regions is also influenced by global climate change.

Oceans

The oceans are plagued by chemicals of concern. The conference attendees listed POPs, current-use pesticides, PAHs, petroleum, metals, radionuclides, PPCP, plastics, and nanoparticles as chemicals that needed additional study. More than one-third of all marine mammals are threatened species, with pollution and accidents ranking as the top two mortality factors. Research is critically needed on exposures of marine mammals to contaminants, especially in regards to multiple stressor scenarios. The most pressing issues they agreed on included harmful algae blooms, BFR, plastics, and coastal pollution.

In summary, many contaminant problems continue to confront the world. It is given that POPs will be around for many more decades, so we need to keep monitoring their concentrations and distributions. The most pressing issues, however, are what are called emerging contaminants. This term has many meanings, but it essentially refers to chemicals that are new and have received little study; thus, their threat to the environment is not well characterized. The chemicals of greatest concern at the present include BFRs, particularly PBDEs, nanoparticles, and PPCPs. Note, however, that new chemicals are being developed constantly and can enter the environment at any time. In addition, each section of the planet has its own problems that need attention.

NEEDS AND SUGGESTIONS FOR THE FUTURE OF RISK ASSESSMENT AND REGULATION

The United States and several other developed countries have rigorous sets of laws covering most aspects of the environment, but many experts have argued that our regulatory process needs some help. Regulations in some other countries are insufficient or lack enforcement. In Chapter 15, we mentioned that more than 62,000 chemicals that could get into the environment were grandfathered in without any toxicological data on them when the Toxic Substances Control Act (TSCA) was enacted in 1976. Clearly, many of these chemicals are produced in small quantities sequestered in laboratories or medical facilities, or can be judged safe from our current knowledge of chemistry, but many also are undoubtedly toxic. We even have 3000 chemicals in the United States whose production exceeds one million pounds per year, again, most of which have no toxicity data. Having only a few hundred chemicals with any toxicity data at all and that can get into the environment should be a matter of great concern. I suspect that the grandfathering of all these chemicals was done to be expedient. It would take decades and billions of dollars to run risk assessments on all of these chemicals. Nevertheless, Congress should take a serious look at the TSCA and determine if the US EPA should demand, on a selective basis, that more chemicals be tested. Perhaps some of these could be screened using QSARs. Methods should also be developed to incorporate multiple chemicals or stressors into these regulatory frameworks.

Newly developed pesticides, especially those with environmental applications such as pesticides, require registration through the US EPA. During this process, the US EPA relies solely on data presented to them by the manufacturer of the proposed chemical. This strongly suggests that a conflict of interest could occur on the part of the manufacturer (Boone et al., 2014). To avoid a potential conflict of interest, the US EPA could require substantiating evidence from sources other than the manufacturers, such as university or nonaffiliated labs.

Currently, the registration process for new pesticides under the Federal Insecticide, Fungicide, and Rodenticide Act is sorely inadequate. The basic framework is to examine toxicity under single chemical laboratory conditions with no consideration for multiple stressors or mixtures of chemicals that are likely to be present in exposures to organisms. This does not mean that the toxicity tests should involve every possible chemical that could be in a mixture, but that if there is a high likelihood that pesticides A and B will be used together the registration process should include a study with both pesticides used jointly.

At present, the US EPA uses risk assessment techniques to evaluate risks to nontarget organisms and other species based on the responses of a handful of organisms generally accepted for toxicity tests. These select few often have little relation to the organisms that will most likely be affected and risk assessment models, as good as they may be, often do not reflect actual exposures under natural conditions. In general, the registration process should be more realistic and inclusive. Greater effort should also be expended on doing community- or ecosystem-level risk assessments.

ACKNOWLEDGMENTS

It is with a true and sincere feeling of gratitude that I thank Greg Linder as a contributing author. Greg stepped in when my first choice of coauthor broke her contractual agreement. Without Greg's assistance, either this book would have not been completed or it would have taken a much longer time to do it. I also wish to thank the editorial staff that I worked with at Academic Press. Kristi, Pat, Susan, and the rest were very professional and very helpful. It goes without saying that I owe a debt of gratitude to my wife, Paulette, who has stood by me for almost five decades.

REFERENCES

Anonymous, 1889. Insecticides and their application. Science 13, 393–396.

Boone, M.D., Bishop, C.A., Boswell, L.A., Brodman, R.D., Burger, J., Davidson, C., et al., 2014. Pesticide regulation amid the influence of industry. Bioscience 64, 917–922.

Carson, R., 1962. Silent Spring. Houghton Mifflin Harcourt, Boston, MA, p. 400.

Cothran, R.D., Brown, J.M., Jenise, M., 2013. Proximity to agriculture is correlated with pesticide tolerance: evidence for the evolution of amphibian resistance to modern pesticides. Evol. Appl. 6, 832–841.

Frear, D.E.H., 1943. A catalogue of insecticides and fungicides. Science 98, 585.

Lorber, M., 2008. Exposure of Americans to polybrominated diphenyl ethers. J. Expos. Sci. Environ. Epidem. 18, 2–19.

Smith, P.N., Afzal, M., Al-Hasan, R., Bouwman, H., Castillo, L.E., Depledge, M.H., et al., 2010. Global perspectives on wildlife toxicology. In: Kendall, R.J., Lacher, T.E., Cobb, G.P., Cox, S.R. (Eds.), Wildlife Toxicology: Emerging Contaminant and Biodiversity Issues. CRC Press, Boca Raton, FL, pp. 197–255.

Union of Concerned Scientists, 2013. The rise of superweeds—and what to do about it. <http://www.ucsusa.org/sites/default/files/legacy/assets/documents/food_and_agriculture/rise-of-superweeds.pdf> (accessed 07.01.16).

US EPA, 2015. Environmental measurement. <http://www.epa.gov/measurements> (accessed 01.04.16).

Index

Note: Page numbers followed by "*f*" and "*t*" refer to figures and tables, respectively.

Printed in the United States
By Bookmasters